Real-Time PCR in Microbiology

From Diagnosis to Characterization

Ian M. Mackay

Caister Academic Press

Caister Academic Press
Norfolk, UK

www.caister.com

British Library Cataloguing-in-Publication Data

A catalogue record for this book is available from the British Library
ISBN: 978-1-904455-18-9

Description or mention of instrumentation, software, or other products in this book does not imply endorsement by the author or publisher. The author and publisher do not assume responsibility for the validity of any products or procedures mentioned or described in this book or for the consequences of their use.

Printed and bound in Great Britain

Contents

Contributors

José Manuel Andrade
Dept of Analytical Chemistry
University of A Coruna
A Coruna
Spain

Katherine E. Arden
Sir Albert Sakzewski Virus Research Centre
Royal Children's Hospital
Herston, Qld, and
Clinical Medical Virology Centre
University of Queensland
St Lucia, Qld
Australia

M.G.H.M. Beld
Dept. Clinical Virology
Academic Medical Center
Amsterdam
The Netherlands

Chris Birch and Julian Druce
Victorian Infectious Diseases Reference Laboratory
Victoria
Australia

Adriaan J.C. van den Brule
PAMM Laboratories
Department of Medical Microbiology
Eindhoven
The Netherlands

a.van.den.brule@pamm.nl

Stephen Andrew Bustin
3rd Floor Alexandra Wing
Royal London Hospital
London
UK

s.a.bustin@qmul.ac.uk

Patricia A. Cane
Health Protection Agency
Porton Down
Salisbury
UK

William Carman
West of Scotland Specialist Virology Centre
Gartnavel General Hospital
Glasgow
UK

Jonathan P. Clewley
Virus Reference Department
Centre for Infections
Health Protection Agency
London
UK

Nick M. Cirino
Biodefense Laboratory
Wadsworth Center
New York State Department of Health
Albany, NY
USA

ncirino@wadsworth.org

Eric C.J. Claas
Leiden University Medical Centre
Department of Medical Microbiology
Leiden
The Netherlands

e.claas@lumc.nl

Esteban Domingo and Cristina Escarmís
Centro De Biología Molecular "Severo Ochoa" (CSIC-UAM)
Universidad Autónoma De Madrid
Cantoblanco
Madrid
Spain

Christina Egan
Biodefense Laboratory
Wadsworth Center
New York State Department of Health
Albany, NY
USA

eganc@wadsworth.org

Ron A. M. Fouchier
Dept. Virology and National Influenza Center
Erasmus MC
Rotterdam
The Netherlands

Vincent Foulongne and Michel Segondy
Department of Virology
Montpellier University Hospital
St Eloi
Montpellier
France

Michael G. Ison
Transplant and Immunocompromised Host Infectious Diseases Service Division of Infectious Diseases
Northwestern University Feinberg School of Medicine
Chicago, IL
USA

Lance C. Jennings
Virology and Molecular Microbiology
Canterbury Health Laboratories
Christchurch
New Zealand

Bernhard Kaltenboeck
Department of Pathobiology
College of Veterinary Medicine
Auburn University
Auburn, AL
USA

Ian D. Kay and Silvano Palladino
Department of Microbiology and Infectious Diseases
Pathwest Laboratory Medicine WA
Royal Perth Hospital
Perth, WA
Australia

Mikael Kubista
TATAA Biocenter
Göteborg
Sweden

Olfert Landt
TIB MOLBIOL Syntheselabor Gmbh
Berlin
Germany

John F. Mackay
Linnaeus Laboratory
Gisborne
New Zealand

Ian Maxwell Mackay
SASVRC
c/o Royal Children's Hospital
Herston, Qld, and
Clinical Medical Virology Centre
University of Queensland
St Lucia, Qld
Australia

ian.mackay@uq.edu.au

Susan Madison-Antenucci
David Axelrod Institute
Wadsworth Center
New York State Department of Health
Albany, NY

susanma@wadsworth.org

Willem JG Melchers
Radboud University Nijmegen Medical Centre
Department of Medical Microbiology
Nijmegen
The Netherlands

w.melchers@mmb.umcn.nl

Reinhold Dietrich Mueller
Gen-Probe Incorporated
San Diego, CA
USA

reinholdm@gen-probe.com

Amy L. Muska
Roche Molecular Biosystems, Inc.
Branchburg, NJ
USA

amy.muska@roche.com

Hubert G.M. Niesters
ErasmusMC
Department of Virology
Rotterdam
The Netherlands

Michael D. Nissen
Sir Albert Sakzewski Virus Research Centre
Royal Children's Hospital
Herston, Qld, and
Clinical Medical Virology Centre
University of Queensland
St Lucia, Qld
Australia

Andreas Nitsche
Robert Koch-Institut
Nordufer
Berlin

nitschea@rki.de

Tania Nolan
Sigma Aldrich
Haverhill
UK

Tania.Nolan@sial.com

Stuart M. Palmer
Roche Molecular Biosystems, Inc.
Branchburg, NJ
USA

stuart.palmer@roche.com

Nikolaos G. Papadopoulos
Allergy Research Center
University of Athens
Goudi
Greece

Edith N. Peck
Roche Molecular Biosystems, Inc.
Pleasanton, CA
USA

edith.peck@roche.com

Astrid Petrich
St. Joseph's Healthcare
Dept. of Pathology and Molecular Medicine
Mcmaster University Microbiology
Hamilton, ON
Canada

Michael W. Pfaffl
Lehrstuhl Fuer Physiologie
Life Science Zentrum Weihenstephan
Technische Universitaet München
Freising-Weihenstephan
Germany

William Rawlinson and Gillian M. Scott
SEH and University of New South Wales
Virology Division
SEALS Microbiology
Prince of Wales Hospital
Randwick, NSW
and
Virology Research
Prince of Wales Hospital
School of Medical Sciences
UNSW, NSW
Australia

Udo Reischl
Institute of Medical Microbiology and Hygiene
University of Regensburg
Regensburg
Germany

Nick A. Saunders
Centre for Infections
HPA (Colindale)
London
UK

Carita Savolainen-Kopra
Enterovirus Laboratory
National Public Health Institute (KTL)
Helsinki
Finland

Oliver Schildgen
Institute for Medical Microbiology, Immunology, and Parasitology
Department of Virology
University of Bonn
Bonn
Germany

Rudolf Seibl
Roche Diagnostics Gmbh
Penzberg
Germany

Theo P Sloots
Sir Albert Sakzewski Virus Research Centre
Royal Children's Hospital
Herston. Qld, and
Clinical Medical Virology Centre
University of Queensland
St Lucia, Qld
Australia

Yi-Wei Tang
Departments of Medicine and Pathology
Vanderbilt University School of Medicine
Nashville, TN
USA

Norma P. Tavakoli
Wadsworth Center
New York State Department of Health
Albany, NY
USA

Raymond Tellier
Hospital for Sick Children
Toronto
and
Dept of Laboratory Medicine and Pathobiology
University of Toronto
Canada

Patrick C. Y. Woo
Department of Microbiology
The University of Hong Kong
University Pathology Building
Queen Mary Hospital
Hong Kong

Preface

Preface

Research workers in more than one field have likened their experience to walking across a darkened room, constantly bumping into various objects. Only later, when eyes adjust to the gloom, can they look back and see exactly how the furniture is arranged. But that is only part of the story. When they switch on the light they discover another door and another darkened room. And the whole process is repeated.

David Tyrell, 2002

Microorganisms have survived, evolved and expanded within the human population without pause. Sometimes they run afoul of the host immune response or a particularly effective drug or vaccine, but they thrive nonetheless. Today there seems to be renewed vigor among researchers to better understand these invaders and the interactions they have with our systems at the molecular level. There is also a renaissance of sorts in the discovery of "new" microorganisms which were previously unknown, although probably cohabiting with us for a very long time. Because of these advances there is an ongoing need for laboratory tools that work quickly, simply and effectively to detect and characterize these tiny terrors. PCR is one of these tools and it quickly found its place in both the clinical and the research microbiology laboratory. The development of real-time PCR heralded the arrival of a tool that could provide results in a clinically relevant timeframe while virtually abolishing the dreaded risk of carrying-over PCR product into subsequent PCRs. All it took to reach these goals was indirect detection of the product, removing the need to open up the reaction vessel. However, as many of us know, the reality of any PCR is that on occasion one must make sure what looks to be happening is what actually is happening.

This book seeks to present some further home truths, asking some tough questions and in so doing, describing and explaining some of the more complex aspects of real-time PCR as clearly as possible. Furthermore, this book aims to provide background for the novice, present a theoretical reference for the experienced user and hint at future developments we can expect to become more familiar with. The early chapters address the basics of PCR history, oligonucleotide design and standardization. The complex issue of quantification in microbiology is touched upon in many chapters but is addressed in particular in the middle section. The remainder of the book focuses on some applications of real-time PCR and the challenges faced, and sometimes overcome, by use of this technology. Finally, in a bid to get experts to address those questions we always want answered, but feel unworthy asking, I have dedicated a chapter to create a virtual roundtable discussion delving into the minds of real-time PCR authorities and experts in affiliated fields. The chapter makes for an insightful, sometimes topical and often humorous read, which I highly recommend.

In today's world of newly identified viruses, imminent pandemics and possible bioterror, methods for microbiological application need to be as simple and robust as we can make them. We're not there yet but we will be soon.

My own research, supported by the Royal Children's Hospital Foundation, the Sir Albert Sakzewski Virus Research Centre and the University of Queensland investigates respiratory viruses, and that bias will thankfully be apparent in several chapters.

Preparing a work such as this marks the passage of many events in one's life, and the lives of others; events that seldom recur but always leave lasting impressions. Sometimes it's important to have somebody point them out so one can wave at them as they pass by. For me, that person is my wife, Katherine Arden, who deserves credit for much of this work on top of my heartfelt thanks for her insightful questions, unswerving support and Herculean endurance throughout this project.

May I sincerely thank all the authors who completed work for this book; your endeavors have provided a diverse and interesting read and have hopefully stimulated some minds to answer the many questions we have posed and must continue to pose in order to open more doors, bump into more objects but ultimately illuminate more rooms in what would seem to be an enormously large house.

Ian M. Mackay
June, 2007

Real-time PCR: History and Fluorogenic Chemistries

1

Ian M. Mackay, John F. Mackay, Michael D. Nissen, and Theo P. Sloots

Abstract

The use of real-time PCR in microbial diagnostics has increased to the point where it has evolved from a novelty into a mature and essential technology for the field. In doing so, real-time PCR has driven significant changes in the way we detect microbes. The predominantly phenotype-related methods of culture and antigen detection, while still of considerable value, are being supplanted by the detection, characterization and quantification of microbial nucleic acids. Real-time PCR has engendered wider acceptance of the PCR technique due to its improved rapidity, sensitivity, reproducibility and the considerably reduced risk of carry-over contamination. There are many fluorogenic chemistries that can detect PCR product as it accumulates but only a few are useful, popular or exciting enough to be the subject of publication in the field of microbiology. We review how real-time PCR has come to be; especially the essential role of fluorescence and we critically review the plethora of detection chemistries available to the end user.

Introduction

The development and introduction of new investigative molecular technologies have begun to change the way we think about virulence, virus variation and pathogenicity. Amplification methods are projected to represent 75 to 90% of the diagnostic nucleic acid test market in 2006 (Kubista, 2004). The most prominent of these methods has been the polymerase chain reaction (PCR; Mullis and Faloona, 1987), increasingly in its latest form of real-time PCR. In the context of infectious disease, diagnostic microbiology applications can be divided into two broad sections; qualitative screening (e.g. routine diagnostics or epidemiology research) and quantitative investigations of the interplay between microbe and host (e.g. microbial load variation in response to therapeutics or cytokine quantification as an indirect measure of the impact of microbial infection).

Many research and diagnostic microbiology laboratories contain a jumble of traditional and leading edge, in-house (home-brew) and commercial kit assays for the detection of microbial antigen or DNA. Furthermore, comparison of any two laboratories might reveal quite disparate jumbles. The wide variety of assays reflects many factors including the perceived reliability of familiar techniques and technologies, the existence of useful quality assurance programs, the scope of a test to perform the required task, the existence and cost of commercial tests able to detect microbes of interest, the experience a laboratory has acquired with a particular technique, the skills of laboratory staff and the degree of support offered by the assay manufacturer. Additionally, the cost of a new diagnostic system may be over-estimated prior to its use in the laboratory, adding to the slow uptake of new technologies.

PCR assays have undergone significant change over the last decade, to the extent that some scientists may be struggling to stay abreast of the latest offerings. This chapter will seek to bring the reader up-to-date with the third-generation of PCR technology; the so-called real-time PCR, a maturing tool widely accepted by many scientific disciplines but especially the field of microbiology (Whelan and Persing, 1996; Ginzinger, 2002; Cockerill III, 2003; Mackay, 2004).

Polymerase chain reaction: just the basics

Sensitivity, specificity and speed: these should be the driving factors behind choosing a microbial detection method, and real-time PCR applications can address all these needs. It has become common to relegate viral culture to specialized virology laboratories rather than include it in the routine diagnostic laboratory, where the focus is on rapid result turnaround (Carman, 2001). The speed with which a negative result is provided is often as important as the timely return of a positive result and when a result from *in vitro* microbial culture can require weeks, PCR assays offer an attractive alternative (Carman, 2001). Of course, speed is the result of a number of factors, and these must all be taken into account when assessing the benefits of any new PCR-based assay. Faster assays should be described in terms of the final optimized and validated test performed on real-world specimens that have followed a normal specimen collection, transport and template extraction process. However what is all too often presented in the literature are preliminary or "one-off" studies using idealized templates (e.g. cloned targets) and conditions. Unfortunately, incomplete PCR assays missing vital information are frequently rushed into publication resulting in technical problems reproducing data in other laboratories.

According to the literature, PCR assays rate as the most commonly used molecular technique to detect and study microbes, appealing more widely than specific commercial template amplification technologies such as Abbott's ligase chain reaction (LCR; Barany, 1991), bioMérieux's nucleic acid sequence based amplification (NASBA; Kievits *et al.*, 1991; Compton, 1991), Becton Dickinson's strand-displacement amplification (SDA; Walker *et al.*, 1992) and Gen-Probe's transcription-mediated amplification (TMA). PCR has also been applied more widely than the signal amplification methods such as Bayer's branched DNA (bDNA) and Digene's Hybrid Capture (Persing, 1993; Whelan and Persing, 1996; Fredricks and Relman, 1999). This apparent popularity may simply reflect the more frequent publication of in-house or home-brew assay results compared to kit-generated data. However PCR is virtually the sole amplification method employed by research laboratories and it is here that routine diagnostics approaches are formulated and eventually transferred to the diagnostic arena.

PCR methods utilize a pair of synthetic oligonucleotides (short, single-stranded pieces of DNA) called primers, each designed to hybridize in a 5′ to 3′ orientation to one particular strand of a double-stranded DNA (dsDNA) target. Each hybridized primer forms a starting point for the production of a complementary DNA strand via the sequential addition of deoxynucleotides using recombinant DNA polymerases derived from thermophilic bacteria. The primer pair spans a target region that is exponentially and precisely replicated during the subsequent reaction. We will refer to the product of this duplicated region as the "amplicon" to indicate that it is amplified material of a single species. The PCR method can be summarized in three steps as shown in Figure 1.1: (i) dsDNA separation at temperatures above 90°C, (ii) primer annealing at a design-specific temperature (commonly between 50°C to 60°C) and (iii), optimal extension of the primed template at 70°C to 75°C (Park *et al.*, 1993). A compact PCR format requiring only two steps is also popular; dsDNA denaturation and a combined annealing and extension step making use of the DNA polymerase's broad thermal range of activity.

The melting temperature (T_M) designates the temperature at which 50% of the oligonucleotide-target duplexes remain hybridized. The actual annealing temperature used for an oligonucleotide (T_A) is typically 5°C to 10°C below the T_M. The T_M is dependent upon the concentration of the DNA, its length, nucleotide sequence and the composition of the solvent in which the DNA is suspended (Ririe *et al.*, 1997). In practice the optimal T_A for a given primer pair is best determined empirically by testing a small range of temperatures around the calculated T_M.

Existing combinations of PCR and non-fluorescent amplicon detection assays will be called "conventional" PCR throughout this chapter. These detection systems include analytical agarose gel electrophoresis (Kidd *et al.*, 2000), Southern blot and ELISA-like

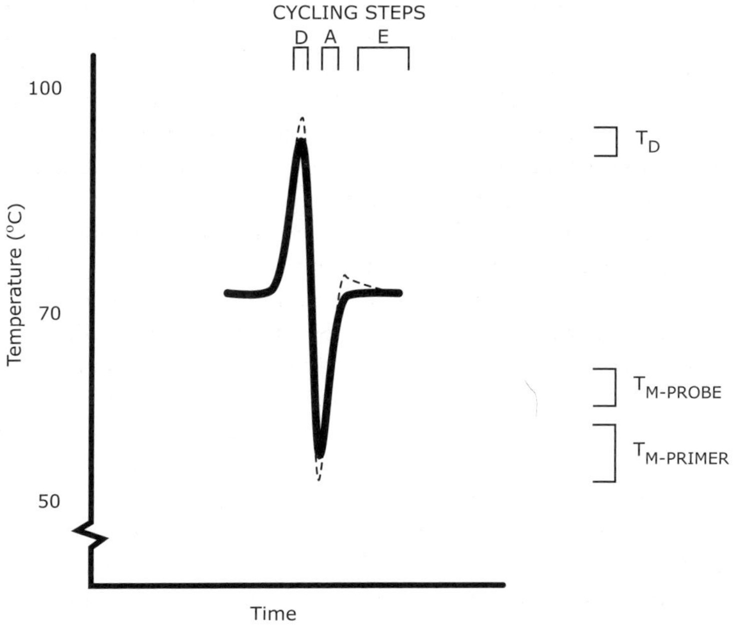

Figure 1.1 A stylized time versus temperature plot during a single PCR cycle. The denaturation (D), primer and probe annealing (A) and primer extension (E) steps are shown. At the indicated optimal temperature ranges, dsDNA denatures (T_D) then oligoprobes anneal ($T_{M\text{-}PROBE}$) followed by the primers ($T_{M\text{-}PRIMER}$) as a precursor to their extension. The actual thermal cycler incubation temperature (dashed line), may overshoot the desired temperature to varying degrees depending on the instrument.

systems such as the enzyme-linked amplicon hybridization assay (ELAHA; (Mackay *et al.*, 2001; Mackay *et al.*, 2006)) and enzyme-linked oligosorbent assay (ELOSA; (van der Vliet *et al.*, 1993; Chandelier *et al.*, 2001). Traditional agarose electrophoresis of amplified DNA in the presence of ethidium bromide is followed by visualization of the bands during irradiation by ultraviolet light (Kidd *et al.*, 2000). The Southern blot detection of amplicon occurs after hybridization of immobilized amplicon to a labeled oligonucleotide probe (oligoprobe). This increases specificity but at the cost of significant time, labor and multiple amplicon handling steps, increasing the risk of spreading amplicon throughout the laboratory. In the past, the Southern blot also depended on the use of radioactive labels (Holland *et al.*, 1991). In the second generation of amplicon detection, PCR-ELISA was used to capture amplicon onto a solid phase via biotin or digoxigenin-labeled primers, oligoprobes or by direct capture after incorporation of the biotin or digoxigenin into the amplicon (Keller *et al.*, 1990; Kemp *et al.*, 1990; van der Vliet *et al.*, 1993; Kox *et al.*, 1996; Dekonenko *et al.*, 1997; Watzinger *et al.*, 2001; Mackay *et al.*, 2003). Once captured, amplicon detection utilized an enzyme-labeled avidin or anti-digoxigenin reporter molecule in a similar manner to a standard ELISA format. PCR-ELISA had the added benefit of employing hardware and techniques commonly available to the ELISA-enabled research and diagnostic microbiology laboratory. These probe detection techniques have now been largely superseded by real-time PCR—a technique which may be thought of in this digital age as an "on-line" Southern blot; the third generation of amplicon detection.

The possibility that, in contrast to conventional PCR, amplicon could be detected as it accumulated, in "real" time was welcomed by researchers and diagnostic analysts alike. The term originally coined for this technique was "kinetic PCR" (Higuchi *et al.*, 1993) but it is now more commonly known as real-time PCR and we shall refer to it as such whether the microbial template is DNA or RNA. Real-time PCR was essentially based on a scientific mistake; preferably thought of as an "unplanned experiment." In the case of real-time PCR, the experiment meant ethidium bromide, usually added to a PCR reaction after thermal cycling to detect amplicon in a UV-irradiated agarose gel, was inadvertently added before the reaction was cycled (Gingeras *et al.*, 2005). Such an act could simply have resulted in amplification inhibition. After all, negative results during the assay development process come as no surprise to those familiar with assay evaluation and validation. However, thanks to the low concentration of ethidium bromide used, the amplicon was both successfully produced and detected while in the tube. From such a simple accident the role of PCR was dramatically changed from that of a specialist research tool to one of a versatile diagnostic technology, providing advanced and powerful systems capable of routine diagnostic applications for the clinical microbiology laboratory (Cockerill III and Smith, 2002). The minimizing of a major concern to the diagnostic laboratory; contamination of new (otherwise negative) reactions with the amplicon generated from previous reactions, was a driving factor in the renewed enthusiasm for PCR. These systems are also called "closed" or "homogeneous" since they combine template amplification and amplicon detection in a single shut-and-forget reaction vessel.

A more controversial factor is cost. Reportedly, real-time PCR is cost effective when implemented in a high throughput laboratory (Martell *et al.*, 1999; Espy *et al.*, 2006), particularly when replacing traditional, culture-based approaches to microbial detection. When calculating costs, it also important to account for the less quantifiable physical and emotional costs to ill patients, caused by unnecessary and sometimes invasive tests, and toxic therapies necessitated by a delay in the provision of laboratory results because of exceedingly lengthy assay times.

A quick review of the literature confirms that, while diagnostic microbiology laboratories are embracing a change to the testing paradigms and determining the best chemistries and assays for peak diagnostic efficiency, research laboratories have already embraced the new technology and real-time PCR applications are entrenched as useful tools. It is interesting to note that the time between technology development and diagnostic utility is continually decreasing. Conventional PCR was the domain of research laboratories for many years before finding diagnostic favor whereas real-time PCR has been rapidly adopted as a diagnostic tool. A similar pattern is appearing in the adoption of microarray technology in diagnostics. Rapid nucleic acid amplification and detection technologies are quickly displacing the traditional assays which are based on pathogen phenotype rather than genotype. The traditional diagnostic microbiological assays including microscopy, microbial culture, antigen detection and serology may be adversely affected by poor sensitivity, slow completion times, the inability to isolate or poor viability of organisms, complex or subjective result interpretation, previous antimicrobial therapy (leaving microbes present but unable to replicate in culture), and non-specific cross-reactions (Whelan and Persing, 1996; Carman *et al.*, 2000). Nonetheless, microbial culture produces valuable data sometimes revealing new, uncharacterized or atypical microbes and yielding intact or infectious organisms for further research (Ogilvie, 2001). It is therefore clear that the role of the traditional assay continues to be an important one (Sintchenko *et al.*, 1999; Johnson, 2000; Biel and Madeley, 2001; Clarke *et al.*, 2002; Ellis and Zambon, 2002). Of course PCR assays, be they real-time or conventional, have their own significant limitations. Our ability to design oligonucleotide primers only extends to our knowledge of a microorganism's genome, as well as the ability of publicly available sequence databases to suitably reflect that knowledge by housing all sequences from a microbe and related variants. Microbial genomes quite commonly contain unexpected sequence polymorphisms which reduce or abolish the function of a PCR and cause falsely negative results. We have already discussed carry-over contamination and the principle approaches to defend against it have resulted in strict guidelines for the design of diagnostic laboratories dedicated to performing PCR. Additionally, PCR is sometimes criticized for offering too high a sensitivity for some applications i.e. detecting a microbe that is present at what are considered non-pathogenic levels. Thus great care is required when designing a PCR assay and interpreting its results.

The increased speed attributed to real-time PCR is mostly due to the removal of post-amplification detection procedures and the combined use of fluorogenic dyes and sensitive methods of directly detecting their emissions *in situ* as they exit the reaction vessel (Wittwer *et al.*, 1990; Wittwer *et al.*, 1997b). The time to reach each temperature of a PCR cycle (ramp rate), the length of the incubation at each temperature and the number of times each cycle of temperatures is repeated are all controlled by a programmable thermal cycler. Advanced instrument designs using electronically controlled heating blocks or fan-forced heated airflows have significantly shortened the ramp rates and therefore the total assay times. Apart from the time saved by amplifying and detecting template in a single tube, the fluorescence generating portion of the assay's performance can be closely scrutinized without introducing errors due to handling of the amplicon (Higuchi *et al.*, 1993).

Today in microbiology, real-time PCR is used to detect and quantify nucleic acids from widely diverse targets including food, viral and non-viral vectors used in gene therapy protocols, genetically modified organisms and to study human and veterinary microbiology, oncology and immunology (Kruse *et al.*, 1997; Böhm *et al.*, 1999; Härtel *et al.*, 1999; Nogva *et al.*, 2000; Fraaije *et al.*, 2001; Mhlanga and Malmberg, 2001; Rudi *et al.*, 2002; Stordeur *et al.*, 2002; Mackay *et al.*, 2002; Klein, 2002; Ahmed, 2002; Barzon *et al.*, 2003).

Shining light upon the accumulation of amplicon

Simply put, it is the detection processes that discriminate real-time PCR from conventional PCR assays. The currently used fluorogenic chemistries can be broadly categorized as interacting specifically or non-specifically with the amplicon's nucleotide sequence (Whitcombe *et al.*, 1999). To clarify, while the amplicon itself is still commonly produced from a PCR using highly sequence-specific primers we are referring here to the specificity for detection, rather than for production of that amplicon.

The monitoring of accumulating amplicon in real-time has been made possible by the attachment to primers, oligoprobes or amplicon of molecules with fluorescent potential. The success of these dyes and the assays employing them is due to a relatively rapid and measurable change in signal produced after the direct or indirect interaction of fluorescent dye and amplicon (Morrison *et al.*, 1989). Some molecules are able to absorb light energy of a specific wavelength and within nanoseconds, emit that energy as light (Figure 1.2a). We have chosen to use the terms "dye" (a fluorogenic entity or fluorochrome in an unbound state) or "label" (when the chemistries may be labeled with a mix of fluorogenic and non-fluorogenic molecules) throughout this chapter. Additionally we employ the term "fluorophore" to describe the oligonucleotide-attached dye which imbues the oligonucleotide with fluorogenic properties. Fluorophores can repeat the absorption-emission cycle hundreds to thousands of times (Behlke *et al.*, 2005).

Because energy is lost to heat and other processes, the emitted photon has less energy than was imparted by the energizing light source, which can be identified by the longer wavelength of the emitted photon. This change in wavelength is called the Stoke's shift (Figure 1.2b) and is defined as the difference between the peak absorbance and the peak emission, where the peaks represent the single wavelength of a spectral curve at which the molecule most efficiently absorbs or emits energy, respectively. A wider shift was a preferred characteristic for the choice of early reporter and quencher dye pairs but improved fluorescence detection optics and software have enhanced the collection and interpretation of fluorescence permitting the use of dye pairs with smaller Stoke's shifts.

The glow of success

The physical transitions from electrophoresis tanks containing a mutagenic liquid to 8-well polystyrene strips, to a comfortable seat and a computer screen are significant, but they can be attributed to a simple process—the direct or indirect fluorescent labeling of newly synthesized DNA. Whether labeled primers, oligoprobes or amplicon are used, real-time PCR is made possible because of a rapid and measurable signal change after the interaction of amplicon and fluorescent dye (in one of many forms) which is identified by sensitive detection platforms (Morrison *et al.*, 1989). The signal is most commonly measured at

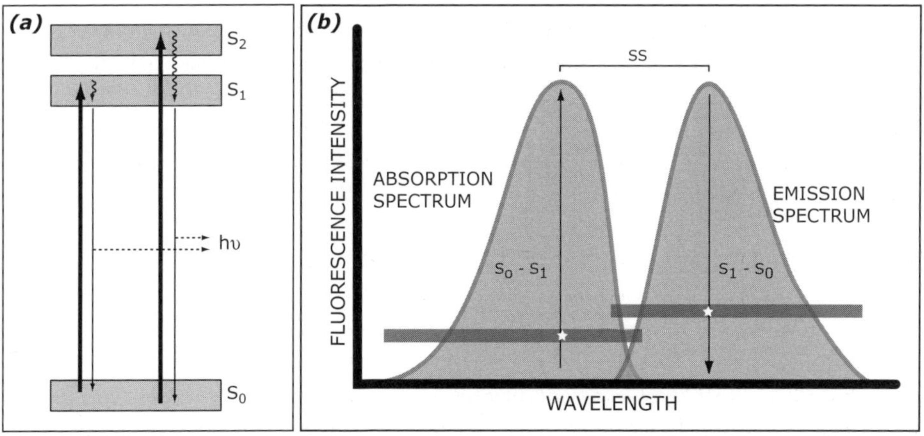

Figure 1.2 (A) The electrons of a given dye molecule may exist in one of several states as demonstrated by the ground state (S_0, grey box), and first (S_1) and second (S_2) excited states. Absorption of a quantum of energy "charges" the dye (bold line) and as the molecule relaxes back to the ground state (thin line) a photon is emitted (where h is Planck's constant and v is the frequency of the electromagnetic radiation), the process of fluorescence. The excited state is maintained for nanoseconds. (B) A diagram of the range of wavelengths over which a fluorogenic molecule can be affected by incoming energy (absorption spectrum) and the range of wavelengths over which that molecule can emit energy after excitation (emission spectrum). Real-time PCR aims to stimulate fluorescence and collect emission data at the peak wavelengths of these ranges (marked by stars) and relies upon the Stoke's shift (SS) to get "clean," noise-free emission data.

some point during each PCR cycle and is related to the amount of amplicon present at that time, increasing in intensity as the DNA is replicated. Radiogenic labels were the technology of the day at the dawn of real-time PCR but easy disposal, an extended shelf life and the removal of radioactive emissions quickly catapulted fluorescent chemistries to the position of label of choice (Matthews and Kricka, 1988).

Most of the popular real-time PCR chemistries depend upon the hybridization of one or more fluorescent oligoprobes to a complementary sequence on one of the amplicon strands. To further improve the odds of a likely meeting, the inclusion of more of the primer that creates the strand complementary to the oligoprobe, a process called asymmetric PCR, can lead to generation of an increased fluorescent signal (Gyllensten and Erlich, 1988; Barratt and Mackay, 2002). This has been shown to produce improved fluorescence from a hairpin oligoprobe PCR (Poddar, 2000) and the method is also applicable to other oligoprobe-hybridization assays especially in variant (e.g. single nucleotide polymorphism; SNP) typing. However, several features require forethought. One must determine which strand to target, particularly in the case of single stranded genomes like the negative-sense, single-stranded paramyxoviruses or positive-sense, single-stranded RNA picornaviruses. It is also better to mediate asymmetric PCR using oligonucleotide design in combination with the amount of primer, rather than amount alone. Studies have shown that the T_M of a primer will change in parallel with its concentration (Sanchez *et al.*, 2004). By designing one primer (A) so that it will have a T_M close to the other primer (B) even though primer

A is present in significantly fewer copies compared to B, amplification can still progress efficiently, but the outcome is a predominance of one strand. This is the process of linear-after-the-exponential (LATE) PCR (Pierce *et al.*, 2005).

The most commonly used fluorogenic oligoprobes rely upon fluorescence resonance energy transfer (FRET; Figure 1.3) between fluorophores or between one fluorophore and a dark, or black-hole, non-fluorescent quencher (NFQ), which disperses energy as heat rather than fluorescence (Didenko, 2001). FRET is a non-radiative process in which energy is passed between permissive molecules that are spatially separated by 10–100 Å and which have overlapping emission and absorption spectra (Stryer and Haugland, 1967; Heller and Morrison, 1985; Clegg, 1992). Förster primarily developed the theory behind this process (Förster, 1948). The energy transfer reduces the lifetime of the excited state of electrons in the original molecule by taking the emitted excess energy and expelling it as fluorescence or heat. The efficiency of energy transfer is proportional to the inverse sixth power of the distance (R) between the donor and acceptor ($1/R^6$) fluorophores (Selvin, 1995; Didenko, 2001). FRET permits the determination of nucleic acid hybridization, without the prior removal of unbound probe (Cardullo *et al.*, 1988). However all dyes

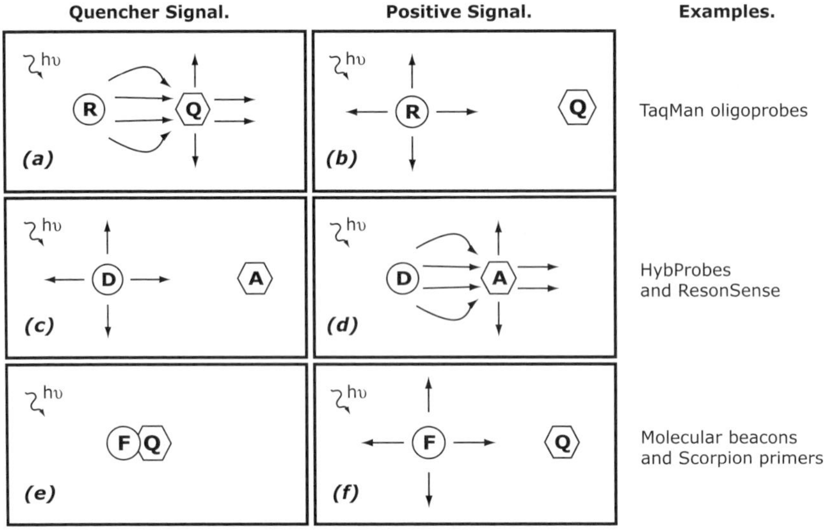

Figure 1.3 Mechanisms of FRET. (A) When the reporter [R] and quencher [Q] of an hydrolysis oligoprobe are in close proximity and illuminated by discharge from an instrument's light source (hv), the quencher "hijacks" the emissions from excitation of the reporter. The quencher, then emits this energy as fluorescence or, if a non-fluorescent or "dark" quencher, heat. When the fluorophores are separated, as occurs upon oligoprobe hydrolysis as depicted in (b), the quencher can no longer influence the reporter, which now fluoresces at a distinctive wavelength recorded by the instrument. In the reverse process using adjacent oligoprobes (c), the fluorophores begin the cycle as separated entities. Whilst the emission of the donor [D] is monitored, it is the signal from the acceptor [A] produced when in close proximity to the donor that indicates a positive reaction (d). In (e), another form of quenching is shown, caused by the intimate contact of labels attached to hairpin oligonucleotides. The fluorophore and quencher interact more by collision than FRET, disrupting each other's electronic structure and directly passing on the excitation energy. When the labels are separated, as is the case in (f), the fluorophore is free to fluoresce.

are not created equal. Marras and colleagues eloquently described experiments to indicate that some fluorophore-quencher pairs interact in such a way that they affect the T_M of the oligoprobe-target hybrid, and those providing the duplex with increased stability also demonstrated higher quenching efficiencies (Marras *et al.*, 2002). Such outcomes are not surprising since many fluorophores contain aromatic rings which imbue intercalating abilities (Mullan *et al.*, 2004), but the effects do complicate our *in silico* efforts to determine the T_M of an oligonucleotide and enforces the importance of empirical studies.

Useful fluorescence data generated by real-time PCR assays are generally collected from early in the amplification process because amplification conditions are optimal and the fluorescence accumulates in proportion to the amplicon (Figure 1.4). Signal detection

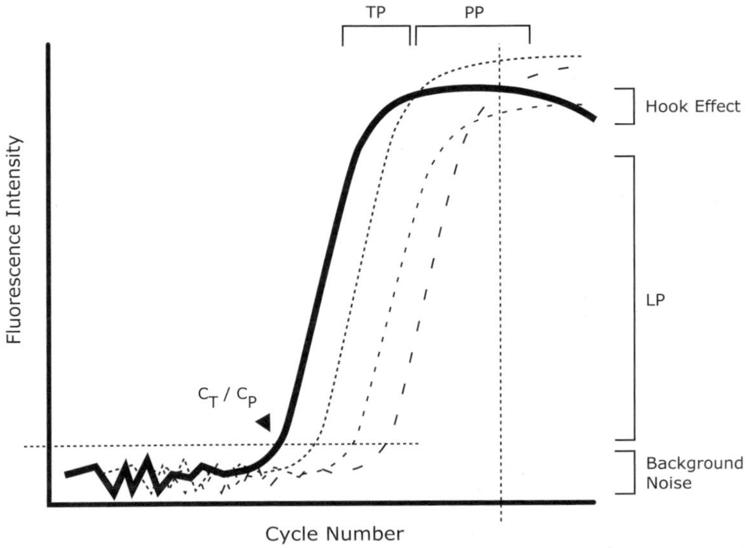

Figure 1.4 Kinetic analysis of fluorescent changes. Indirect charting of the duplication of a template by real-time PCR (solid line) ideally takes on the form of a sigmoidal curve when plotted as cycle number versus fluorescence emission intensity. Early exponential amplification cannot be viewed because the signal is below the sensitivity of the instrument's detector. However, when enough amplicon is present, the assay's exponential progress can be monitored as the rate of amplification enters a log-linear phase (LP). Under theoretically ideal conditions, the amount of amplicon increases at a rate of one $\log_{10}$ every 3.32 cycles, i.e. it would double (increases by one $\log_2$) every cycle. As primers and enzyme become limiting and products accumulate that are inhibitory to PCR assays or compete for hybridization with an oligoprobe, the reaction slows, entering a transition phase (TP). Eventually a plateau phase (PP) is reached in which there is little or no increase in fluorescent signal although amplicon may continue to accumulate. Some fluorescent detection chemistries display an overall reduction in fluorescence intensity after the plateau phase (Hook). The point at which fluorescence surpasses a pre-defined background noise (dashed horizontal line) is called the threshold cycle or crossing point (C_T or C_P; indicated by an arrow) which is proportional to starting template concentration. Construction of a standard curve using a range of known starting concentrations permits the calculation of the quantity in the original specimen. Traditional PCR data collection is performed at the end of the assay (dashed vertical line) by electrophoresis, Southern blot or PCR-ELISA. Also shown are curves representing a serial titration of template (dashed curves), consisting of decreasing starting template concentrations, which produce increasing numerical C_T or C_P values.

occurring at the end-point of the reaction is problematic since amplicon accumulation may have been adversely affected by inhibitors, poorly optimized reaction conditions or saturation effects due to excess double-stranded amplicon. In fact at the end-point there may be no relationship between the initial template and final amplicon concentrations (see Figure 1.4). Because the emissions from fluorescent chemistries are temperature-dependent, data are generally acquired only once per cycle, at the same temperature (Wittwer *et al.*, 1997a). Data acquisition is best performed after the annealing step (Reynisson *et al.*, 2006).

The fractional cycle number at which the real-time fluorescence signal mirrors progression of the amplification reaction above the background noise level is used as an indicator of successful target amplification (Wilhelm *et al.*, 2001a). Most commonly, this is called the threshold cycle (C_T) but the same value is described for use with the LightCycler™ (Roche Applied Science) where the fractional cycle is called the crossing point (C_P). The C_T is defined as the PCR cycle at which the gain in fluorescence generated by the accumulating amplicon exceeds 10 standard deviations of the mean baseline fluorescence, using data taken from cycles 3 to 15 (Jung *et al.*, 2000). The C_T is proportional to the number of template copies present in the sample (Gibson *et al.*, 1996). The values are assumed to result from equal amounts of amplicon in each vessel since the C_T/C_P values are acquired from a single fluorescence intensity value (see Figure 1.4). In practice the C_T is calculated after the definition of a noise band which permits the exclusion of data from early PCR cycles that cannot be distinguished from background fluorescence. The C_P used on the LightCycler system is most frequently obtained using algorithms derived from each fluorescent curve. Therefore the final C_T and C_P value is the fractional cycle at which a reaction's plotted amplicon accumulation curve intersects a single fluorescence value (usually at or close to the noise band but within the LP; Wilhelm *et al.*, 2001a). The accuracy of the C_T or C_P depends upon the concentration and nature of the fluorescence-generating system, the amount of template initially present, the sensitivity of the fluorescence detector and its ability to discriminate specific fluorescence from background noise. Given a C_T value (T_n), we can roughly estimate the amount of amplicon present using the equation 1.1.

$$T_n = T_0(E)^n \qquad 1.1$$

We also need to know the assay's exponential amplification (Equation 1.2; Wilkening and Bader, 2004) and amplification efficiencies (E; Equation 1.3 is the most frequently used; Gingeras *et al.*, 2005).

$$10^{-1/\text{slope}} \qquad 1.2$$

$$[10^{-1/\text{slope}}] - 1 \times 100\% \qquad 1.3$$

The slope is determined from a plot of the decreasing C_T values obtained from amplifying, in ascending order, template dilution series of starting from a known amount (standard curve). To solve equation 1.1 we also need an understanding of the real-time PCR fluorescence detection system's ability to first discriminate specific fluorescence from background (generally 10^{10} to 10^{11} copies (T_0) of labeled DNA). Equation 1.1 is the basis for quantitative real-time PCR. For example, if a standard with 10^4 copies was amplified in an assay with an efficiency of 1.8, the expected C_T would be 23.1 as shown below,

$$10^{10} \text{ copies} = 10^4 \text{ copies } (1.8 \text{ copies/cycle})^{n \text{ cycles}} \qquad 1.4$$

Therefore, log (10^6 copies) = n cycles.log(1.8 copies/cycle),
which solving for n gives us n = 6/0.26 cycles (= 23.1).

Forearmed with some knowledge about your assay's efficiency and the real-time C_T data from a dilution series, it is possible to quickly identify problems with a dilution series used as a standard curve. Moreover, if one uses the same variables, one can also identify where a single copy's amplification plot ($T_n = 1$) should first appear, in this case using our assay efficiency and T_0 value, at a C_T of 38.5 cycles. Therefore one can quickly see if an unknown has amplified too late suggesting the presence of amplification inhibitors, non-specific amplification or even low-level contamination which may only be present at 1 copy per several microliters and prime inefficiently during the early cycles.

Real-time instrumentation

Broadly speaking, real-time PCR instruments can be divided into two classes based on the method used to heat the reaction vessel; the solid heating block thermal cyclers and the heated air thermal cyclers. Block cyclers generally ramp at 1 to 10°C/sec whereas air cyclers can ramp as quickly at 20°C/sec. The faster rates permit 10 to 20 min assays (Wittwer *et al.*, 2003; excluding hands-on time) however these times are infrequently applied to microbial real-time PCR assays, probably due to technical difficulties when clinical specimens are used under real-life conditions as well as the need for cycling parameters that provide maximum sensitivity (e.g. lengthy PCR enzyme activation times for hot-start PCR). An off-shoot of the block cyclers are miniaturized thermal cyclers capable of performing real-time PCR but constructed of sturdy, solid-state and relatively low-cost materials, permitting transport of the PCR laboratory to the site of testing (Northrup, 1998; Ibrahim *et al.*, 1998).

The quality of the instrument has a significant role in the reproducibility of results, which subsequently influences how well a described protocol transports between laboratories and countries. A common discrepancy between systems is the temperature profile of a PCR cycle (Schoder *et al.*, 2003). Since temperature affects enzyme function, fluorescence and oligonucleotide binding, the instrument's ability to obtain and hold the desired temperature should be an important consideration when purchasing a real-time PCR system.

The majority of PCR thermal cyclers employ capped plastic reaction vessels to contain the amplification reaction. However the LightCycler employs glass capillaries (Wittwer *et al.*, 1997a). The LightCycler's plastic and glass composite capillaries are optically clear and act as cuvettes for fluorescence analysis, as well as facilitating rapid heat transfer. However, they are relatively fragile and require some experience to handle, which should be considered along with the laboratory layout when employing such a system for infectious diseases diagnosis (Schalasta *et al.*, 2000). The LightCycler and the RotorGene™ (Corbett life science) instruments vary their temperature by rapidly heating and cooling air using a heating element and fan which produce ramp rates of 20°C and 2.5°C per second, respectively. Faster rates prolong polymerase survival and significantly shorten the assay's completion time but existing assays may require some fine-tuning in the transition between instruments (Weis *et al.*, 1992). There are several reviews which include the features of real-time PCR instrumentation, and it does not fall within the scope of this chapter to reproduce these further (Bustin, 2000; Bustin, 2002; Cockerill III, 2003; Herrmann *et al.*, 2006; Espy *et al.*, 2006; Kubista *et al.*, 2006).

The fluorescent options in the real-time PCR toolbox

Two categories of chemistry exist because of (i) the use of dyes that interact with any and all dsDNA (e.g. SYBR® green I), and (ii) the use of sequence-specific, fluorogenic oligoprobes (e.g. TaqMan® oligoprobes). In general however, the specific and non-specific fluorogenic chemistries detect amplicon with equal sensitivity (Wittwer *et al.*, 1997b; Newby *et al.*, 2003; Fernández *et al.*, 2006), despite the common misconception that adding an oligoprobe to a reaction will automatically make it more sensitive. The appearance of new fluorogenic chemistries has slowed in recent times and few applications using the most recently described chemistries have been reported for the specific detection, quantification and genotyping of microbes. Undoubtedly some of the newest chemistries are the result of pure research projects however it is tempting to propose that at least some of these chemistries have been developed in order to circumvent existing patents (and generate new ones) rather than convincingly demonstrate any significant advance to the field of real-time PCR.

DNA-associating dyes

The DNA-associating dyes are the basis of the non-specific, real-time PCR amplicon detection methods. Many of these fluorogenic molecules interact with dsDNA by associating with the minor groove of the DNA duplex. As a group these dyes display minimal fluorescence when free in solution but emit strongly when associated with dsDNA (which they do with high affinity) and exposed to a wavelength of light capable of exciting that dye. These are the simplest real-time PCR reporter systems. Ethidium bromide (Higuchi *et al.*, 1992), as described earlier, was the original dye used in this context but now this class includes YO-PRO-1 (Ishiguro *et al.*, 1995; Tseng *et al.*, 1997), SYBR® green I (Wittwer *et al.*, 1997a), SYBR Gold (Tuma *et al.*, 1999), BEBO (Bengtsson *et al.*, 2003), BOXTO (Lind *et al.*, 2006), LCGreen (Wittwer *et al.*, 2003; Herrmann *et al.*, 2006) and SYTO9 (Monis *et al.*, 2005). The asymmetric, minor groove-binding (MGB) cyanine dyes are very well represented amongst this list and are a subject of ongoing research and development (Kubista *et al.*, 2004). The non-specific chemistries are relatively inexpensive, do not require additional oligonucleotide design or chemical conjugation and are minimally affected by small changes in template sequence which may abolish hybridization of an oligoprobe, even when primers have successfully amplified the template (Komurian-Pradel *et al.*, 2001). A recent study showed that nearly half of all possible single nucleotide mismatches in an oligoprobe-binding site could not be detected by a TaqMan oligoprobe but were detectable using SYBR Green I (Papin *et al.*, 2004). This provides a perfect example of how we can now see "into" the workings of the PCR thanks to fluorogenic detection methods and obtain more reliable data than we ever could have using end-point methods such as PCR-ELISA which is more likely to report that mismatches in an oligonucleotide have no significant effect on product yield (Christopherson *et al.*, 1997). However, recent evidence suggests that SYBR green I preferentially binds to amplicon species that melt at higher temperatures, indicating a preference for G+C-rich regions (Giglio *et al.*, 2003). Possibly, SYBR Green I binds to high-temperature duplexes after melting off lower-temperature duplexes due to its use at non-saturating concentrations; resulting in a hierarchy of melting peak heights that may have no relation to starting template concentration or sequence (Wittwer *et al.*, 2003). Such unreliability precludes efficient microbiological multiplex real-time PCR applications which employ this dye to discriminate amplicon based on melting curve analysis. Furthermore, apart from its direct toxicity (especially if used at saturating concentrations), SYBR green I can degrade into products capable of inhibiting the PCR (Karsai *et al.*, 2002). A recent addition to this chemistry, LCGreen™ I, displays a higher sensitivity for lower-temperature duplexes (Wittwer *et al.*, 2003) and is the favored dye for High Resolution Melting applications (Wittwer *et al.*, 2003; Herrmann *et al.*, 2006).

Primer–dimer formation is common during PCR and is strongly associated with entry of the PCR into a plateau phase of amplification (Figure 1.4; (Chou *et al.*, 1992; Halford, 1999; Halford *et al.*, 1999)). Association of a DNA-binding fluorophore with primer–dimer, or with another non-specific amplification product, can confuse interpretation of the PCR results. The formation of primer–dimers is also more common in the presence of little or no specific template when idle primers become the dimers' play-thing. Unfortunately, limited template can be a frequent occurrence when analyzing nucleic acids purified from patient specimens for the presence of microbial targets. The presence of primer–dimer can sometimes be revealed using software capable of melting curve analysis (Ririe *et al.*, 1997; Wittwer *et al.*, 1997a) producing fluorescent dissociation curves. In the context of the non-specific dyes, melting curve analysis can be completed in minutes, requires no amplicon manipulation and utilizes the temperature at which a dsDNA amplicon denatures or "melts" (T_D; Figure 1.1). The shorter primer–dimer can theoretically be discriminated by its reduced T_D compared to the full-length amplicon. Practically, this discrimination is heavily reliant upon the comparative G+C content of the specific amplicon and of the primer–dimer or larger non-specific products, as well as the length of the amplicon. The contribution of non-specific products to the accumulating fluorescence signal during the PCR can reportedly be reduced by selecting a data collection temperature above the known T_D of the unwanted products (Morrison *et al.*, 1998; Ball *et al.*, 2003) but only if those products melt at a discernibly different T_D compared to the specific amplicon and after some preliminary melting experiments to determine the T_D of all possible unwanted products. Of course this does not prevent the production and amplification of primer–dimers, merely the detection of them. Unfortunately this approach may reduce the total signal since primer–dimers may contribute significantly to the overall fluorescence, while the higher data acquisition temperature may also reduce fluorescence levels. This approach is not a substitute for well designed primers (and perhaps a titration of magnesium ions).

The melting of dsDNA in parallel with constant monitoring of the changes in fluorescence can be prone to error and is reliant upon the abilities of the instrument to resolve small differences in temperature (Herrmann *et al.*, 2006). Also, analysis of the melting character of an amplicon in the presence of SYBR green I, initially observed by experimental accident, has shown that the sensitivity of DNA-binding fluorophores is limited by non-specific amplification products which exert greater influence when a limited amount of the specific template is present (Wittwer *et al.*, 1997a; Gingeras *et al.*, 2005). Unfortunately there is no reliable real-time PCR solution to discriminate non-specific amplicon when it has a similar or higher T_D to the target amplicon species. Nonetheless low-resolution confirmation of amplicon identity is the most common application of the non-specific fluorophores (Herrmann *et al.*, 2006) although species discrimination is also possible with well-designed assays (Merien *et al.*, 2005).

DNA binding fluorophores increase the T_D and broaden the melting transition of dsDNA, requiring substantial sequence change to produce a noticeable shift in the T_D compared to using the T_M to discriminate nucleotide polymorphisms. Oligoprobes permit a much clearer discrimination of single point sequence variations using the T_M (Wetmur, 1991). They are especially useful for discriminating related microbes and are our method of choice for rapid microbial genotyping.

Fluorogenic oligoprobe and primer chemistries

The adoption of oligoprobes to detect amplicon has added an additional layer of specificity to the PCR by indirectly confirming the sequence of the amplicon, in addition to employing a pair of sequence-specific primers. By using an excess of oligoprobe, the time required for it to hybridize with its target is significantly reduced, especially after PCR amplification (Morrison *et al.*, 1989; Wetmur, 1991). Similar benefits are not enjoyed by the PCR-ELISA and Southern blot probe-based methods since neither of these can fully exploit the speed aspect.

A candidate fluorescent dye should efficiently absorb and emit energy, releasing those emissions at dissimilar wavelengths so that excitation and emission can occur concurrently. An optimal label must also easily attach to DNA and be detectable at low concentrations. The label or labels should produce an altered signal upon specific hybridization or shortly thereafter, remain biologically innocuous and remain functional after periods at elevated temperatures and after long-term storage on the shelf. Additionally, the label should not interfere with the activity of the polymerase (Matthews and Kricka, 1988; Holland *et al.*, 1991). The relationship between a fluorescence signal indicating positive hybridization and a signal from unwanted fluorescence is often referred to as the signal to noise ratio. A high signal to noise ratio is preferred (i.e. more signal, less noise) Generally speaking, this can best be achieved by oligoprobe chemistries that utilize a non-fluorescent quencher and require the lowest number of molecular interactions to generate a signal.

While the most common oligoprobes are based on conventional nucleic acid chemistries, peptide nucleic acid (PNA) is becoming an increasingly popular choice for oligonucleotide backbones. PNA is a DNA analogue formed from neutral, repeated N-(2-aminoethyl) glycine units instead of the negatively charged sugar-phosphates of DNA (Egholm *et al.*, 1993). The PNA oligoprobe retains the same sequence recognition properties of DNA but it cannot be extended or hydrolyzed by a DNA polymerase. A more recent family of DNA analogues are the locked nucleic acids (LNA; 2′-O,4′-C-methylene-β-D-ribofuranosyl) (Obika *et al.*, 1998; Koshkin *et al.*, 1998; Singh *et al.*, 1998; Kumar *et al.*, 1998; Petersen and Wengel, 2003). LNA are modified nucleic acids in which the sugar has been conformationally "locked," imparting unprecedented hybridization affinity towards DNA and RNA. LNA monomers can be incorporated into a synthetic oligonucleotide as desired using conventional phosphoramidite chemistry. Their addition increases the thermal stability of the oligonucleotide, permitting the construction of very short 8–11 nt oligoprobes for real-time PCR applications, even shorter than TaqMan-MGB oligoprobes (Simeonov and Nikiforov, 2002; Goldenberg *et al.*, 2005). This is extremely useful when only short regions of suitably conserved or discriminating sequence are available and these bases have been used in several oligoprobe formats described below.

Other modified nucleotides include SuperA™ and SuperT™. Each are designed to bind more strongly to their complementary base and can be included in oligonucleotides (Epoch Biosciences). SuperG™ has been modified to reduce self-association and permit oligonucleotide designs with runs of Gs. SuperN™ is designed to bind to any base with equal stringency.

General considerations for the design of a fluorogenic oligoprobe should include a length of 15 to 35 nt with a G+C content of 40 to 60%. The oligoprobe should not contain clusters of a single nucleotide, particularly G, neither should it have repeated sequence patterns or hybridize with the sense or antisense primers. An oligoprobe should also have a T_M at least 5°C higher than that of the primers, to ensure the oligoprobe(s) hybridizes with its template before extension of the primers begins (Landt, 2001). This caveat also applies to the components of a multi-oligonucleotide fluorogenic chemistry requiring two or more hybridization events for effective signaling such as the tripartite molecular beacons, universal template primer or duplex Scorpion® primers.

Deoxyguanosine nucleotides (G) naturally quench some dyes (e.g. FITC) in a position-dependent manner (Crockett and Wittwer, 2001). This quenching varies linearly with a defined concentration range of template. The level of quenching can be increased if more Gs are present or if a single G is located in the first non-hybridized position of the oligoprobe and amplicon duplex. In this position, the G will be located one nucleotide beyond ("overhanging") the dye-labeled terminus of the probe, where it will remain free once the oligoprobe has hybridized to the target amplicon. A hybridized G does not quench to the same degree as a free G. Since a single labeled oligonucleotide is easy to design and use and relatively simple to synthesize, this knowledge of natural quenching has been applied to some amplicon detection approaches (e.g. LUX™ primers described below). In addition, natural quenching does not require a DNA polymerase with nuclease activity (Crockett and Wittwer, 2001). However, this system should be applied cautiously as the reduction in fluorescence is not as efficient as with a dedicated quencher and is more favorable for fluorophores emitting in the green and yellow wavelengths (Marras *et al.*, 2002). This application has typically been described for SNP analysis where template amounts are not as much of an issue as they are in microbiology (Leman *et al.*, 2006).

All the non-incorporating, nucleotide-based, oligoprobe chemistries described in the following sections include a 3′ phosphate or similar moiety (Cradic *et al.*, 2004), which blocks their extension by the DNA polymerase thus preventing the oligoprobe acting as a primer, but with no impact on the amplicon yield. These fluorogenic chemistries can be further divided into two classes; those which are destroyed to produce fluorescence and those which are not. Oligoprobes depending on a destructive process for signal generation are usually located close to the primer that hybridizes to the same strand in order to maximize the chance that polymerase will make early contact with the bound oligoprobe.

Non-destructive oligoprobes are usually located as far as possible from the primer (unless they are themselves the primer) on the same strand to ensure signal is produced before the polymerase dislodges the duplex.

Destructive oligonucleotide systems

In 1991, Holland *et al.* described a technique that was to form the foundation for homogeneous PCR using fluorogenic oligoprobes. The 5′ to 3′ nuclease activity of Taq DNA polymerase on specific oligoprobe and target DNA duplexes was monitored using radiolabel incorporated into the amplicon (Holland *et al.*, 1991). The products were examined using thin layer chromatography and the presence of hydrolysis was used as an indicator of specific duplex formation.

Lee *et al.* (1993) later reported an innovative approach using nick-translation PCR in combination with a dual-fluorophore labeled oligoprobe (Lee *et al.*, 1993). In the first truly homogeneous assay of its kind, one potentially fluorescent dye was added to the 5′ terminus and one to the middle of a sequence specific oligoprobe. When in such close proximity, the 5′ reporter fluorophore (6-carboxy-fluorescein; FAM) transferred laser-induced excitation energy by FRET to the 3′ quencher fluorophore (6-carboxy-tetramethyl-rhodamine; TAMRA). TAMRA emitted the new energy at a wavelength that was monitored but not specifically utilized in the presentation of data. However, when the oligoprobe hybridized to its template, the nuclease activity of the DNA polymerase destroyed the oligoprobe and released the fluorophores. Once the fluorophores were separated, the reporter's emissions were no longer quenched and the instrument detected and presented the resulting fluorescence data. These oligoprobes have been called hydrolysis, 5′ nuclease, dual-labeled or most commonly, the trade name has been used; TaqMan® oligoprobes (Figure 1.5a). Except where specific examples are provided, we will use the first term.

To make the hydrolysis oligoprobes more user-friendly, a platform was required that could both excite and detect fluorescence in addition to thermal cycling. In 1993 a charge-coupled device described the year before was combined with a thermal cycler, resulting in the first real-time PCR fluorescence excitation and detection instrument (Nakayama *et al.*, 1992; Higuchi *et al.*, 1993). In 1996 the first commercial platform was released, the ABI

Figure 1.5 (a) Following primer hybridization, the DNA polymerase (Pol) progresses along the relevant strand during the extension step of the PCR, displacing and then hydrolyzing the oligoprobe. Once the reporter (R) is removed from the extinguishing influence of the quencher (Q), it is able to release excitation energy. Inset shows the NFQ (Q, filled hexagon) and minor groove binding molecule (grey diamonds) which make up the TaqMan®-MGB oligoprobes. These reporter systems acquire data from the reporter's emissions. Data can be collected during the annealing or extension steps of the PCR; (b) After the universal template (UT) primer is extended, the nascent strand acts as the template for the second primer. The polymerase encounters and hydrolyses the UT-oligoprobe (whilst extending the second primer; dashed line) in the same fashion as an hydrolysis oligoprobe. This system can produce fluorescence data from the emissions of the released fluorophore during the annealing or extension steps; (c) When the strand incorporating the DzyNA primer is duplicated by a complementary strand (dashed line), a DNAzyme is created. A complementary, dual-labeled oligonucleotide substrate will be specifically cleaved by the DNAzyme releasing the fluorophore (F; circle) from its proximity to the quencher, releasing the labels and permitting fluorescence. Data can be collected during the annealing or extension steps of the PCR; (d) Adjacent hybridization probes (HybProbe®

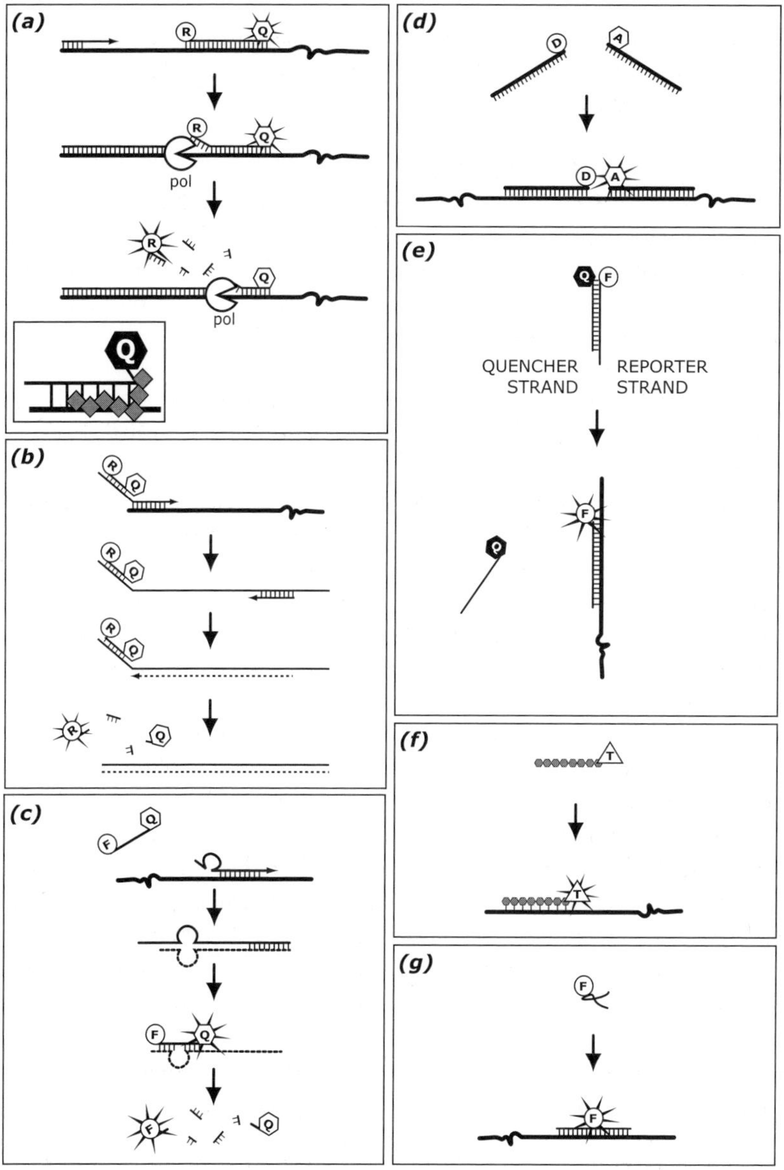

probes) can produce a FRET signal due to interaction between the donor (D) and acceptor (A) spectra which is detected during the annealing step of the PCR. This system acquires its data from the acceptor's emissions; the opposite of the hydrolysis oligoprobe chemistry; (e) The yin-yang oligoprobes consist of two hybridized oligonucleotides. The shorter NFQ-labeled strand is displaced when the fluorophore-labeled (F; circle) strand preferentially hybridizes to the longer specific amplicon strand. Data are collected from this system during the annealing step of the PCR; (f) The Light-Up™ Probe has a PNA backbone and generates a fluorescent signal when the asymmetric thiazole orange fluorophore (T; open triangle) hybridizes to the amplicon strand. Data are collected from this system during the annealing step of the PCR; (g) The fluorophore of the HyBeacon oligoprobe emits fluorescence when in close proximity to DNA as occurs upon hybridization with the specific amplicon strand. Data are collected from this system during the annealing step of the PCR.

Prism® 7700 sequence detection system (Perkin Elmer Corporation/Applied Biosystems, USA), and it remains in use today. Its accompanying chemistry, the hydrolysis oligoprobes, remain the most popular specific chemistry, followed by SYBR® Green I which was released on the next commercial platform, the LightCycler (Wittwer *et al.*, 1997b). Evaluation of TaqMan oligoprobes with LNA substitutions resulted in real-time PCR curves with a better linear phase and earlier C_T values than conventional TaqMan oligoprobes (Reynisson *et al.*, 2006). This may have been caused by their slightly decreased length, resulting in an enhanced level of quenching due to the changed proximity of reporter and quencher and thus an improved signal to noise ratio.

A modification to the hydrolysis oligoprobe chemistry resulted in the MGB oligoprobes usually known as TaqMan-MGB probes (Figure 1.5a, inset; (Afonina *et al.*, 2002a)). This chemistry has a fluorescent reporter dye at the 5′ end and an NFQ at the 3′ end (although they were first reported with a TAMRA quencher). In addition the oligoprobe has an MGB molecule at the 3′ end that further stabilizes an oligoprobe-target duplex by folding into the minor groove of the dsDNA (Kutyavin *et al.*, 2000; Afonina *et al.*, 2002b). In the unbound state the oligoprobe assumes a random coil configuration that is quenched. When hybridized, the stretched oligoprobe is able to fluoresce. However, as with many dual-labeled oligoprobe systems, this relaxed state can "leak" fluorescence at higher temperatures when the oligoprobe is prone to partial unfolding. We have, accidentally, experienced such leaky fluorescence when developing a TaqMan assay for human metapneumovirus (HMPV) in which we employed asymmetric primer concentrations that may have created suitably inefficient conditions for us to demonstrate leakage in the form of a melting peak under controlled conditions (Mackay *et al.*, 2003).

The TaqMan-MGB chemistry permits the use of very short (12–17 nt) oligoprobes because of a 15 to 30°C rise in their T_M as a result of the interaction of the MGB and the DNA helix, in particular its stabilization of A:T bonds. These short oligoprobes are theoretically ideal for detecting SNPs, since a short oligoprobe is more significantly destabilized by changes within the hybridization site than a longer oligoprobe. However, in practice the degree of discrimination depends on the base of interest, since some nucleotides can significantly influence fluorescence intensity and because different mismatches impart different degrees of instability. Originally this approach was designed for the allelic discrimination of genomic DNA where the ratio of heterozygote to wild-type sequence was 1:1. In situations where this ratio is significantly skewed, the discriminatory power of polymorphism detection by these oligoprobes may be reduced.

A chemistry similar to the TaqMan-MGB is the MGB-Eclipse™ oligoprobe, which replaces the standard TAMRA or DABCYL (4-[4′-dimethylamino-phenylazo]-benzoic acid) quencher with a proprietary NFQ (Eclipse dark quencher) and places it at the 5′ end along with the MGB molecule, with the fluorophore at the 3′ end (opposite to the TaqMan-MGB; also known as QuantiProbes™ from QIAGEN). The MGB molecule protects the oligoprobe from degradation by the polymerase and instead the signal is generated by stretching and relaxing of the oligoprobe during annealing and melting which also permits melting curve analyses. While stringent hybridization is useful for polymorphism analysis, this is not a generally desirable feature for routine detection of viruses unless an extremely stable hybridization site can be determined (Whiley and Sloots, 2006).

A recent addition to the destructive chemistry set is the universal template (UT) primer (Figure 1.5b; (Zhang *et al.*, 2003)). This system adds a generic or "universal" sequence to the 5′ end of one PCR primer with which a common hydrolysis oligoprobe can hybridize permitting the use of the same oligoprobe for different amplicons. The remainder of the primer provides assay specificity so that multiple primers can be used to amplify numerous targets. This chemistry may be useful for detecting viruses consisting of many genotypes such as the picornaviruses. During the second PCR cycle, the nascent strand is copied and the polymerase encounters and hydrolyses the bound oligoprobe, releasing the reporter and permitting fluorescence to be generated. Conceivably, locating the oligoprobe as far as possible from the primer on the same strand may lead to inefficient hydrolysis. Another universal system called common reporter real-time PCR takes a slightly different approach (Rickert *et al.*, 2004). In this system two sense primers are employed. The first sense primer incorporates two non-target sequences; one for an assay specific hydrolysis oligoprobe preceded by one for a second, universal sense primer as well as a 3′ target sequence. The sense primers and a target-specific reverse primer are included in the reaction but the universal sense primer only begins to amplify target after the antisense strand has been produced (which will include the complement of the universal sense primer's and the hydrolysis oligoprobe's sequence). Using the same approach, a distinctively labeled hydrolysis oligoprobe can be employed to detect a housekeeping gene within the same reaction. Because the oligoprobes in these two systems rely upon hybridization with a primer, it is possible that this chemistry will also produce fluorescence from the formation of non-specific amplicon.

The padlock probe is a new class of ligation-dependent oligoprobe designed primarily for human genotyping analysis by microarray but it has also proven useful to detect plant pathogens (Nilsson, 2006). The padlock probe contains target complementary sequence at each terminus of a ~100 nucleotide linear DNA strand, bracketing the hybridization sites for two primers and an hydrolysis oligoprobe. During hybridization to specific template the termini are brought close enough together to permit ligation and circularization (Szemes *et al.*, 2005). This linear DNA molecule has no fluorogenic potential by itself but the exonuclease-resistant, circularized molecules can be amplified by PCR along with the oligoprobe binding site permitting a real-time PCR assay.

Another TaqMan-like oligonucleotide sequence has been used as the signal-generating portion of the DzyNA-PCR system, now in a commercial variant entitled the Qzyme™ (Figure 1.5c; Todd *et al.*, 2000; Applegate *et al.*, 2002). The reporter and quencher are attached to the termini of an oligonucleotide substrate which is essentially an hydrolysis oligoprobe. Cleavage of the oligosubstrate is performed by a DNAzyme, which is created during the PCR. This creation is the result of the PCR duplicating an antisense DNAzyme sequence included in the 5′ tail of one of the primers. The duplicated sequence is the "functional" form of the DNAzyme. Upon cleavage of the substrate, the fluorophores are released permitting fluorescence in an identical manner to an hydrolysis oligoprobe. A similar approach is employed by catalytic molecular beacons which consist of a hammerhead deoxyribozyme appended to a traditional molecular beacon structure (Stojanovic *et al.*, 2001). These oligonucleotides are all quite complex requiring careful design and, because they contain non-target sequences, may perform with difficulty when employed to discriminate microbes within human genomic DNA-laden matrices where the existence of potentially similar, but non-target sequences abound.

To increase the specificity of the hydrolysis oligoprobe chemistry whilst retaining the process of probe destruction and fluorophore release a hairpin modification was created essentially replicating the structure of a molecular beacon (described below), but permitting the oligoprobe's hydrolysis (Kong *et al.*, 2002). This chemistry was first trialed on hepatitis B virus DNA but comprehensive sensitivity or specificity comparisons were not performed and subsequent use of this chemistry has not been reported for microbial studies.

Non-destructive oligonucleotide systems

Since these oligoprobes are not destroyed to produce a fluorescent signal, they can be re-used to examine the nature of the amplicon(s). Briefly, a suitable instrument denatures and anneals the dsDNA amplicon at the end of the amplification and then slowly increases the temperature beyond that required for denaturation. Fluorescence data are continually collected throughout and the point at which one or both oligoprobes melts off the template is apparent as a precipitous change in the fluorescence signal. Plotting the first derivative of this data (change in fluorescence with temperature) versus temperature results in a melting "peak" that can be used to identify differences between microbial strains, identify sequence drift and even function as an alternative approach to multi-probe multiplex real-time PCR.

Linear chemistries

The majority of fluorogenic oligoprobes fall into the class of linear oligoprobes. In fact, the use of a pair of adjacent, fluorogenic oligoprobes was first described in the mid 1980s, predating the hydrolysis oligoprobe chemistry but taking longer to achieve the same degree of commercial support. The pair of oligonucleotides were used to identify the distance between fluorophores on a complementary nucleic acid template and the system held promise for similar detection within living cells, hinting at further possibilities for diagnostic use (Heller and Morrison, 1985; Cardullo *et al.*, 1988). This chemistry is now known commercially as the hybridization probes (HybProbe® probes), having become the manufacturer's chemistry of choice for the LightCycler (Wittwer *et al.*, 1997b). The upstream oligoprobe is typically labeled with a 3′ donor fluorophore (FITC) and the downstream probe with either a proprietary LightCycler Red 500, 610, 640, 670 or Red 705 acceptor fluorophore at the 5′ terminus, with a phosphate at the 3′ terminus (or similar moiety; Cradic *et al.*, 2004) to prevent extension by the DNA polymerase. Ideally when both oligoprobes are hybridized to the amplicon, the two fluorophores are located within 5 nt of each other (Figure 1.5d) to allow FRET between the excited FITC dye and the particular red dye employed; and it is the red dye signal that is measured. The ratio of acceptor to donor emissions can also be used as an internal reference signal rendering the results independent of absolute fluorescence (Huang *et al.*, 2001).

The double-stranded or "Ying-Yang" oligoprobes function by displacement hybridization (Figure 1.5e; (Li *et al.*, 2002)). In this process a 5′ fluorophore-labeled oligonucleotide is, in its resting state, hybridized with a complementary, but shorter, quenching DNA strand that is 3′ end-labeled with an NFQ. Quenching reportedly occurs via a collisional mechanism similar to the hairpin chemistries (to be described below). When the full-length complementary sequence in the form of an amplicon is generated by PCR, the reporter strand preferentially hybridizes to the longer target amplicon strand, disrupting the quenched oligoprobe duplex and permitting the fluorophore to emit its excitation energy directly. This emission accumulates due to nuclease-mediated destruction of the reporter strand. This approach has been used for the detection of hepatitis B virus DNA (Kong *et al.*, 2003) and the quantification of plasmid-derived hepatitis C virus (HCV) sequences demonstrating an 8-$\log_{10}$ dynamic range of template concentration (Shengqi *et al.*, 2002). Modification of the chemistry via the addition of LNA bases has made hybridization more stringent, but at the expense of increased production cost without a significant improvement in performance (Kennedy *et al.*, 2006). The Ying-Yang chemistry is made more complex by its need for the quenching strand to re-anneal after dissociating with amplicon in order to complete the signaling process, failure of which could see a rise in background fluorescence.

The next two chemistries are very simple, single-fluorophore oligoprobes making them relatively easy to both design and use. The LightUp® probe (LightUp Technologies) is also a linear PNA, and is labeled with an asymmetric cyanine fluorophore, thiazole orange (TO; Figure 1.5f (Svanvik *et al.*, 2001)). The combination of these two design features conveys extraordinary shelf-life upon this chemistry making it a good candidate for ready-to-use kit-based assays (Leijon *et al.*, 2006). Because PNA does not interact with other proteins, this chemistry may be useful for direct detection of microbes in crude specimens. When hybridized with a nucleic acid target, either as a duplex or triplex, the fluorophore emits fluorescence due to proximity to the amplicon (Svanvik *et al.*, 2000). The LightUp probes do not interfere with the PCR, do not require conformational change and because they bind more strongly to DNA than do DNA oligonucleotides, they can be shorter than similar oligoprobes based on a DNA backbone. These probes are very sensitive to single nucleotide mismatches, permitting further amplicon characterization but creating similar concerns to those brought up for the TaqMan-MGB chemistry (Egholm *et al.*, 1993; Leijon *et al.*, 2006). Because a single reporter is used, a direct measurement of fluorescence can be made instead of the measurement of a change in fluorescence between two fluorophores thus bringing our detection system closer to direct analysis of amplicon accumulation (Isacsson *et al.*, 2000; Svanvik *et al.*, 2001). However, non-specific fluorescence has been reported during extended cycling (Svanvik *et al.*, 2000). The chemistry has matured to the commercial arena (ReSSQ®, LightUp Technologies) where it has been used to monitor human cytomegalovirus (CMV) and severe acute respiratory syndrome coronavirus (SARS-CoV) viral load in clinical and *in vitro* studies respectively (Wirgart *et al.*, 2005; Åkerstrom *et al.*, 2005).

The HyBeacon™ (LGC, United Kingdom) is a single linear oligonucleotide internally labeled with a fluorophore that emits an increased signal upon formation of a duplex between the target DNA strand and the HyBeacon (Figure 1.5g; French *et al.*, 2001; French *et al.*, 2002; Marks *et al.*, 2005). HyBeacons are principally applied for SNP detection where they perform comparably to the HybProbes. Recent improvements to the chemistry have also improved fluorescence emission intensity.

A dual-labeled PNA hybridization probe (Seitz, 2000) was also briefly known as the Lightspeed probe or linear PNA beacon (Stender *et al.*, 2002). The oligoprobe is analogous to the hydrolysis probe chemistry terminally labeled with a fluorophore and quencher, but it differs in its backbone which is PNA (Figure 1.6a). DNA (containing 2′-O-methylribonucleotides to prevent nuclease destruction) and PNA versions called "stemless molecular beacons" have been reported for the detection of synthetic HIV-1 templates where they incorporate an NFQ (Kuhn *et al.*, 2001; Kuhn *et al.*, 2002). The DNA versions suffer from high background fluorescence (Kuhn *et al.*, 2002). This format also includes a pair of PNA "openers" which require two short, closely located oligopurine stretches spanning the oligoprobe target region. The openers bind and hold apart the region of the DNA duplex, permitting oligoprobe insertion without the need for dsDNA denaturation (Kuhn *et al.*, 2001). In an aqueous solution the PNA backbone brings the fluorophore and quencher into close proximity, quenching the systems. When the probes hybridize to a specific target, they flex open and fluorescence is possible.

The result of combining a single sequence-specific, Cy5-labeled linear oligoprobe with SYBR green I created the Bi-probe system. This functions via a variation of FRET termed induced FRET (iFRET; (Cardullo *et al.*, 1988; Howell *et al.*, 2002)). Bi-probes are more specific than using SYBR green I alone and are simpler to design and manufacture than other oligoprobes but retain an enhanced signal to noise ratio (Brechtbuehl *et al.*, 2001; Walker *et al.*, 2001). Interestingly this approach functioned on the Idaho Technologies LightCycler but not the closely related Roche LightCycler version 1.0. The disparity was due to the narrow band pass filter sets employed by the latter instrument, which prohibited FRET between these two particular fluorophores. A similar technical problem rendered some proprietary LightCycler fluorophores unusable on the ABI PRISM 7700 (Nitsche *et al.*, 1999). The commercial form of the Bi-probe chemistry, called the ResonSense® probe

---

Figure 1.6 (a) In aqueous solution the PNA backbone (grey hexagons) of the Lightspeed probe forms a random coil conformation that quenches fluorescence due to the proximity of the fluorophore and quencher (Q, hexagon). Upon hybridization this system is stretched open sufficiently to permit fluorescence which can be recorded during the annealing step of the PCR; (b) Hybridization of the loop sequence of the molecular beacon to the target separates the fluorophore and NFQ (Q, filled hexagon) allowing fluorescence. Data from these systems are collected during the annealing step of the PCR. Inset shows a wavelength-shifting hairpin oligoprobe incorporating a harvester molecule (H, filled circle); (c) Upon binding of the tripartite molecular beacon to specific amplicon the fluorophore is removed from the influence of the NFQ which permits fluorescence. Data are collected during the annealing step of the PCR. (d) The Amplifluor™ primer's fluorophore is separated from the NFQ during disruption of the primer's hairpin structure. The disruption occurs during creation of the complementary strand of the nascent DNA incorporating the Amplifluor primer (i) and whenever nascent dsDNA duplexes form during reannealing. An intermediate primer may also be employed for SNP studies, and then the Uniprimer is designed to target the 5′ sequence appended to the primer (ii). The data from this system are collected during the annealing step of the PCR; (e) The Scorpion® primer is blocked from being extended by a hexethylene glycol molecule (B; diamond) so that the hairpin can only be disrupted by specific hybridization and not by the extension of a complementary amplicon strand. The fluorophore is separated from a 3′ methyl red quencher during self-hybridization of the loop portion of the Scorpion with a complementary region on the nascent amplicon strand. Inset shows a duplex Scorpion. Data are collected from this unimolecular system during the annealing step of the PCR; (f) The IntraTaq primer is very similar to the Scorpion primer except that the signaling portion functions when the polymerase disturbs and hydrolyses the 5′ end of the primer during the process of creating

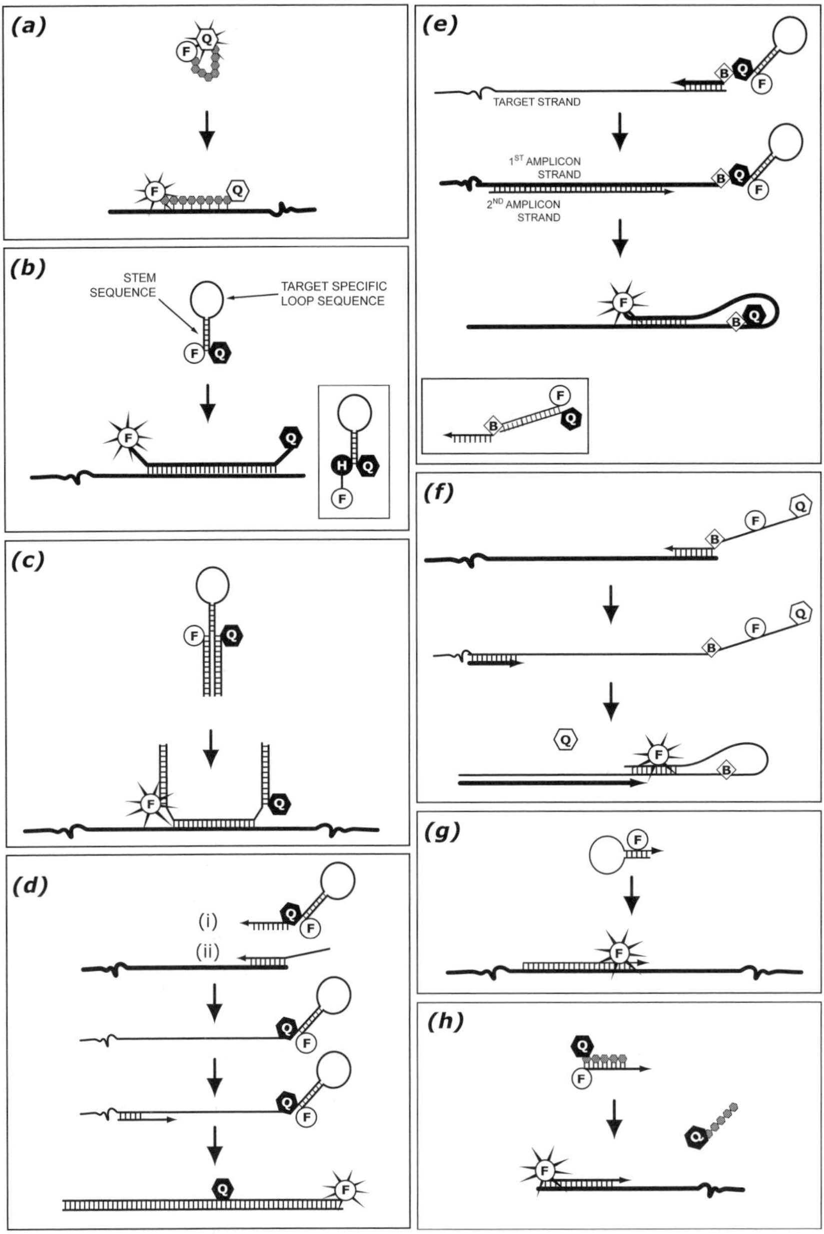

the 2nd nascent strand; (g) The LUX™ primer is labeled with a single fluorophore positioned next to a guanine nucleotide when the hairpin portion of the primer is intact. The G naturally quenches the fluorophore. In the presence of the specific target strand the primer hybridizes, disrupts the hairpin and is extended. The fluorophore is now free to fluorescence and data can be collected from this signaling system during the annealing or extension step of the PCR. (h) In the absence of the longer specific amplicon, quenching of the Q-PNA displacement primer chemistry is achieved by a short NFQ-labeled PNA backbone (grey hexagons) designed to hybridize with the fluorophore-labeled primer. Fluorescence data can be collected during the annealing and extension steps of the PCR.

(Defence Science and Technology Laboratory, United Kingdom; (Lee *et al.*, 2002)), has overcome the platform-specific incompatibility, possibly due to the substitution of SYBR Gold for SYBR Green I and the use of Cy5.5.

Single-fluorophore systems have also been described for melting curve analysis to discriminate homozygous from heterozygous DNA without the need for FRET (Kurata *et al.*, 2001; Gundry *et al.*, 2003). In one format, the fluorophore is carefully chosen and positioned so that its emissions are quenched by proximity to a complementary guanine (Crockett and Wittwer, 2001; Leman *et al.*, 2006). This approach suffers if used for the detection of nucleotide polymorphisms in a guanine-rich sequence. A commercial form, called the SimpleProbe® probe is used on the LightCycler and the genotyping platform, the LightTyper (Roche Diagnostics). The SimpleProbe targets a single 5′ or 3′ fluorescein labeled hybridization oligoprobe to a polymorphism and produces an increased fluorescence when hybridized. Genotyping is performed by melting curve analysis (Chase *et al.*, 2005).

Hairpin oligoprobes

Molecular beacons were the first fluorogenic hairpin oligoprobes described for real-time PCR applications. The hairpin oligoprobe's fluorophore and quencher are positioned at the termini of the oligoprobe (Figure 1.6b). Fluorophores have included EDNAS (5-[2′-aminoethyl] aminonapthalene-1-sulfonic acid), FAM, TET (tetrachloro-6-carboxyfluorescein), HEX (hexaclhloro-6-carboxy fluorescein), TAMRA and ROX (5-carboxyrhodamine-X) while the most commonly used quencher has been DABCYL. Gold nanoparticles have also been used successfully as quencher molecules (Dubertret *et al.*, 2001; Goel *et al.*, 2005). The labels are held in close proximity by an adjacent region of homologous base pairing deliberately designed to create a hairpin structure encompassing a target-specific sequence that forms a bulging loop. The intimate proximity of the label molecules results in quenching by direct energy transfer via a collisional mechanism (Tyagi *et al.*, 1998; Marras *et al.*, 2002). In the presence of a sequence complementary to the molecular beacon's loop, the oligoprobe is shifted into an open configuration. The reporter is then separated from the quencher's influence and fluorescence can be detected (Tyagi and Kramer, 1996). Molecular beacons are not intended to be hydrolyzed during the PCR. Hydrolysis is avoided by careful design to ensure that the specific hybridization occurs during annealing but does not continue during extension. Studies have employed different approaches to demonstrate that mostly intact molecular beacon oligoprobe remains following amplification indicating they are not destroyed in the reaction. Experiments have included the use of polymerases which lack endonuclease activity and gel electrophoresis of reaction products to directly determine that there is little destruction of oligoprobe. Additionally, similar fluorescent signaling patterns have resulted when employing non-cleavable backbone chemistries in a molecular beacon structure while the use of non-polymerase related amplification methods such as NASBA and SDA have produced the same fluorescence patterns.

Wavelength-shifting hairpin oligoprobes were introduced to improve the hairpin oligoprobe chemistry, making use of a second, "harvesting" fluorophore (Inset, Figure 1.6b). The harvester passes on excitation energy acquired from a blue light source as fluorescent energy in the far-red spectrum. A receptive "emitter" fluorophore can then be selected which uses the energy to produce light at characteristic wavelengths. This approach offers the potential for improved multiplex real-time PCR and nucleotide polymorphism analysis by increasing the number of emitters that can be excited using a single energizing wavelength (Tyagi *et al.*, 2000). This is a useful workaround for instruments with a limited energizing light source although it may increase the manufacturing costs of the oligoprobes.

The multicomponent tripartite molecular beacons (TMB) are also hairpin oligoprobes (Figure 1.6c; (Nutiu and Li, 2002). These highly complex oligoprobes combine a molecular beacon's hairpin with longer, unlabeled, single-stranded arms. Each arm is designed to hybridize to an oligonucleotide labeled with either a fluorophore or an NFQ. The system is quenched in the hairpin state due to the close proximity of the labels but fluorescent when hybridized to the specific amplicon strand. However, for quenching to occur the arms must re-hybridize after the denaturation step. The TMB chemistry may be subject to high background fluorescence in practical applications where genomic nucleic acids and accumulating amplicon interfere with re-annealing.

Because the function of all the hairpin oligoprobes depends upon correct hybridization of the stem, accurate design is crucial to their function and is considerably more challenging than for other oligoprobe chemistries.

Self-priming, fluorogenic amplicon

The self-priming amplicon is similar in concept to the hairpin oligoprobe, except that the label(s) becomes irreversibly incorporated into the nascent amplicon. While these systems have only been described with a single primer, it is conceivable that both primers could be labeled in a similar way to provide a stronger, if more costly, fluorescent signal. The first contact between primer and template is a bimolecular event, but from then on the signaling is unimolecular. Intramolecular hybridization is extremely fast and kinetically favorable since it does not rely upon the chance meeting of oligoprobe and amplicon each cycle (Bustin, 2002). Fast cycling conditions appear to better suit the chemistries in this group (Thelwell *et al.*, 2000). Three approaches have been described: sunrise primers (Amplifluor™ hairpin primers, Chemicon International), Scorpion primers (Nazarenko *et al.*, 1997), and the light upon extension primers (LUX; Whitcombe *et al.*, 1999).

Amplifluor primers consist of a 5′ fluorophore and a 3′ DABCYL NFQ (Figure 1.6d). The labels are brought together by complementary sequences that create a stem, closing the primer. A target-specific primer sequence is located at the 3′ terminus downstream of the NFQ. The sunrise primer's sequence is intended to be duplicated by the nascent complementary strand and during subsequent annealing and extension steps, the stem is destabilized, the two fluorophores are forced approximately 20 nucleotides (70 Å) apart and the fluorophore can emit excitation energy (Nazarenko *et al.*, 1997). The Uniprimer or universal energy transfer primer is a variant of the Amplifluor primer which improves this chemistry's power to discriminate nucleotide polymorphisms as well as reduces costs by re-using a single pair of fluorogenic primers for multiple SNP targets. Uniprimers are described for the identification of which of two possible nucleotides may be at the site of interest, and in this role are most useful for human SNP studies (Bengra *et al.*, 2002). Two primers are designed such that the 3′ terminal nucleotide sits on the site of polymorphism. A "universal" label sequence (which is the complement of the 3′ sequence of one of two detecting Uniprimers) is included at the 5′ end of these primers. In this way, multiple primer sets can be designed to detect different variable targets, but only a single pair of fluorogenic primers is required to detect which of two nucleotide variant are present in the template. This approach may be extremely useful for the development of multiplex microbial screening assays in which many primer sets are included, but only one Uniprimer. This would overcome the added expense inherent to the need for multiple oligoprobes in multiplex real-time PCR assays (Rickert *et al.*, 2004). Uniprimer positives could then be further examined with individual, pre-existing real-time PCR assays. It is possible however, that the Amplifluor's duplication could also occur during the formation of non-specific amplicon, producing reduced signal to noise ratios or false positive fluorescence.

The Scorpion primer is similar in design to the sunrise primer except for a hexethylene glycol molecule that blocks duplication of the signaling portion of the Scorpion, which increases manufacturing complexity and costs (Figure 1.6e). In addition to the difference in structure, the function of the Scorpion primer varies in that the 5′ region of the oligonucleotide is designed to hybridize to a complementary region within the nascent amplicon strand creating an intramolecular signaling system which combines hairpin and self-priming mechanisms. This approach to signal generation separates the labels disrupting the hairpin and permitting fluorescence in the same way that hairpin oligoprobes function (Whitcombe *et al.*, 1999). The duplex Scorpion primer requires the primer, a quencher and the target amplicon to interact. The stem-loop is exchanged for a separate, complementary oligonucleotide labeled with a quencher at the 5′ terminus. The additional oligonucleotide interacts with the primer element which is terminally labeled with the fluorophore (Figure 1.5e, inset; Solinas *et al.*, 2001). When it exists as a duplex, the chemistry is quenched but becomes fluorescent after hybridization to the longer specific amplicon strand. Interestingly, because the quencher is not part of the same molecule, brighter fluorescence can be achieved than with the unimolecluar version where the labels are neither absolutely quenched, nor freely fluorescent. Uniplex Scorpion primers have been used to quantify HIV-1 load (Saha *et al.*, 2001) and can be used for nucleotide polymorphism detection. In the latter role either the primer component or the probe component covers the polymorphism (Whitcombe *et al.*, 1999; Thelwell *et al.*, 2000). As with any microbial genotyping, signal intensity and thus results, may suffer if the target region is subject to any nucleotide variation.

A combination of the hydrolysis probe and a Scorpion primer was been presented as the intramolecular TaqMan, or IntraTaq probe (Figure 1.6f; Solinas *et al.*, 2002). This chemistry exhibits the proprieties of an hydrolysis oligoprobe combined with a self-probing chemistry since it is also part primer. Only when the probing portion has annealed to the newly formed amplicon, will the exonuclease activity of the polymerase be able to cleave the link between the reporter and quencher.

Another variation of the hairpin and self-priming fluorophore chemistries is the self-quenching hairpin oligonucleotide primer which has been commercially denoted the LUX fluorogenic primer (Figure 1.6g; Nazarenko *et al.*, 2002). At annealing temperatures, this chemistry is non-fluorescent in the absence of specific amplicon through the natural quenching ability of a carefully placed guanosine nucleotide. The natural quencher is brought into close proximity with responsive FAM (carboxyfluorescein) or JOE (carboxy-4′,5′-dichloro-2′,7′-dimethoxyfluorescein) dyes via stretches of 5′ and 3′ complementary sequences. In the presence of the complementary target strand, a nascent strand is extended, which incorporates the LUX primer. The dsDNA opens the hairpin, permitting fluorescence from the fluorophore. This non-destructive chemistry is simple to design and use, relatively inexpensive and it does not require the inclusion of an additional oligoprobe. However, the presence of primer–dimer and non-specific amplicon could, as for SYBR Green I, the sunrise primers and the UT-oligoprobe, be displayed as a fluorescent signal. The products can be discriminated using their differing T_D however there is no reliable way to account for their signal in quantitative real-time PCR. A variant of these chemistries is the Angler® oligoprobe which closely resembles a Scorpion primer, except for the absence of a quenching moiety (Lee *et al.*, 2002). Angler oligoprobes consist of a 3′ specific primer sequence linked via a hexethylene glycol molecule to a 5′, Cy5-labeled tail. The tail portion of the Angler is designed to self-hybridize to downstream sequences in the nascent strand, producing a unimolecular signaling system. In contrast to the dual-labeled Scorpion, the Angler uses FRET between the terminal Cy5 and SYBR gold incorporated into the self-annealed duplex to produce a fluorescent signal.

The technique of hybridization displacement can also be used with a fluorophore-labeled primer and because additional stringency is provided by the longer complementary strand, the system performs its own "hot-start" as exemplified by an NFQ-labeled PNA strand (Q-PNA; Figure 1.6h; Fiandaca *et al.*, 2001; Stender *et al.*, 2002). In the Q-PNA approach, a short quenching PNA oligoprobe is bound to an unincorporated fluorogenic primer such that the NFQ and fluorophore are adjacent, resulting in a quenched system. However, once the dsDNA amplicon is created by primer extension, the shorter Q-PNA is displaced in favor of the longer target amplicon, after which the fluorophore can fluoresce. During cycling, increasing amplicon concentrations will compete with reannealing of the quencher strands and could cause a non-specific, cumulative rise in background fluorescence. This issue should be considered for all oligoprobe systems, particularly the multicomponent chemistries which require some "assembly" to complete each cycle of reporting and quenching.

A different approach to the self-quenching amplicon resulted from the creation of a completely new nucleotide base pair (Switzer *et al.*, 1989) formed by isoguanine (iG) and 5′-methylisocytosine (iC) which were labeled with a fluorophore and quencher (Johnson *et al.*, 2004; Sherrill *et al.*, 2004). In this example the sense primer was 5′ labeled with fluorophore-modified diCTP, and free diGTP conjugated to DABCYL was incorporated into the nascent amplicon using a nuclease-deficient *Taq* DNA polymerase. This approach was able to detect 1 to 10 copies of DNA or RNA derived from HIV-1 in a real-time PCR format. Another new base pair (Piccirilli *et al.*, 1990) was employed to create a universal real-time or end-point PCR signaling system (Moser *et al.*, 2003). Similar to the Uniprimer system, a universal label sequence is incorporated into the 5′ terminus of the sense primer, separated from the target specific sequence by iG and iC bases. This separation provides an entry-point for nuclease activity which hydrolyses an oligoprobe labeled with an internal DABCYL quencher and a 3′ fluorophore.

Comparison of fluorogenic chemistries

While reviewing the previous signaling options one should note the number of hybridization events required to generate and quench the fluorescent signal (a signaling "event").

Each fluorogenic oligonucleotide must meet and hybridize with its specific amplicon (a bimolecular event for TaqMan, TaqMan-MGB, HyBeacon, Lightspeed probe, SimpleProbe and LightUp probes). If there is a second oligoprobe required, either to provide a quenching moiety or as a partner for FRET, the likelihood of a chance encounter between all three molecules theoretically decreases (a trimolecular event, e.g. HybProbes, Yin-Yang oligoprobes and the Q-PNA), especially as the amplicon concentration increases favoring dsDNA strand re-annealing but also generally interfering with hybridization via random collision. This should also be extended to hairpin oligoprobes which signal via reversible hybridization (e.g. molecular beacons). But the fluorogenic primers complicate this classification. They are incorporated into each nascent amplicon strand only once and do not need to meet and hybridize anew with amplicon each cycle. Nonetheless that first hybridization between primer and template is a bimolecular event. The incorporated primer then generates a signal principally because it has been forced open (e.g. Sunrise, LUX and IntraTaq primers), or because of intramolecular hybridization (a unimolecular event, e.g. Scorpion primers). One should also remember that despite the complexity of multimolecular interactions, systems employing them should theoretically function with higher specificity driven by the necessity for multiple, sequence-specific interactions to generate the signaling event.

The most popular chemistries can often display unique signal-generation quirks. The cyclical destruction of hydrolysis oligoprobes will theoretically continue despite a plateau in amplicon accumulation whereas SYBR green I fluorescence can start to increase non-specifically during later cycles, even without the intended template, due to the presence of primer–dimers. HybProbe fluorescence can begin to decrease in later cycles as the rate of collision between the amplicon strands increases, resulting in a phenomenon called the "hook effect" (Figure 1.4). At this stage, the formation of dsDNA is favored over the hybridization of oligoprobe to its target DNA strand, adversely affecting total fluorescence. Whilst these end-point effects are mainly cosmetic (since they don't affect the C_T), they may interfere with genotyping via melting curve analysis and require primer ratio optimization (Barratt and Mackay, 2002). The possibility exists that some linear oligoprobes are consumed by sequence-related nuclease activity, which may also contribute to the hook effect as has been reported for HybProbes (Lyamichev *et al.*, 1993; Wittwer *et al.*, 1997a; Wilhelm *et al.*, 2001b). Molecular beacons, TaqMan-MGB, Scorpion primers and LNA-containing oligoprobes are extremely sensitive to nucleotide mismatch, more so than conventional TaqMan oligoprobes (Täpp *et al.*, 2000). This is a positive feature for their use in SNP analysis, but a burden for their use in detecting viral templates especially those obtained from RNA viruses which exhibit more genetic variation than DNA viruses.

It is worth pointing out that functional comparison between chemistries is complicated by the different requirements for amplicon length, primer location relative to oligoprobe placement, the number of molecular events required to generate a signal and oligonucleotide sequence composition due to the different needs of each type of chemistry. This means that direct comparison between two or more different fluorogenic chemistries employing the same conditions is sometimes impossible, making a clear-cut hierarchy of chemistries extremely difficult to compile, even when they are examined using the same microbial template. To get the most useful data from a published comparison the reader should ensure that the same primer sequences are used throughout and that the same amplicon length has been used. But often we rely on commercial "experts" to design our chemistries, especially so if we are trialing a novel chemistry or one we have no prior experience with. For the microbiologist this reliance can be risky since the majority of expert designers developed their skills creating assays for human genomic targets and the pitfalls encountered when designing real-time PCR assays for the microbiologist are perhaps, at best, considered an oddity. This reliance upon out-sourcing the designs of assays employing more exotic chemistries may be a major reason why more of the chemistries which are harder to design have not gained purchase in microbiology, even if they could occupy useful niches. Microbiologists are generally more familiar with designing hydrolysis or HybProbe oligoprobes which means they can maintain control over the finer design aspects that are essential to accurate and efficient assay performance. We can also expect to pay more for the more complex or less popular chemistries, and it takes a considerably longer period of time (days versus weeks) to manufacture these chemistries. The latter is an important consideration for urgent research studies or for developing assays to diagnose putative pandemic pathogens.

In one of few comparative oligoprobe studies it was found that hydrolysis chemistries produced a greater change in emission intensity after hydrolysis compared with the "open" state of a molecular beacon however there is a link between the stem-loop chemistries, reduced background fluorescence and a corresponding increased detection sensitivity, presumably due to a reduction in the amount of signal lost to noise. (Wang *et al.*, 2005; Gunson *et al.*, 2006). Another recent comparison of hydrolysis chemistries investigated the efficiency of multiple reporter and quencher pairings, the impact of LNA base substitution in parallel with reduced length and a comparison to Scorpion primers (Reynisson *et al.*, 2006). The results indicated that FAM-BHQ1 and Cy5-BHQ3 gave the highest fluorescence yields while Scorpion primers were less sensitive and resulted in a lower PCR efficiency than conventional or LNA-substituted hydrolysis oligoprobes for the detection of *Salmonella* DNA. When the same target was examined, the Amplifluor chemistry performed with similar efficiency to a TaqMan oligoprobe, with only slightly reduced sensitivity when employed to quantify *Listeria* species (Rodriguez-Lázaro *et al.*, 2004). When hydrolysis oligoprobes of identical sequence but with either LNA substitutions or an MGB moiety attached were compared for *Staphylococcus aureus* detection, the two chemistries performed with equivalent sensitivity and specificity (Letertre *et al.*, 2003). However, TaqMan-MGB oligoprobes are reportedly slightly more sensitive than an LNA Yin-Yang oligoprobe for a human target (Kennedy *et al.*, 2006). Yin-Yang oligoprobes are also relatively easy to manufacture and therefore less expensive and this factor must also be considered when choosing chemistries (Kong *et al.*, 2003). MGB-Eclipse and SimpleProbe oligoprobes perform with equal sensitivity when compared using the same primers for the identification of *Yersinia pestis* (Chase *et al.*, 2005). In a comparison of hydrolysis probes, HybProbes and SYBR Green I all using the same amplicons; the three chemistries displayed similar sensitivities but the hybridization probes exhibited the highest specificity (Newby *et al.*, 2003), most likely due to the trimolecular signaling required. Others report hydrolysis and HybProbes can be used for quantification with similar results when using the same viral target, although targeting different regions of the genome (Hesselink *et al.*, 2005).

Future trends

Chemistries are no longer the hot developmental topic they once were and applications are now driving better understanding of the use of those chemistries we have. Improved instrument evaluation studies and better understanding of the benefits and limitations of new and existing fluorophores are also providing current and future users with the background information they need to implement real-time PCR in their microbiology laboratories. While implementation has not progressed as quickly in microbiology as it has in human gene research, the following chapters will make it clear that our very diverse field is quickly progressing. The horizon is crowded with intriguing developments such as the possibility of analyzing hundreds of liters of human specimens for viruses after concentrating the material into tiny volumes (Anderson *et al.*, 2003), or the implementation of biochips derived from small microfluidic devices capable of concentration, lysis and purification of template as well as detection of amplicon all within a few hours (Kubista, 2004). The immediate future of real-time PCR appears to be secure and fluorogenic oligonucleotides will be with us in some capacity, for decades to come.

Conclusions

While real-time PCR may have come about by scientific good luck, its arrival was seized upon by microbiology laboratories excited by the promise of PCR diagnostics but disappointed because of the looming specter of assay contamination by amplicon from previous reactions. In the years since its introduction many aspects of real-time PCR and the instruments used to perform it have been investigated and enhanced, most notably among these, the fluorogenic chemistries which make the technique possible. Microbiology thrives on the benefits of sensitivity, speed and containment afforded by real-time PCR. In the following chapters we will see how this essential research aid has matured into a principle tool for clinical and research microbiology laboratories around the world and why it is finally fulfilling the expectations of its core business—identifying otherwise undetectable amounts of viral, bacterial, fungal and parasitic RNA or DNA from ill human subjects.

Acknowledgments

A number of kind people took time to contribute information for the construction of this chapter. Thanks go to Andreas Nitsche, David McDowell, Jane Theaker, Michael Zuker, Brian Erich Caplin, Vadim Demidov, Olfert Landt, David Spodick and Sanjay Tyagi. As always special thanks go to Katherine Arden for honest critical editing of this work and support of its completion.

Web sites

Premier biosoft
www.premierbiosoft.com/index.html
Software developers that provide many programs to help create of oligonucleotide applications including real-time PCR and arrays.

Molecular beacons.org
www.molecular-beacons.org/

Information of the structure, function, design and applications of molecular beacons from the home of their discovery.

TIB MOLBIOL
www.tib-molbiol.com/index.html
A vendor of and design-house for real-time PCR oligonucleotides since 1990.

Applied Biosystems
www.appliedbiosystems.com/
Home to the TaqMan (hydrolysis) oligoprobe and the first of a continuing line of real-time PCR instruments and support world-wide.

Idaho technologies
www.idahotech.com/index.html
A privately held company not actually based in Idaho. The developers of the LightCycler instrument and trailblazers in high resolution melting curve analysis.

LightCycler® online
www.lightcycler-online.com
Information on the fastest instrument capable of combined thermal cycling and fluorescence detection.

Gene quantification
http://www.gene-quantification.info/
A one-stop site for quantitative real-time PCR with information on topics ranging from efficiency to quantification and events to microarrays.

Black Hole Quenchers
http://www.eurogentec.be/code/en/what.asp?pk_id_what=103
An informative site describing the quirks of the BHQs.

Fluorescence lectures
http://probes.invitrogen.com/resources/education/
An exceptional series of fluorescence tutorials from Invitrogen™ and Molecular Probes™.

References

Afonina, I.A., Reed, M.W., Lusby, E., Shishkina, I.G., and Belousov, Y.S. (2002a). Minor groove binder-conjugated DNA probes for quantitative DNA detection by hybridization-triggered fluorescence. Biotechniques 32, 940–949.

Afonina, I.A., Sanders, S., Walburger, D., and Belousov, Y.S. (2002b). Accurate SNP typing by real-time PCR. Pharmagenomics *Jan/Feb*, 48–54.

Ahmed, F.E. (2002). Detection of genetically modified organisms in foods. Trend Biotechnol. *20*, 215–223.

Anderson, N.G., Gerin, J.L., and Anderson, N.L. (2003). Global screening for human viral pathogens. Emerg. Infect. Dis. 9, 768–773.

Applegate, T.L., Iland, H.J., Mokany, E., and Todd, A.V. (2002). Diagnosis and molecular monitoring of acute promyelocytic leukemia using DzyNA reverse transcription-PCR to quantify PML/RARα fusion transcripts. Clin. Chem. *48*, 1338–1343.

Åkerstrom, S., Mousavi-Jazi, M., Klingström, J., Leijon, M., Lundkvist, A., and Mirazimi, A. (2005). Nitric oxide inhibits the replication cycle of severe acute respiratory syndrome coronavirus. J. Virol. *79*, 1966–1969.

Ball, T.B., Plummer, F.A., and HayGlass, K.T. (2003). Improved mRNA quantitation in LightCycler RT-PCR. Int. Arch. Allergy Immunol.*130*, 82–86.

Barany, F. (1991). The ligase chain reaction in a PCR world. PCR Methods Appl. *1*, 5–16.

Barratt, K. and Mackay, J.F. (2002). Improving real-time PCR genotyping assays by asymmetric amplification. J. Clin. Microbiol. *40*, 1571–1572.

Barzon, L., Bonaguro, R., Castagliuolo, I., Chilosi, M., Franchin, E., Del Vecchio, C., Giaretta, I., Boscaro, M., and Palù, G. (2003). Gene therapy of thyroid cancer via retrovirally-driven combined expression of human interleukin-2 and herpes simplex virus thymidine kinase. Eur. J. Endocrinol. *148*, 73–80.

Behlke, M.A., Huang, L., Bogh, L., Rose, S., and Devor, E.J. Fluorescence and fluorescence applications. 2005. Ref Type: Pamphlet

Bengra, C., Mifflin, T.E., Khripin, Y., Manunta, P., Williams, S.M., Jose, P.A., and Felder, R.A. (2002). Genotyping of essential hypertension single-nucleotide polymorphisms by a homogenous PCR method with universal energy transfer primers. Clin. Chem. *48*, 2131–2140.

Bengtsson, M., Karlsson, H.J., Westman, G., and Kubista, M. (2003). A new minor groove binding asymmetric cyanine reporter dye for real-time PCR. Nucleic Acids Res. *31*, e45.

Biel, S.S. and Madeley, D. (2001). Diagnostic virology-the need for electron microscopy: a discussion paper. J. Clin. Virol. *22*, 1–9.

Böhm, J., Hahn, A., Schubert, R., Bahnweg, G., Adler, N., Nechwatal, J., Oehlmann, R., and Oswald, W. (1999). Real-time quantitative PCR: DNA determination in isolated spores of the mycorrhizal fungus *Glomus mosseae* and monitoring of *Phytophthora infestans* and *Phytophthora citricola* in their respective host plants. J. Phytopathol. *147*, 409–416.

Brechtbuehl, K., Whalley, S.A., Dusheiko, G.M., and Saunders, N.A. (2001). A rapid real-time quantitative polymerase chain reaction for hepatitis B virus. J. Virol. Methods *93*, 105–113.

Bustin, S.A. (2000). Absolute quantification of mRNA using real-time reverse transcription polymerase chain reaction assays. J. Mol. Endocrinol. *25*, 169–193.

Bustin, S.A. (2002). Quantification of mRNA using real-time reverse transcription PCR (RT-PCR): trends and problems. J. Mol. Endocrinol. *29*, 23–39.

Cardullo, R.A., Agrawai, S., Flores, C., Zamecnik, P.C., and Wolf, D.E. (1988). Detection of nucleic acid hybridization by nonradiative fluorescence resonance energy transfer. Proc. Natl. Acad. Sci. USA *85*, 8790–8794.

Carman, B. (2001). Molecular techniques should now replace cell culture in diagnostic virology laboratories. Rev. Med. Virol. *11*, 347–349.

Carman, W.F., Wallace, L.A., Walker, J., McIntyre, S., Noone, A., Christie, P., Millar, J., and Douglas, J.D. (2000). Rapid virological surveillance of community influenza infection in general practice. Br. Med. J. *321*, 736–737.

Chandelier, A., Dubois, N., Baelen, F., De Leener, F., Warnon, S., Remacle, J., and Lepoivre, P. (2001). RT-PCR-ELOSA tests on pooled sample units for the detection of virus Y in potato tubers. J. Virol. Methods *91*, 99–108.

Chase, C.J., Ulrich, M.P., Wasieloski, L.P., Kondig, J.P., Garrison, J., Lindler, L.E., and Kulesh, D.A. (2005). Real-time PR assays targeting a unique chromosomal sequence of *Yersinia pestis*. Clin. Chem. *51*, 1778–1785.

Chou, Q., Russell, M., Birch, D.E., Raymond, J., and Bloch, W. (1992). Prevention of pre-PCR mispriming and primer dimerization improves low-copy-number amplifications. Nucleic Acids Res. *20*, 1717–1723.

Christopherson, C., Sninsky, J., and Kwok, S. (1997). The effects of internal-primer-template mismatches on RT-PCR: HIV-1 model studies. Nucleic Acids Res. *25*, 654–658.

Clarke, S.C., Reid, J., Thom, L., Denham, B.C., and Edwards, G.F.S. (2002). Laboratory confirmation of meningococcal disease in Scotland, 1993–9. J. Clin. Pathol. *55*, 32–36.

Clegg, R.M. (1992). Fluorescence resonance energy transfer and nucleic acids. Methods Enzymol. *211*, 353–388.

Cockerill III, F.R. (2003). Application of rapid-cycle real-time polymerase chain reaction for diagnostic testing in the clinical microbiology laboratory. Arch. Pathol. Lab. Med. *127*, 1112–1120.

Cockerill III, F.R. and Smith, T.F. (2002). Rapid-cycle real-time PCR: a revolution for clinical microbiology. ASM News *68*, 77–83.

Compton, J. (1991). Nucleic acid sequence-based amplification. Nature *350*, 91–92.
Cradic, K.W., Wells, J.E., Allen, L., Kruckeberg, K.E., Singh, R.J., and Grebe, S.K.G. (2004). Substitution of 3′-phosphate cap with a carbon-based blocker reduces the possibility of fluorescence resonance energy transfer probe failure in real-time PCR. Clin. Chem. *50*, 1080–1082.
Crockett, A.O. and Wittwer, C.T. (2001). Fluorescein-labelled oligonucleotides for real-time PCR: using the inherent quenching of deoxyguanosine nucleotides. Anal. Biochem. *290*, 89–97.
Dekonenko, A., Ibrahim, M.S., and Schmaljohn, C.S. (1997). A colorimetric PCR-enzyme immunoassay to identify hantaviruses. Clin. Diag. Virol. *8*, 113–121.
Didenko, V.V. (2001). DNA probes using fluorescence resonance energy transfer (FRET): designs and applications. Biotechniques *31*, 1106–1121.
Dubertret, B., Calame, M., and Libchaber, A.J. (2001). Single-mismatch detection using gold-quenched fluorescent oligonucleotides. Nat. Biotechnol. *19*, 365–370.
Egholm, M., Buchardt, O., Christensen, L., Behrens, C., Freier, S.M., Driver, D.A., Berg, R.H., Kim, S.K., Norden, B., and Nielsen, P.E. (1993). PNA hybridizes to complementary oligonucleotides obeying the Watson–Crick hydrogen bonding rules. Nature *365*, 566–568.
Ellis, J.S. and Zambon, M.C. (2002). Molecular diagnosis of influenza. Rev. Med. Virol. *12*, 375–389.
Espy, M.J., Uhl, J.R., Sloan, S.P., Buckwalter, S.P., Jones, M.F., Vetter, E.A., Yao, J.D.C., Wengenack, N.L., Rosenblatt, J.E., Cockerill III, F.R., and Smith, T.F. (2006). Real-time PCR in clinical microbiology: Applications for routine laboratory testing. Clin. Microbiol. Rev. *19*, 165–256.
Fernández, F., Gutiérrez, J., Sorlózano, A., Romero, J.M., Soto, M.J., and Ruiz-Cabello, F. (2006). Comparison of the SYBR Green and the hybridization probe format for real-time PCR detction of HHV-6. Microbiol. Res. *161*, 158–163.
Fiandaca, M.J., Hyldig-Nielsen, J.J., Gildea, B.D., and Coull, J.M. (2001). Self-reporting PNA/DNA primers for PCR analysis. Genome Res. *11*, 609–611.
Förster, T. (1948). Zwischenmolekulare energiewanderung und fluoreszenz. Ann. Phys. *6*, 55–75.
Fraaije, B.A., Lovell, D.J., Coelho, J.M., Baldwin, S., and Hollomon, D.W. (2001). PCR-based assays to assess wheat varietal resistance to blotch (*Septoria tritici* and *Stagonospora nodorum*) and rust (*Puccinia striiformis* and *Puccinia recondita*) diseases. Eur. J. Plant Pathol. *107*, 905–917.
Fredricks, D.N. and Relman, D.A. (1999). Application of polymerase chain reaction to the diagnosis of infectious disease. Clin. Inf. Dis. *29*, 475–488.
French, D.J., Archard, C.L., Andersen, M.T., and McDowell, D.G. (2002). Ultra-rapid analysis using HyBeacon™ probes and direct PCR amplification from saliva. Mol. Cell. Probes *16*, 319–326.
French, D.J., Archard, C.L., Brown, T., and McDowell, D.G. (2001). HyBeacon™ probes: a new tool for DNA sequence detection and allele discrimination. Mol. Cell. Probes *15*, 363–374.
Gibson, U.E.M., Heid, C.A., and Williams, P.M. (1996). A novel method for real time quantitative RT-PCR. Genome Res. *6*, 995–1001.
Giglio, S., Monis, P.T., and Saint, C.P. (2003). Demonstration of preferential binding of SYBR Green I to specific DNA fragments in real-time multiplex PCR. Nucleic Acids Res. *31*, e136.
Gingeras, T.R., Higuchi, R., Kricka, L.J., Lo, Y.M.D., and Wittwer, C.T. (2005). Fifty years of molecular (DNA/RNA) diagnostics. Clin. Chem. *51*, 661–671.
Ginzinger, D.G. (2002). Gene quantification using real-time quantitative PCR: An emerging technology hits mainstream. Exp. Hematol. *30*, 503–512.
Goel, A., Kumar, A., Puniya, A.K., Chen, W., and Singh, K. (2005). Molecular beacon: a multitask probe. J. Appl. Microbiol. *99*, 435–442.
Goldenberg, O., Landt, O., Schumann, R.R., Göbel, U.B., and Hamann, L. (2005). Use of locked nucleic acid oligonucleotides as hybridization/FRET probes for quantification of 16S rDNA by real-time PCR. Biotechniques *38*, 29–32.
Gundry, C.N., Vandersteen, J.G., Reed, G.H., Pryor, R.J., Chen, J., and Wittwer, C.T. (2003). Amplicon melting analysis with labeled primers: A closed-tube method for differentiating homozygotes and heterozygotes. Clin. Chem. *49*, 396–406.
Gunson, R.N., Collins, T.C., and Carman, W.F. (2006). Practical experience of high throughput real time PCR in the routine diagnostic virology setting. J. Clin. Virol. *35*, 355–367.
Gyllensten, U.B. and Erlich, H.A. (1988). Generation of single-stranded DNA by the polymerase chain reaction and its application to direct sequencing of the HLA-DQA locus. Proc. Natl. Acad. Sci. USA *85*, 7652–7656.
Halford, W.P. (1999). The essential prerequisites for quantitative RT-PCR. Nat. Biotechnol. *17*, 835.

Halford, W.P., Falco, V.C., Gebhardt, B.M., and Carr, D.J.J. (1999). The inherent quantitative capacity of the reverse transcription-polymerase chain reaction. Anal. Biochem. *266*, 181–191.

Härtel, C., Bein, G., Kirchner, H., and Klüter, H. (1999). A human whole-blood assay for analysis of T-cell function by quantification of cytokine mRNA. Scand. J. Immunol. *49*, 649–654.

Heller, M.J. and Morrison, L.E. (1985). Chemiluminescent and fluorescent probes for DNA hybridization. In Rapid detection and identification of infectious agents., D.T.Kingsbury and, S.Falkow, eds. (New York: Academic Press), pp. 245–256.

Herrmann, M.G., Durtschi, J.D., Bromley, L.K., Wittwer, C.T., and Voelkerding, K.V. (2006). Amplicon melting analysis for mutation scanning and genotyping: Cross-platform comparison of instruments and dyes. Clin. Chem. 52, 494–503.

Hesselink, A.T., van den Brule, A.J.C., Groothuismink, Z.M.A., Molano, M., Berkhof, J., Meijer, C.J.L.M., and Snijders, P.J.F. (2005). Comparison of three different PCR methods for quantifying human papillomavirus type 16 DNA in cervical scrape specimens. J. Clin. Microbiol. *43*, 4868–4871.

Higuchi, R., Dollinger, G., Walsh, P.S., and Griffith, R. (1992). Simultaneous amplification and detection of specific DNA sequences. Biotechnology (N Y) *10*, 413–417.

Higuchi, R., Fockler, C., Dollinger, G., and Watson, R. (1993). Kinetic PCR analysis: real-time monitoring of DNA amplification reactions. Biotechnology (N Y) *11*, 1026–1030.

Holland, P.M., Abramson, R.D., Watson, R., and Gelfand, D.H. (1991). Detection of specific polymerase chain reaction product by utilizing the 5′-3′ exonuclease activity of *Thermus aquaticus*. Proc. Natl. Acad. Sci. USA *88*, 7276–7280.

Howell, W.M., Jobs, M., and Brookes, A.J. (2002). iFRET: An improved fluorescence system for DNA-melting analysis. Genome Res. *12*, 1401–1407.

Huang, J., DeGraves, F.J., Gao, D., Feng, P., Schlapp, T., and Kaltenboeck, B. (2001). Quantitative detection of *Chlamydia* spp. by fluorescent PCR in the LightCycler. Biotechniques *30*, 150–157.

Ibrahim, M.S., Lofts, R.S., Jahrling, P.B., Henchal, E.A., Weedn, V.W., Northrup, M.A., and Belgrader, P. (1998). Real-time microchip PCR for detecting single-base differences in viral and human DNA. Anal. Chem. *70*, 2013–2017.

Isacsson, J., Cao, H., Ohlsson, L., Nordgren, S., Svanvik, N., Westman, G., Kubista, M., Sjöback, R., and Sehlstedt, U. (2000). Rapid and specific detection of PCR products using light-up probes. Mol. Cell. Probes *14*, 321–328.

Ishiguro, T., Saitoh, J., Yawata, H., Yamagishi, H., Iwasaki, S., and Mitoma, Y. (1995). Homogeneous quantitative assay of hepatitis C virus RNA by polymerase chain reaction in the presence of a fluorescent intercalater. Anal. Biochem. *229*, 207–213.

Johnson, J.R. (2000). Development of polymerase chain reaction-based assays for bacterial gene detection. J. Microbiol. Methods *41*, 201–209.

Johnson, S.C., Sherrill, C.B., Marshall, D.J., Moser, M.J., and Prudent, J.R. (2004). A third base pair for the polymerase chain reaction: inserting isoC and isoG. Nucleic Acids Res. 32, 1937–1941.

Jung, R., Soondrum, K., and Neumaier, M. (2000). Quantitative PCR. Clin. Chem. Lab. Med. *38*, 833–836.

Karsai, A., Müller, S., Platz, S., and Hauser, M.-T. (2002). Evaluation of a Homemade SYBR® Green I Reaction Mixture for Real-Time PCR Quantification of Gene Expression. Biotechniques 32, 790–796.

Keller, G.H., Huang, D.-P., Shih, J.W.-K., and Manak, M.M. (1990). Detection of hepatitis B virus DNA in serum by polymerase chain reaction amplification and microtiter sandwich hybridization. J. Clin. Microbiol. *28*, 1411–1416.

Kemp, D.J., Churchill, M.J., Smith, D.B., Biggs, B.A., Foote, S.J., Peterson, M.G., Samaras, N., Deacon, N.J., and Doherty, R. (1990). Simplified colorimetric analysis of polymerase chain reactions: detection of HIV sequences in AIDS patients. Gene *94*, 223–228.

Kennedy, B., Arar, K., Reja, V., and Henry, R.J. (2006). Locken nucleic acids for optimizing displacement probes for quantitative real-time PCR. Anal. Biochem. *348*, 294–299.

Kidd, I.M., Clark, D.A., and Emery, V.C. (2000). A non-radioisotopic quantitative competitive polymerase chain reaction method: application in measurement of human herpesvirus 7 load. J. Virol. Methods *87*, 177–181.

Kievits, T., van Gemen, B., van Strijp, D., Schukkink, R., Dircks, M., Adriaanse, H., Malek, L., Sooknanan, R., and Lens, P. (1991). NASBA isothermal enzymatic *in vitro* nucleic acid amplification optimized for the diagnosis of HIV-1 infection. J. Virol. Methods 35, 273–286.

Klein, D. (2002). Quantification using real-time PCR technology: applications and limitations. Trends Mol. Med. *8*, 257–260.

Komurian-Pradel, F., Paranhos-Baccalà, G., Sodoyer, M., Chevallier, P., Mandrand, B., Lotteau, V., and André, P. (2001). Quantitation of HCV RNA using real-time PCR and fluorimetry. J. Virol. Methods *95*, 111–119.

Kong, D.-M., G.u., L., Shen, H.-X., and Mi., H.-F. (2002). A modified molecular beacon combining the properties of TaqMan probe. Chem. Commun. 854–855.

Kong, D.-M., Huang, Y.-P., Zhang, X.-Y., Yang, W.-H., Shen, H.-X., and Mi., H.-F. (2003). Duplex probes: a new approach for the detection of specific nucleic acids in homogenous assays. Anal. Chim. Acta *491*, 135–143.

Koshkin, A.A., Singh, S.K., Nielsen, P., Rajwanshi, V.K., Kumar, R., Meldgaard, M., Olsen, C.E., and Wengel, J. (1998). LNA (locked nucliec acids): Synthesis of the adenine, cytosine, guanine, 5-methylcytosine, thymine and uracil bicyclonucleoside monomers, oligomerisation, and unprecedented nucleic acid recognition. Tetrahedron *54*, 3607–3630.

Kox, L.F.F., Noordhoek, G.T., Kunakorn, M., Mulder, S., Sterrenburg, M., and Kolk, A.H.J. (1996). Microwell hybridization assay for detection of PCR products from *Mycobacterium tuberculosis* complex and the recombinant *Mycobacterium smegmatis* strain 1008 used as an internal control. J. Clin. Microbiol. *34*, 2117–2120.

Kruse, N., Pette, M., Toyka, K., and Rieckmann, P. (1997). Quantification of cytokine mRNA expression by RT PCR in samples of previously frozen blood. J. Immunol. Methods *210*, 195–203.

Kubista, M. (2004). Nucleic acid-based technologies: application amplified. Pharmagenomics 5, 1–7.

Kubista, M., Andrade, J.M., Bengtsson, M., Forootan, A., Jonák, J., Lind, K., Sindelka, R., Sjöback, R., Sjögreen, B., Strömbom, L., Ståhlberg, A., and Zoric, N. (2006). The real-time polymerase chain reaction. Mol. Aspects Med. *27*, 95–125.

Kubista, M., Karlsson, J., Bengtsson, M., Zoric, N., and Westman, G. (2004). DNA binding fluorophores. In Encyclopedia of Medical Genomics and Proteomics, J.Fuchs and, M.Podda, eds. Dekker), pp. 351–355.

Kuhn, H., Demidov, V.V., Coull, J.M., Fiandaca, M.J., Gildea, B.D., and Frank-Kamenetskii, M.D. (2002). Hybridization of DNA and PNA molecular beacons to single-stranded and double-stranded DNA targets. J. Am. Chem. Soc. *124*, 1097–1103.

Kuhn, H., Demidov, V.V., Gildea, B.D., Fiandaca, M.J., Coull, J.C., and Frank-Kamenetskii, M.D. (2001). PNA beacons for duplex DNA. Antisense Nucleic Acid Drug Dev. *11*, 265–270.

Kumar, R., Singh, S.K., Koshkin, A.A., Rajwanshi, V.K., Meldgaard, M., and Wengel, J. (1998). The first analogues of LNA (locked nucleic acids): Phosphorothioate-LNA and 2′-thio-LNA. Bioorg. Med. Chem. Lett. *8*, 2219–2222.

Kurata, S., Kanagawa, T., Yamada, K., Torimura, M., Yokomaku, T., Kamagata, Y., and Kurane, R. (2001). Fluorescent quenching-based quantitative detection of specific DNA/RNA using a BODIPY®FL-labeled probe or primer. Nucleic Acids Res. *29*, e34.

Kutyavin, I.V., Afonina, I.A., Mills, A., Gorn, V.V., Lukhtanov, E.A., Belousov, E.S., Singer, M.J., Walburger, D.K., Lokhov, S.G., Gall, A.A., Dempcy, R., Reed, M.W., Meyer, R.B., and Hedgpeth, J. (2000). 3′-Minor groove binder-DNA probes increase sequence specificity at PCR extension temperatures. Nucleic Acids Res. *28*, 655–661.

Landt, O. (2001). Selection of hybridization probes for real-time quantification and genetic analysis. In Rapid cycle real-time PCR: methods and applications., S.Meuer, C.Wittwer, and, K.Nakagawara, eds. (Germany: Springer Verlag), pp. 35–41.

Lee, L.G., Connell, C.R., and Bloch, W. (1993). Allelic discrimination by nick-translation PCR with fluorogenic probes. Nucleic Acids Res. *21*, 3761–3766.

Lee, M.A., Siddle, A.L., and Page, R.H. (2002). ResonSense®: simple linear fluorescent probes for quantitative homogeneous rapid polymerase chain reaction. Anal. Chim. Acta *457*, 61–70.

Leijon, M., Mousavi-Jazi, M., and Kubista, M. (2006). LightUp® probes in clinical diagnostics. Mol. Aspects Med. *27*, 160–175.

Leman, A.R., Polochock, S., Mole, S.E., Pearce, D.A., and Rothberg, P.G. (2006). Homogeneous PCR nucleobase quenching assays to detect four mutations that cause neuronal ceroid lipofuscinosis: T75P and R151X in *CLN1*, and IVS5–1G> C and R208X in *CLN2*. J. Neurosci. Methods.

Letertre, C., Perelle, S., Dilasser, F., Arar, K., and Fach, P. (2003). Evaluation of the performance of LNA and MGB probes in 5′-nuclease PCR assays. Mol. Cell. Probes *17*, 307–311.

Li, Q., Luan, G., Guo, Q., and Liang, J. (2002). A new class of homogeneous nucleic acid probes based on specific displacement hybridization. Nucleic Acids Res. *30*, e5.

Lind, K., Ståhlberg, A., Zoric, N., and Kubista, M. (2006). Combining sequence-specific probes and DNA binding dyes in real-time in real-time PCR for specific nucleic acid quantification and melting analysis. Biotechniques *40*, 315–319.

Lyamichev, V., Brow, M.A.D., and Dahlberg, J.E. (1993). Structure-specific endonucleolytic cleavage of nucleic acids by eubacterial DNA polymerases. Science *260*, 778–783.

Mackay, I.M. (2004). Real-time PCR in the microbiology laboratory. Clin. Microbiol. Infect. *10*, 190–212.

Mackay, I.M., Arden, K.E., and Nitsche, A. (2002). Real-time PCR in virology. Nucleic Acids Res. *30*, 1292–1305.

Mackay, I.M., Harnett, G.B., Jeoffreys, N., Bastian, I., Sriprakash, K.S., Siebert, D., and Sloots, T. (2006). Detection and discrimination of herpes simplex viruses, Haemophilus ducreyi, Treponema pallidum, and Calymmatobacterium (Klebsiella) granulomatis from genital ulcers. Clin. Inf. Dis. *42*, 1431–1438.

Mackay, I.M., Jacob, K.C., Woolhouse, D., Waller, K., Syrmis, M.W., Whiley, D.M., Siebert, D.J., Nissen, M.N., and Sloots, T.P. (2003). Molecular assays for the detection of human metapneumovirus. J. Clin. Microbiol. *41*, 100–105.

Mackay, I.M., Metharom, P., Sloots, T.P., and Wei, M.Q. (2001). Quantitative PCR-ELAHA for the determination of retroviral vector transduction efficiency. Mol. Ther. *3*, 801–808.

Marks, A.H.R., Bhadra, P.K., McDowell, D.G., French, D.J., Douglas, K.T., Bichenkova, E.V., and Bryce, R.A. (2055). Molecular basis of action of HyBeacon™ fluorogenic probes: A spectroscopic and molecular dynamics study. J. Biomol. Struct. Dyn. *23*, 49–62.

Marras, S.A.E., Kramer, F.R., and Tyagi, S. (2002). Efficiencies of fluorescence resonance energy transfer and contact-mediated quenching in oligonucleotide probes. Nucleic Acids Res. *30*, e122.

Martell, M., Gómez, J., Esteban, J.I., Sauleda, S., Quer, J., Cabot, B., Esteban, R., and Guardia, J. (1999). High-throughput real-time reverse transcription-PCR quantitation of hepatitis C virus RNA. J. Clin. Microbiol. *37*, 327–332.

Matthews, J.A. and Kricka, L.J. (1988). Analytical strategies for the use of DNA probes. Anal. Biochem. *169*, 1–25.

Merien, F., Portnoi, D., Bourhy, P., Charavay, F., Berlioz-Arthaud, A., and Baranton, G. (2005). A rapid and quantitative method for the detection of Leptospira species in human leptospirosis. FEMS Microbiol. Lett. *249*, 139–147.

Mhlanga, M.M. and Malmberg, L. (2001). Using molecular beacons to detect single-nucleotide polymorphisms with real-time PCR. Methods *25*, 463–471.

Monis, P.T., Giglio, S., and Saint, C.P. (2005). Comparison of SYTO9 and SYBR Green I for real-time polymerase chain reaction and investigation of the effect of dye concentration on amplification and DNA melting curve analysis. Anal. Biochem. *340*, 24–34.

Morrison, L.E., Halder, T.C., and Stols, L.M. (1989). Solution-phase detection of polynucleotides using interacting fluorescent labels and competitive hybridization. Anal. Biochem. *183*, 231–244.

Morrison, T.B., Weis, J.J., and Wittwer, C.T. (1998). Quantification of low-copy transcripts by continuous SYBR® green I monitoring during amplification. Biotechniques *24*, 954–962.

Moser, M.J., Marshall, D.J., Grenier, J.K., Kieffer, C.D., Killeen, A.A., Ptacin, J.L., Richmond, C.S., Roesch, E.B., Scherrer, C.W., Sherrill, C.B., van Hout, C.V., Zanton, S.J., and Prudent, J.R. (2003). Exploiting the enzymatic recognition of an unnatural base pair to develop a universal genetic analysis system. Clin. Chem. *49*, 407–414.

Mullan, B., Sheehy, P., Shanahan, F., and Fanning, L. (2004). Do *Taq*-generated RT-PCR products from RNA viruses accurately reflect viral genetic heterogeneity? J. Viral Hepatitis *11*, 108–114.

Mullis, K.B. and Faloona, F. (1987). Specific synthesis of DNA *in vitro* via a polymerase-catalysed chain reaction. Methods Enzymol. *155*, 335–350.

Nakayama, H., Yokoi, H., and Fujita, J. (1992). Quantification of mRNA by non-radioactive RT-PCR and CCD imaging system. Nucleic Acids Res. *20*, 4939.

Nazarenko, I., Lowe, B., Darfler, M., Ikonomi, P., Schuster, D., and Raschtchian, A. (2002). Multiplex quantitative PCR using self-quenched primers labeled with a single fluorophore. Nucleic Acids Res. *30*, e37.

Nazarenko, I.A., Bhatnager, S.K., and Hohman, R.J. (1997). A closed tube format for amplification and detection of DNA based on energy transfer. Nucleic Acids Res. *25*, 2516–2521.

Newby, D.T., Hadfield, T.L., and Roberto, F.F. (2003). Real-Time PCR Detection of Brucella abortus: a Comparative Study of SYBR Green, I., 5′-Exonuclease, and Hybridization Probe Assays. Appl. Environ. Micro. *69*, 4753–4759.

Nilsson, M. (2006). Lock and roll: single-molecular genotyping in situ using padlock probes and rolling-circle amplification. Histochem. Cell Biol. *126*, 159–164.

Nitsche, A., Steuer, N., Schmidt, C.A., Landt, O., and Siegert, W. (1999). Different real-time PCR formats compared for the quantitative detection of human cytomegalovirus DNA. Clin. Chem. *45*, 1932–1937.

Nogva, H.K., Rudi, K., Naterstad, K., Holck, A., and Lillehaug, D. (2000). Application of 5′-nuclease PCR for quantitative detection of *Listeria monocytogenes* in pure cultures, water, skim milk and unpasteurized whole milk. Appl. Environ. Micro. *66*, 4266–4271.

Northrup, M.A. (1998). A miniature analytical instrument for nucleic acids based on micromachined silicon reaction chambers. Anal. Chem. *70*, 918–922.

Nutiu, R. and, L.i., Y. (2002). Tripartite molecular beacons. Nucleic Acids Res. *30*, e94.

Obika, S., Nanbu, D., Hari, Y., Andoh, J.-I., Morio, K.-I., Doi, T., and Imanishi, T. (1998). Stability and structural features of the duplexes containing nucleoside analogues with a fixed N-type conformation, 2′-O,4′-C-methyleneribonucleosides. Tetrahedron Lett. *39*, 5401–5404.

Ogilvie, M. (2001). Molecular techniques should not now replace cell culture in diagnostic virology laboratories. Rev. Med. Virol. *11*, 351–354.

Papin, J.F., Vahrson, W., and Dittmer, D.P. (2004). SYBR green-based real-time quantitative PCR assay for detection of West Nile Virus circumvents false-negative results due to strain variability. J. Clin. Microbiol. *42*, 1511–1518.

Park, J.H., Kim, J.S., Kwon, S.-T., and Lee, D.-S. (1993). Purification and characterization of *Thermus caldophilus* GK24 DNA polymerase. Eur. J. Biochem. *214*, 135–140.

Persing, D.H. (1993). *In vitro* nucleic acid amplification techniques. In Diagnostic molecular microbiology: Principles and applications., D.H.Persing, T.F.Smith, F.C.Tenover, and, T.J.White, eds. (Washington, D.C.: ASM), pp. 51–87.

Petersen, M. and Wengel, J. (2003). LNA: a versatile tool for therapeutics and genomics. Trend Biotechnol. *21*, 74–81.

Piccirilli, J.A., Krauch, T., Moroney, S.E., and Benner, S.A. (1990). Enzymatic incorporation of a new base pair into DNA and RNA extends the genetic alphabet. Nature *343*, 33–37.

Pierce, K.E., Sanchez, J.A., Rice, J.E., and Wangh, L.J. (2005). Linear-after-the-exponential (LATE)-PCR: Primer design criteria for high yields of specific single-stranded DNA and improved real-time detection. Proc. Natl. Acad. Sci. USA *102*, 8609–8614.

Poddar, S.K. (2000). Symmetric vs asymmetric PCR and molecular beacon probe in the detection of a target gene of adenovirus. Mol. Cell. Probes *14*, 25–32.

Reynisson, E., Josefsen, M.H., Krause, M., and Hoorfar, J. (2006). Evaluation of probe chemistries and platforms to improve the detection limit of real-time PCR. J. Microbiol. Methods *66*, 206–216.

Rickert, A.M., Lehrach, H., and Sperling, S. (2004). Multiplexed real-time PCR using universal reporters. Clin. Chem. *50*, 1680–1683.

Ririe, K.M., Rasmussen, R.P., and Wittwer, C.T. (1997). Product differentiation by analysis of DNA melting curves during the polymerase chain reaction. Anal. Biochem. *245*, 154–160.

Rodriguez-Lázaro, D., Hernández, M., Scortti, M., Esteve, T., Vázquez-Boland, J.A., and Pla, M. (2004). Quantitative detection of *Listeria monocytogenes* and *Listera innocua* by real-time PCR: Assessment of *hly*, *iap*, and *lin02483* targets and AmpliFluor technology. Appl. Environ. Micro. *70*, 1366–1377.

Rudi, K., Nogva, H.K., Moen, B., Nissen, H., Bredholt, S., Møretrø, T., Naterstad, K., and Holck, A. (2002). Development and application of new nucleic acid-based technologies for microbial community analyses in food. Int. J. Food Microbiol. *78*, 171–180.

Saha, B.K., Tian, B., and Bucy, R.P. (2001). Quantitation of HIV-1 by real-time PCR with a unique fluorogenic probe. J. Virol. Methods *93*, 33–42.

Sanchez, J.A., Pierce, K.E., Rice, J.E., and Wangh, L.J. (2004). Linear-after-the-exponential (LATE)-PCR: An advanced method of asymmetric PCR and its uses in quantitative real-time analysis. Proc. Natl. Acad. Sci. USA *101*, 1933–1938.

Schalasta, G., Arents, A., Schmid, M., Braun, R.W., and Enders, G. (2000). Fast and type-specific analysis of herpes simplex virus types 1 and 2 by rapid PCR and fluorescence melting-curve-analysis. Infection *28*, 85–91.

Schoder, D., Schmalwieser, A., Schauberger, G., Kuhn, M., Hoorfar, J., and Wagner, M. (2003). Physical characteristics of six new thermocyclers. Clin. Chem. *49*, 960–963.

Seitz, O. (2000). Solid-phase synthesis of doubly labeled peptide nucleic acids as probes for the real-time detection of hybridization. Angew Chem. Int. Ed. Engl. *39*, 3249–3252.

Selvin, P. (1995). Fluorescence resonance energy transfer. Methods Enzymol. *246*, 300–334.

Shengqi, W., Xiaohong, W., Suhong, C., and Wei, G. (2002). A new fluorescent quantitative polymerase chain reaction technique. Anal. Biochem. *309*, 206–211.

Sherrill, C.B., Marshall, D.J., Moser, M.J., Larsen, C.A., Daudé-Snow, L., and Prudent, J.R. (2004). Nucleic acid analysis using an expanded genetic alphabet to quench fluorescence. J. Am. Chem. Soc. *126*, 4550–4556.

Simeonov, A. and Nikiforov, T.T. (2002). Single nucleotide polymorphism genotyping using short, fluorescently labeled locked nucleic acid (LNA) probes and fluorescence polarization detection. Nucleic Acids Res. *30*, e91.

Singh, S.K., Nielsen, P., Koshkin, A.A., and Wengel, J. (1998). LNA (locked nucleic acids): synthesis and high-affinity nucleic acid recognition. Chem. Commun. 455–456.

Sintchenko, V., Iredell, J.R., and Gilbert, G.L. (1999). Is it time to replace the petri dish with PCR? Application of culture-independent nucleic acid amplification in diagnostic bacteriology: Expectations and reality. Pathology *31*, 436–439.

Solinas, A., Brown, L.J., McKeen, C., Mellor, J.M., Nicol, J.T.G., Thelwell, N., and Brown, T. (2001). Duplex scorpion primers in SNP analysis and FRET applications. Nucleic Acids Res. *29*, e96.

Solinas, A., Thelwell, N., and Brown, T. (2002). Intramolecular TaqMan probes for genetic analysis. Chem. Commun. 2272–2273.

Stender, H., Fiandaca, M., Hyldig-Nielsen, J.J., and Coull, J. (2002). PNA for rapid microbiology. J. Microbiol. Methods *48*, 1–17.

Stojanovic, M.N., de Prada, P., and Landry, D.W. (2001). Catalytic molecular beacons. Chembiochem *2*, 411–415.

Stordeur, P., Poulin, L.F., Craciun, L., Zhou, L., Schandené, L., de Lavareille, A., Goriely, S., and Goldman, M. (2002). Cytokine mRNA quantification by real-time PCR. J. Immunol. Methods *259*, 55–64.

Stryer, L. and Haugland, R.P. (1967). Energy transfer: A spectroscopic ruler. Proc. Natl. Acad. Sci. USA *58*, 719–726.

Svanvik, N., Sehlstedt, U., Sjöback, R., and Kubista, M. (2000). Detection of PCR products in real time using light-up probes. Anal. Biochem. *287*, 179–182.

Svanvik, N., Westman, G., Wang, D., and Kubista, M. (2001). Light-up probes: Thiazole orange-conjugated peptide nucleic acid for the detection of target nucleic acid in homogeneous solution. Anal. Biochem. *281*, 26–35.

Switzer, C., Moroney, S.E., and Benner, S.A. (1989). Enzymatic incorporation of a new base pair into DNA and RNA. J. Am. Chem. Soc. *111*, 8322–8323.

Szemes, M., Bonants, P., de Weerdt, M., Baner, J., Landergren, U., and Schoen, C.D. (2005). Diagnostic application of padlock probes-multiplex detection of plant pathogens using universal microarrays. Nucleic Acids Res. 33, e70.

Täpp, I., Malmberg, L., Wik, M., and Syvänen, A.-C. (2000). Homogeneous scoring of single-nucleotide polymorphisms: Comparison of the 5′-nuclease TaqMan® assay and molecular beacon probes. Biotechniques *28*, 732–738.

Thelwell, N., Millington, S., Solinas, A., Booth, J., and Brown, T. (2000). Mode of action and application of Scorpion primers to mutation detection. Nucleic Acids Res. *28*, 3752–3761.

Todd, A.V., Fuery, C.J., Impey, H.L., Applegate, T.L., and Haughton, M.A. (2000). DzyNA-PCR: Use of DNAzymes to detect and quantify nucleic acid sequences in a real-time fluorescent format. Clin. Chem. *46*, 625–630.

Tseng, S.Y., Macool, D., Elliott, V., Tice, G., Jackson, R., Barbour, M., and Amorese, D. (1997). An homogeneous fluorescence polymerase chain reaction assay to identify *Salmonella*. Anal. Biochem. *245*, 207–212.

Tuma, R.S., Beaudet, M.P., Jin, X., Jones, L.J., Cheung, C.-Y., Yue, S., and Singer, V.L. (1999). Characterization of SYBR gold nucleic acid gel stain: A dye optimised for use with 300-nm ultraviolet transilluminators. Anal. Biochem. *268*, 278–288.

Tyagi, S., Bratu, D.P., and Kramer, F.R. (1998). Multicolor molecular beacons for allele discrimination. Nat. Biotechnol. *16*, 49–53.

Tyagi, S. and Kramer, F.R. (1996). Molecular beacons: Probes that fluoresce upon hybridization. Nat. Biotechnol. *14*, 303–308.

Tyagi, S., Marras, S.A.E., and Kramer, F.R. (2000). Wavelength-shifting molecular beacons. Nat. Biotechnol. *18*, 1191–1196.

van der Vliet, G.M.E., Hermans, C.J., and Klatser, P.R. (1993). Simple colorimetric microtiter plate hybridization assay for detection of amplified *Mycobacterium leprae* DNA. J. Clin. Microbiol. *31*, 665–670.

Walker, G.T., Little, M.C., Nadeau, J.G., and Shank, D.D. (1992). Isothermal *in vitro* amplification of DNA by a restriction enzyme/DNA polymerase system. Proc. Natl. Acad. Sci. USA *89*, 392–396.

Walker, R.A., Saunders, N., Lawson, A.J., Lindsay, E.A., Dassama, M., Ward, L.R., Woodward, M.J., Davies, R.H., Liebana, E., and Threlfall, E.J. (2001). Use of a LightCycler *gyrA* mutation assay for rapid identification of mutations conferring decreased susceptibility to ciprofloxacin in multiresistant *Salmonella enterica* serotype typhimurium DT104 isolates. J. Clin. Microbiol. *39*, 1443–1448.

Wang, L., Blasic, J.r., J.R., Holden, M.J., and Pires, R. (2005). Sensitivity comparison of real-time PCR probe designs on a model DNA plasmid. Anal. Biochem. *344*, 257–265.

Watzinger, F., Hörth, E., and Lion, T. (2001). Quantitation of mRNA expression by competitive PCR using non-homologous competitors containing a shifted restriction site. Nucleic Acids Res. *29*, e52.

Weis, J.H., Tan, S.S., Martin, B.K., and Wittwer, C.T. (1992). Detection of rare mRNAs via quantitative RT-PCR. TIG *8*, 263–264.

Wetmur, J.G. (1991). DNA Probes: Applications of the principles of nucleic acid hybridization. Crit. Rev. Biochem. Mol. Biol. *26*, 227–259.

Whelan, A.C. and Persing, D.H. (1996). The role of nucleic acid amplification and detection in the clinical microbiology laboratory. Annu. Rev. Gen. *50*, 349–373.

Whiley, D.M. and Sloots, T.P. (2006). Sequence variation can affect the performance of minor groove binder TaqMan probes in viral diagnostic assays. J. Clin. Virol. 35, 81–83.

Whitcombe, D., Theaker, J., Guy, S.P., Brown, T., and Little, S. (1999). Detection of PCR products using self-probing amplicons and fluorescence. Nat. Biotechnol. *17*, 804–807.

Wilhelm, J., Hahn, M., and Pingoud, A. (2001a). Influence of DNA target melting behavior on real-time PCR quantification. Clin. Chem. *46*, 1738–1743.

Wilhelm, J., Pingoud, A., and Hahn, M. (2001b). Comparison between *Taq* DNA polymerase and its Stoffel fragment for quantitative real-time PCR with hybridization probes. Biotechniques *30*, 1052–1062.

Wilkening, S. and Bader, A. (2004). Quantitative real-time polymerase chain reaction: Methodological analysis and mathematical model. J. Biomol. Tech. *15*, 107–111.

Wirgart, B.Z., Andersson, P., and Grillner, L. (2005). Evaluation of the ReSSQ assay in relation to the COBAS AMPLICOR CMV MONITOR test and an in-house nested PCR method for detection of cytomegalovirus DNA. J. Clin. Microbiol. *43*, 4057–4063.

Wittwer, C.T., Fillmore, G.C., and Garling, D.J. (1990). Minimizing the time required for DNA amplification by efficient heat transfer to small samples. Anal. Biochem. *186*, 328–331.

Wittwer, C.T., Herrmann, M.G., Moss, A.A., and Rasmussen, R.P. (1997a). Continuous fluorescence monitoring of rapid cycle DNA amplification. Biotechniques *22*, 130–138.

Wittwer, C.T., Reed, G.H., Gundry, C.N., Vandersteen, J.G., and Pryor, R.J. (2003). High-resolution genotyping by amplicon melting analysis using LCGreen. Clin. Chem. *49*, 853–860.

Wittwer, C.T., Ririe, K.M., Andrew, R.V., David, D.A., Gundry, R.A., and Balis, U.J. (1997b). The LightCycler™: A microvolume multisample fluorimeter with rapid temperature control. Biotechniques *22*, 176–181.

Zhang, Y., Zhang, D., L.i., W., Chen, J., Peng, Y., and Cao, W. (2003). A novel real-time quantitative PCR method using attached universal template probe. Nucleic Acids Res. *31*, e123.

Oligonucleotide Design for In-house Real-time PCR Applications in Microbiology

2

Andreas Nitsche

Abstract

Today, real-time PCR has entered nearly every field in the biosciences. Numerous applications of real-time PCR have been described for basic scientific purposes and for diagnosis of hereditary and infectious diseases. Ready-to-use assays for profitable targets are commercially available and can be easily implemented in the laboratory's methods repertoire. However, for special situations or just to reduce expenses, it is often desirable to design a real-time PCR assays by oneself. Although often considered a daunting task, a manageable number of design rules permit the generation of reproducible real-time PCR assays. The basic guidelines for successful real-time PCR assay design are presented here.

Introduction

The first functional real-time PCR assays based on fluorescence signal production were published about 15 years ago (Holland *et al.*, 1991). Since then, numerous different real-time PCR fluorescence generating chemistries have been developed some appearing useful for only a single, new real-time PCR platform while others have broad applicability to multiple platforms.

The first real-time PCR assays simply added intercalating dyes to primers used in previously established conventional PCR (Higuchi *et al.*, 1993) reactions but soon the introduction of specific probes required the definition of additional design rules for these oligonucleotides. Besides giving a significant sequence-specific fluorescence signal the probes have to be designed to work in combination with the primer pair used. The primers are also designed to meet special criteria for real-time PCR applications.

Nowadays there are three ways to add a real-time PCR assay to the laboratory methods panel. You buy one as a kit, you copy one from the relevant literature or you design one by yourself.

For prominent nucleic acid targets well-evaluated, commercially available kits can be used and these perform well when applied according to the manufacturer's instructions. Applied Biosystems (Foster City, USA), Roche Applied Science (Penzberg, Germany) and Qiagen (Hilden, Germany) are among the most popular vendors of ready-to-use real-time PCR kits. Real-time PCR kits are purchased for diagnosis of infectious diseases, oncology, metabolic diseases, gene expression analysis and many more applications. For licensing reasons, the sequences of the primers and detection probes are typically not provided in most kits. Similar to a kit, Applied Biosystems offers pre-developed "assay-on-demand" reagents that are MGB hydrolysis probe-based and designed for the detection of numerous human, rat and mouse genes. In addition, Applied Biosystems designs real-time PCR assays for individual customers who can afford this expensive option. Moreover, these assays use also always hydrolysis probes and detailed sequence information is not provided to the customer. Other companies like TIB MOLBIOL (Berlin, Germany) offer excellent free custom-design services and support for several real-time PCR chemistries as long as the designed oligonucleotides are purchased from this company.

One important issue that has arisen from the use of kits by clinical microbiology laboratories is whether the scope of the evaluation has been sufficient to permit detection and/or discrimination of all strains of the target microorganism. Recent years have seen kits redesigned or recalled due to poor performance on this front.

Alternatively, various real-time PCR assays for numerous targets can be taken from the literature (Espy *et al.*, 2006; Mackay, 2004; Mackay *et al.*, 2002), or public databases (Pattyn *et al.*, 2003; Pattyn *et al.*, 2006). The number of publications dealing with real-time PCR is still exponentially increasing and in general the sequences of the oligonucleotides used as well as the reaction conditions employed are well-described by the better publications. Readiness to invest a little work in the adaptation of the reaction conditions nearly always permits successful transfer of a published real-time PCR assay to one's own real-time PCR platform (Nitsche *et al.*, 1999).

If no assay for the desired target exists or special demands must be met for the individual application, a new assay has to be set up from scratch. Each real-time PCR run for the evaluation of a self-designed assay incurs costs by consumption of enzymes, oligonucleotides and probably most of all due to personnel. These efforts can be minimized by a good working knowledge of several design rules which are specific for the intention of the PCR application. The main applications of real-time PCR include:

- detection (diagnostics)
- quantification (microbial load)
- genotyping (microbial characterization).

In certain situations real-time PCR may only be required for the detection of a target. In addition to speed, the minimized risk of carry-over contamination, the option of on-line analysis (available on some real-time PCR instruments) and the additional specificity brought about by the binding of a third oligonucleotide probe makes a real-time PCR diagnostic assay much more reliable than a conventional PCR assay without probes (Mackay *et al.*, 2002; Watzinger *et al.*, 2006). For example, in diagnosing an infection, the detection of a pathogen of interest is commonly of higher importance than its quantification. Therefore, the detection probe has to be located in a target region justifying specific identification of the target of interest which may require compromises with the design rules we will describe below.

Quantification can be relevant for several purposes. For example, gene expression analysis is nearly always a question of quantification (Bustin, 2000; Bustin, 2002; Klein, 2002). Since the sequence of the gene of interest is usually precisely known, the detection probe can be designed according to existing design rules. However, some real-time

PCR assays intend to detect exclusively expressed RNA and not genomic DNA and again, special assay design rules may apply.

The third classical application of real-time PCR in microbiology is genotyping (Gibson, 2006). For microbial genotyping assays the variable sequence position must necessarily be well defined and not be subject to "drift" over time or across distance. Moreover, not every real-time PCR chemistry is suitable for genotyping, which may restrict the assay's application to certain real-time PCR platforms (Bustin, 2004; Bustin, 2004).

However, in the real-life clinical laboratory most assays have to meet more than one of the listed criteria and a combination of numerous design rules is required. A complex example comes from the need to quantify the viral concentration or "viral load" of particular strains when a patient is undergoing antiviral treatment. Such a real-time PCR assay needs a very low detection limit to prove the total clearance of the virus after antiviral treatment. The analytical sensitivity, allowing the quantitative demonstration of very small changes in viral load, must be of sufficient accuracy to verify any desired response to the drug early on during treatment. Finally, if individual virus variants are of interest, a subsequent virus genotyping using the same assay can be applied to confirm the drug's efficacy against the right virus strain. This example demonstrates just how challenging today's real-time PCR design requirements can be.

Beside the basic experimental conditions which predefine certain design strategies, individual technical rules have to be followed according to the real-time platform available. These technical rules are implemented in various software packages that either come with the PCR instrument, can be purchased separately or are available as freeware. A selection of the available software is given in the web resources section although it is not exhaustive. However, there is no software that considers all requirements for a robust assay design and even if such software is developed, it will always make sense to check if the results are in accordance with the basic rules.

During recent years the number of real-time PCR chemistries as well as the number of real-time PCR applications has increased significantly (Bustin, 2004; Bustin, 2004). Some are just modifications of existing chemistries; others are completely new constructions that afford specific signal production however, all recent chemistries are based on the production of a fluorescent signal. Interestingly, as can be followed in the literature, not every published real-time PCR chemistry finds reasonable application, because of complex design requirements or complex and therefore expensive synthesis. It is not the intention of this chapter to cite every single sophisticated real-time PCR chemistry precisely. This chapter will give a brief review of the design of the most popular real-time PCR chemistries, which are sufficient to establish assays for common real-time PCR applications in microbiology. The design rules are presented as an instructional list that can be applied to design and subsequently judge well-performing PCR oligonucleotides. By using these lists, the reader should be able to design basic real-time PCR assays for themselves.

Selecting the best real-time PCR chemistry

Several different real-time PCR chemistries can be applied for most applications in microbiology (Mackay, 2004) and the choice of which chemistry to use depends on only a few criteria. The first consideration should be the purpose of the designed real-time PCR assay as outlined above.

Nearly all known chemistries can be applied for precise quantification and even SYBR® Green I can produce accurate quantification results so long as the target amount is not close to the detection limit of the assay (Queipo-Ortuno *et al.*, 2005; Shu *et al.*, 2003). For genotyping purposes hybridization probes (HybProbes) and molecular beacons have proven useful (Nitsche *et al.*, 2004; Nitsche *et al.*, 2005; Olson *et al.*, 2004; Lindler *et al.*, 2001; Vet *et al.*, 1999; Szuhai *et al.*, 2001). Minor grove-binding (MGB) hydrolysis probes and MGB-Eclipse™ probes can also be used for genotyping (Belousov *et al.*, 2004), but it may be harder to obtain a reliably differentiating assay compared to HybProbes where considerable design experience has accumulated over time.

The available real-time PCR instrument is also a crucial factor when deciding on which is the best chemistry. Special features or quirks occur for the excitation or detection of different fluorophores and these must be considered. For example, fluorescein derivatives that are usually used as reporter fluorophores for hydrolysis probes can be excited and their emission can be detected with every real-time PCR instrument on the market. Therefore fluorescein-coupled hydrolysis probes can be applied universally. On the other hand, classical fluorophores coupled to HybProbes include LightCycler™ Red 640 and Red 705. These can be detected best with Roche LightCycler instruments while other instruments may fail to detect these long wavelength dyes.

Finally the costs can contribute to reaching a decision on which real-time PCR chemistry to use (Speers, 2006). While SYBR Green is definitely the cheapest way to perform quantitative real-time PCR, it has its drawbacks in terms of specificity. Any additional hybridizing oligonucleotide contributes to a higher degree of specificity and also to increased costs (Yeung *et al.*, 2004). Costs for hydrolysis probes are generally comparable to HybProbes, although the synthesis of HybProbes is restricted to Roche-licensed companies while the production of universally available hydrolysis probes is affected by commercial competition which reduces pricing. However, some special chemistries, like MGB or LNA are restricted to selected manufacturers and therefore pricing and synthesis times for oligonucleotides modified with these items may be predetermined and considerable.

Oligonucleotide design

General considerations

For the design of a real-time PCR assay several constituent parts have to be coordinated. Elements that are of highest importance are

- the DNA fragment that is flanked by the two primers and subsequently amplified, here referred to as "amplicon"
- the two primers
- the detection probes.

All of these elements contribute to the PCR efficiency, which describes the idealized property of a PCR assay to double the amount of PCR product with each cycle (Kubista *et al.*, 2006). When establishing a new real-time PCR it is of highest importance to include controls in every run. Samples that contain the targets be they genomic DNA, cDNA, plasmids or just artificially synthesized DNA, can be good positive controls. In addition, negative controls should exclude the possibility for interactions between primers and/or probes that may result in an product that resembles the desired amplicon size in agarose gel analysis.

The amplicon

The amplicon is a fragment of the larger DNA or cDNA strand that is flanked by, and incorporates two primers and is exponentially amplified by duplication during each PCR cycle. The length and structure of the amplicon are fundamental to a good PCR. The length is defined as the distance between the two flanking primers, although it should also be noted that an amplicon includes the primer sequences at each end and not necessarily the original sequence of the target nucleic acid template. Sequence representing the original template is only found between the primers. The optimal length of the amplicon depends on the real-time PCR chemistry; in general, real-time PCR amplicons are very short compared to conventional PCR products. Shorter amplicons result in more efficient PCR reactions; however, highly efficient PCR reactions with exceptionally long amplicons are also described. A smaller amplicon will particularly lead to more efficient detection when the nucleic acid template may be fragmented during the preparation step, for example when extracted from paraffin-embedded tissue.

The three-dimensional structure of the amplicon depends on the sequence, environmental temperature and ionic conditions of the solvent and describes the possible folding an amplicon may adopt. Currently our best estimate of amplicon configuration (although still only predictive) is of secondary structure. The critical DNA region is not only the amplicon itself, but also the template sequences immediately adjacent to the amplicon. Because in the first PCR cycles these surrounding sequence stretches may fold back into the primer or probe-binding regions, they could inhibit efficient oligonucleotide binding and subsequent DNA amplification. In later PCR cycles only the folding of the amplicon is critical. However, all possible structures should be checked for the actual reaction conditions including the likely annealing temperature (T_A) and the concentration of salt ions and magnesium ions. Under the chosen reaction conditions, the binding regions for primers and probes should be completely accessible to or else the oligonucleotides must be moved to an unstructured target region. An appropriate tool to evaluate the folding of relatively short nucleic acid fragments can be found at www.bioinfo.rpi.edu/applications/mfold/.

If possible, GC rich sequences which melt at high temperature should be excluded from the amplicon, since they may hamper strand dissociation at the PCR denaturation step making primer elongation by the Taq DNA polymerase difficult and reducing the PCR efficiency.

Further aspects of amplicon selection relate to special applications including gene expression analysis or the generic or specific detection of microbial variants.

The role of amplicon in transcript analysis

For gene expression analysis the amplicon can be located according to the method of RNA preparation. RNA preparations may contain total RNA or polyA mRNA that is free of ribosomal RNA or tRNA molecules. If polyA RNA is used as the template for reverse-transcription PCR (RT-PCR), amplicons located near to the polyA (3′) end of the RNA promise a higher sensitivity because fragments near the 5′ end may be lost during the preparation step. Examples of this issue can be found among the single-stranded RNA genomes of viruses belonging to the family *Picornaviridae* or *Coronaviridae*.

Another aspect to consider during design is whether the assay is intending to exclusively detect mRNA but not contaminating genomic DNA (Bustin, 2002; Bustin and Mueller, 2005; Bustin, 2005; Bustin *et al.*, 2005). The determination of cellular gene expression in response to virus or bacterial infection is one relevant example, where genomic DNA must be clearly differentiated from mRNA. In this scenario primers are selected to span an intron (that intervening sequences excised during the transcription process) therefore enabling primer binding only to spliced mRNA and not to the genomic DNA (Moore *et al.*, 1990). This approach can be employed for one primer, both primers or even for a detection probe. Intron-spanning primers should extend from the upstream exon with at least four to five nucleotides into the adjacent downstream exon. Usually an intron starts with GU and ends with AG nucleotides. Additionally, the first nucleotides of the spanned intron should be taken into account and similarities to the adjacent downstream exon should be excluded to avoid the 3′ terminus of the primer also binding to genomic DNA. On the other hand, including too many nucleotides from the adjacent downstream exon will encourage binding to genomic DNA despite incomplete hybridization of the oligonucleotide to the template. However, if the detection probe alone spans the intron, the primers will also amplify genomic DNA, but the detection probe will exclusively bind to cDNA. Amplification of genomic DNA in parallel with the mRNA will reduce PCR efficiency.

Alternatively, very large introns can be included in the amplicon, flanking primers will discriminate genomic DNA from cDNA by out-competing the amplification of the much larger DNA amplicon under the selected reaction conditions. In cases when introns are not present, incompletely characterized or the gene consists of only one exon, an additional DNA digestion step before PCR is essential to eliminate contaminating traces of genomic DNA that could bias the results. The mRNA expression analysis of DNA viruses is one example that needs additional DNA digests especially when high titers of virus have been used for infection and huge amounts of viral genomic DNA are present. Also the detection of HIV RNA and not HIV DNA integrated within the human genome requires special primer selection (Kollmann *et al.*, 1992).

Because either very specific or generic "catch-all" assays may also be a design requirement, the positioning of the amplicon will obviously also depend on the suitability of the target sequence required to fulfill the assay's specificity requirements.

Oligonucleotide primers

Good primers have to do only one job: amplify the desired amplicon. But this can be achieved in very different ways. It can be in a highly specific PCR assay that exclusively detects the target of interest and omits amplification of close relatives, or it can be a broadly specific or generic PCR assay that detects a group of targets and catches as many variants as possible in a single amplification reaction (Nadkarni *et al.*, 2002). The specificity of a primer is defined by its complete sequence, but mostly by the 3′ end; the region elongated by the Taq DNA polymerase (Rychlik, 1995). The 3′ end should contain at least five to seven unique nucleotides not found anywhere in the target sequence, thereby reducing false-priming and the creation of side-products. Usually "T" should be avoided as the terminal 3′ nucleotide due to an increased risk of mispriming. Realistically it is extremely difficult to find a sequence of five to seven nucleotides that do not also hybridize with human genomic DNA.

For variant-specific amplification the differentiating nucleotide should be located at the 3′ end. This will not completely stop amplification when the unintended base is present at this position, but it will severely impair amplification efficiency. Sometimes it is also recommended to use the second-last nucleotide in this fashion. The best solution must be proven by amplification of both genotypes with each primer.

In any case, the specificity of a selected primer should first be evaluated *in silico* which should be done during the design. Most commonly used is the "Basic Local Alignment Search Tool" (BLAST) found at www.ncbi.nlm.nih.gov. A primer that displays homology to undesired pathogen variants or to human genomic DNA already *in silico* should not be considered further for use. However, when designing assays to detect, quantify or characterize microorganisms, it quickly becomes clear that these ideal rules can only rarely be strictly adhered to and compromises have to be made. For example, as long as the primers are definitely specific for the desired target, then the detection system may display some *in silico* cross-reactivity with unexpected organisms, since a detection probe will only hybridize with the amplicon which was produced from the specific template the primers were designed for. To carry these considerations to extremes, a real-time PCR assay can be highly specific due to the specificity of only one primer that prevents amplification of target sequences displaying homology to the second primer and/or the detection probe. In any case, the design of a new real-time PCR assay should always aim for the desired specificity of each applied oligonucleotide. At the end this "theoretically" evaluated specificity must always be confirmed by application to as many targets as possible.

Finally, there are a few more technical considerations that define a good primer. First of all, primers should "play well" with each other (Rychlik, 1995). They should not be complementary to themselves (forming homodimers) or to each other (forming heterodimers) and if this is achieved you will avoid dimerization (Brownie *et al.*, 1997). In certain chemistries dimerization causes trouble for the detection of weak positives (see below), but the efficiency of any PCR reaction is significantly reduced when primer–dimers are formed due to irrelevant consumption of oligonucleotides. The formation of hairpins occurs when self-complementary regions in a primer molecule are present, and the conditions are right. In the worst case, the 3′ end of the primer folds back on its 5′ end and elongation at the 3′ end is hampered. As a rule of thumb primer–dimers and secondary structures should remain below a binding energy of 5 kcal/mol which can be calculated by appropriate software. The "Integrated DNA Technologies" homepage offers a package of useful tools including the calculation of primer–dimer binding energies (www.idtdna.com).

Both primers should also bind with similar or identical efficiency to the target. This can be achieved by the adjustment of the annealing temperature (T_A) or the concentration of the primer. As a rule of thumb, doubling the concentration of one primer increases its T_A by 1–2°C. The primer melting temperature (T_M) is that temperature at which half of the primer-target duplex is dissociated and becomes single stranded, indicating the duplex stability. The T_A is usually 5 to 10°C below the T_M. The T_M of a primer can be calculated by several different methods but depends on the primer length, the number of tight-binding G and C nucleotides compared to weaker binding A and T nucleotides, the distribution of the nucleotides and the reaction conditions. The easiest way to estimate the T_M is provided by the "G+C-rule" which adds 4°C for each G or C nucleotide and 2°C for each A or T nucleotide. A primer of 20 nucleotides consisting of 5 As, Cs, Gs and Ts each would have a T_M of roughly 60°C. It seems obvious that the distribution of the nucleotides will influence the T_M value significantly. Therefore, one of the best T_M calculation methods for short oligonucleotides is the "nearest neighbor" method that considers not only the number of each nucleotide, but also the surrounding nucleotides (Breslauer *et al.*, 1986). Numerous tools are available that calculate the T_M value based on this method, however, as long as the result of any method can be interpreted and transferred to the cycling conditions any calculation method is suitable. For example, a primer that has a T_M value of approximately 57°C according to the nearest neighbor calculation method, will perform well with a PCR T_A of 60°C on most real-time PCR platforms. Nevertheless, as long as the same calculation method is applied for all oligos and all designs from one laboratory, the results will always be comparable. The pure value is not that important. More important is how we use it. Nonetheless, the average length of good primers should be 18–22 nucleotides which is long enough for adequate specificity and short enough for efficient binding. These annealing temperature considerations also apply to any kind of oligonucleotide detection probes.

For generic PCR applications, the sequence of the primers cannot be unique. To catch more than one possible variant of the target microorganism, primers with mixed nucleotides at one primer position can be used. These primers are called degenerate primers. The degenerate positions of the primer should not be located directly at the 3′ end, but can be anywhere else in the primer. Efficiently binding primers are described that contain a degeneracy in up to 25% of nucleotide positions (Aitichou *et al.*, 2005). To describe degenerate nucleotides, which actually represent a mixture of oligonucleotide with the different nucleotides at the desired position, special acronyms were defined. For example, "S" (strong) describes either a G or a C, because these nucleotides form three Hydrogen-bonds in Watson–Crick base pairs which are strong compared to the two H-bonds of an A-T base pair. Therefore, having an A or a T at a position is described by a "W" (weak). In contrast to "N" (any), which means the equimolar incorporation of all four regular nucleotides A, C, G and T, at one position, Inosine is a nucleotide that is able to pair well with C and A, less strongly with T and a little with G. A complete summary of the acronyms used for the description of degenerate primers is given in Table 2.1.

To shorten the establishment of a PCR assay, it is recommended to design four or six primers that can be used in four or nine different combinations, respectively. Testing all possible combinations will lead rapidly to a functional PCR assay and saves time and money compared to the laborious optimization of one primer pair.

But what if the newly designed primer(s) does not work as expected? The worst case can be that a primer or primers does not amplify the target at all. Agarose gel analysis should always be applied to prove the presence of a single distinct PCR product band even for real-time PCR assays. If no amplicon is formed the primers should be checked for orientation (the reverse primer should be synthesized as the reverse-complement of the strand the forward primer is designed from), for T_M and for production quality. Low-melting primers will not bind to the target strand at high annealing temperatures.

Table 2.1 Nucleotide acronyms

Acronym	Nucleotide(s)	Explanation
A		Adenosine
C		Cytosine
G		Guanine
T		Thymine
U		Uracil
N	A C G T	aNy
Y	C T	pYrimidine
R	A G	puRine
W	A T	Weak
S	C G	Strong
M	A C	aMino
K	G T	Keto
B	C G T	not A
D	A G T	not C
H	A C T	not G
V	A C G	not T

If several product bands or a product of unexpected size are shown by gel analysis, the specificity of the primers should be checked. Most important is to check if the last five to seven 3′ nucleotides are unique to the target sequence. Sometimes moving the primer a few nucleotides can be helpful. However, a distinct band is a pre-requisite for an efficient PCR assay and real-time PCR should always be based on that. Basic primer-design rules to obtain good assay results are summarized in Table 2.2.

Oligonucleotide detection probes

Detection probes can add specificity to an assay or they can be used for genotyping or for quantification. Common detection probes are hydrolysis probes (also known as TaqMan® probes, 5′ nuclease probes or dual-labeled probes) (Livak *et al.*, 1995; Holland *et al.*, 1991; Lie and Petropoulos, 1998), hybridization probes (Wittwer *et al.*, 1997), molecular beacons (Tyagi and Kramer, 1996) and in the broadest sense also Scorpion® primers (Scorpions) (Whitcombe *et al.*, 1999) which combine primer and detection probe in one molecule. Several more chemistries exist but their benefit compared to those listed above is unclear to date. The design of each probe can vary with the intention of the PCR assay. In any case the detection probe(s) has to be compatible with the primers, i.e. primers and probes should not bind to each other due to complementarities or compete for the same

Table 2.2 Basic primer design rules

Design rule	Reason
Avoid secondary structures	Enhances primer binding
Avoid primer-dimers	Increases PCR efficiency
Use balanced sequences	Increases specificity
Avoid >4 identical nucleotides (G,C>A,T)	Increases likelihood of specific binding
Avoid repetitive sequences	Prevents primer from slipping
Avoid G-rich primers	Avoids formation of undesired complexes through non-Watson–Crick interactions
Adapt GC-content to the target sequence	Increases specificity
Incorporate at least 5 unique 3' nucleotides	Determines primer specificity
Adjust annealing temperatures	Allows simultaneous binding of both primers

stretch of target sequence. Usually detection probes are designed to have a higher T_M than the primers to enable efficient signal production. The increased T_M of the detection probe guarantees that the oligonucleotide has bound before the primers are elongated by the polymerase. Concomitant binding of primers and detection probe due to similar T_M values could lead to blockage of the probe binding site by the rapidly elongated primer. However, special rules for individual detection probe chemistries have to be applied for successful assay design.

Because detection probes are crucial for the outcome of a real-time PCR assay, a new design always creates the detection probes first and then adds suitable primers latter.

Specific oligonucleotide design rules

SYBR® Green I and other intercalating dyes

Intercalating dyes emit a strong fluorescence signal due to interaction with double stranded DNA, such as PCR amplicons. However, these dyes cannot discriminate between amplicons; they will recognize primer–dimers or larger non-specific side-products of the PCR reaction. Consequently, this chemistry is not more specific than conventional PCR but it is less specific than chemistries that rely on binding of additional sequence specific probes. Beside SYBR Green I (Morrison *et al.*, 1998; Karlsen *et al.*, 1995), fluorescent dyes including the more stable SYBR Gold® (Tuma *et al.*, 1999), ethidium bromide (Higuchi *et al.*, 1993), BEBO® (Bengtsson *et al.*, 2003), YOYO® (Srinivasan *et al.*, 1993; Ogura *et al.*, 1994) YO-PRO® or BOXTO® (Lind *et al.*, 2006) have been introduced. However, SYBR Green I is the dye most frequently applied. Many SYBR Green I or intercalating dye based real-time PCR assays give weak signals in negative controls because of binding to single stranded DNA and primer–dimers. Reliable differentiation of signal due to primer–dimers from that due to weak positive samples is impossible without post-PCR dissociation curve analysis. Most software tools for oligonucleotide analysis permit a primer–dimer check, e.g. Oligo 6.0 (Molecular Biology Insights).

On the other hand, SYBR Green I is less expensive than any detection probe and can be detected by every real-time PCR instrument, making this chemistry highly flexible. Numerous publications describe the use of intercalating dyes for the detection of viral (Aldea *et al.*, 2002; Komurian-Pradel *et al.*, 2001; Manaresi *et al.*, 2002; Drosten *et al.*, 2002; Escutenaire *et al.*, 2006) and bacterial pathogens (Aarts *et al.*, 2001);(Lee *et al.*, 1999) (Bystrom *et al.*, 2005; Debeaumont *et al.*, 2005; Brennan and Samuel, 2003)

After a SYBR Green I PCR run the amplicons can be subjected to a melting curve analysis (Ririe *et al.*, 1997). If only one specific amplicon was produced a distinct melting peak is observed. This peak denaturation temperature (T_D) depends on the amplicon length, sequence and ionic environment and is usually between 75°C and 90°C. Primer–dimers are generally of lower length and display a lower T_D. A PCR that produces primer–dimers, and almost every PCR does, will show up two distinct peaks, one for the amplicon and one for the primer–dimers; the latter peak will be at a lower temperature. To circumvent the detection of primer–dimers, an additional measuring step can be added at a temperature for which the primers are already molten and the amplicon is still fluorescing. If primer–dimers and amplicon display similar melting temperatures, finding the right measuring temperature which does not also reduce the intensity of the specific signal, and therefore the sensitivity of amplicon detection, may be difficult. A first practical approach is to reduce the annealing incubation to the shortest time that will still produce sensitive results. General rules for the design of SYBR® Green I assays are described in Table 2.3.

Hydrolysis, TaqMan or 5′ nuclease probes

Hydrolysis probes were the first convenient and sequence-specific real-time PCR chemistry (Holland *et al.*, 1991). An hydrolysis probe assay consists of two primers and one detection probe that binds to the amplified target strand and is dually labeled with different molecules. When excited by a LASER or other light source, one molecule, a fluorophore at the 5′ end, reports signal accumulation after degradation of the probe by the DNA polymerase, while the other molecule, near or at the 3′ end, quenches the reporter dye

Table 2.3 Assay design rules when using SYBR® Green I

Design rule	Reason
Amplicon	
Create amplicon length of 200-300 bp	Generally facilitates differentiation from primer-dimers by melting analysis and produces high fluorescence signals due to incorporation of more SYBR Green Shorter amplicons can give better PCR efficiency
Primers	
Avoid primer-dimers!!!	Helps reliability of negative results
Detection	
Always perform melting analysis	Generally permits estimation of primer-dimer formation

as long as the two molecules are in close spatial proximity, e.g. the hydrolysis probe is intact. The DNA polymerase used must therefore have a 5′ exonuclease activity which is essential for hydrolysis of the probe (Kreuzer *et al.*, 2000). The distance between reporter and quencher should not exceed 25–30 nucleotides which is about 100 Å in length. If longer hydrolysis probes are designed, the quencher should be placed somewhere in the middle of the hydrolysis probe to enable proper quenching. To keep the fluorophores in the same spatial plane, 22 nucleotides are an ideal distance between reporter and quencher fluorophore. Commonly suitable reporter fluorophores are fluorescein derivatives like 6-carboxy-fluorescein (FAM) which emit strong fluorescence signals at a wavelength of 518 nm when excited at 494 nm. Several other fluorophores have been reported including TET (6-carboxy-2′,4,7,7′-tetrachlorofluorescein), JOE (6-carboxy-4′,5′-dichloro-2′,7′-dimethoxyfluorescein), HEX (6-carboxy-2′,4,4′,5′,7,7′-hexachlorofluorescein), VIC and Yakima Yellow which display longer emission wavelengths. Although they yield good detectable signals, fluorescein derivatives are the first choice, for uniplex assays as well as for one of the assays in a multiplex format. The first quencher fluorophore used was TAMRA (6-carboxy-tetramethyl-rhodamine), but recently dark quenchers have been introduced including DABCYL (4-((4-(dimethylamino)phenyl)azo)benzoic acid) that has an absorption range that overlaps hardly with fluorescent dyes emitting above 500 nm and must be in closer proximity to the reporter fluorophore, and the Black hole quenchers (BHQ, Biosearch Technology, USA), which emit energy as heat instead of visible light. For multiplexing purposes the selected reporter dyes should be spectrally separated as much as possible and dark quenchers should be used. However, the outcome of a multiplex detection depends strongly on the excitation source employed and the detection capacity of the real-time PCR instrument. These considerations are addressed in more detail in chapter 6.

Hydrolysis probes can be used for sequence-specific detection or for quantification (Mackay *et al.*, 2002) but are only rarely described for genotyping (Livak, 1999). Depending on the reporter dye, which is in most applications FAM, hydrolysis probes are able to be used on many different real-time PCR instruments. For a well designed and functioning assay, one hydrolysis probe molecule is degraded per amplicon, each cycle. Continuous excitation by light allows the online monitoring of the fluorescence signal in "real" time. Since a hydrolyzed probe molecule remains in pieces for the remainder of the reaction, the signal produced can be read at any time point during the amplification process.

Although it is possible to plan hydrolysis probes to fit between an existing primer pair, the design of completely new assays is usually more beneficial although this will depend on the situation. Amplicons for sensitive detection are usually as short as possible. Between the 3′ end of the primer and the hydrolysis probe binding to the same strand a gap of one nucleotide is satisfactory. The primer on the other strand can be placed at any position behind the hydrolysis probe; however, a one or two base overlap with the hydrolysis probe is fine. Additional overlapping nucleotides will lead to complementary binding which could produce an artificial signal and should be avoided. For longer amplicons place the hydrolysis probe close to the primer on the same strand to allow maximal signal generation. A "G" at the first 5′ position of the probe should be avoided, since unpaired "G" quenches the fluorescence of fluorescein derivatives (Crockett and Wittwer, 2001). In general, whenever possible the strand with the fewest "Gs" should be selected for the hydrolysis probe target to reduce secondary structures formed by non-Watson–Crick base pairing.

If a hydrolysis probe does not give a signal, the first step is to verify generation of a PCR product by agarose gel analysis. No PCR product indicates problems with the primers. A PCR product of the expected size points to a problem with the hydrolysis probe. There are two possible explanations: The hydrolysis probe has not bound to the target strand and could not be hydrolyzed by the polymerase, or the signal is too weak to be detected. To bind to the target strand tightly, the 5′ nucleotides should match the target strand perfectly, an issue that is much less important for the middle position nucleotides. However, too many mismatches, wherever they are located, can impede probe binding at the annealing temperature resulting in incomplete hydrolysis. Possible formation of stable secondary structures within the probe must also be avoided since they will compete with binding to the target. Finally, the quality of the hydrolysis probe should be considered. Poor quality production can be controlled by nuclease digestion of the probe in a fluorimeter that should result in a clear fluorescence increase over time. It is recommended to digest a good performing hydrolysis probe in parallel as standard. Basic rules for successful hydrolysis probe design are given in Table 2.4.

MGB probes have proven to be a helpful modification of regular hydrolysis probes. These have a tricyclic organic molecule attached to the 3′ end that can fold back into the minor groove of the perfectly matching probe-target duplex (Afonina *et al.*, 2002). This results in an increase of the T_A by 10 to 15°C, allowing the design of hydrolysis probes that are only between 13 and 18 nucleotides in length. A temperature calculation tool is incorporated into the "Primer Express" software included with all real-time PCR instruments sold by Applied Biosystems. This can be used to estimate the final T_M of an MGB hydrolysis probe which is still usually designed according to regular design rules for hydrolysis probes but can also be designed using this software.

Very short hydrolysis probes can be useful tools for the generic detection of sequences of low homology such as different virus species. Originally, a pair of differently reporter labeled MGB probes was designed for allelic discrimination (Kutyavin *et al.*, 2000) but this application has also been useful for genotyping microorganisms (Campsall *et al.*, 2004). Longer regular hydrolysis probes tolerate mismatches located in the middle of the molecule but not mismatches in the first two or three 5′ terminal nucleotides; these lead to reduced signal production. In contrast, shorter probes need to match perfectly for stable binding at common PCR temperatures and a single mismatch may lead to destabilization and poor signal generation. Since the MGB moiety can only function with perfectly matching probes, one mismatch in the MGB binding region results in a significant decrease of the annealing temperature. The actual T_A of such a mismatched probe will be close to that of the probe without an MGB modification, which is usually below 40°C. Consequently the probe cannot be degraded by the polymerase and no signal is generated. The use of two differently labeled MGB probes, each perfectly matching one of the two genotypes, allows differentiation of single nucleotide polymorphisms. However, background signal production for the non-matching genotype can be observed creating difficulties when interpreting the data in certain cases when genotypes have to be excluded definitively.

HybProbes, adjacent hybridization probes or LightCycler probes

HybProbes were first described for use with the Roche LightCycler instrument but can be used on other instruments depending on the available emission detection channels (Wittwer *et al.*, 1997). This real-time PCR detection chemistry consists of two primers

Table 2.4 Assay design rules for hydrolysis probes

Design rule	Reason
Amplicon	
Create amplicon as short as possible	Higher efficiency and sensitivity
Primers	
Avoid complementarities to hydrolysis probe	Minimizes poor signal generation due to primer–probe binding
Hydrolysis probe	
Place probe close to the primer on the same strand	Maximal degradation of probes molecules per cycle, equating to maximum signal generation
Choose the purine-poor strand	Improves hybridization stability
Avoid complementarity to primers	Minimizes poor signal generation by primer–probe binding
Length 20–30 nucleotides (up to 35)	Permits efficient probe binding
T_M 5 to10°C above primer T_M	Permits efficient signal generation while avoiding blocking of target region by elongated primer
T_M should not be higher than elongation temperature	Ensures probe does not block primer elongation
No G at the 5′ end (for FAM reporters)	Prevents unpaired Gs from quenching fluorescence leading to poor signal generation
Block 3′ end by phosphate or quencher fluorophore	Prevents elongation of the probe
Do not introduce mismatches at 5′ end	When probe and target match perfectly optimal degradation can occur
Avoid GC-rich stretches	Minimizes the non-specific binding of the probe to GC-rich target islands
Avoid G-rich probes	Excessive Gs can form undesired complexes through non-Watson–Crick interactions
Avoid sequence palindromes	Prevents the formation of self-complementary stem-loop structures
Detection	
Performed at annealing or polymerization step	Signal is permanent

and two additional oligonucleotides that are designed to bind specifically to the same target strand. Both probes are selected to bind adjacently, within a distance of only a few nucleotides. One of the probes, the donor, is labeled with a fluorescein at its 3′ end, preventing elongation of the molecule by the DNA polymerase. The downstream probe, or acceptor, is 5′ end-labeled with a fluorophore and the 3′ terminus is blocked by a phosphate moiety (Nitsche *et al.*, 1999; Landt O and Nitsche A, 1999). Common acceptor fluorophores are LC Red 640, LC Red 705 and Cy5, but more recent LightCycler instruments can detect additional LC Red fluorophores that are also clearly distinguished when employed for multiplex detection assays. If both probes have bound to the target strand, the donor and the acceptor dye come in close proximity. While the donor fluorophore can be excited by the light source, the acceptor cannot. However, following excitation the donor emits energy that is of an appropriate wavelength for excitation of the acceptor dye, a process called fluorescence resonance energy transfer (FRET). FRET can only happen if both fluorophores are in close proximity which occurs exclusively during the annealing phase of the PCR reaction. The probes do not require degradation for signal generation, so this chemistry can be regarded as non-destructive, in contrast to the hydrolysis probe chemistry. However, the DNA polymerase cannot differentiate between the different real-time PCR chemistries and will therefore also degrade HybProbes that are annealed to the target strand during primer elongation and signal production is aborted. Therefore, HybProbes should always be detected during the annealing phase of the PCR reaction, since most of the probes are already bound to the target strand and primers are not significantly elongated at this stage. This chemistry requires special considerations for assay design (see Table 2.5). If all these criteria are met, each single amplicon is displayed by the FRET signal and the accumulation of PCR product can be precisely followed. HybProbes can therefore be used for sequence-specific detection and quantification.

Beside general considerations for successful probe design, an important criterion of HybProbe design is their positioning on the target strand. If the pair of probes is located too close to the primer on the same strand, the probes will be displaced by the DNA polymerase before a signal is detected. Agarose gel analysis will reveal PCR product generation while subsequent fluorescent melting curve analysis should produce a clear signal, but the amplification cannot be monitored during the PCR run. Similar symptoms will be observed if the melting temperature of the probes is too low to bind efficiently at the annealing temperature. Poor complementarities with the target strand or poor production quality of either the oligonucleotides or the attachment of the fluorescence labels should also be considered as sources of poor fluorescence. Basic rules for the design of HybProbes are given in Table 2.5.

Interestingly, since the HybProbe chemistry is non-destructive, the probes can be used for fluorescence melting curve analysis, a key feature of the HybProbes (Nitsche *et al.*, 2004; Nitsche *et al.*, 2005; Olson *et al.*, 2004; Panning *et al.*, 2004; Lindler *et al.*, 2001). Fluorescence curve melting analysis is a useful method for the typing of microbial mutations, polymorphisms or other sequence variants. Assuming that DNA polymerase-mediated probe degradation is not a feature of the chemistry, then the intact HybProbes remain after the PCR run is finished. A melting analysis includes a denaturation step at 95°C that dissociates the double-stranded DNA amplicon, an annealing step at approximately 40°C where the HybProbes can bind to perfect or mismatched target sequences and finally a controlled incremental increase in temperature to above 80°C which melts the HybProbes from their target. A perfectly matched HybProbe will melt off its target at a higher T_M compared to a HybProbe partly annealed to a mismatched target which is destabilized. The resulting characteristic T_M for the binding of a HybProbe to a target can be used for genotyping. The probe covering the polymorphism is therefore called the "sensor" probe and should be selected first. One should aimed at placing the polymorphism in the middle of the sensor probe. Since the second HybProbe functions only for signal generation by

Table 2.5 Assay design rules for HybProbes used for detection and quantification

Design rule	Reason
Amplicon	
Amplicon should be between 150–200 bp	Compromise between PCR efficiency and sufficient signal generation (also see below)
Primers	
Avoid complementarities to hydrolysis probe	Minimizes reduced signal due to primer to probe binding
HybProbes	
Place probes far away from the primer on the same strand	Increase the time for signal detection (until the polymerase displaces the probes)
Choose the purine-poor strand	Improves hybridization stability
Avoid complementarity to primers	Minimizes reduced signal due to primer to probe binding
Avoid complementarities of HybProbes	Prevents probe-to-probe hybridization which can produce target-independent signal
Length of 20 to 30 nucleotides (up to 35)	Permits efficient probe binding
T_M 5 to 10°C above primer T_M and annealing	Permits efficient signal generation while avoiding blocking of target region by elongated primer
T_M should not be higher than elongation temperature	Ensures probes do not block primer elongation
T_M of both probes within 2°C	Permits concurrent binding of the probes
Leave a gap of 1 to 5 nucleotides between probes	Ideal distance for reliable FRET
No Gs should be located in the target strand between the annealed probes	Overcomes tendency for unpaired Gs to quench fluorescence reducing signal strength
Label adjacent probe ends with 3′ fluorescein and 5′ LC red dye	Avoids difficult synthesis of vice versa labeling, although still possible
Block 3′ end by phosphate or fluorescein	Prevents elongation by the polymerase
Avoid GC-rich stretches	Minimizes the non-specific binding of the probe to GC-rich target islands
Avoid G-rich probes	Excessive Gs can form undesired complexes through non-Watson–Crick interactions
Avoid sequence palindromes	Prevents the formation of self-complementary stem-loop structures
Detection	
At annealing step	Adjacent probe hybridization is require for signal generation

FRET, its T_M should be approximately 5°C higher so that the melting temperature is only dependent on the sensor probe. The second probe is therefore called the "anchor" probe. Either the anchor or the sensor probe can be fluorescein labeled or LC Red labeled.

For the perfect match, the T_M depends on the length and composition of the probe and this value determines the annealing temperature. The lowered duplex-stability of a mismatch is reflected by a decrease in the T_A that cannot be predicted exactly. The T_A will depend on the position and the type of the mismatch. While G:C and A:T binding is characterized by formation of three and two stabilizing hydrogen bonds, respectively, G:T still forms one H-bond leading to less destabilization and smaller differences of the melting temperature. On the other hand, any mismatch that includes a C leads to a bigger difference in the melting temperatures due to efficient destabilization of the probe-target duplex (von *et al.*, 2001). No H-bonds are formed, which applies to several mismatch situations, but sterical considerations lead to maximal duplex destabilization when C is included. Since the target region is defined by the position of the polymorphism to be detected, the location of the probes is clearly predetermined.

If the discrimination of two genotypes is not significant, the reverse strand should be used to create a different mismatch situation. General rules for the design of genotyping HybProbes are given in Table 2.6.

A modification of the HybProbe concept is called the "internal LightCycler" approach. This chemistry is based on the use of one regular primer, one primer that is fluorophore-labeled which additionally functions as a FRET probe, and a single dedicated HybProbe. The HybProbe binds to the strand that is produced by elongation of the labeled primer, and in close proximity to it permitting FRET from one strand to the other. Consequently, both quantification and melting curve analysis can be performed (Nitsche *et al.*, 2004). The internal LightCycler can be applied for very short amplicons or when the amplicon sequence offers no suitable stretches for the placement of a pair of HybProbes. The design of an internal LightCycler assay proceeds according to the above rules; however, the HybProbe must be close to the labeled primer.

Other chemistries

Beside hydrolysis probes and HybProbes there are many more chemistries published that are not listed here in detail. Among the useful alternative chemistries are molecular beacons (Tyagi and Kramer, 1996). A molecular beacon consists of a sequence specific loop and a stem (www.molecular-beacons.org). The stem is formed by complementary binding of an artificial sequence of approximately six nucleotides which is coupled to the 5′ end and in the reverse orientation attached to the 3′ end. The reporter fluorophore is attached to the 5′ end and the quencher fluorophore to the 3′ end. By forming a duplex the stem brings both fluorophores close together and if excited by a light source the reporter is quenched. At higher temperatures the beacon opens by melting of the stem and the reporter fluorophore can emit light after excitation by a light source. Cooling down to the annealing temperature, the beacon has two options: it can either close again by forming the stem or it can bind to the target strand that is complementary to the loop region. As long as no matching target sequence is present, the beacon will close and the reporter fluorescence is quenched. Otherwise the beacon will bind to the target, remaining open and permitting its fluorescence to be detected during the annealing phase of the PCR reaction. As mentioned

Table 2.6 Assay design rules for HybProbes used for genotyping

Design rule	Reason
Amplicon	
Amplicon can be of any length	Because this application of HybProbes is about small sequence changes so length is not a concern so long as overly long amplicons do not form secondary structures
Primers	
Avoid complementarities to hydrolysis probe	Minimizes reduced signal due to primer to probe binding
HybProbes	
Place probes anywhere on the target strand	The important signal is generated during melting analysis, not during PCR cycling
Choose the purine-poor strand	Improves hybridization stability
Avoid complementarity to primers	Minimizes reduced signal due to primer to probe binding
Avoid complementarities of HybProbes	Prevents probe-to-probe hybridization which can produce target-independent signal
Length of 20 to 30 nucleotides (up to 35)	Permits efficient probe binding
T_M 5–10°C above primer T_M and annealing	Permits efficient signal generation while avoiding blocking of target region by elongated primer
Probe T_M can be below primer T_M	Still retains the end-point analysis application if amplification does not need to be followed
T_M should not be higher than elongation temperature	Ensures probes do not block primer elongation
T_M of anchor probe is > 5°C higher than T_M of sensor probe	Ensures FRET can occur as soon the sensor probe hybridizes
Leave a gap of 1 to 5 nucleotides between probes	Ideal distance for reliable FRET
No Gs should be located in the target strand between the annealed probes	Overcomes tendency for unpaired Gs to quench fluorescence reducing signal strength
Label adjacent probe ends with 3′ fluorescein and 5′ LC red dye	Avoids difficult synthesis of vice versa labeling, although still possible
Block 3′ end by phosphate or fluorescein	Prevents elongation by the polymerase
Create C:A mismatches instead of A:T mismatches	Permits better discrimination between mismatches than G:T than C:A mismatches
The variable position can be anywhere in the sensor probe, but not closer than 3 nucleotides from either end	Avoids less destabilizing terminal positions

Table 2.6 continued

Design rule	Reason
Avoid GC-rich stretches	Minimizes the non-specific binding of the probe to GC-rich target islands
Avoid G-rich probes	Excessive Gs can form undesired complexes through non-Watson–Crick interactions
Design fluorescein sensor probes	Produce a strong emission signal
Avoid sequence palindromes	Prevents the formation of self-complementary stem–loop structures
Detection	
If amplification is to be followed, detect in the annealing phase. If only genotyping results are required, detect signal during melting analysis	Depends on the intention of the assay
Consider asymmetric PCR when signal is low	May produce excess of the target strand binding the sensor probe and thus increase he signal

above, the DNA polymerase can only degrade an oligonucleotide that perfectly matches the target strand at its 5′ end. Since the stem of the beacon is not designed to hybridize with the target, the 5′ end of the beacon is never hybridized to the target and the polymerase cannot degrade the beacon, which is instead pushed off the target strand. This chemistry is non-destructive and can therefore be used for quantification as well as for genotyping (Marras *et al.*, 1999; Marras *et al.*, 2003; Mhlanga and Malmberg, 2001; Tapp *et al.*, 2000). The design of a beacon can be difficult, because the right balance between opening of the beacon at relevant temperatures and optional closing or target-binding can be dependent on a single nucleotide in the stem (Monroe and Haselton, 2003). If the beacon closes before it gets the chance to bind to the target, signal will never be detected. On the other hand, the beacon must close if no matching target sequences are present, otherwise it will produce a false signal. However, apart from the design of the stem which must work in tandem with the target-specific loop, common rules for oligonucleotide design still apply.

Vet *et al.* demonstrated the potential of molecular beacons presenting a real-time PCR assay that detected and discriminated four pathogenic retroviruses in a multiplex PCR (Vet *et al.*, 1999). An advancement of molecular beacons that enables multiplexing with highly efficient excitation of other reporter dyes than FAM by a LASER is the wavelength-shifting beacon (Tyagi *et al.*, 2000). This molecule contains beside a quencher a FAM fluorophore that harvests the excitation energy and transmits it to an emitter fluorophore in close proximity. Therefore, the excitation wavelength can be optimized for FAM and several different emitter fluorophores can be used as emitters permitting the use of multiple beacons on older instruments.

Similar to MGB hydrolysis probes, Eclipse probes utilize an MGB moiety, but in contrast to hydrolysis probes the labeling orientation is changed. The MGB moiety and the quencher are coupled to the 5′ end and the reporter fluorophore is coupled to the 3′ end

(Afonina *et al.*, 2002). Spatial proximity of reporter and quencher in the free probe leads to signal suppression after excitation as already described. Signal production happens by stretching the probe during its binding to the target strand increasing the distance between reporter and quencher fluorophore. Due to the MGB at the 5′ end, the detection probe cannot be degraded by the DNA polymerase so signal production is reversible. This implies that Eclipse probes can be used for melting analysis. A recent application in microbiology describes the detection of *Yersinia pestis* using MGB Eclipse probes (Chase *et al.*, 2005).

Additional real-time PCR chemistries often resemble combinations of previously established chemistries. For example Scorpion primers are a combination of a primer and a beacon in one molecule that primes elongation and detects the generated sequence using a single molecule (Whitcombe *et al.*, 1999). Scorpions are ideally suited for genotyping as shown for the typing of human papillomaviruses (Hart *et al.*, 2001). ResonSense® probes consist of a Cy5 labeled hybridization probe that is excited by SYBR Green, emitting fluorescence after binding to the probe and target (Lee *et al.*, 2002).

A clever combination of two different chemistries is the use of a hydrolysis probe (or MGB hydrolysis probe) to function as a donor for a single HybProbe that is labeled with a LC Red fluorophore. The hydrolysis probe is usually labeled with a 5′ fluorescein reporter and 3′ quencher. The HybProbe is located upstream of the hydrolysis probe, requiring the LC Red fluorophore label to be coupled to its 3′ end, a modification which can be synthesized at higher expense. This combination allows the detection of the hydrolysis probe like a regular hydrolysis probe in the fluorescein channel of most real-time instruments, giving information about amplification. The LC Red fluorophore can be monitored additionally to follow amplification, but, most importantly, it can be used for melting analysis in suitable instruments (Olson *et al.*, 2004).

Several more chemistries and modifications are published (see Chapter 1), but most of them lack actual applications in microbiology. Finally, none of these chemistries offers crucial features that cannot be accomplished by the design of hydrolysis probes or HybProbes.

Oligonucleotide modifiers

LNA

A further promising approach to create tightly binding oligonucleotides is the introduction of locked nucleic acid (LNA) nucleotides into primers or probes (Kumar *et al.*, 1998). LNA are modifications of DNA nucleotides that are characterized by an altered ring structure in the sugar moiety (a 2′-O, 4′-C-methylene bridge) which induces a conformational change in the regular B-DNA double helix to an A-DNA double helix structure (Nielsen *et al.*, 1999). Thereby, the duplex stability is significantly increased. Introduction of one single LNA base into an oligonucleotide leads to an increase in T_M of 2 to 6°C. Recently, LNA nucleotides were used to improve oligonucleotide binding in terms of a better discrimination in melting analyses (Simeonov and Nikiforov, 2002). However, apart from the knowledge that LNA nucleotides should not be introduced to the 3′ end of a primer since this leads to poor elongation, additional critical factors for the application of LNA in detection probes are lacking.

Superbases
Additional modifications, the so called "superbases A™, G™ and T™" increase binding stability, mismatch discrimination and reduce self-annealing. While super A and super T stabilize AT-bonds, super G is modified to reduce secondary structures by non-Watson–Crick binding and does not quench adjacent fluorescein when unpaired. However, only a few applications of Eclipse probes containing superbases have been described to date.

Oligonucleotide quality
Whatever the chosen real-time PCR chemistry, the oligonucleotides must be manufactured to a high quality as a prerequisite for a reproducible real-time PCR. Unlabeled oligonucleotides of high quality are provided by several companies today. Most companies offer various purification grades and these may be important for special applications, however for most PCR assays a gel filtration of the synthesized oligonucleotide is satisfactory. Fluorescently labeled oligonucleotides are usually purified by High Performance Liquid Chromatography (HPLC) to remove uncoupled fluorophore molecules and provide a homogenous molecule population. A uniform length is crucial for genotyping assays employing HybProbes since a probe population of mixed length would lead to a broad shoulder in the peak fluorescence melting curve and mean data were extremely difficult or impossible to interpret. For any fluorophore-labeled oligonucleotide, the yield of fluorophore coupled to the oligonucleotide is a very important quality criterion. Unlabeled probe molecules compete with labeled probe molecules for target binding and consequently reduce the maximum possible fluorescence signal. For hydrolysis probes, quantitative labeling is essential to reduce the background originating from insufficiently quenched hydrolysis probes.

However, the oligonucleotide manufacturing quality of most vendors is at least acceptable and one should be able to find a compromise between price, quality and delivery time to choose a reliable provider. A certain level of quality can easily be maintained by staying with a company for all oligonucleotides of the same type. Beside these more technical considerations, the support provided by a vendor can be a decisive factor. Several companies offer online or telephone support for PCR-related questions; however, competent specialists are rare.

Troubleshooting
Presuming that the oligonucleotides are of high quality, there are still manifold possibilities for a real-time PCR assay to fail. The following table (Table 2.7) may be understood as a flow-chart for basic troubleshooting of newly established real-time PCR assays.

Future trends

Well designed assays based on hydrolysis probes or HybProbes provide reliable sensitivity and reproducibility. The introduction of further chemistries does not seem to have improved the outcome of basic applications in the microbiology laboratory and the most recently presented complex chemistries have not found their way into the diagnostic laboratory yet.

Table 2.7 Basic real-time PCR troubleshooting

No amplification occurs as shown by agarose gel analysis		
check	*for*	*because*
Reaction mixture	Buffer dNTPs Mg^{2+} ions Taq DNA polymerase Two primers Template DNA	All ingredients are essential for enzymatic elongation
Primer orientation	Reverse complementation of one primer	Each primer needs to bind to complementary DNA strands
Amplicon	Length, and use longer annealing/extension steps or shorten the amplicon	Extremely long amplicons may not be amplified under common real-time PCR conditions
T_A	The relationship to the primer T_M and adjust T_A to T_M	With too high T_A high primers cannot bind efficiently and with too low T_A primers bind non-specifically producing several side products that hide the correct amplicon
Incorrect amplification occurs as shown by agarose gel analysis		
check	*for*	*because*
Primer	Specificity by BLAST search and the position of the specific nucleotide	Primer specificity is determined by the 3′ end
T_A	Maximum possible T_A	Specificity requires high stringency
Five to seven 3′ nucleotides of the primers	Redundancy in the target sequence	Primers may bind falsely
Amplification occurs as shown by agarose gel analysis but no florescent signal is detectable		
check	*for*	*because*
Amplicon	Secondary structures in the probe binding region	Hairpins reduce binding efficiency of the probe(s)
Detection probe (s)	Correct sequence	Probes with mismatches may show extremely reduced binding efficiency
Detection probe (s)	G content and G positions	G quenches fluorescein
Detection probe (s)	Placement in the amplicon	Hydrolysis probes work best close to the primer on the same strand and HybProbes when placed far away from the primer on the same strand

Table 2.7 continued

The fluorescence signal is also present in negative controls		
check	*for*	*because*
Laboratory facilities	Precautions against contamination	DNA and PCR products are very stable and easily cause contamination
Wrong positives	Amplification by agarose gel analysis	Signals may occur from interactions of complementary primers and/or detection probes
The real-time PCR assay is insensitive		
check	*for*	*because*
Amplicon	length	Short amplicon promises higher PCR efficiency and lower detection limits
Primers	Tendency to form dimers	Primer–dimers compete with primer binding to low target amounts
Reaction conditions	Primer binding efficiency	Improper T_A or Mg^{2+}-concentration reduce primer binding
Genotyping temperature shift is not significant		
check	*for*	*because*
Polymorphism	Nucleotides in mismatching duplex	Different mismatches evoke different duplex destabilization
Differentiating probe	Location of the polymorphic nucleotide	HybProbes work best when the polymorphic nucleotide is in the middle of the probe and MGB probe when it is close to the 3′ end were the MGB moiety resides

Today a lot of experience has developed for the design hydrolysis probes and HybProbes and relevant software has been developed and delivered with certain PCR instruments to facilitate the straight forward design of individual real-time PCR assays. However, these software solutions are usually restricted to the supported real-time PCR chemistry and may fail to solve complex problems. Numerous freely available software packages also exist, which aid the design of oligonucleotides by calculating melting temperatures, predicting secondary structures or proposing primer or probe sequences. A complete package that combines the design of primers and detection probes for various chemistries that also consider different sequence variants for the creation of highly specific assays would be ideal. Such an ideal design tool would also propose reaction conditions for the selected oligonucleotides that lead to accurate PCR results from the first run. However, such an idealized tool does not exist yet.

Conclusions

There are various ways to establish good real-time PCR assays. For prominent targets, like pathogens or diagnostic markers of clinical chemistry or oncology, ready-to-use kits can be purchased. These kits make guarantees for successful real-time PCR and do not require specialist knowledge for assay design. On the other hand, troubleshooting is often hampered by the absence of detailed information about the oligonucleotides employed.

A second option is to order a complete assay design from a company. Numerous companies offer the design of real-time PCR assays according to individual requirements. The companies apply either software solutions or very rarely make use of specialists that do the design manually, employing various computer-based tools for the control of temperatures and secondary structures etc. These customized designs can be expensive but sometimes a complete evaluation has been performed and ready-to-use reaction conditions are supplied with the oligonucleotides sequences. In the case of unexpected results, troubleshooting in cooperation with the support division of the company is desirable. Nonetheless. if the oligonucleotide sequences are known, the easiest approach is to ensure that they meet the general design guidelines outlined in this chapter.

The third option is to design a real-time PCR assay by oneself using tried and tested, popular, robust and relatively inexpensive chemistries such as the hydrolysis probes or the HybProbes. This can be facilitated by the help of several online software tools or complete software packages that come either with a real-time PCR machine or can be purchased individually. Some tools are public domain and freely available on the internet. A selection of oligonucleotide design-related software is listed below. For simple real-time assays that detect microbial DNA or RNA, most of these tools will generate acceptable results. However, knowledge of the basic design rules is pivotal to improve a sub-optimal assay. For more complex assays, the highly specific detection of minority target variants or for the design of generic real-time PCR assays, most of the programs will fail to create robust results. In these cases "hand-made" real-time PCR assays will usually be superior to "machine-made" ones.

When looking at the literature it is interesting to note that sometimes there are real-time PCR assays that reportedly perform well despite offending several of the design guidelines presented here. In other cases, a design accommodates all the established guidelines but results in an unacceptable assay. This indicates that there are factors influencing the real-time PCR assay performance that we still do not understand.

The listed guidelines do not guarantee an easy path to success when designing real-time PCR assays. Instead, they should be understood as helpful guidelines to judge primers and detection probes used for real-time PCR.

Web resources

Ready-to-use real-time PCR assays

A useful real-time PCR oligonucleotide database (Pattyn *et al.*, 2006):
Center for Medical Genetics, Ghent University Hospital, Belgium http://medgen.ugent.be/rtprimerdb/

Companies that offer real-time PCR kits or customer designs

Roche Applied Science:	www.roche-applied-science.com
Applied Biosystems:	www.appliedbiosystems.com
Qiagen:	www1.qiagen.com
TIB Molbiol:	www.tib-molbiol.com

Primer/probe design tools

A list of oligo design software at Bioinformatics.Net:
www.bioinformatics.vg/biolinks/bioinformatics

Commercial design tools

Oligo 6.0	www.oligo.net
Vector NTI	www.invitrogen.com
DNAStar	www.dnastar.com
ProbeITy	www.celadonlabs.com/StoreFront/default.aspx
Primer Premier	www.premierbiosoft.com/primerdesign
Scorpio	www.dnasoftware.com

Freeware

Primer Quest	www.idtdna.com/SciTools/SciTools.aspx
Primer 3.0	www.frodo.wi.mit.edu/cgi-bin/primer3/primer3_www.cgi
FastPCR	www.biocenter.helsinki.fi/bi/Programs/fastpcr.htm
PrimerD	www.genes.cs.wustl.edu/cgi-bin/license.cgi?software = primerD
Oligo Explorer	www.genelink.com/tools/gl-oe.asp
Autoprime	www.autoprime.de
Oligo Factory	www.ueg.ulb.ac.be/oligofaktory/index.jsp
PerlPrimer	perlprimer.sourceforge.net/
PrimerX	http://bioinformatics.org/primerx/

Additional useful design tools

BLAST	www.ncbi.nlm.nih.gov
BioEdit	www.mbio.ncsu.edu/BioEdit/bioedit.html
Mfold	www.bioinfo.rpi.edu

References

Aarts, H.J., Joosten, R.G., Henkens, M.H., Stegeman, H., and van Hoek, A.H. (2001). Rapid duplex PCR assay for the detection of pathogenic Yersinia enterocolitica strains. J. Microbiol. Methods *47*, 209–217.

Afonina, I.A., Reed, M.W., Lusby, E., Shishkina, I.G., and Belousov, Y.S. (2002). Minor groove binder-conjugated DNA probes for quantitative DNA detection by hybridization-triggered fluorescence. Biotechniques 32, 940–949.

Aitichou, M., Saleh, S.S., McElroy, A.K., Schmaljohn, C., and Ibrahim, M.S. (2005). Identification of Dobrava, Hantaan, Seoul, and Puumala viruses by one-step real-time RT-PCR. J. Virol. Methods. *124*, 21–26.

Aldea, C., Alvarez, C.P., Folgueira, L., Delgado, R., and Otero, J.R. (2002). Rapid detection of herpes simplex virus DNA in genital ulcers by real-time PCR using SYBR green I dye as the detection signal. J. Clin. Microbiol. *40*, 1060–1062.

Belousov, Y.S., Welch, R.A., Sanders, S., Mills, A., Kulchenko, A., Dempcy, R., Afonina, I.A., Walburger, D.K., Glaser, C.L., Yadavalli, S., Vermeulen, N.M., and Mahoney, W. (2004). Single nucleotide polymorphism genotyping by two colour melting curve analysis using the MGB Eclipse Probe System in challenging sequence environment. Hum. Genomics *1*, 209–217.

Bengtsson, M., Karlsson, H.J., Westman, G., and Kubista, M. (2003). A new minor groove binding asymmetric cyanine reporter dye for real-time PCR. Nucleic Acids Res. *31*, e45.

Brennan, R.E. and Samuel, J.E. (2003). Evaluation of Coxiella burnetii antibiotic susceptibilities by real-time PCR assay. J. Clin. Microbiol. *41*, 1869–1874.

Breslauer, K.J., Frank, R., Blocker, H., and Marky, L.A. (1986). Predicting DNA duplex stability from the base sequence. Proc. Natl. Acad. Sci. USA *83*, 3746–3750.

Brownie, J., Shawcross, S., Theaker, J., Whitcombe, D., Ferrie, R., Newton, C., and Little, S. (1997). The elimination of primer–dimer accumulation in PCR. Nucleic Acids Res. *25*, 3235–3241.

Bustin, S.A. (2000). Absolute quantification of mRNA using real-time reverse transcription polymerase chain reaction assays. J. Mol. Endocrinol. *25*, 169–193.

Bustin, S.A. (2002). Quantification of mRNA using real-time reverse transcription PCR (RT-PCR): trends and problems. J. Mol. Endocrinol. *29*, 23–39.

Bustin, S.A. (2004). A-Z of Quantitative PCR. IUL Biotechnology.

Bustin, S.A. (2005). Real-time, fluorescence-based quantitative PCR: a snapshot of current procedures and preferences. Expert. Rev. Mol. Diagn. *5*, 493–498.

Bustin, S.A., Benes, V., Nolan, T., and Pfaffl, M.W. (2005). Quantitative real-time RT-PCR—a perspective. J. Mol. Endocrinol. *34*, 597–601.

Bustin, S.A. and Mueller, R. (2005). Real-time reverse transcription PCR (qRT-PCR) and its potential use in clinical diagnosis. Clin. Sci. (Lond) *109*, 365–379.

Bystrom, M., Bocher, S., Magnusson, A., Prag, J., and Johansson, A. (2005). Tularemia in Denmark: identification of a Francisella tularensis subsp. holarctica strain by real-time PCR and high-resolution typing by multiple-locus variable-number tandem repeat analysis. J. Clin. Microbiol. *43*, 5355–5358.

Campsall, P.A., A.u., N.H., Prendiville, J.S., Speert, D.P., Tan, R., and Thomas, E.E. (2004). Detection and genotyping of varicella-zoster virus by TaqMan allelic discrimination real-time PCR. J. Clin. Microbiol. *42*, 1409–1413.

Chase, C.J., Ulrich, M.P., Wasieloski, L.P., Jr., Kondig, J.P., Garrison, J., Lindler, L.E., and Kulesh, D.A. (2005). Real-time PCR assays targeting a unique chromosomal sequence of Yersinia pestis. Clin. Chem. *51*, 1778–1785.

Crockett, A.O. and Wittwer, C.T. (2001). Fluorescein-labeled oligonucleotides for real-time pcr: using the inherent quenching of deoxyguanosine nucleotides. Anal. Biochem. *290*, 89–97.

Debeaumont, C., Falconnet, P.A., and Maurin, M. (2005). Real-time PCR for detection of Brucella spp. DNA in human serum samples. Eur. J. Clin. Microbiol. Infect. Dis. *24*, 842–845.

Drosten, C., Gottig, S., Schilling, S., Asper, M., Panning, M., Schmitz, H., and Gunther, S. (2002). Rapid detection and quantification of RNA of Ebola and Marburg viruses, Lassa virus, Crimean-Congo hemorrhagic fever virus, Rift Valley fever virus, dengue virus, and yellow fever virus by real-time reverse transcription-PCR. J. Clin. Microbiol. *40*, 2323–2330.

Escutenaire, S., Mohamed, N., Isaksson, M., Thoren, P., Klingeborn, B., Belak, S., Berg, M., and Blomberg, J. (2006). SYBR Green real-time reverse transcription-polymerase chain reaction assay for the generic detection of coronaviruses. Arch. Virol.

Espy, M.J., Uhl, J.R., Sloan, L.M., Buckwalter, S.P., Jones, M.F., Vetter, E.A., Yao, J.D., Wengenack, N.L., Rosenblatt, J.E., Cockerill, F.R., III, and Smith, T.F. (2006). Real-time PCR in clinical microbiology: applications for routine laboratory testing. Clin. Microbiol. Rev. *19*, 165–256.

Gibson, N.J. (2006). The use of real-time PCR methods in DNA sequence variation analysis. Clin. Chim. Acta *363*, 32–47.

Hart, K.W., Williams, O.M., Thelwell, N., Fiander, A.N., Brown, T., Borysiewicz, L.K., and Gelder, C.M. (2001). Novel method for detection, typing, and quantification of human papillomaviruses in clinical samples. J. Clin. Microbiol. 39, 3204–3212.

Higuchi, R., Fockler, C., Dollinger, G., and Watson, R. (1993). Kinetic PCR analysis: real-time monitoring of DNA amplification reactions. Biotechnology (NY) *11*, 1026–1030.

Holland, P.M., Abramson, R.D., Watson, R., and Gelfand, D.H. (1991). Detection of specific polymerase chain reaction product by utilizing the 5′–3′ exonuclease activity of Thermus aquaticus DNA polymerase. Proc. Natl. Acad. Sci. USA *88*, 7276–7280.

Karlsen, F., Steen, H.B., and Nesland, J.M. (1995). SYBR green I DNA staining increases the detection sensitivity of viruses by polymerase chain reaction. J. Virol. Methods 55, 153–156.

Klein, D. (2002). Quantification using real-time PCR technology: applications and limitations. Trends Mol. Med. *8*, 257–260.

Kollmann, T.R., Zhuang, X., Rubinstein, A., and Goldstein, H. (1992). Design of polymerase chain reaction primers for the selective amplification of HIV-1 RNA in the presence of HIV-1 DNA. AIDS 6, 547–552.

Komurian-Pradel, F., Paranhos-Baccala, G., Sodoyer, M., Chevallier, P., Mandrand, B., Lotteau, V., and Andre, P. (2001). Quantitation of HCV RNA using real-time PCR and fluorimetry. J. Virol. Methods *95*, 111–119.

Kreuzer, K.A., Bohn, A., Lass, U., Peters, U.R., and Schmidt, C.A. (2000). Influence of DNA polymerases on quantitative PCR results using TaqMan probe format in the LightCycler instrument. Mol. Cell Probes *14*, 57–60.

Kubista, M., Andrade, J.M., Bengtsson, M., Forootan, A., Jonak, J., Lind, K., Sindelka, R., Sjoback, R., Sjogreen, B., Strombom, L., Stahlberg, A., and Zoric, N. (2006). The real-time polymerase chain reaction. Mol. Aspects Med. *27*, 95–125.

Kumar, R., Singh, S.K., Koshkin, A.A., Rajwanshi, V.K., Meldgaard, M., and Wengel, J. (1998). The first analogues of LNA (locked nucleic acids): phosphorothioate-LNA and 2′-thio-LNA. Bioorg. Med. Chem Lett. *8*, 2219–2222.

Kutyavin, I.V., Afonina, I.A., Mills, A., Gorn, V.V., Lukhtanov, E.A., Belousov, E.S., Singer, M.J., Walburger, D.K., Lokhov, S.G., Gall, A.A., Dempcy, R., Reed, M.W., Meyer, R.B., and Hedgpeth, J. (2000). 3′-minor groove binder-DNA probes increase sequence specificity at PCR extension temperatures. Nucleic Acids Res. *28*, 655–661.

Landt O and Nitsche A. (1999). Selection of Hybridization Probes for use with the LightCycler. LC Technical Notes *LC6/99*, 1–7 .

Lee, M.A., Siddle, A.L., and Page, R.H. (2002). ResonSense: simple linear fluorescent probes for quantitative homogeneous rapid polymerase chain reaction. Analytica Chimica Acta *457*, 61–70.

Lee, M.A., Brightwell, G., Leslie, D., Bird, H., and Hamilton, A. (1999). Fluorescent detection techniques for real-time multiplex strand specific detection of Bacillus anthracis using rapid PCR. J. Appl. Microbiol. *87*, 218–223.

Lie, Y.S. and Petropoulos, C.J. (1998). Advances in quantitative PCR technology: 5′ nuclease assays. Curr. Opin. Biotechnol. *9*, 43–48.

Lind, K., Stahlberg, A., Zoric, N., and Kubista, M. (2006). Combining sequence-specific probes and DNA binding dyes in real-time PCR for specific nucleic acid quantification and melting curve analysis. Biotechniques *40*, 315–319.

Lindler, L.E., Fan, W., and Jahan, N. (2001). Detection of ciprofloxacin-resistant Yersinia pestis by fluorogenic PCR using the LightCycler. J. Clin. Microbiol. *39*, 3649–3655.

Livak, K.J. (1999). Allelic discrimination using fluorogenic probes and the 5′ nuclease assay. Genet. Anal. *14*, 143–149.

Livak, K.J., Flood, S.J., Marmaro, J., Giusti, W., and Deetz, K. (1995). Oligonucleotides with fluorescent dyes at opposite ends provide a quenched probe system useful for detecting PCR product and nucleic acid hybridization. PCR Methods Appl. *4*, 357–362.

Mackay, I.M. (2004). Real-time PCR in the microbiology laboratory. Clin. Microbiol. Infect. *10*, 190–212.

Mackay, I.M., Arden, K.E., and Nitsche, A. (2002). Real-time PCR in virology. Nucleic Acids Res. *30*, 1292–1305.

Manaresi, E., Gallinella, G., Zuffi, E., Bonvicini, F., Zerbini, M., and Musiani, M. (2002). Diagnosis and quantitative evaluation of parvovirus B19 infections by real-time PCR in the clinical laboratory. J. Med. Virol. *67*, 275–281.

Marras, S.A., Kramer, F.R., and Tyagi, S. (1999). Multiplex detection of single-nucleotide variations using molecular beacons. Genet. Anal. *14*, 151–156.

Marras, S.A., Kramer, F.R., and Tyagi, S. (2003). Genotyping SNPs with molecular beacons. Methods Mol. Biol. *212*, 111–128.

Mhlanga, M.M. and Malmberg, L. (2001). Using molecular beacons to detect single-nucleotide polymorphisms with real-time PCR. Methods *25*, 463–471.

Monroe, W.T. and Haselton, F.R. (2003). Molecular beacon sequence design algorithm. Biotechniques *34*, 68–3.

Moore, R.E., Shepherd, J.W., and Hoskins, J. (1990). Design of PCR primers that detect only mRNA in the presence of DNA. Nucleic Acids Res. *18*, 1921.

Morrison, T.B., Weis, J.J., and Wittwer, C.T. (1998). Quantification of low-copy transcripts by continuous SYBR Green I monitoring during amplification. Biotechniques *24*, 954–8, 960, 962.

Nadkarni, M.A., Martin, F.E., Jacques, N.A., and Hunter, N. (2002). Determination of bacterial load by real-time PCR using a broad-range (universal) probe and primers set. Microbiology *148*, 257–266.

Nielsen, C.B., Singh, S.K., Wengel, J., and Jacobsen, J.P. (1999). The solution structure of a locked nucleic acid (LNA) hybridized to DNA. J. Biomol. Struct. Dyn. *17*, 175–191.
Nitsche, A., Ellerbrok, H., and Pauli, G. (2004). Detection of orthopoxvirus DNA by real-time PCR and identification of variola virus DNA by melting analysis. J. Clin. Microbiol. *42*, 1207–1213.
Nitsche, A., Steger, B., Ellerbrok, H., and Pauli, G. (2005). Detection of vaccinia virus DNA on the LightCycler by fluorescence melting curve analysis. J. Virol. Methods. *126*, 187–195.
Nitsche, A., Steuer, N., Schmidt, C.A., Landt, O., and Siegert, W. (1999). Different real-time PCR formats compared for the quantitative detection of human cytomegalovirus DNA. Clin. Chem *45*, 1932–1937.
Ogura, M., Keller, C., Koo, K., and Mitsuhashi, M. (1994). Use of the fluorescent dye YOYO-1 to quantify oligonucleotides immobilized on plastic plates. Biotechniques *16*, 1032–1034.
Olson, V.A., Laue, T., Laker, M.T., Babkin, I.V., Drosten, C., Shchelkunov, S.N., Niedrig, M., Damon, I.K., and Meyer, H. (2004). Real-time PCR system for detection of orthopoxviruses and simultaneous identification of smallpox virus. J. Clin. Microbiol. *42*, 1940–1946.
Panning, M., Asper, M., Kramme, S., Schmitz, H., and Drosten, C. (2004). Rapid detection and differentiation of human pathogenic orthopox viruses by a fluorescence resonance energy transfer real-time PCR assay. Clin. Chem. *50*, 702–708.
Pattyn, F., Robbrecht, P., D.e., P.A., Speleman, F., and Vandesompele, J. (2006). RTPrimerDB: the real-time PCR primer and probe database, major update 2006. Nucleic Acids Res. *34*, D684-D688.
Pattyn, F., Speleman, F., D.e., P.A., and Vandesompele, J. (2003). RTPrimerDB: the real-time PCR primer and probe database. Nucleic Acids Res. *31*, 122–123.
Queipo-Ortuno, M.I., Colmenero, J.D., Reguera, J.M., Garcia-Ordonez, M.A., Pachon, M.E., Gonzalez, M., and Morata, P. (2005). Rapid diagnosis of human brucellosis by SYBR Green I-based real-time PCR assay and melting curve analysis in serum samples. Clin. Microbiol. Infect. *11*, 713–718.
Ririe, K.M., Rasmussen, R.P., and Wittwer, C.T. (1997). Product differentiation by analysis of DNA melting curves during the polymerase chain reaction. Anal. Biochem. *245*, 154–160.
Rychlik, W. (1995). Priming efficiency in PCR. Biotechniques *18*, 84–90.
Shu, P.Y., Chang, S.F., Kuo, Y.C., Yueh, Y.Y., Chien, L.J., Sue, C.L., Lin, T.H., and Huang, J.H. (2003). Development of group- and serotype-specific one-step SYBR green I-based real-time reverse transcription-PCR assay for dengue virus. J. Clin. Microbiol. *41*, 2408–2416.
Simeonov, A. and Nikiforov, T.T. (2002). Single nucleotide polymorphism genotyping using short, fluorescently labeled locked nucleic acid (LNA) probes and fluorescence polarization detection. Nucleic Acids Res. *30*, e91.
Speers, D.J. (2006). Clinical applications of molecular biology for infectious diseases. Clin. Biochem. Rev. *27*, 39–51.
Srinivasan, K., Morris, S.C., Girard, J.E., Kline, M.C., and Reeder, D.J. (1993). Enhanced detection of PCR products through use of TOTO and YOYO intercalating dyes with laser induced fluorescence—capillary electrophoresis. Appl. Theor. Electrophor. *3*, 235–239.
Szuhai, K., Sandhaus, E., Kolkman-Uljee, S.M., Lemaitre, M., Truffert, J.C., Dirks, R.W., Tanke, H.J., Fleuren, G.J., Schuuring, E., and Raap, A.K. (2001). A novel strategy for human papillomavirus detection and genotyping with SybrGreen and molecular beacon polymerase chain reaction. Am. J. Pathol. *159*, 1651–1660.
Tapp, I., Malmberg, L., Rennel, E., Wik, M., and Syvanen, A.C. (2000). Homogeneous scoring of single-nucleotide polymorphisms: comparison of the 5′-nuclease TaqMan assay and Molecular Beacon probes. Biotechniques *28*, 732–738.
Tuma, R.S., Beaudet, M.P., Jin, X., Jones, L.J., Cheung, C.Y., Yue, S., and Singer, V.L. (1999). Characterization of SYBR Gold nucleic acid gel stain: a dye optimized for use with 300-nm ultraviolet transilluminators. Anal. Biochem. *268*, 278–288.
Tyagi, S. and Kramer, F.R. (1996). Molecular beacons: probes that fluoresce upon hybridization. Nat. Biotechnol. *14*, 303–308.
Tyagi, S., Marras, S.A., and Kramer, F.R. (2000). Wavelength-shifting molecular beacons. Nat. Biotechnol. *18*, 1191–1196.
Vet, J.A., Majithia, A.R., Marras, S.A., Tyagi, S., Dube, S., Poiesz, B.J., and Kramer, F.R. (1999). Multiplex detection of four pathogenic retroviruses using molecular beacons. Proc. Natl. Acad. Sci. USA *96*, 6394–6399.

von, A.N., Wittwer, C.T., and Schutz, E. (2001). Oligonucleotide melting temperatures under PCR conditions: nearest-neighbor corrections for mg(2+), deoxynucleotide triphosphate, and dimethyl sulfoxide concentrations with comparison to alternative empirical formulas. Clin. Chem. *47*, 1956–1961.

Watzinger, F., Ebner, K., and Lion, T. (2006). Detection and monitoring of virus infections by real-time PCR. Mol. Aspects Med. *27*, 254–298.

Whitcombe, D., Theaker, J., Guy, S.P., Brown, T., and Little, S. (1999). Detection of PCR products using self-probing amplicons and fluorescence. Nat. Biotechnol. *17*, 804–807.

Wittwer, C.T., Ririe, K.M., Andrew, R.V., David, D.A., Gundry, R.A., and Balis, U.J. (1997). The LightCycler: a microvolume multisample fluorimeter with rapid temperature control. Biotechniques *22*, 176–181.

Yeung, A.T., Holloway, B.P., Adams, P.S., and Shipley, G.L. (2004). Evaluation of dual-labeled fluorescent DNA probe purity versus performance in real-time PCR. Biotechniques *36*, 266–5.

QPCR: Target Preparation

3

Tania Nolan, Reinhold Mueller, and Stephen Bustin

Abstract

Application of modern molecular biology techniques, for the detection of pathogens has resulted in a radical change in the methodology adopted by many diagnostic microbiology laboratories. Use of real-time quantitative PCR technology has resulted in much more sensitive and specific detection. It is becoming apparent that the most vulnerable stages of the quantitative PCR (qPCR) and reverse transcription (RT)-qPCR protocols are those concerned with target preparation. Each specimen, organism and nucleic acid combination requires a different set of considerations for collection, transport and nucleic acid purification and the most appropriate combination should be ascertained for each project and laboratory.

In order to ensure that the data collected are reliable the relevant controls should also be included. In this way a set of standard operating procedures can be defined that ensure reliable, reproducible and robust data.

Introduction

When detecting microorganisms for research or diagnostic purposes, the goal is quite clear: Determine accurately, reproducibly and with high specificity the presence and, in some cases, also the quantity or activity of the organism(s) of interest. It is becoming apparent that many conventional approaches lack sensitivity and specificity, are labor intensive and in some cases expensive to perform. Molecular techniques are becoming adopted as potential alternatives. A recent addition to the molecular analysis portfolio is quantification using real-time PCR (real-time qPCR). This is a technology that utilizes the amplification of a specific DNA (qPCR) or cDNA (RNA) sequence (RT-qPCR) and concurrently detects the amplified product via the generation of an associated fluorescent signal. Specialized qPCR instruments detect changes in target-associated fluorescence and plot these as a function of PCR cycle number. Various algorithms are now used to determine the cycle at which the fluorescence signal is most closely associated with the initial copy number and these are reported as threshold cycle, C_t (Higuchi 1992, although derived using different algorithms, for the purposes of this discussion, the Ct can be considered to be synonymous to the "Take off point" and "Crossing point," Cp used by the software of some instruments). Under optimal conditions the fluorescent signal changes are directly proportional to the initial target input, and therefore the C_t values are logarithmically distributed, over a range of 5 to 10 orders of magnitude.

Conventional techniques for detection of the common food borne pathogen *Campylobacter jejuni* rely upon culture of the organism however, this is difficult because it requires special growth conditions and it appears that the bacterium can enter a phase where it remains viable but cannot be cultured. Nogva (2000) demonstrated the power of a real-time qPCR assay for quantitation of *C. jejuni* that could even be used to distinguish between living and dead bacteria in food samples. Although this is an exciting new technology, an important step in the evaluation of any new technique is a comparison to the established procedures. Uhl (2003) clearly demonstrated the use of qPCR as a suitable diagnostic technique for the detection of group A streptococci isolated from throat swabs and established that the data were comparable to those determined after use of established rapid antigen immunoassay and culture. Application of real-time qPCR technology to detection of pathogens has resulted in a radical change in the diagnostic methodology adopted by many microbiology laboratories for the diagnosis of pathogens (Cockerill, 2002, 2003; Bankowski 2004; Mackay 2004; Smith 2004; Espy, 2006).

Real-time qPCR can be very sensitive because the target is amplified permitting the detection of a single genome (Raghunathan, 2005). The high sensitivity of qPCR assays may be compromised at one or more of the following steps:

1. sampling
2. sample preparation and storage
3. nucleic acid isolation
4. conversion of RNA into cDNA, when the target is RNA
5. PCR (target amplification)
6. detection of amplicon, and
7. data analysis and reporting.

In this review we are mainly concerned with the preanalytical steps, also known as pre-PCR processing (Radstrom 2004; Bustin, 2004), for qPCR assays, which are the processes performed during the first three steps described above; sampling, preparing the sample and purifying the target nucleic acid. These general considerations apply regardless of the source of the target material and the downstream application of the experiment, encompassing all experimental aims from relatively straightforward genomic sequence analysis to more challenging transcript quantification in single cells. However, specific examples will also be provided since some extraction methods work very well for isolation of a specific organism from a given sample but do not work for different organisms or specimen material (Espy, 2006). It is pertinent to note that the effect on the final data and interpretation of each of these influencing factors must be considered in the context of the required application. For example, when the aim of the test is to genotype the target material, the specificity of the assay is paramount whereas sensitivity is of less importance than when a diagnostic assay detecting the presence or absence of a pathogenic organism is being developed, and visa versa.

Initially let us consider a Utopian qPCR experiment: The presence and quantity of a specific pathogen is to be determined in a number of samples of a defined type, e.g. blood, a swab or bronchoalveolar lavage, biopsy, laser dissection, food or environmental sample.

These samples would be acquired from all sources under standard conditions and the number of host cells (or the sample volume) would be determined immediately. If an RNA pathogen is to be identified the target RNA of the pathogen would remain perfectly intact and be reverse transcribed with 100% efficiency into cDNA with no sequence bias. The target DNA could be amplified without purification or would be purified from the sample with 100% efficiency. The subsequent PCR amplification would also occur at maximum efficiency, be absolutely reproducible within and between tests and all amplicons would be accurately detected. This perfect, standardized analysis would also include a reproducible and representative standard curve for target quantification consisting of template material that amplifies in exactly the same way as the sample material (Rosenstraus 1998; Luebeck 2003; Malorny, 2004).

The everyday challenge in the real world is that the scientific method is never perfect. All stages of our Utopian assay are vulnerable to variation. These variables introduce a degree of uncertainly to the interpretation of the final data. It is for these reasons that experimental controls and defined reference standards and calibrators are required. It is appropriate to define the standard and control elements that are required for each stage of the process and incorporate this into a Standard Operating Procedure (SOP) specific for each sample type and organism. Examples of these are described below as each process of the workflow is defined.

Sample collection methods

Pathogens are diverse in origin and belonging to all phylogenetic groups: virus, bacteria, fungi and protozoa and are found in and on host animal bodies, in air, soil, foodstuffs and water. Each of these require specific consideration of the diversity of the biological system including variation in the cell wall (presence, absence and composition), membrane biochemistry, DNA and RNA binding proteins and cytoplasmic biology. As a fundamental example; extraction of nucleic acid from fungi or Gram-positive bacteria requires inclusion of procedures to disrupt the cell wall that are relatively more severe than those required when working with Gram negative bacteria. For some pathogens the stage of growth has an effect on microbial cell morphology and physiology, e.g. *Plasmodium* sp.

Therefore, sampling protocols are diverse and are adapted to the biology of the organism, the environment of origin and the nucleic acid to be analyzed. The main goal is to establish consistency in the sampling procedure and to maintain the integrity of the sample material until the nucleic acid is examined. This requires inactivation of nucleic acid hydrolyzing enzymes as soon as possible. Often samples are acquired from clinical settings where the major focus of the medical team is to preserve the life or health of a patient. Under these circumstances the integrity of the sample in the time between acquisition and transfer to the waiting scientist is not a priority consideration so a simple procedure must already be implemented and adhered to in order for the laboratory to derive reliable data. The results from recent studies indicate that the treatment of samples during this period (i.e. from sampling to nucleic acid preparation) can be critical, especially if the target nucleic acid is RNA. The target organisms are often pathogenic and the stages of nucleic acid preparation at which the sample can be considered non-pathogenic have not been defined. Until these recommendations have been reported conclusively, it should be assumed that all specimens are infectious and the required precaution adopted into the SOP (Espy 2006, CDC, 1999).

The FTA® range of products (Whatman) have been developed to collect, store and extract nucleic acid from a range of sample types, including blood and buccal smears, plant material and solid tissue. All procedures are optimized for room temperature operation and the stability of sample is exemplified by the claim that human genomic DNA (gDNA) has been amplified in a PCR after 14 years storage on a FTA card. After application of the sample to the FTA card, microorganisms are inactivated, cell membranes lysed and nucleic acids are entrapped onto the FTA matrix. Similarly, many companies have developed kits to address the specific challenges of the target specimen, although an optimal extraction procedure would be highly efficient over a wide range of specimen types, require a simple protocol and be cost effective. Fahle (2000) compared the extraction of *Cytomegalovirus* sp. (CMV) DNA that had been spiked into human specimens using six commercially available DNA extraction kits. They selected specimens presenting different challenges; bronchoalveolar lavage, cerebral spinal fluid, plasma and whole blood. The tested extraction kits were; Puregene DNA isolation kit, Generation Capture Column kit (both Gentra Systems Inc,), MasterPure DNA purification kit (Epicentre Technologies) IsoQuick nucleic acid extraction kit (MicroProbe Corp), QIAamp blood kit (Qiagen) and NucliSens isolation kit (Organon Teknika Corp). Extraction using the NucliSens isolation kit resulted in the most sensitive and also the most expensive detection from all sample types when all kits were used according to recommendations. It is clear that it is important to validate an extraction procedure for the specific specimen and pathogen type of interest (Petrich, 2006)

Sampling from blood

For identification of blood borne parasites and viruses, the usual sample material is either whole blood, plasma or peripheral blood mononuclear cells (PBMCs). The key to the successful isolation of high quality RNA and to the reliable and meaningful comparison of RT-qPCR data is to ensure that neither endogenous nor exogenous RNAses are introduced during the extraction procedure. Two related problems are that different mRNAs display differential stability and that mRNA expression profiles can change rapidly after cells or tissue samples have been collected, but before they have been frozen. This implies that quantification of a single transcript or determination of relative transcript levels could be seriously affected by the sampling procedure adopted. This is a particular problem with cells extracted from whole blood (Keilholz, 1998) and mRNA profiles can change over several orders of magnitude even in the short time it can take from collecting the blood to processing it in the laboratory. Causes are RNA degradation, the induction of genes, as well as the method of RNA preparation. Cox-2 is a very unstable protein and its transcripts have been implicated in inflammation as well as pathogenic processes. This serves as an example of a target which is particularly sensitive to sampling and it has been demonstrated that *ex vivo* analyses are dependent on the time between tissue extraction and analysis (Learn 2000, Lukiw, 1997). In addition, blood is notorious for containing numerous inhibitors of the PCR reaction (Akane 1994; Al Soud 2001; Fredricks, 1998). Therefore, appropriate sample acquisition has a major influence on the quality of the RNA and subsequently on any result of RT-qPCR assays (Vlems, 2002). Once a biological sample has been obtained, immediate stabilization of RNA is the most important consideration (Madejon, 2000).

Whole blood is commonly collected in either EDTA treated tubes (BD, New Jersey) or PAXgene™ tubes (PreAnalytiX, GmbH). EDTA is a metal complexing agent that has been widely used commercially since the 1950s (Bersworth Chemical Company, 1954), the disodium salt being commonly used as an anticoagulant which prevents the formation of fibrin by calcium chelation (Seegers, 1967). Tubes contain either K_2EDTA that is sprayed dried onto the side of a plastic tube or K_3EDTA as a solution in a glass tube. The International Council for Standardization in Haematology Expert Panel on Cytology (1993) recommended the preferential use of K_2EDTA collection tubes because these allow more accurate white blood cell counts and reduce both the shrinkage effect of red blood cells and the increase in cellular volume observed when using K_3EDTA. Interestingly the concentration and source of EDTA in forensic blood samples was pertinent to the interpretation of data in the murder case of the State of California vs. Orenthal James Simpson (Sheppard and Henion, 1997).

The PAXgene system is a combined approach to sample collection and nucleic acid extraction. Blood is drawn directly into specialized Vacutainer™ tubes that contain a proprietary buffer designed to stabilize the sample during storage and subsequent cell lysis and nucleic acid purification. Blood samples in PAXgene tubes are stable for 14 days at room temperature or for 28 days at 2–8°C. For long-term storage the samples can be frozen at –70°C. The PAXgene system incorporates a nucleic acid purification procedure to then extract either RNA or DNA from whole blood. After storage of specimen blood in PAXgene tubes, nucleic acids are isolated using specific preparation columns to isolate either RNA or DNA. Chai (2005) compared the use of PAXgene tubes and columns to the standard erythrocyte lysis method for the extraction of RNA to be used for downstream RT-PCR quantification of the relative levels of ribonuclease reductase R1 and R2 subunits. They showed conclusively that the PAXgene approach resulted in intact RNA of high quality whereas conventional lysis approaches resulted in seriously degraded RNA. However, they also observed a high level of gDNA contamination in the PAXgene isolated RNA that could cause erroneous results. In this report the authors suggested a modification of the standard protocol to include an off column DNase digestion step for further gDNA removal. Coyne (2004) compared the isolation of *Bacillus anthracis* Sterne vegetative cells and spores (Gram positive) and *Yersinia pestis* (Gram negative) from spiked samples of whole blood, serum and buffer using either the IsoCode Stix or QIAamp DNA minikit DNA extraction procedures. Overall the conclusion was that both approaches provide comparable detection limit but that the IsoCode Stix method of template preparation was faster with a simpler protocol making this amenable to higher throughput requirements. Although these recommendations are generally accepted and this system is regarded as the method of choice for handling blood samples it is important to report some caution. Kagedal (2005) were studying malignant melanoma and inflammatory bowel disease in a multicenter study. Blood samples were stored for several days at room temperature, according to manufacturer's recommendation and the authors report that some target transcripts were not stable in the PAXgene storage system and therefore, advise that the stability of a given target should be established prior to incorporating any storage and extraction procedure into an SOP.

While these methods of collection are simple for those of us fortunate enough to work in well-equipped laboratories, in some areas of the world even the basic protocols for processing of blood and plasma prior to shipping to a central diagnostic laboratory are unavailable. Dried blood spots on filter paper have been used for screening of metabolic disorders and Fiscus (1998) demonstrated the use of dried blood spots on filter paper (Schleicher and Schuell 903 and IsoCode filter paper) for quantification of HIV RNA load in plasma. Their preliminary results demonstrated that they could reliably use filter paper blood samples to quantify HIV, although they experienced stability issues which require addressing before this approach could be adapted for field use. However, the FTA card (see above) and IsoCode Stix system (Schleicher and Schuell, Keene, NH) have been developed to be used to preserve and stabilize blood samples when long periods of transportation are required. The IsoCode Stix system has been shown to be effective for downstream malarial diagnosis using qPCR (Zhong, 2001).

Sampling from solid tissue

Generally, solid tissue biopsies are snap-frozen in liquid nitrogen and can be subsequently processed to extract either RNA or DNA. When sampling for RNA organisms it is particularly important to pre-chill sample tubes on dry ice/ethanol and collect tissue immediately into the cold tubes. When multiple samples of the same tissue are available they should be processed into individual tubes of similar sample sizes. This is to avoid attempting to break sections from a solid frozen lump for subsequent extractions. It also helps to prevent the sample from defrosting during homogenization. The tubes are then snap frozen and stored in liquid nitrogen or at −80°C. Labeling and cataloguing of frozen samples must be rigorous to ensure rapid and accurate retrieval and that tube markings are not removed during freezing!

An alternative approach is to immerse tissues into aqueous sulfate salt solutions (such as ammonium sulfate) at controlled pH and room temperature. This treatment results in precipitation of RNAses and other solubilized proteins and protects tissue RNA. Tissues can be stored at −80°C prior to processing using standard RNA preparation techniques (Lader, 2001). This forms the basis of the commercially available RNAlater® solution (Ambion, Inc). RNA, DNA or even denatured proteins can be isolated from samples of fresh tissues, cultured cells, bacteria or yeast after storage in RNAlater. Tissue samples should be prepared as slices less than 0.5 cm, preferably 2 mm (Mutter, 2004). A larger relative surface area facilitates diffusion of the solution into the tissue. Ideally ten volumes of RNAlater solution are added to one volume of tissue section. Since the RNA would be degrading while the sample size was being determined the recommendation to use a standard protocol of 1.5 ml for samples up to 150 mg and 5 ml for samples between 1.5 mg and 500 mg is a very useful practical tip (Qiagen RNAlater handbook). Samples can be stored in RNAlater at room temperature for up to 1 week. For longer term storage, store overnight and for up to 4 weeks at 4°C and then transfer to −20°C or −80°C. RNAlater is not suitable for use with some cells or tissues such as those that cannot be penetrated by the solution, e.g. waxy plant tissue or bone and so it is wise to check suitability with the manufacturer before use. RNAlater-ICE has been developed to aid tissue processing of previously frozen material.

Mutter *et al.* (2004) compared the RNA profiles of identical uterine myometrial samples after immediate extraction from fresh tissue. Samples were either snap frozen in liquid nitrogen or stored in RNAlater for 24 hours or 72 hours at room temperature. They concluded that identical profiles were produced when analyzed using microarray analysis and determined that the storage conditions did not influence the RNA profile of 14, 639 gene targets. However, they noted that tissue sections stored in RNAlater should be a maximum of 2 mm thick to allow passive diffusion of salt solution and that agitation is recommended to prevent tissue pieces from forming clumps. Storage of tissues in RNAlater at temperature above or below 25°C was not examined.

Sampling from stool, swab, aspirate, and lavage samples

Pathogens are often isolated from samples collected using apparently less invasive techniques; stool, urine, swab, aspirate and lavage samples. Extractions from stool samples are notoriously challenging but this process is facilitated by the STAR system (Stool Transport and Recovery Buffer; Roche Diagnostics Corporation, Indianapolis, IN). STAR has been formulated to render infectious organisms non-pathogenic and prevent nucleic acid degradation during magnetic bead extraction.

Gouvea (1990) developed a protocol for the extraction, amplification and equivalent serotype identification of the dsRNA rotavirus molecule from stool samples. In this report they describe a relatively crude SDS/proteinase K disruption procedure followed by phenol chloroform extraction. The essential step to optimize was that of the first round of PCR to accommodate reverse transcription of dsRNA and the process was shown to be highly dependent on DMSO, $MgCl_2$ and viral template concentration. The data from this study clearly illustrate that the reaction component balance required for these reactions is much more critical than when relatively pure, conventional samples are processed. Petrich (2006) conducted a multicenter comparison of nucleic acid extraction methods to identify coronavirus from stool samples in order to facilitate the development of a molecular diagnostic test for Severe Acute Respiratory Syndrome (SARS). They observed a discrepancy in the recorded rates of infection detected in stool samples from different centers and hypothesized that this was due to differences in RNA extraction procedures. In order to determine the most effective extraction procedure they conducted an international, multicenter comparison of manual and automated purification methods. When RNA was extracted using manual techniques the sensitivity of detection ranged from 50% to 100% when using the magnetic bead systems, Cortex Biochem Magnazorb being the most sensitive of the protocols tested. The range of sensitivities of detection after automated extraction was 57% to 100% with both the bioMerieux NucliSens and miniMag systems achieving maximum sensitivity. The study revealed that a number of the protocols used resulted in retention of inhibitory substances that reduced subsequent amplification efficiency. In this way they confirmed that the observed differences in apparent infection were as a result of sample processing in some part attributable to the yield of RNA and in some part to the efficiency of amplification of the extracted material.

Microorganisms collected on swabs are typically immersed in a transport solution, shipped and processed in a reference laboratory or lysed immediately (Bustin, 2004). Surface sampling is routinely done by use of a swipe (Speci-sponge; Nasco, Fort Atkinson, WI), Heavy Wipe (Handy Wipes Heavy Wipes; First Branch Corp. Danbury, CT) or swab sample processing kit (ASD; Fort Lauderdale, FL). Pathogens trapped on the sampling device are then subjected to elution into extraction buffer (Buttner, 2004).

The SETS kit (Swab Extraction Tube System; Roche Diagnostics Corporation, Indianapolis, IN) is a simple alternative swab processing system, specifically designed for recovery of organisms collected on swab fibers. The SETS method consists of a specialized swab collection spin tube. The process simply requires 1 min to break the shaft of the specimen collection swab leaving the head in the collection tube, a 2 min centrifugation step and a 2 min heating step (to 100°C). Issa (2005) used this system to isolate HSV DNA from dermal and genital swabs and demonstrate that it is faster and less expensive than comparable sample preparation using the MagNA Pure system. Importantly application of the SETS provided a more sensitive assay, possibly because the sample handling allowed for a 10 fold lower extraction buffer to be used with the SETS and so a corresponding increase in sample DNA concentration.

Identification of respiratory pathogens has rapidly improved due to development of techniques for both sampling and organism identification. Traditionally causative agents of lung and associated infections have been identified after lengthy culture or by highly invasive lung tissue examination either of biopsy or postmortem samples. Most respiratory pathogens are RNA virus but there are also many which are fungal and cause pneumonia resulting in serious health issues in humans and economic losses when livestock are infected. Culture of the infectious organism can be challenging. *Mycoplasma hyopneumoniae* causes pneumonia in pigs but attempts to culture often result in growth of other *Mycoplasma* sp. present in the pig lung preventing accurate diagnosis. Immunological detection methods cross react with other *Mycoplasma* sp. resulting in reduced specificity. PCR techniques were developed for detection of *M. hyopneumoniae* in nasal swabs (Mattson, 1995) and tracheobroncheolar washes (Verdin, 1996) but one of the challenges when using bronchoalveolar washes in PCR is that these samples contain high concentrations of compounds that inhibit the reaction. Baumeister (1998) compared sampling methods for the detection of *M. hyopneumoniae* in pig bronchoalveolar lavage fluid in order to select an approach that minimized the presence of PCR inhibitors. The bronchoalveolar lavage washes were in 0.15 M phosphate buffered saline and collected after immediate instillation and aspiration from the bronchus trachealis. Various methods for removal of inhibitors were tested and found to be ineffective; these included boiling and centrifugation, proteinase K and detergent treatment. The highest sensitivity was reported after DNA was extracted using either the Nucleon II DNA extraction kit (Scotlab) or the CTAB (cetyltrimethylammonium bromide) method (Maass and Dalhoff, 1994).

Pneumocystis carinii is a causative agent of pneumonia in humans and is associated with infection of immunocompromised patients. As HIV infection and AIDS have become more prevalent the incidence of *P. carinii* pneumonia has also increased. *P. carinii* had been alternatively described as a protozoan until its classification as a fungus in 1988 (Edman). The diagnosis of *P. carinii* pneumonia was originally dependent upon clinical observation and histological examination of open lung biopsy specimen or autopsy tissue. Collection of bronchoalveolar lavage fluid or induced sputa has replaced these invasive techniques and has assumed a major role in diagnosis protocols for this and other respiratory infections. Ribes (1996) analyzed the sensitivity and specificity of PCR techniques for detection of *P. carinii* and compared these to conventional immunological detection methods used to detect cysts or trophozoites. Current staining techniques were inefficient because the samples were dilute and so microscopic detection was difficult and different stains were required to detect cysts or trophozoites.

Samples were collected into sterile saline and DNA was purified using the IsoQuick extraction procedure (Orca Research Inc). The extraction of pathogen nucleic acid was estimated to be equivalent to concentrating the sample by 100 fold. PCR was sensitive enough to detect single cysts and both cysts and trophozoites simultaneously and was around 100 fold more sensitive than direct examination. An improvement in the DNA extraction procedure of samples from bronchoalveolar lavage fluid and a nested PCR assay were developed for detection of *Aspergillus* sp. (Hayette, 2001). The assay was used to confirm the presence of fungus but could not be used to discriminate between infection and colonization.

Sampling from environmental samples

When attempting to sample microorganisms from water, sensitivity is a major issue. Although the microorganisms are present at very low concentration, often around 1 organism per liter, this is still sufficient to present a health hazard. Prior to nucleic acid extract a concentration step is usually required. These include; hollow fiber filtration that is used to capture microorganisms based on molecular weight, immunological techniques such as affinity chromatographic separation approaches using carbohydrate or hydroxyapatite columns for extraction of a broad range of organisms while immumomagnetic separation using specific antibodies is used for capture of target microorganisms.

Extraction of nucleic acid from soil and sediment presents particular challenges since these samples usually contain high concentrations of compounds that are difficult to separate from nucleic acids due to similar chemistry but will inhibit downstream enzyme mediated reactions (Schreiner and Laptrop, 1912; Bottomley, 1917). A number of publications address improvements to specific stages of the nucleic acid purification procedure to enhance the DNA yield. One approach has been to lyse all cells directly by bead beating (Miller, 1999), or heating and freezing (Bruns and Buckley, 2002). The alternatives target enhanced purification by addition of hexadecyltrimethylammonium (CTAB) or polyvinylpolypyrrolidone PVPP (Zhou, 1996), or by chemical flocculation using multivalent cations such as $AINH_4(SO_4)_2$ (Braid, 2003). Chandler (1996) demonstrated that electroelution could be used to remove inhibitory compounds and that the sensitivity of detection of organisms from surface soils and sediments using PCR was increased by up to a factor of 10^4 when detecting DNA targets and 10^3 when targeting RNA, relative to non-electroelution treated samples.

The SoilMaster DNA Extraction Kit has been developed to address removal of organic inhibitors such as fumic and fulvic acid while providing high quality, intact DNA. This protocol uses a heating lysis procedure in the presence of detergent and employs chromatography to remove organic inhibitor compounds. TruRNA (Atom Sciences Inc) present a system for purification of RNA and DNA (independently or by copurification) from soil samples and this has been used for recovery of nucleic acids from actinomycetes and Gram positive cocci including *Rhodococcus* sp. and *Micrococcus* sp.

Felnagle (2003) compared the sensitivity, bias and feasibility of use of commercial DNA extraction kits (Isoquick, ORCA Research; Soil DNA extraction kit, Mo Bio Laboratories, Inc; Dneasy Tissue kit Qiagen Inc; Genomic Tip 20/G Qiagen and application of FTA filter, Whatman Bioscience). Samples of water and soil were spiked with known quantities of *E. coli*, *Bacillus megaterium (Gram positive)*, *Haloarcula marismortui (archaean)* and both PCR and real-time qPCR were used for detection. The results of this comparison indicated that the FTA card was suitable for use in planktonic environmental microbiology investigations and that the MoBio kit was efficient and sensitive for extraction of nucleic acids from soil.

Nucleic acid extraction

When the pathogen of interest is associated with the host, varying amounts of host cells or inhibitory components can be combined with the pathogen sample, possibly causing different degrees of amplification inhibition or a reduction in specificity. Commercially produced kits are available for purification of pathogenic sequences from the host DNA or RNA. In general, the pathogen is separated from the host before nucleic acid extraction, typically by magnetic beads coated with anti-pathogen antibody.

Pathogen DNA and RNA need to be accessible for purification and analysis, and one goal is to recover 100% of the target nucleic acid without fragmentation. Initially, cells must be disrupted. This is achieved by various methods including mechanical shearing, homogenization, sonication and enzymatic disruption. When isolating RNA from frozen specimens, the samples must be maintained completely frozen until RNAses have been inactivated. Hence if disruption is carried out using mechanical procedures e.g. in a mortar and pestle, the processing should be carried out in the presence of liquid nitrogen to keep the specimen frozen. The alternative is to use the RNAlater-ICE solution described previously.

Bacterial cell disruption is often enzyme mediated using lysozyme or lysostaphin followed by passing the digest through a syringe with needle attached, or by sonication. Yeast cell walls are digested with lyticase or zymolyase and the resulting spheroplasts can be lyzed in lysis buffer containing proteinase K or disrupted with 0.5 mm glass beads in a bead mill. Plant tissue and cells are typically ground in a mortar and pestle or with 5 mm tungsten carbide beads again using a bead mill. Tissue culture cells can be broken open with common cell lysis buffer, with a Rotor-Stator (Omni International Inc, Marietta GA, PRO Scientific Inc., Oxford CT) or equivalent using disposable probes. Simple to use spin columns (included in some nucleic acid extraction kits e.g. Sigma Aldrich or purchased separately e.g. the Qiashredder, Qiagen) can be added as a simple step prior to standard extraction protocols. Tissue biopsies can be disrupted by using an Omni Rotor, a mortar and pestle or in a bead mill with 5 mm steel beads. Mammalian red blood cells do not contain a nucleus and are often removed to help reduce PCR inhibition. Red blood cells are frequently removed by hypotonic lysis, Ficoll density centrifugation or by plain centrifugation for 10 minutes at 3300 × g, thus forming a top plasma layer, a middle buffy coat containing nucleated blood cells, and an erythrocyte pellet (Bustin, 2004).

Radstrom (2004) has listed four sample preparation methods: (1) biochemical (adsorption and DNA extraction), (2) immunological adsorption, (3) physical (two-phase systems, centrifugation systems, filtration and dilution), and (4) enrichment cultures. The latter is a way to increase the amount of target, rather than a technique to isolate microbial genomes. The biochemical sample preparation method was shown to remove more of the common inhibitors than the other three methods, but at a higher cost. Traditionally either DNA or RNA is extracted, although protocols and kits for the co-extraction of

DNA and RNA are available and gaining popularity because of their flexibility. Phenol based protocols, including the phenol based compound alternatively marketed under the trade names of TriZol (Invitrogen) and TriReagent (Sigma-Aldrich) are extensively and successfully used to extract nucleic acid. However care must be taken to perform the procedure accurately since not only will residual phenol inhibit downstream enzyme mediated reactions but this is a hazardous reagent, both caustic and corrosive, causing both severe burns and damage to the human nervous system. It is prudent to sacrifice yield for purity when removing the nucleic acid aqueous phase from phenol and interface layers to avoid contaminating the nucleic acid sample with plasmid.

Many commercial companies have developed cost effective kits using simple and rapid protocols incorporating less hazardous compounds (Table 3.1).

Commercial systems aim to purify DNA, RNA or both (Espy, 2006), that are intact and free of inhibitors that would compromise downstream enzymatic reactions. Ideally the nucleic acids are delivered at high concentration (preferably eliminating final ethanol precipitation steps) in storage buffer this is compatible with subsequent qPCR. Kits can be divided into either specific target purification kits, usually used on fully automated or semi-automated purification platforms, conventional purification kits or laboratory developed protocols. A detailed laboratory developed protocol can be found in Bustin (2004). Several authors have performed and published side-by-side comparisons of a select number of nucleic acid purification kits where yield, quality, hands-on time, ease-of-use and price per reaction have been analyzed. When setting up a new assay it is wise to perform an independent comparison of as many kits as possible before selecting the one to be adopted as the standard for the specific procedure. The nucleic acid yield, protocol details and manipulation steps vary and the most appropriate kit for the experimental conditions (sample source and target organism) should be determined. While the protocols for these kits are relatively simple it is recognized that these extractions require technical competence and constant attention during template preparation (Clinical Laboratory Improvement Amendments, 1988) high variability in yield and nucleic acid quality have been observed. In a recent training course at EMBL, post doctoral scientists studying qPCR techniques were tasked with extraction of total cellular RNA from identical samples and provided with identical extraction kits and laboratory facilities. The resulting extraction ranged in yield from 0 to 450 ng (Figure 3.1) and clearly demonstrated that these extraction procedures can lead to variability until personnel are fully trained and competent. Many of the nucleic acid preparation kits use ethanol either to ensure specific binding of nucleic acid to column material or for precipitation of nucleic acid when eluted at low concentration. An additional and familiar phenomenon is the incomplete removal of ethanol leading to samples floating out of agarose gels when attempting to load. While this is irritating, a more significant consequence is that remaining ethanol contamination inhibits PCR.

When processing a high volume of samples an automated extraction procedure is usually adopted. The major advantage of automation is that hands on time and the potential opportunities for operator error are significantly reduced. Adoption of automated systems requires substantial financial investment but also laboratory space, initial set up time and effort. The development of automated systems and the procedures for extraction requires highly skilled personnel but it is clear that this is a valuable investment because once developed, less skilled assistants can carry out subsequent operations.

Table 3.1 Comparison of commercially available DNA preparation systems

Vendor	Item	Catalog number	Quantity	Max yield	Time	Method of purification	Amt starting material	Types of starting material
Bioneer	DNA PrepMateTM -M	K-3011	100		40 min			
BD Biosciences Clontech	NucleoBond Blood and Cell Culture Mini Kits	635949	20			NucleoBond AX Resin		
	NucleoBond Blood and Cell Culture Midi Kits	635950	20			NucleoBond AX Resin		
	NucleoBond Blood and Cell Culture Maxi Kits	635951	10			NucleoBond AX Resin		
	NucleoSpin Tissue Kit	K3053–1	50	25 mg		Spin Column		
		K3053–2	250	25 mg		Spin Column		
	NucleoSpin Multi-96 Tissue Kit	K3081–1	4 × 96		192 preps in 3 hours	Spin Column		
		K3081–2	24 × 96		192 preps in 3 hours	Spin Column		

BioRad	AquaPure Genomic DNA Isolation Kit	732–6340	100		30 min		0.5 ml	Does not purify Gram positive DNA
	AquaPure Genomic DNA Tissue Kit	732–6343	100		30 min		0.5 ml	Does not purify Gram positive DNA
	InstaGene Matrix	732–6030	100			Matrix		
Dynal Biotech	Dynabeads DNA DIRECT Universal	630.06	300	600 ng–1 mg	10 min	Magnetic Bead separation	up to 30 ml	
Epicentre Technologies	MasterPure DNA Purification Kit	MCD85201	200	> 90% of theoretical DNA yield from *E. coli*	45 min	Modified salt and enzyme based technology	1.5 ml or approx 2 × 106 cells	Tissue, bacteria, yeast
	MasterPure Complete DNA and RNA Purification Kit	MC89010	10	> 90% of theoretical DNA yield from *E. coli*, can be used for BOTH DNA and RNA by splitting sample after initial isolation of NA	46 min	Centrifugation	1.5 ml or approx 2 × 106 cells	Tissue, bacteria, yeast
		MC85200	200	> 90% of theoretical DNA yield from *E. coli*, can be used for BOTH DNA and RNA by splitting sample after initial isolation of NA	45 min	Centrifugation	1.5 ml or approx 2 × 106 cells	Tissue, bacteria, yeast

Table 3.1 continued

Vendor	Item	Catalog number	Quantity	Max yield	Time	Method of purification	Amt starting material	Types of starting material
	MasterPure™ Gram Positive DNA Purification Kit	MGP04020	20	specific for G+ can be used for any microbial application	2 hours	DNA precipitation	1 ml ON culture	Gram-positive bacteria
		MGP04100	100	specific for G+ can be used for any microbial application	2 hours	DNA precipitation	1 ml ON culture	Gram-positive bacteria
GE Healthcare Life Sciences	GFX Genomic Blood DNA Purification Kit	27–9603–01	100	15 mg	15–20 min	Glass Fiber Matrix		Blood, Gram(-), Gram (+), yeast
Idaho Technology	IT 1–2-3TM FLOW Sample Purification Kit	ASAY-ASY-0004	40			Spin Column		Whole blood, air, ware, milk, tuna salad, mixed greens, ground beef, gastric wash
	IT 1–2-3TM SWIPE Sample Purification Kit	ASAY-ASY-0005	40			Spin Column		Nasal swabs, pus, surface, powder, culture, lymph node
	IT 1–2-3TM VIBE Sample Purification Kit	ASAY-ASY-0500	40			Spin Column		Whole blood, nasal swabs, sputum

	IT 1–2-3TM SCOOP Sample Purification Kit	ASAY-ASY-0502	40			Spin Column		Stool, soil
Intron Biotechnology	G-spinTM lip Genomic DNA Extraction Kit	17271L	200		60 min	Spin Column		Plant material, food
		17271M	100			Spin Column		Plant material, food
		17271	50			Spin Column		Plant material, food
Invitrogen	ChargeSwitch gDNA Mini Bacteria Kit	CS11301	50	12 mg	40 min Gram(–), 2 hours Gram(+)	Magnetic Beads	one colony or 0.5 ml ON culture	Gram(-) and Gram(+) bacteria
Maxim Biotech	Bdtract Genomic Isolation Kit	SA-40002	50	30–40 mg		EtOH precipitation	1 × 10^6 cells	Whole blood, buffy coat, cultured cells, tissues and bacteria
		SA40001	100	30–40 mg		EtOH precipitation	1 × 10^6 cells	Whole blood, buffy coat, cultured cells, tissues and bacteria
Mo Bio Laboratories	UltraClean Microbial DNA Isolation Kit	12224–50	50	20 mg	20 min	Bead beating/ spin filter	1.8 ml	Yeast, fungi, Gram(-) and Gram(+) bacteria

Table 3.1 continued

Vendor	Item	Catalog number	Quantity	Max yield	Time	Method of purification	Amt starting material	Types of starting material
		12224–250	250	20 mg	20 min	Bead beating/ spin filter	1.8 ml	Yeast, fungi, Gram(-) and Gram(+) bacteria
	UltraClean-htp 96 Well Microbial DNA Isolation Kit	10196–4	384	20 mg	1.5 hours for 192 samples	96 well format bead beating/ silica spin filter plate method	1.8 ml	Yeast, fungi, Gram(-) and Gram(+) bacteria
		10196–12	1152	20 mg	1.5 hours for 192 samples	96 well format bead beating/ silica spin filter plate method	1.8 ml	Yeast, fungi, Gram(-) and Gram(+) bacteria
Norgen Biotek Corp.	Bacterial Genomic DNA Isolation Kit	17900	50	20 mg	1 hour	Spin Column	0.5 to 1 ml culture, 2 × 109 cells	Gram(-) and Gram(+) bacteria
Promega	Wizard Genomic DNA Purification Kit	A1120	100	18 mg Gram(-), 6–13 mg Gram (+)	60 min	Solution based method	300 ml blood	White blood cells, tissue culture cells, animal and plant tissues, yeast, Gram(-) and Gram(+) bacteria
		A1125	500	19 mg Gram(-), 6–13 mg Gram (+)	61 min	Solution based method	300 ml blood	White blood cells, tissue culture cells, animal and plant tissues, yeast, Gram(-) and Gram(+) bacteria

		A1620	100	20 mg Gram(-), 6–13 mg Gram (+)	62 min	Solution based method	10 ml blood	White blood cells, tissue culture cells, animal and plant tissues, yeast, Gram(-) and Gram(+) bacteria
	Maxwell 16 Tissue kit	AS1030	48	21 mg Gram(-), 6–13 mg Gram (+)	16 samples in 30 minutes	Solution based Instrument system	50–400μl of blood	Requires Maxwell instrument
	Maxwell 16 Cell kit	AS1029	48	22 mg Gram(-), 6–13 mg Gram (+)	17 samples in 30 minutes		up to 1.2cm of rodent tail	Requires Maxwell instrument
Qbiogene	GNOME DNA Kit	2010–200	10	100 mg	< 1 hour	Proprietary	Varies	Bacteria, yeast, animal cells and tissues
		2010–400	25	100 mg	< 1 hour	Proprietary	Varies	Bacteria, yeast, animal cells and tissues
		2010–600	100	100 mg	< 1 hour	Proprietary	Varies	Bacteria, yeast, animal cells and tissues
Qiagen	QIAamp DNA Mini Kit	51304	50		20 min after lysis	Spin Column	Varies	For isolation of genomic, mitochondrial, bacterial, parasite or viral DNA

Table 3.1 continued

Vendor	Item	Catalog number	Quantity	Max yield	Time	Method of purification	Amt starting material	Types of starting material
		51306	250		20 min after lysis	Spin Column	Varies	For isolation of genomic, mitochondrial, bacterial, parasite or viral DNA
	DNeasy Tissue Kit	69504	50	20–40 mg	0.33 hours	Silica gel membrane technology	25 mg tissue or 1.2 cm tail	Animal tissues, blood, cultured cells, rodent tails, insects, bacteria, yeast.
		69506	250	20–40 mg	0.33 hours	Silica gel membrane technology	25 mg tissue or 1.2 cm tail	Animal tissues, blood, cultured cells, rodent tails, insects, bacteria, yeast.
	DNeasy 96 Tissue Kit, 4 and 12 plate kits	69581	384	20–30 mg	1 hour after lysis		20 mg tissue or 1.2 cm tail	Animal tissues and cells.
		69582	1152	20–30 mg	1 hour after lysis		20 mg tissue or 1.2 cm tail	Animal tissues and cells.
	AllPrep DNA/RNA Mini Kit	80204	50		35 min			Easy to use cells only (not gram +)

	QIAGEN DNA/ RNA Maxi Kit	14162	10	500 mg gDNA, 1 mg RNA				
Roche Applied Science	DNA Isolation Kit for Cells and Tissues	11814770001	10		2.5 hours plus resuspension time	Salting out proteins, DNA precipitation	1g tissue or 5 × 10^7 cultured cells, 1011 Gram(-) bacteria	Tissues, cultured cells, Gram(-) bacteria, mouse tail, yeast
	MagNA Pure LC DNA Isolation Kit II (Bacteria, Fungi)	3264785001	192		95 min for 32 samples	Magnetic glass particles	100 ml	Requires use of MagNA Pure LC Instrument
Sigma	GenElute Bacterial Genomic DNA kit	NA2100	10	20 μg	< 75 min Gram(–), < 120 min Gram(+)	Silica Bind and elute	1.5 ml culture	
		NA2110	70					
		NA2120	350					
Takara Mirus Bio	Dr GenTLE	TAK9081	200		40 min	Precipitation	100 ml blood	Whole blood, bacteria
Trevigen	Genomic DNA Isolation Kit	4850–20-GD	20			Biphasic Extraction	100 mg tissue, 105 to 107 cells	Tissue, cultured cells

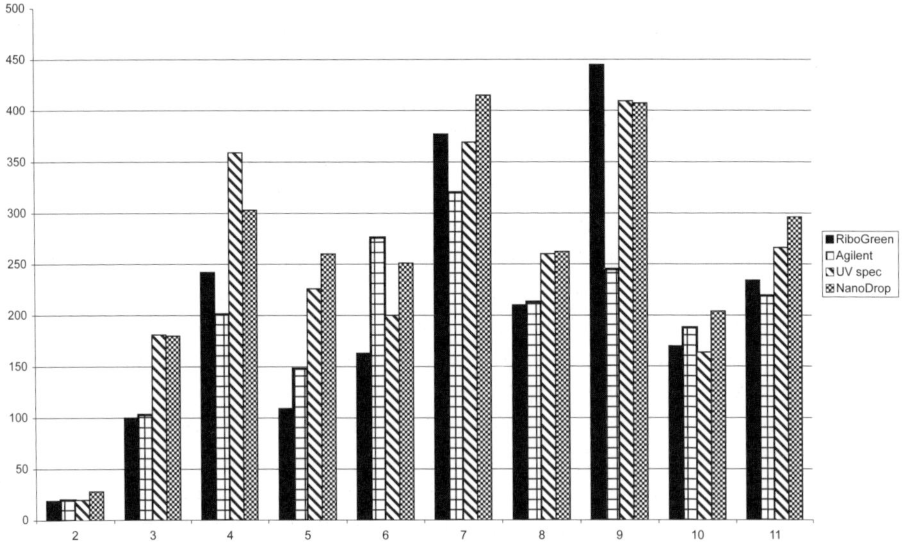

Figure 3.1 Groups of students extracted RNA from identical cell pellets and the yield measured using RiboGreen, Agilent, UV spectrophotometry and NanoDrop. Measurements of yield varied from zero (group 1) to 450 ng (group 9) and measurement values were also highly variable and dependent on the technique used.

When automated and manual procedures are directly compared, the automated approach may not always give the highest yield but it invariably results in higher consistency when identical procedures are repeated.

Beuselinck (2005) compared the performance of an automated system (m1000, Abbott Laboratories, IL) with a manual extraction system for purification of viral RNA and DNA. The m1000 can simultaneously extract nucleic acid from up to 48 samples using guanidinium isothiocyanate for lysis and enzyme deactivation and uncoated iron particles for capture of RNA and silica-coated magnetic particles for capturing DNA. The report describes the extraction of hepatitis C virus (HCV) RNA, hepatitis B virus DNA, CMV and Epstein-Barr virus DNA. When purifying nucleic acid from 24 samples the total time for manual and automated plasma/serum extraction is approximately equivalent (between 120 and 132 min.) but the operator hands-on time required for the automated system is reduced by about 65%. Accuracy was determined by comparison of automated extraction of the HCV 2004 QCMD proficiency program panel and detection with the laboratory developed HCV-PCR. No significant differences were revealed with almost all results within 0.5 $\log_{10}$ HCV IU/ml.

In order to compare the reliability of each extraction procedure, Beuselinck extracted nucleic acids from forty clinical samples using both automated and manual protocols and assayed these for HBV. Of the 40 samples, 25 samples were found positive in both extractions but nine samples assayed for HBV gave different results after automated and manual extraction; six samples that were negative with manual extraction were positive in automated extraction. Three weakly positive manually extracted samples were negative when extracted on the m1000. For four of these samples enough material was available for the extraction to be repeated; the second time samples that had been positive (automated) and negative (manual) in the first round, were all positive in the second round of extractions indicating that the manual extraction would have lead to a false negative result after the first extraction. While this report focused on the comparison between manual and automated extraction procedures, these data reveal the absolute requirement for technical replicates or extraction controls to be included in any diagnostic SOP.

Aldous and coworkers (2005) described a comparison of six ways of extracting *Mycobacterium tuberculosis* DNA from respiratory specimens. Digested and decontaminated sputum was spiked with *M. tuberculosis* (ATCC 27294) and adjusted to the standard concentration of 1.0 McFarland, divided into aliquots and stored at –70°C. Each of the six DNA purification protocols were repeated eight times. The thawed samples were centrifuged at 6000 × g for 1 min and the pellet was saved for extraction.

The first procedure relied upon boiling of the pellet in TE (10 mM Tris-HCl pH 7.5, 0.1 mm EDTA) Kocagoz, 1993) for 15 min and analysis of the subsequent supernatant. The second technique used was the PrepMan Ultra sample preparation reagent (Applied Biosystems, Foster City, CA), which is a commercial reagent that also uses a boiling procedure. The IDI extraction protocol (Infectio Diagnostics, Inc., Quebec, Canada) uses a lysis system employing a proprietary lysis tube containing glass beads used for cell disruption in conjunction with a boiling step. The Qiagen QIAamp DNA mini kit tissue protocol (Qiagen, Valencia, CA) was also used with the following modifications; an enzymatic lysis step using 30 mg/ml lysozyme added to the tissue lysis buffer was followed by boiling for 15 min. and a proteinase K incubation at 56°C for 1 hour. An SDS-Triton X extraction (Khan, 2004) was performed in which non-ionic detergent mix (2% SDS, 10% Triton X-100) replaced the TE in a standard boiling protocol. The final protocol consisted of re-suspension of the bacterial pellet was in the non-ionic detergent mix followed by sonicated for 15 min and then a boil extraction protocol.

Each extract was analyzed on two platforms, the SmartCycler® II (Cepheid, Sunnyvale, CA) and the Rotor-Gene 3000 (Corbett Research, Sydney, Australia). After optimizing the assay conditions on each instrument, 25 µl reactions were run using the same cycling conditions for both platforms. Ten-fold dilutions of *M. tuberculosis* DNA (2.5 fg to 25 pg) were run in triplicate on both platforms and standard curves were generated. For each assay the standard curve was included in the run or imported for calculation of *Mycobacterium* DNA concentration. In presence of internal control DNA the limit of detection for *Mycobacterium* standard DNA was 0.5 fg for both platforms. The utility of the incorporated internal control was clearly evident when DNA was extracted using non-ionic detergents (protocols 5 and 6) because the no template controls (NTCs) showed amplification of internal controls while there was no template detected in the samples As expected, 2% SDS strongly inhibited the PCR. There was a difference in the DNA extractions (as determined by C_t) with the IDI extracted sample amplifying around three C_ts earlier than the other techniques indicating either a higher yield or lower inhibitor concentration. Interestingly, the authors report that the differences in C_t between the recorded C_t for samples amplified after extraction using each technique (IDI and TE, QIAGEN and TE and PrepMan and TE) are significant ($P = 0.05$) when determined on the Smart-Cycler but not on the Rotor-Gene leading to the suggestion that statistical analysis on differences in Ct with differing standard deviations should be performed with caution! These differences are most likely due to higher standard deviation recorded on the Rotor-Gene than on the SmartCycler®. The DNA yield was calculated from the mean Ct for each successful extraction and ranged from 51.7 pg when using the IDI approach to 7.4 pg after using TE boil extraction protocol. These extractions were performed in less than 1 hour when using the IDI, PrepMan and TE approaches, but 2.5 to 3 hours using the Qiagen kit. The IDI approach is the most expensive, and, predictably, the TE the least (Aldous, 2005).

Pre-analytical nucleic acid quantification and quality assessment

Once the nucleic acid has been purified and is in a suitable storage buffer, usually in TE or equivalent, the concentration needs to be determined. One common method for estimation of concentration is UV absorbance at 260 nm (A260). This is most accurate when the nucleic acid is at a concentration of greater than 250 ng/ml in the absence of nucleotides and proteins and at neutral pH. Conventionally the conversion factors for A260 = 1 are: 50 μg/ml (double stranded DNA), 33 μg/ml (single stranded DNA), 40 μg/ml (single stranded RNA) and 20–30 μg/ml (oligonucleotides). As proteins and phenol absorb at a wavelength of 280 nm, the ratio A260/A280 is an indicator of the purity of the sample and should be greater than 1.8. The NanoDrop® spectrophotometer (NanoDrop Technologies, Wilmington, DE) is the preferred instrument for spectrophotometric measurements when a relatively small number of samples is being analyzed. This is because it can measure across a remarkably large dynamic range of concentrations (from 2 to 3700 ng/μl) with surprising accuracy and simply requires a microliter of sample to be applied to the detector. More recently, fluorescent dyes such as PicoGreen (Invitrogen, Carlsbad, CA) and OliGreen (Invitrogen, Carlsbad, CA) have been used for determination of DNA concentration and Ribogreen for RNA concentration. PicoGreen binds preferentially to double stranded DNA and an increase of fluorescence at 535 nm is observed. The linear range of response when using PicoGreen is greater and sensitivity is higher with respect to UV spectrometry. The method of choice for analysis of total RNA is either the Agilent 2100 Analyzer (Agilent, Santa Clara, CA) with an RNA 6000 LabChip or the BioRad Experion System (BioRad Hercules, CA). Using these systems an estimate of RNA integrity and quantity can be determined. As little as 1 μl of RNA is loaded into the cartridge that includes a capillary electrophoresis system, an RNA binding fluorescent dye and a fluorescent detector. After an 18S rRNA and 28S rRNA (16S rRNA and 23S rRNA for bacteria, resp.) electropherogram has been established the area below the ribosomal RNAs peaks is calculated (BioRad). The ratio between the two areas is calculated by the instrument's software and is an indicator of the integrity of the RNA preparation. Alternatively, an RNA Integrity Number (RIN) is calculated using the Agilent algorithm (Schroeder, 2006). The RIN may also be used cautiously to estimate RNA integrity (Fleige, 2006).

More recently, a real-time qPCR based approach to determine RNA integrity was reported (Nolan, 2006a). This method relies on quantifying the relative degradation of one or more transcripts of interest. It is prudent to select the targets that are to be investigated as well as a reference gene. The system can be adapted to determine the degradation of any target transcript, although the internal reference gene, Glyceraldehyde 3 phosphate dehydrogenase is used as an example. If the targets of interest are poly A tailed, representative cDNA is prepared using oligo dT priming. This approach is used for this procedure regardless of reverse transcription protocol to be adopted for subsequent experimental analysis. For analysis of transcripts lacking a polyA tail, a target specific RT primer can be designed that is positioned at the extreme 3′ of the RNA sequence. Two (or more) real-time qPCR assays are designed for each target such that these are spatially separated along the length of the RNA molecule. In the example given three assays were used and referred to as the 5′, centre and 3′ assays. The quantity of the target detected using each assay was compared. If the RNA is intact an equal quantity should be detected using either assay whereas, if the RNA is degraded a higher quantity would be recorded when using the 3′ assay than when using the 5′ assay. This provides a rapid, cost effective system to interrogate RNA integrity that also uses little of the sample RNA.

Sample storage

Purified DNA and RNA are stored as aliquots in TE buffer at −20°C or at −80°C, respectively. For long term storage, nucleic acids should be stored in 70% ethanol at −80°C.

Contamination of template and inhibition of amplification

Since it is possible to amplify genomic DNA from a single bacterium and use the amplicon for speciation (Raghunathan, 2005) it is also clear that it is possible to contaminate samples and detect these targets as false positive signals. Because of the high sensitivity of qPCR, great care must be taken to avoid contamination of the reaction. In amplification laboratories basic rules must be obeyed. These include control of air flow, separation of set-up, amplification and analysis locales, and the use of dedicated micropipettes with barrier tips. The micropipettes need to be calibrated at least twice yearly to ensure accuracy. When working with RNA targets, contaminating reactions with RNase results in variable or erroneous results as much as contaminating with unwanted target would do. Working with aliquots of buffer, enzyme, primers and probe stock solutions helps to reduce the loss of expensive ingredients and aids troubleshooting. For a detailed guide to successful amplification results consult Bustin and Nolan (2004).

Assay inhibitors

Real-time qPCR and RT-qPCR efficiency can be determined by reference to a serial dilution of specific target or by measurement of the amplification efficiency in a single tube. In order to estimate amplification in a single tube, fluorescent changes are measured during subsequent cycles and these changes fitted to one of a number of curve fitting algorithms. These methods are generally more effective when using DNA binding dyes for detection of the amplicon because they give a larger dynamic range of signal detection. The serial dilution or standard curve method is still the most common: A standard curve is constructed from a serial dilution of a representative target material. The specific target is amplified and the product detected with a characteristic C_t. Subsequent dilutions are detected with higher C_t values because more cycles are required to reach the same DNA (and detection chemistry) concentration. Typically a target dilution series is run, and the $\log_{10}$ of target concentration is plotted against the C_t. Using linear regression analysis of the standard curve the slope and the Y-intercept of the curve are calculated. The slope of the best-fit line is a measure of PCR efficiency. Since the C_t measurements are logarithmically distributed, in theory 3.323 cycles are required to amplify a target by ten fold. Hence when an assay is 100% efficient it results in a plot of log target concentration against C_t with a gradient of −3.32. The Y-intercept is a measure of the sensitivity of the assay, indicating the number of cycles required to detect a single unit of target, where the unit corresponds to the units specified for the standard curve. In addition the R^2-value is calculated, which is a measure of the closeness of the replicate data points to the best-fit line and is indicative of the reproducibility of the assay. A value below 0.98 indicates that the assay is highly variable and probably requires optimization. For reliable quantification it is important to use only the linear range of the standard curve (Muller, 2004).

Factors affecting the PCR efficiency include PCR inhibitors, PCR enhancers (facilitators), tissue decay, target and background DNA/RNA concentration, DNA/RNA degradation, reverse transcription (if applicable), the source and batch variability of PCR enzymes and buffer components, non-specific PCR products, detection chemistry, qPCR platforms, reaction plates/tubes, cycling conditions, and differences between operators.

Inhibition may act upon one or more steps of the PCR assays including the cell lysis, nucleic acid purification, DNA polymerase and reverse transcriptase. Nucleic acid is mainly degraded by hydrolysis, non-enzymatic methylation, oxidative damage and enzymatic cleavage. If the template is fragmented, PCR will not be efficient, especially when the nick in the target occurs between the primers. DNA polymerase will not be able to function in PCR if an inhibitor is bound directly to the polymerase and inactivates the 5′- to 3′ exonuclease function (for TaqMan® assays), or interferes with the DNA binding domain or the polymerization site of the enzyme. This is a common phenomenon with many reverse transcriptase enzymes. DNA polymerase can also be inhibited indirectly when the inhibitor(s) compete with or totally block the polymerase binding sites on the target or the primers (Wilson, 1997). Two sources of PCR inhibitors are known: (1) The original sample or (2) the nucleic acid extraction reagents. A few inhibitory agents have been identified including bile salts and complex polysaccharides (feces), heme, hemoglobin, lactoferrin and immunoglobulin G (blood), urea (urine), melanin and myoglobin (tissue), proteinases and calcium ions (milk), collagen (food), polysaccharides (plants), and humic acid (soil; Radstrom, 2004).

One method to detect PCR inefficiency is to include an internal positive control into the reaction. These can be included at the specimen processing stage and be monitored throughout nucleic acid extraction to identify the efficiency of the entire extraction and amplification procedure. Alternatively a control molecule may be included with the extracted sample to control for PCR and RT inhibitors. Some controls are designed to be amplified with the same primers as the target molecule and are differentiated on the basis of size when using gel visualization or qPCR using DNA binding dyes or alternative probe binding sites (Roche Amplicor). Carefully quantified internal amplification controls (IACs), or internal controls (ICs) are added to the reaction to detect the presence of inhibitors. The IAC can be a synthetic oligonucleotide, a purified amplicon, *in vitro* transcribed RNA, purified viral nucleic acid or a cloned fragment. A distinct number of IAC molecules are added to an assay tube before the purification. The controlled reaction consists of two assay tubes, both containing an identical number of external control molecules (e.g. 100 000 copies) and the identical amount of material to be purified (e.g.

10 μl of whole blood). The purification procedure is carried out on one of the samples but not the other. The number of external control copies is determined in both reactions. The difference in external control copy number between the non-purified and the purified control is equal to the loss that has occurred.

These approaches require the design of specific controls for each assay. The SPUD assay was developed as an alternative, universal and rapid protocol for detection of inhibitors (Nolan, 2006). This assay consisted of an artificial target amplicon, specific primers and target probe. In the absence of template, a characteristic C_t was recorded which depended on the concentration of the universal (SPUD) amplicon included (water control). In the presence of template that did not contain inhibitors, an identical C_t was recorded to the water control. However, in the presence of inhibitors, a shift to higher C_t value was observed. Detection of inhibition was sensitive when 20 000 copies were included in reactions containing EDTA, phenol or ethanol.

The preferred approach is to relieve inhibition by removal of the offending agent during target extraction. However the alternative is the addition of PCR enhancers or facilitators. PCR enhancers can be grouped into proteins (bovine serum albumin, T4 gene protein 32, *E. coli* single stranded binding protein), organic solvents (formamide, Dimethylsulphoxide, TEAC, TMAC), non-ionic detergents (Tween-20 and Triton X), biologically active solutes (glycerol and betaine) and polymers (Polyethyleneglycol and dextran).

Unfortunately it is impossible to predict which of these will be effective in relieving a specific case of inhibition and most protocols are determined after a process of trial and error. The final approach to overcoming the effects of inhibitors is to perform limiting dilution of the sample. This is commonly used when samples contain inhibitors with similar biochemical makeup to nucleic acid, such as fumic acid in samples extracted from soil but this is only useful when the target nucleic acid is present at a sufficiently high concentration to tolerate at least 100 or 1000-fold dilutions.

Nucleic acid amplification

One of the major challenges for extraction of microbial nucleic acid is preparing sufficient material to perform downstream tests. This has been addressed by optimization of the sampling and extraction procedures as well as by enhancing the detection systems. When as little as a single cell is available for analysis, but it is desirable to perform multiple analyses there is now the further option of performing a procedure that amplifies the total nucleic acid in the sample such that sufficient material is prepared to be used in numerous downstream reactions, including qPCR analysis; gDNA can be amplified using a variety of applications of the Phi29 enzyme used for plasmid amplification in bacteria and adapted to continuous, large molecule replication (Qiagen). Alternatively a library of representative fragments can be generated using random priming and consensus oligonucleotides (Rubicon Genomplex). This approach has been developed for amplification of all sample DNA from many specimens including clinical samples, environmental, plant and tissue material. Kits and protocols are also available for amplification of single cells and for formalin fixed paraffin embedded material (Rubicon, Sigma Aldrich). A similar, random-primed approach is also used for amplification of the entire RNA complement of a sample (Rubicon).

Conclusion

The entire process of specimen acquisition through to nucleic acid amplification is vulnerable to experimental variation. It is essential to include control samples to monitor each stage of the real-time qPCR or RT-qPCR in order to easily identify aberrant samples. It is clear that different specimen types, pathogens and target nucleic acid samples require different extraction procedures and it is essential to determine the most appropriate extraction procedure for the desired outcome. Target specific IACs can be used to monitor the sample extraction RT procedures; alternatively inhibitors can be identified using the SPUD assay. RNA integrity can be determined using the Agilent BioAnalyser algorithm and the 3′/5′ assay can be used to ensure that the specific targets of interest are not degraded. When the target is to be determined from microbial mRNA, complex procedures for accurate normalization of host cell and pathogen quantity are required. When possible this should include quantification of both the microorganism and a host reference gene target. In addition a reverse transcription control, such as an accurately quantified, stable RNA is included when assaying RNA targets. When virulence plasmids are assayed, a chromosomal gene target from the same microorganism needs to be included for the purpose of calculating the number of virulent and the number of non-virulent bacteria, as well as for control of successful DNA extraction. Positive and negative reverse transcriptase and PCR controls allow a rapid check to ensure that these steps have performed correctly. In this ideal assay set-up the template preparation, the reverse transcription, the target amplification and detection, as well as the data analysis are controlled, allowing for a true comparison of results from day-to-day (e.g. residual disease assays) and lab-to-lab (e.g. pandemic outbreaks).

Without the appropriate standard operating procedure, controls and validation assays, quantitative assay results are not reproducible at best and at worst are totally meaningless.

Acknowledgment

We would like to thank Julie Robinson (MoBio laboratories) for kindly providing information about different extraction kits.

References

Akane, A., Matsubara, K., Nakamura, H., Takahashi, S., Kimura, K. (1994). Identification of the heme compound copurified with deoxyribonucleic acid (DNA) from blood stains, a major inhibitor of polymerase chain reaction (PCR) amplification. J. Forensic Sci. 39, 362–372.

Aldous, W.K., Pounder, J.I., Cloud, J.L., Woods G.L. (2005). Comparison of six methods of extracting *Mycobacterium tuberculosis* DNA from processed sputum for testing by quantitative real-time PCR. J. Clin. Microbiol. 2471–2473.

Al-Soud WA and Radstrom P. (2001). Purification and characterization of PCR-inhibitory components in blood cells. J. Clin. Microbiol. 485–493.

Baumeister, A.K., Runge, M., Ganter, M., Feenstra, A.A., Delbeck, F., Kirchhoff, H. (1998). Detection of Mycoplasma hyponeumoniae in bronchoalveolar lavage fluids of pigs by PCR. J. Clin. Microbiol. 1984–1988.

Bankowski, M.J., and Anderson, S.M. (2004). Real time nucleic acid amplification in clinical microbiology. Clin. Microbiol. Newsl. 26, 9–15.

Bersworth chemical company (1954). The Versenes, Powerful organic chelating agents for the exacting control of cations. Framingham, MA.

Beuselinck, K., van Ranst, M., van Eldere, J. (2005). Automated extraction of viral-pathogen RNA and DANN for high-throughput quantitative real-time PCR. J. Clin. Microbiol, 5541–5546.

Bottomley, W.B. (1917). The Isolation from peat of Certain Nucleic Acid Derivatives. Proc. Royal Soc. Lon. *90*, 39–44

Braid, M.D., Daniels, L.M., Kitts, C.L. (2003). Removal of PCR inhibitors from soil DNA by chemical flocculation. J. Microbial Meth. *52*, 389–393.

Bruns, M.A., and Buckley, D.H., (2002). Isolation and purification of microbial community nucleic acids from environmental samples pp564–572 in CJ Hurst (ed) Manual of Environmental Biology, 2nd Ed ASM Press Washington, DC.

Bustin, S.A. (2004). A-Z Quantitative PCR (IUL Press, La Jolla California USA).

Bustin, S.A., and Nolan, T. (2004). Pitfalls of quantitative real-time reverse transcription polymerase chain reaction. J. Biomol. Tech. *15*, 155–166.

Buttner, M.P., Cruz, P., Stetzenbach, L.D., Klima-Comba, A.K., Stevens, V.L., Emanuel, P.A. (2004). Evaluation of the biological sampling kit (BiSKit) for large area surface sampling. Appl. Environ. Microbiol. 7040–7045.

Centre for Disease Control (1999). Biosafety in microbiological and medical laboratories. Centres for Disease Control and Prevention, Atlanta, Ga.

Chai, V., Vassilakos, A., lee, Y., Wright, J.A., Young, A.H. (2005). Optimisation of the PAXgene blood RNA extraction system for gene expression analysis of clinical samples. J. Clin. Lab. Anal. *19*, 182–188.

Chandler, D.P., Schreckhise, R.W., Smith, J.L., Bolton, H., Jnr (1996). Electroelution to remove humic acid compounds from soil DNA and RNA extracts. J. Microbiol. Methods *28*, 11–19.

Cockerill, F.R. (2003). Application of rapid real-time polymerase chain reaction for diagnostic testing in the clinical microbiology laboratory. Arch. Pathol. Lab. Med. *127*, 1112–1120.

Cockerill FR and Uhl, J.R. (2002). Applications and challenges of real-time PCR for the clinical microbiology laboratory. In Reischl, U., Wittwer C and Cockerill, F.R. (ed) Rapid cycle real-time PCR methods and applications, Springer Verlag, Berlin, Germany.

Coyne, S.R., Craw, P.D., Norwood, D.A., Ulrich, M.P. (2004). Comparative analysis of the Schleicher and Schuell IsoCode Stix DNA isolation device and the Qiagen QIAamp DNA mini kit. J. Clin. Microbiol, 4859–4862.

Edman, J.C., Kovacs, J.A., Masur, H., Santi, D.V., Elwood, H.J., Sogin, M.L. (1988). Ribosomal RNA sequences show Pneumocystis carinii to be a member of the fungi. Nature *334*, 519–522.

Espy, M.J., Rys, P.N., Wold, A.D., Uhl, J.R., Sloan, L.M., Jenkins, G.D., Ilstrup, D.M., Cockerill III, F.R., Patel, R., Rosenblatt, J.E., Smith, T.F. (2001). detection of herpes simplex virus DNA in genital and dermal specimens by LightCycler PCR after extraction using IsoQuick, MagNA Pure and BioRobot 9604 methods. J. Clin. Microbiol, *39*, 2233–2236.

Espy, M.J., Uhl, J.R., Sloan, L.M., Buckwalter, S.P., Jones, M.F., Vetter, E.A., Yao, J.D.C., Wengenack, N.L., Rosenblatt, J.E., Cockerill, F.R., Smith, T.F. (2006). Real-time PCR in clinical microbiology: applications for routine laboratory testing. Clin. Microbiol. Revs. 165–256.

Fahle, G.A., and Fischer, S.H. (2000). Comparison of six commercial DNA extraction kits for recovery of cytomegalovirus DNA from spiked human specimens. J. Clin. Microbiol, 3860–3863.

Fenagle, E. (2003). An investigation into the efficacy and bias of nucleic acid extraction techniques on mixed microbial populations from environmental samples. California Institute of Technology, Intern Report

Fiscus, F.A., Brambilla, D., Grosso, L., Schock, J., Cronin, M. (1998). Quantitation of human immunodeficiency virus type I RNA in plasma by using blood dried on filter paper. J. Clin. Microbiol, 258–260.

Fleige, S., Pfaffl, M.W. (2006). RNA integrity and the effect on the real time qRT-PCR performance. Mol. Aspects Med. *27*, 126–139

Fredricks, D.N., Relman, D.A. (1998). Improved amplification of microbial DNA from blood cultures by removal of the PCR inhibitor sodium polyanetholesulfonate. J. Clin. Microbiol, *36*, 2810–2816.

Gouvea, V., Glass, R.I., Woods, P., Taniguchi, K., Clark, H.F., Forrester, B., Fang Z-Y (1990). Poymerase chain reaction amplification and typing of rotavirus nucleic acid from stool specimens. J. Clin. Microbiol, 276–282.

Hayette, M.P., Vaire, D., Susin, F., Boland, P., Christiaens, G., Melin, P., de Mol, P. (2001). Detection of *Aspergillus* species DNA by PCR in bronchoalveolar lavage fluid. J. Clin. Microbiol, 2338–2340.

Higuchi, R., Dollinger, G., Walsh, P.S., Griffith, R. (1992). Simultaneous amplification and detection of specific DNA sequences. Nat. Biotechnol. *10*, 413–417.

International Council for Standardization in Haematology Expert Panel on Cytometry (1993) Recommendation of the International Council in Haematology for ethylenediaminetetraactic acid anticoagulation for blood cell counting and sizing. Am. J. Clin. Pathol. *100*, 371–372.

Issa, N.C., Espy, M.J., Uhl, J.R., Harmsen, W.S., Mandrekar, J.N., Gullerud, R.E., Davis, M.D., Smith, T.F. (2005). Comparison of specimen processing and nucleic acid extraction by the Swab Extraction Tube System versus the MagNA Pure LC system for laboratory diagnosis of Herpes Simplex Virus infections by LightCycler PCR. J. Clin. Microbiol, 1059–1063.

Kagedal, B., Linqvist, M., Farnebäck, M., Lenner, L., Peterson, C. (2005). Failure of the PAXgene Blood RNA system to maintain mRNA stability in whole blood. Clin. Chem. Lab. Med. *43*, 1190–1192.

Keilholz, U., Willhauck, M., Rimoldi, D., Brasseur, F., Dummer, W., Rass, K., de Vries, T., Blaheta, J., Voit, C., Lethe, B., and Burchill S. (1998) Reliability of reverse transcription-polymerase chain reaction (RT-PCR)-based assays for the detection of circulating tumour cells: a quality-assurance initiative of the EORTC Melanoma Cooperative Group. Eur. J. Cancer *34*, 750–753

Khan, I.U., and, J.S. Yadav. (2004). Development of a single-tube, cell lysis-based, genus-specific PCR method for rapid identification of mycobacteria: optimization of cell lysis, PCR primers and conditions, and restriction pattern analysis. J. Clin. Microbiol. *42*, 453–457.

Kocagöz, T., E. Yilmaz, S. Ozkara, S. Kocagöz, M. Hayran, M. Sachedeva, and H.F. Chambers (1993). Detection of Mycobacterium tuberculosis in sputum samples by polymerase chain reaction using a simplified procedure. J. Clin. Microbiol. *31*, 1435–1438

Kreader, C.A. (1996). Relief of amplification inhibition in PCR with bovine serum albumin or T4 gene 32 protein. App. Envir. Microbiol. 1102–1106.

Lader, E.S. (2001) Methods and reagents for preserving RNA in cell and tissue samples. US patent 6, 204,375

Learn, C.A., Mizel, S.B., McCall, C.E. (2000). mRNA and protein stability regulate the differential expression of pro- and anti-inflammatory genes in endotoxin-tolerant THP-1 cells. J. Biol. Chem. *275*, 12185–12193.

Luebeck, P.S., Cook, N., Wagner, M., Fach, P., Hoorfar, J. (2003). Toward and International standard for PCR-based detection of food borne thermotolerant Campylobacters: Validation in a multicenter collaborative trial. App. Envir. Microbiol. 5670–5672.

Lukiw, W.J., Bazan, M.G. (1997). Cyclooxygenase 2 RNA message abundance, stability and hypervariability in sporadic Alzheimer neocortex. J. Neurosci. Res. *50*, 937–945.

Maass, M., and Dalhoff, K. (1994). Comparison of sample preparation methods for the detection of *Chlamydia pneumoniae* in bronchoalveolar lavage fluids by PCR. J. Clin. Microbiol. *32*, 2616–2619.

Mackay, I.M. (2004). Real time PCR in the microbiology laboratory. Clin. Microbiol. Infect. *10*, 190–212.

Madejon, A., Manzano, M.L., Arocena, C., Castillo, I., Carreno, V. (2000). Effects of delayed freezing of liver biopsies on the detection of hepatitis C virus RNA strands. J. Hepatol. *32*, 1019–1025.

Malorny, B., Hoorfar, J. (2004). Toward standardization of diagnostic PCR testing of fecal samples: lessons from the detection of salmonellae in pigs. J. Clin. Microbiol. 3033–3037.

Mattson, J.G., Bergstrom, K., Wallgren, P., Johansson, K.E. (1995). Detection of *Mycoplasma hypopneumoniae* in nose swabs from pigs by *in vitro* amplification of the 16S ribosomal gene. J. Clin. Microbiol. *33*, 893–897.

Mutter, G.L., Zahrieh, Z., Liu, C., Neuberg, D., Finkelstein, D., Baker, H.E., Warrington, J.A. (2004). Comparison of frozen and RNAlater solid tissue storage methods for use in RNA expression microarrays. BMC Genomics 5, 88.

Miller, D.N., Bryant, J.E., Madsen, E.L., Ghiorse, W.C. (1999). Evaluation and optimisation of DANN extraction and purification procedures for soil and sediment samples. App. Environ. Microbiol. *65*, 4715–4724.

Muller, Z., Stelzl, E., Bozic, M., Haas, J., Marth, E., Kessler, H.H. (2004). Evaluation of automated sample preparation and quantitative PCR LCx assay for determination of human immunodeficiency virus type I RNA. J. Clin. Microbiol. 1439–1443.

Nolan, T., Hands, R.E., Bustin, S.A. (2006a) Quantification of mRNA using real time RT-PCR. Nature Protocols 1559–1582.

Nolan, T., Hands, R.E., Ogunkolade, W., Bustin, S.A. (2006). SPUD: A quantitative PCR assay for the detection of inhibitors in nucleic acid preparations. Anal. Biochem. *351*, 308–310.

Nogva, H.K., Bergh, A., Holck, A., Rudi, K. (2000). Application of the 5′ nuclease PCR assay in evaluation and development of methods for quantificative detection of *Campylobacter jejuni*. App. Envir. Microbiol. *66*, 4029–4036.

Petrich, A., mahony, J., Chong, S., Broukhanski, G., Ghjarabaghi, F., Johnson, G., Louie, L., Luinstra, K., Willey, B., Akhaven, P., Chui, L., Jamieson, F., Louie, M., Mazzulli, T., Tellier, R., Smieja, M., Cai, W., Chernesky, M., Richardson, S.E. (2006). Multicenter comparison of nucleic acid extraction methods for detection of severe acute respiratory syndrome coronavirus RNA in stool specimens. J. Clin. Microbiol. 2681–2688.

Radstrom, P., Knuttson, R., Wolffs, P., Lovenklev, M., Lofstrom, C. (2004). pre-PCR processing: strategies to generate PCR compatible samples. Mol. Biotechnol. *26*, 133–146.

Raghunathan, A., Ferguson Jr, H.R., Bornarth, C.J., Song, W., Driscoll, M., Lasken, R.S. (2005). Genomic DNA amplification from a single bacterium. App. Envir. Microbiol. 3342–3347.

Ribes JA., Limper AH., Espey MJ., Smith, T.F. (1996). PCR detection of *Pneumocystis carinii* in bronchoalveolar lavage specimens: analysis of sensitivity and specificity. J. Clin. Microbiol. 830–835.

Rosenstraus, M., Wang, Z., Chang S-Y, DeBonville, D., Spadoro P. (1998). An internal control for routine diagnostic PCR: design properties and effect on clinical performance. J. Clin. Microbiol. 191–197.

Schreiner and Laptrop (1912). US Dept Agric Bureau of Soils Bull 89.

Schroeder *et al.* (2006). The RIN: an RNA integrity number for assigning integrity values to RNA measurements. BMC Mol. Biol. *7*, 3.

Seegers, W.H. (1967). Blood Clotting and Enzymology, Academic Press, New York, 364–366.

Sheppard, R.L., and Henion, J. (1997). Determining EDTA in blood. Analytical Chemistry *69*, 477–480.

Smith, T.F., Uhl, J.R., Espy, M.J., Sloan, L.M., Vetter, E.A., Jones, M.F., Rosenblatt, J.E., Cockerill, F.R. (2004). Development, implementation and trend analysis of real time PCR tests for the clinical microbiology laboratory. Clin. Microbiol. Newlett. *26*, 145–154.

Uhl, J.R., Adamson, S.C., Vetter, E.A., Schleck, C.D., Harsen, W.S., Iverson, L.K., Santrach, P.J., Henry, N.K., Cockerill, F.R. (2003). Comparison of LightCycler PCR, rapid antigen immunoassay, and culture for detection of group A streptococci from throat swabs. J. Clin. Microbiol. *41*, 2127–2136.

Verdin (1996) Use of nested PCR diagnosis test to detect Mycoplasma hypopneumoniae under field conditions IOM Lett. *4*, 101–102.

Vlems, F.A., Soong, R., Diepstra, H., Punt, C., Wobbes, T., Tabiti, K., van Muijen G. (2002). Effect of blood sample handling and reverse transcripase polymerase chain reaction assay sensitivity on detection of CK20 expression in healthy donor blood. Diagn. Mol. Pathol. *11*, 90–97.

Zhong, K.J., Salas, C.J., Shafer, R., Gubanov, A., Gasser, Jnr, R.A., Magill, A.J., Forney, J.R., Kain, K.C. (2001). Comparison of IsoCode Stix and FTA gene Guard collection matrices as whole-blood storage and processing devices for diagnosis of malaria by PCR. J. Clin. Microbiol. *39*, 1195–1196.

Zhou, J., Bruns, M.A., Tiedje, J.M. (1996). Rapid method for recovery of DNA from soils of diverse composition. App. Environ. Microbiol. *62*, 316–322.

Standards and Controls: Concepts for Preparation and Use in Real-time PCR Applications

4

Amy Muska, Edith Peck, and Stuart Palmer

Abstract

The use of real-time PCR for molecular diagnostic applications requires a high degree of assurance that the analytic result reflects accurately the true concentration of the target nucleic acid and is not affected by inhibitors of the reaction. A variety of strategies have been developed to ensure confidence in the resulting assay information: Well-characterized reference standards, established by extensive collaborative studies using a variety of amplification methodologies can provide a basis for accurate calibration and a means for preparation of secondary and tertiary standards while co-amplification of internal controls provides assurance the PCR has functioned without inhibition. When considering the increasing complexity of reagent composition for multiplexed real-time PCR and the associated hardware for kinetic fluorescent signal acquisition, such measures would appear not only desirable but necessary. The preparation methods for reference standards, calibration standards and internal controls often present unique challenges. Consistent, stable formulation of low copy number targets requires stringent control of the laboratory environment to prevent contamination by exogenous RNAse or DNAse and a reliable means of measuring the target at such low concentrations. Depending upon the intended application for the real-time PCR, linkage of the assay calibration to a reference standard may also be desirable. The quantity of information yielded by a single multiplex real-time PCR can now readily discriminate the signal output from four or more individual target probes. This output of enhanced information content from a single reaction will continue to rapidly increase with advances in detection chemistries and hardware capabilities. This has particularly beneficial implications for molecular diagnostics. However, further challenges will be posed for the development of standards and controls that monitor and assure the accuracy of large numbers of potentially related target sequences. In the present chapter, we discuss concepts for the application of standards to assay calibration and discuss the development, preparation and use of reference standards and controls in real-time PCR assays.

Introduction

The use of quantitative molecular methods has seen wide application to the detection and study of microorganisms in the diagnostic laboratory. Included in these methods are a series of related reaction technologies that are collectively referred to as real-time PCR (also known as "homogeneous PCR" or "kinetic PCR") (Higuchi *et al.*, 1992). The advent of real-time PCR methods, in all of its variants, allows the simultaneous amplification and detection of increasing amplicon concentrations by a number of general reaction mechanisms that links target amplification to increasing fluorescent signal.

The major variations for real-time PCR methods that have found routine use in the clinical laboratory include the 5′ nuclease PCR assay, fluorescence resonance energy transfer (FRET) and double-stranded DNA dye-binding. In the case of the 5′ nuclease PCR reaction, the accumulation of specific target nucleic acid is detected by incorporation of a synthetic, dual-labeled fluorescent oligonucleotide probe complementary to sequences downstream from a primer-binding site (Livak *et al.*, 1995). As the 5′ nuclease assay proceeds, the reporter and quencher dye molecules present on the dual-labeled probe are separated by the cleavage of the probe by the 5′ nuclease activity of the Taq polymerase, releasing the reporter molecule and resulting in an increase in detectable fluorescence signal. FRET-based signal generation relies upon the hybridization of two different dye-labeled oligonucleotide probes to an accumulating, specific amplicon; the close proximity of the two fluorescent dyes on the amplicon allows for fluorescent energy transfer from the donor dye to the acceptor which can be measured (Wittwer *et al.*, 1997). Lastly, the method of double-stranded DNA dye-binding using ethidium bromide or SYBR Green I is a useful but less specific method of detecting PCR products continuously during amplification where the direct binding of the dye to DNA increases the fluorescent signal (Morrison *et al.*, 1998). The absence of a target sequence-specific detection probe makes the double-stranded DNA dye-binding method subject to errant results when non-specific PCR products are generated.

The complexity of real-time PCR methods has increased with the optimization of multitarget amplification strategies in single tube reactions (Tettelin *et al.*, 1999) and the application of these to challenging homogeneous or kinetic assays (Du *et al.*, 2006). In order to achieve the level of accurate and reliable results expected in diagnostic settings, such real-time PCR methods often require precise and controlled formulation of reagents, well-defined and controlled sample-processing methods and sophisticated laboratory equipment. While the reliability of such complex test systems remains high, the need to ensure consistency in the quantitative detection of a target microorganism on a run-to-run, lab-to-lab and sample-to-sample basis poses some interesting challenges (National Committee for Clinical Laboratory Standards, 2003). These challenges have led many in the field to develop a variety of approaches to standardize the measurements of real-time PCR products and to incorporate a variety of external and internal controls which assure the integrity of the results (Ullmannova and Haskovec, 2003; Brey *et al.*, 2006). Some factors to be considered in assuring the quality of results and, ultimately, the utility of the information rendered by any real-time PCR method include sample integrity, interfering substances, "trueness" or accuracy of value, specificity and measurement range. Standards, calibrators and controls are crucial components in the reliable use of real-time PCR assays.

The use of standards and controls in real-time PCR assays is both wide-spread and variable in design and application strategy. Therefore, it is not unexpected that the terminology in the literature accompanying the technical descriptions of assay formats is equally wide-spread and variable. For the purposes of the present discussion, the terminology used herein centers on the generally accepted use derived from consensus references (NCCLS, 2003). The term "Standard" or "Reference Standard" will be used to describe a material suitable for real-time PCR applications which is a specimen derived from a known biological source and whose value has been established by consensus means. Standards and Reference Standards are typically procured from independent sources (e.g. World Health Organization or the UK National Institute of Biological Standards and Control) and are intended to provide means for researchers or others in the field to adhere to a common measurement. This terminology usage is in alignment with the International Organization for Standardization (ISO), definition on the topic. The term "Calibrator" will be used to describe a synthetic or natural analytic standard whose concentration value has been established by a defined method which may or may not include comparison to a reference standard. Calibrators are tested externally to the specimen, typically in the same assay run (external calibrator), and are used to calculate the concentration of target nucleic acid in the sample. "Internal Controls" are typically synthetic nucleic acid constructs of naked or encapsidated RNA or DNA which are added to the sample either before specimen preparation (full process internal controls) or just prior to amplification (direct amplification internal controls). The term "Quantitation Standard" will be used interchangeably with "Internal Control" although in true context, it is more properly an "Internal Calibrator." The term "Control", "Run Control" or "Positive Control" will be used to describe natural biologic specimens (e.g. viral or bacterial agents) or synthetic RNA or DNA constructs with nucleic acid sequences which mimic the target of interest and are analyzed as specimens independent of test samples. Controls are typically included in all test method runs to monitor assay variation; this is determined by deviations from the known control concentration). The control samples may also have internal controls added to the reaction to assure extraction or amplification consistency.

Reference standards

A variety of strategies have been developed in recent years to enhance the analytic accuracy of a particular method by relating the assay response of the real-time PCR (Ct) to a series of external calibrator concentrations prepared by serial dilution which are then used to generate the quite familiar standard curve. Provided that one has an accurate or "true" measure of the starting concentration of the highest level of calibration solution or its parent material, highly accurate volumetric or gravimetric dilutions can be prepared. Preparation of a calibration panel by volumetric or gravimetric means firmly establishes the quantitative relationship between the panel members and eliminates variability that may be imparted when using real-time PCR methods. In general, direct comparisons of the fluorescent signal (measured at the threshold or Ct point) for the highest calibrator level may be made directly to the Ct of an external standard and from that ratio, accurate assignment of each calibrator panel member's concentration may be made on the basis of the dilution relationship. An inherent prerequisite of this approach is the availability of a recognized reference standard for the initial comparison.

Since the first description of an international reference standard for Hepatitis C virus (HCV) by the World Health Organization (WHO) in 1999 (Saldanha *et al.*, 1999), several well-characterized, quantified standards have become available for use in molecular diagnostics including HIV and Parvovirus B19 (Saldanha *et al.* 2000, Saldanha *et al.* 2003,

Holmes *et al.*, 2003). As recognized international reference standards, the available inventory of such materials is often quite limited. It is therefore common practice for commercial manufacturers, reference labs and "home brew" users to create secondary or tertiary standards using biologically sourced materials that are related to the available reference standard in their assay response. Such materials have concentration values which are assigned by the real-time PCR method being employed and are traceable to the reference standard concentration in the specific test system. For the purpose of routine use, external calibrators in real-time PCR applications are typically referenced to these secondary or tertiary standards. The approach of preparing such assay calibrators which are linked to defined standards has been utilized in a number of real-time PCR applications, including quantitative assays for Hepatitis B virus (Sum *et al.*, 2004), Respiratory Syncytial virus (Achenbach *et al.*, 2004), Hepatitis C virus (Kim *et al.*, 2005) and Japanese Encephalitis virus (Kim *et al.*, 2004).

The existence of viral or microbial variants creates a number of additional considerations that must be taken into account for the preparation and use of well-characterized, quantitated reference standards that are suitable for broad application. The ability to discriminate between serovars, serotypes, genotypes, clades and other variant types in multiplex real-time PCR applications is limited by a combination of suitably stable genetic variation as well as the availability of fluorescent dye chemistries and instrumentation to perform the required multiplex reactions. The selected reference standards may respond differently relative to the sample target sequence as a result of primer or probe sequence mismatches. As a result of such variants, absolute quantification based simply on the relation of assay response to an input copy number of a reference standard to the assay response of a synthetic calibrator may be somewhat elusive. In some circumstances, secondary analyses of real-time PCR amplification products based on fluorescent probe:amplicon or amplicon melting temperatures (Tm or Td) may allow discrimination between exact and inexact sequence matches (Nicolas *et al.*, 2002). In such a situation, the assay response of a potential reference standard will be expected to differ from the non-homologous target, but may be at least partially compensated for in secondary analyses. It will become increasingly important to consider the complexity introduced by variants when attempting to measure assay response of reference standards and test specimens using real-time PCR methods. Such considerations have been recognized as being important for commercial assay developers by regulatory agencies (Food and Drug Administration, 2005).

Internal controls

The integrity of nucleic acid in samples of biologic origin is highly dependent upon the methods and conditions of sample acquisition, transport and storage (NCCLS, 2003). Defined procedures for sample handling prior to arrival at the laboratory are key steps to assuring that the nucleic acid in the sample is representative of the original biological mater. While important in clinical settings, exacting procedures and use of internal controls are particularly important with field samples that may be collected from more remote, out door or "field" locations (Holman *et al.*, 2004). Such internal controls are pseudo-targets where the pre-defined base sequence utilizes the same primer binding regions as the assay target nucleic acid sequence but differs in the probe hybridization region. These are also known as "competitive" internal controls since there is a true competition between target and internal control for the same primer. The addition of a known quantity of such an internal control to a sample that may be subsequently co-extracted, handled or stored in the normal chain of processing, has been a useful method of compensating for incomplete recovery, sample deterioration, or the presence of PCR inhibitors (Brightwell *et al.*, 1998, Zimmerman *et al.*, 2000). The strategies for optimization of these pseudo-targets for use as internal controls in heterogeneous and RT-PCR formats have been well described in the literature (Henley *et al.*, 1996) and are readily applicable to real-time PCR formats. Similar, but alternative non-competitive internal control strategies have also been described (Dingle *et al.*, 2004). In this approach, two separate primer pairs are directed towards the internal control and the target sequence. The non-competitive internal control sequence may be endogenous RNA or DNA sequences present in the sample or they may be synthetic constructs spiked into the specimen.

Positive reactions in real-time PCR analyses are defined by the cycle (Ct) during amplification for which the target of interest is first detected rather than the measurement of the amount of the PCR product of interest which is accumulated at the end of a PCR reaction. Real-time PCR does this by monitoring the amount of fluorescence emitted during the PCR reaction and this acts as an indicator of the amount of PCR amplification that occurs during each PCR cycle. As such, the kinetics of amplification for both internal control molecule and target molecule must be closely matched in order for the internal control to perform as a true molecular mimic (Lee *et al.*, 2004). The use of a competitive internal control which shares primer sequences with the target of interest simplifies the composition of the PCR reaction mix and allows for a more straightforward assay optimization. An example of this approach to internal control design include the recent report of the use of a *Mycobacterium avium* subspecies *paratuberculosis* synthetic plasmid construct in which a specific deletion of the target sequence created a near-mimic sequence that still allowed discrimination between target and internal control (Brey *et al.* 2006). Despite its simplicity, one has to keep in mind the potential for limiting the overall sensitivity in a competitive internal control-based assay, especially in situations where an infectious agent may be present in very low concentrations such as in donor blood screening. At low concentrations of target nucleic acid (e.g. $\sim 10^1$ to 10^2 copies per mL range) the potential for competition for primer binding to the same sequences of an assay internal control cannot be readily eliminated. Therefore, the impact on assay sensitivity should be assessed (e.g. by performing parallel real-time PCR reactions on low copy number specimens both with and without added internal controls).

An alternative means of measuring a target sequence that may require normalization for inefficient target nucleic acid recovery or direct inhibition of the PCR reaction is to identify and utilize an endogenous sequence of known concentration that is co-extracted and co-measured with the target sequence (Ullmanovnova and Haskovec, 2003). In this non-competitive approach, the reaction is a true real-time PCR duplex reaction that contains separate primer sequences for the internal control sequence and the target under study. This may only be possible with the required degree of confidence in samples where the amount of the control sequence is sufficiently constant to serve as a reference level (e.g. blood samples, cervical or buccal scrapings). Control gene sequences of this type are usually "housekeeping" genes in defined biological systems and often include examples such as β2-microglobin, glyceraldehyde-3-phosphate dehydrogenase, 18S ribosomal RNA and dihydrofolate reductase (Watzinger and Lion, 1998). Expression of different housekeeping genes yields RNA that serves as a control template in RT-PCR applications and therefore knowledge of the stability of RNA (e.g. rRNA vs. mRNA) may be a prerequisite for using a particular control template in any particular situation. Applications of this approach for PCR internal controls in the detection and monitoring of infectious agents includes real-time RT-PCR assays for tick borne encephalitis virus using 16S rRNA (Schwaiger and Cassinotti, 2002), DNA viral load testing in high-risk human papillomavirus type 16 using endogenous β-globin DNA (Lefevre *et al.*, 2003) and RT-PCR detection of severe acute respiratory syndrome (SARS) coronavirus RNA using 18S rRNA (Poon *et al.*, 2004). Real-time PCR assays developed with either the competitive or non-competitive internal control formats, provide a suitable means for determining the validity of a negative or positive result. This is a crucial factor for proper evaluation of the information for diagnostic purposes.

Calibrators

The dynamic range of an assay determines by how much a target sequence concentration can vary and still be readily measured. Real-time PCR assays generally have a much wider dynamic range compared to conventional RT-PCR. Heterogeneous RT-PCR formats routinely achieve dynamic ranges of 10^5 to 10^6 while an overall dynamic range of 10^8 or more for real-time PCR formats is not atypical. A wide dynamic range permits a more accurate quantitation provided there is available a reliable reference sample or standard of known concentration for reference to the assay range. It also reduces the necessity of diluting out-of-range specimens which, aside from the expenditure of laboratory time and effort, is a potential source of error. The selection and characterization of a suitable assay standard is a key determinant in the accuracy of the resulting diagnostic information (NCCLS, 2003, Kim *et al.*, 2004). Accurate quantitation over a broad dynamic range is often a primary consideration for nucleic acid targets of microbial origin where extremely low levels of individual bacteria or viruses are the normal occurrence *in vivo* but can often be propagated to high levels *in vitro* [e.g. in clinical laboratory settings for blood screening (Kleinman *et al.*, 2005)]. The concentration range of external calibrators should be developed keeping in mind the nucleic acid levels expected to be encountered in diagnostic and experimental samples.

The preparation and quantitation of an external calibrator for real-time PCR may be achieved by direct measurement of nucleic acid concentration independent of available reference standards. For example, synthetically transcribed HIV-1 RNA quantified spectrophotometrically at 260 nm (Palmer *et al.*, 2003) and 18S rRNA quantified by RiboGreen™ (Hashimoto *et al.*, 2004) have been used in their respective assay systems as calibration standards. In these applications, the high concentration parent solutions for the calibrators are analyzed due to the limited sensitivity of the methods. Once reliable concentration values are obtained, the parent solutions are diluted by accurate volumetric or gravimetric means to the final concentration range for the external calibrator set. An alternative approach to preparing an external calibrator series based on the measurement of physical/chemical attributes is to measure the external calibrator by the Poisson statistical analysis (Wang and Spadoro, 1998). In this approach, multiple, limiting dilutions of calibrator are analyzed for a positive or negative response by PCR and applying a statistical distribution analysis to the results. The Poisson analysis is also considered a reference-independent approach to quantifying external calibrator concentrations. Once established, the external calibrator set is then run along with the test specimens and the standard curve generated by the assay response for the external calibrators may be used to measure the concentration of specimen.

The choice for the investigator to either prepare external calibrators using physical/chemical methods (e.g. spectrophotometrically at 260 nm or RiboGreen™ binding) or a direct PCR-based method (e.g. Poisson analysis) requires a number of factors for the test system and target be considered. For example, the availability of a reference standard of known concentration and the expected range of target concentration (e.g. 10^5 copies per mL or higher versus 10^2 copies per mL or lower) are particularly relevant. Exotic microbial specimens such as those derived from environmental sources (soil, marine, etc.) are less likely than important medical pathogenic agents to have previously established and quantified reference standards to use for external calibrator assignment. As such, it may be necessary to first attempt to establish a measurement standard for the exotic specimen by propagating the organism and measuring target nucleic acid physically. Once established, that material is used for both optimization of the real-time PCR assay for dynamic range and sensitivity, but also for preparing direct dilution series for use as external calibrators.

External calibrators can also be used in combination with internal controls. By comparing the assay response (concentration) for both external calibrators and internal controls co-amplified with the target nucleic acid, one establishes the quantitative relationship between the calibrators and internal controls. Once this relationship is defined, the internal control can serve the dual purpose of assay validation and single-point reference calibrator for quantitation of the target nucleic acid. By defining the concentration of the internal control, a comparison of Ct values may be made directly (e.g. sample to standard ratio) in order to estimate the target concentration. Standardization of the internal control concentration is particularly useful when comparing results across multiple laboratories (Malorny *et al.*, 2003). This combined approach eliminates the need to run an external calibrator set with each run or series of samples.

Whether the external calibrator is quantified in relation to an established reference standard or whether it is quantified using means that are independent of reference standards, external calibrators should be tested in the assay system with defined frequency and the results compared to pre-defined (expected) assay response. For routine use in quantitative molecular diagnostic assays, the ability to validate the sustained accuracy of a method over time is directly linked to the standardization concept that one has chosen when engaged in assay design (Haberhausen *et al.*, 1998). For such molecular diagnostic applications, this must also be complemented by rigorous adherence to established laboratory procedures.

In order to ensure ongoing accuracy of a laboratory's quantitative molecular methods it is often desirable to participate in an external quality assurance (EQA) or quality control program. Such programs utilize blinded samples that are tested in parallel at participating laboratories and the results analyzed by statistical means. The unblinded, tabulated results are shared amongst the participants so that determinations of accuracy between labs and between methods may be assessed. Recently, reports of such EQA programs have been emerging (Marubini *et al.*, 2004; Raggi *et al.*, 2005). For a similar purpose, the College of American Pathologists offers a series of Nucleic Acid Test-based survey programs for HIV, HCV, HBV, West Nile virus and human papilloma virus (CAP, 2006). It is expected that programs such as these will continue to expand the range of viral or microbial survey targets offered. While not exclusive for use with real-time PCR applications, such programs are essential to ensuring the continued accuracy of analytic results. Because they provide the enhanced ability to compare quantitative results amongst laboratories practicing real-time PCR for diagnostic purposes, the importance of assay standardization to common reference materials becomes heightened. It is expected that reduced operator manipulation associated with single-tube amplification and detection of real-time PCR assays may contribute to lower peer group variation for commercial products.

Main body

Internal controls

For real-time PCR applications in life science and molecular diagnostics, the requirement for high confidence in the resultant data hinges on the correct design and use of internal controls for the detection of target loss during sample preparation or inhibition of the reaction during amplification and detection (Saunders *et al.*, 2004). Both occurrences may lead to false negatives which might otherwise not be detected. Beyond the essential element of controlling for false negative results, the ability to multiplex PCR in real-time mode has expanded the role for internal controls to facilitate the quantitative detection of specific nucleic acid sequences (Madej *et al.*, 2001). Internal controls also validate that the proper manipulations from specimen preparation through to detection have occurred. Both non-competitive and competitive internal control formats are most commonly used in qualitative applications. Whereas competitive, quantified internal controls (sometimes also referred to as quantitation standards or QS) find common use in applications where more precise measures of nucleic acid target may be necessary. In addition, the incorporation of internal controls in the form of a quantitation standard is particularly useful for targets which vary over time, such as the levels of HIV RNA in response to antiviral therapy (Orito *et al.*, 1995).

In real-time PCR, the measurement of viral RNA, for example, may be accomplished by comparing the emerging Ct signal of the viral target probe hydrolysis to the Ct signal for the quantitation standard present in each specimen (Figure 4.1 and Figure 4.2).

The Ct curves depicted in Figures 4.1 and 4.2 for target or QS sequences are typical output data from real-time PCR reactions. Because the fluorescent signal increase with each cycle represents (approximately) a doubling of the target by amplification, the means of measuring concentration will be based on a logarithmic function. This contrasts with the quantitative function observed in heterogeneous PCR tests (e.g. microwell plate-based assays) where the concentration of target nucleic acid is more linear and calculated from absorbance measurements (OD) that relate the concentration of the internal control or quantitation standard. The linear relationship for target and QS OD in heterogeneous

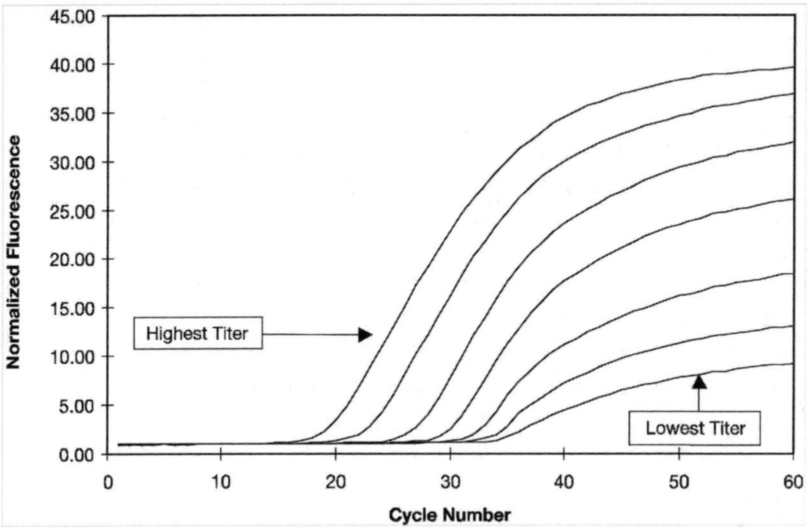

Figure 4.1 Depicts typical viral target growth curves (e.g. HCV) for a dilution series of virus spanning a 5-$\log_{10}$ range. As the concentration of the virus increases, the growth curves shift to earlier cycles. Therefore, the leftmost growth curve corresponds to the highest viral titer level whereas the rightmost growth curve corresponds to the lowest viral titer level.

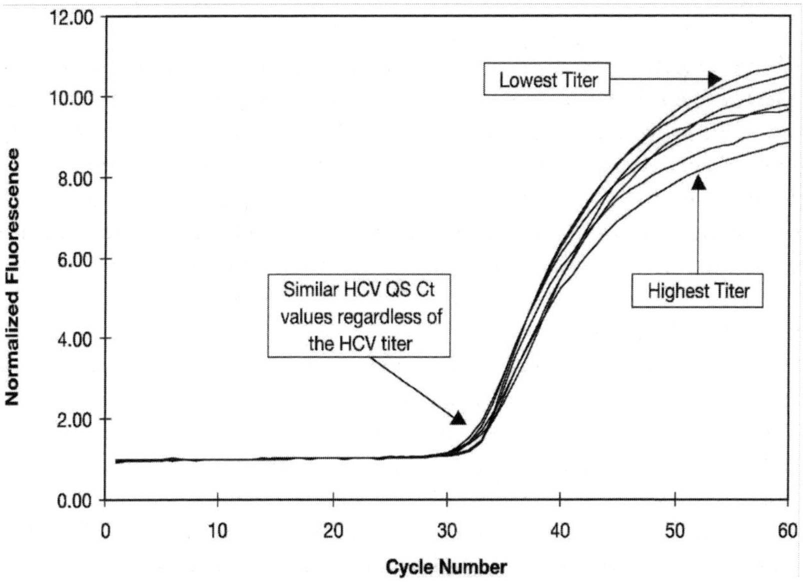

Figure 4.2 Depicts the Quantitation Standard signal growth curves for specimens from the Figure 4.1 viral dilution series. The amount of Quantitation Standard added to each specimen is constant for each reaction. The Ct value of the Quantitation Standard is similar regardless of the viral titer.

assays (expressed in copies per mL or copies per μL) is illustrated by the following calculation:

$$Target\ (Copies/\mu L) = \frac{Target\ OD}{QS\ OD} \cdot (Input\ QS\ copies/PCR) \cdot (SVF) \qquad 4.1$$

Where:
Total Target OD is measured for Target Probe Hydrolysis
Total QS OD is measured for QS Probe Hydrolysis
Input Copies/PCR of QS is determined per batch of QS
SVF (Sample Volume Factor) converts copies/PCR to copies/uL
Note: The sample volume factor is the amount of sample added to the reaction

As applied to real-time PCR applications, the measurement of a target signal Ct is performed in relation to the signal for the quantitation standard Ct for each specimen by a non-linear (second-order polynomial) relationship:

$$T = SVF \cdot Q / R - 10^{(a \cdot \Delta n^2 + b - \Delta n + c)} \qquad 4.2$$

Where:

T	titer (copies or IU/mL)
SVF	Sample Volume Factor
Q	QS input (copies/PCR)
R	recovery (adjustment factor)
A	2nd order polynomial constant
B	1st order polynomial constant
C	0th order polynomial constant
Δn	elbow value of QS minus elbow value of target

The reporter dye selections made during oligonucleotide probe design allow the emission wavelength to be specified. Internal controls or quantitation standards are readily distinguished from the specimen target through the fluorescence emission monitoring of their probes at different wavelengths. Wavelength selection, and therefore dye choice, is further dependent upon the available hardware (including wavelength filter options, detector discrimination and sensitivity capabilities and fluorescent channel "cross talk" reduction capabilities).

In standard TaqMan® probes, the actual "reporter:quencher" dye combination chosen to be used for target detection and the dye choice for internal control detection must have emission spectra that are suitably distinct for the instrumentation to resolve the fluorescent signals. Since the threshold intensity of the florescent signal is related to the amount of target the threshold intensity of the target will be seen to vary while the threshold intensity of the quantitation standard will be relatively constant (Figure 4.2). Other considerations for the choice of dyes for target signal or internal control signal reporting include fluorescent yield (or "brightness") and moieties for linking the dyes to nucleic acids.

Calibration and quantitation standards

Beyond the simple and direct sample:standard ratio approach taken in heterogeneous PCR formats (Equation 4.1), it is desirable to use accurate means for ensuring reliability of the concentration measurements over the broad dynamic range of a particular real-time PCR method. Due to the inherent non-linear relationship of Ct value versus concentration (Equation 4.2), it is important to make compensating adjustments across the dynamic range of the assay. When sample concentrations differ significantly from the concentration of the quantitation standard, one cannot use a simple sample:standard ratio method (Equation 4.1) as there is a likelihood this calculation will not be adequate due to potentially large extrapolations. Therefore, one may use an alternative approach to establish a calibration range for the assay using samples of known concentration (natural or synthetic) and define the assay response for these samples (calibration standards). This will be based on the non-linear relationship described above (Equation 4.2). The plot of a calibration curve prepared in this manner would also be a second-order polynomial equation defined by the formula:

$$y = a(x)^2 + b(x) + c \qquad 4.3$$

In this relationship, the concentration is presented on the y-axis in appropriate units (e.g. copies/mL or copies/μL) and the difference between the Ct values (Δ Ct) of the calibration standards are plotted as values on the x-axis. Because this pre-defined relationship of assay response to concentration will vary in subtle ways with changes due to normal run-to-run variation or due to changes in the assay components (batches of reagents or raw materials), one can use the relationship to compensate for this variation by using a quantitation standard in each sample. By comparing the measured concentrations of the QS to the recovered target concentration, one can readily express the relationship in terms of concentration by transforming Equation 4.3 to the following:

$$Log\,(T/QS\ in\ Copies/\mu L) = a[Ct\ QS - Ct\ Target]^2 + b[Ct\ QS - Ct\ Target] + c \qquad 4.4$$

Where:
T is Target Titer in copies/μL
QS is QS Titer in copies/μL
a,b,c are second order polynomial terms

With this approach, the external reference calibration curve is generated when Ct values from the calibration panel are plotted (Elbow $value_{QS}$ – Elbow $value_{target}$) vs. the Log of Target/QS, thus deriving the calibration coefficients using a second order polynomial fit. In practice, the variables ***a*** and ***b*** do not deviate significantly with the types of changes described above. However, ***c*** will be expected to show some variation. One can conceptualize this as the shape of the curve remaining constant, but exhibiting "offset" due to changes in ***c***. (Figure 4.3).

Coefficient assignment is dependent upon the production and titer determination of calibration panels that are at known concentrations. Ideally, a multiple level panel of calibrators prepared in a processed human plasma matrix such as Normal Human Plasma

(NHP) can be constructed to cover the test dynamic ranges (approximately 7 $\log_{10}$ or more as shown in Figure 4.3) and may be referenced to a known primary standard. The dilution of the individual panel members themselves should be done in a manner which minimizes additive error of serial dilutions. Depending upon requirements for use of the calibration panel, one may select a starting target stock material which is naturally sourced or synthetic. A number of synthetic construct approaches have developed (Donia *et al.*, 2005) and can result in minimizing drift for calibration panels.

Calibration accuracy

Once an external calibration panel has been prepared, the concentration of the individual members can be linked to known concentrations of reference standards (primary and secondary) as a means of anchoring the test system. This may be done using a secondary standard, which itself may be concentration assigned relative to a primary standard such as a WHO International Standard. The multilevel calibrator panel is analyzed with the secondary standard so titers can be reassigned (in either copies per ml or International Units (IU)). When the dose-response curve is generated, using the nominal titers of the series, replicate samples of the secondary standard are included. The nominal value of the secondary standard, determined by the curve, is compared to the known titer for the standard. This comparison results in the calculation of a reassignment factor between nominal titer and known secondary standard titers (IU). The result is a factor to adjust or "normalize" the curve. This reassignment factor is applied to each calibrator level in the series and the

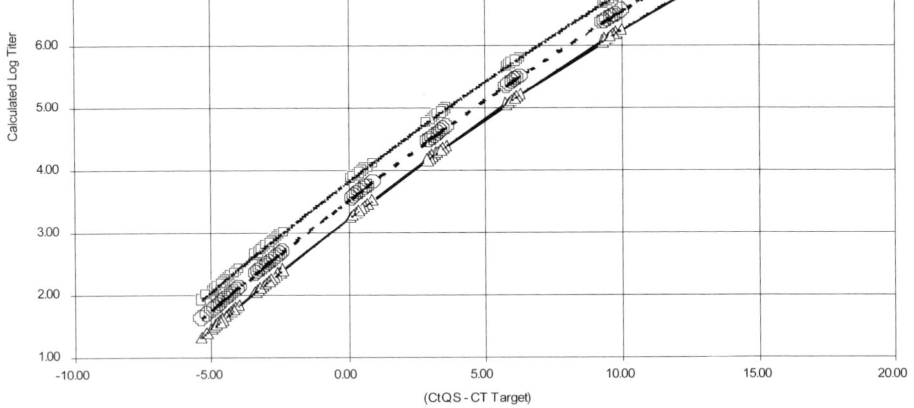

Figure 4.3 Calibration Coefficient (*a*,*b*,*c*) Determination. Variation in non-linear calibration curves for well-developed assay systems is observed to be most influenced by the c term of the polynomial equation (□ = +0.3*c*; ○ = c; △ = –0.3*c*).

calibration curve is recalculated using the newly assigned titer for the calibrators. Figure 4.4 demonstrates this re-calculation.

Positive controls

Assay positive controls differ from internal controls in that they are prepared, amplified and detected separately from specimens. They may be considered characterized specimens and are useful for assessing run-to-run quality control attributes of a particular assay's performance. Positive controls also differ from internal controls in that their sequences have the same primer *and* probe binding regions as the target virus or bacteria. Positive controls are designed to mimic the specimen and are most effective in monitoring continued assay performance when their concentration is set to a level that is approximately three times the standard deviation of the assay's limit of detection (NCCLS, 2004). This ensures that even at the lowest levels of sensitivity, the positive control will continue to validate the assay run with a high degree of confidence. The positive control also allows the analyst to assess the potential effects of run-to-run differences in primer/probe binding or hydrolysis and differences in general amplification efficiency separate from the internal control. Under such circumstances, a "positive" result for a control sample within an assay run, in combination with an appropriately measured internal control signal (i.e. within its expected range for Ct or concentration) for a specimen, provides strong indication that a negative specimen is truly negative and that there has been no interference of the PCR by the sample matrix.

PM#	Nominal Calibrator Titer (cp/mL)	×	CAF (avg.) (IU/cp)	=	Assigned Calibrator Titer (IU/mL)
11	1.0×10^2	×	0.3236	=	3.236×10^1
10	3.0×10^2	×	0.3236	=	9.708×10^1
9	5.0×10^2	×	0.3236	=	1.618×10^2
8	1.0×10^3	×	0.3236	=	3.236×10^2
7	1.0×10^4	×	0.3236	=	3.236×10^3
6	1.0×10^5	×	0.3236	=	3.236×10^4
5	1.0×10^6	×	0.3236	=	3.236×10^5
4	1.0×10^7	×	0.3236	=	3.236×10^6
3	5.0×10^7	×	0.3236	=	1.618×10^7
2	1.0×10^8	×	0.3236	=	3.236×10^7
1	1.0×10^9	×	0.3236	=	3.236×10^8

Figure 4.4 Reassignment of External Calibration Series in International Units. From replicate measurements of known reference materials (e.g. Primary or Secondary Standards) in the test system, the difference between the measured and known values can be determined. This difference is then used as a calibrator assignment factor for adjusting the nominal concentrations of the external calibration series. Because the difference between the measured and known values is generally constant across a wide range of concentrations, each panel member is adjusted by the same factor. PM = Panel member (the sequential number of the calibrator). CAF = Calibration Assignment Factor (difference of Secondary Standard nominal titer vs. known titer).

In many real-time PCR applications, positive controls have also been constructed by synthetic means and have consisted of short segments of single-stranded RNA or double-stranded DNA (Dingle *et al.*, 2004). Such positive controls may be constructed in plasmids containing non-infectious, conserved sequences of the target microorganism designed to permit linearization using restriction endo-nucleases such as *Eco*RI (Wu, 1978). The use of linearized DNA is preferred because super-coiled (and un-nicked) plasmid DNA is a poor template for PCR due to limited strand separation upon thermal denaturation and because linearization ensures consistency in preparation batch-to-batch. The resulting linearized recombinant DNA plasmid or fragment may be used directly (after appropriate dilution) in the PCR or may be the substrate for *in vitro* run-off transcription if an RNA positive control template is desired (Beld *et al.*, 2004 and Eisler *et al.*, 2004). Direct dilution of linearized DNA or *in vitro* transcribed RNA into a buffered aqueous solution is a standard method for positive control preparation. However, such "naked" nucleic acid is generally subject to digestion by RNAses or DNAses which are ubiquitous. These synthetic materials are also not fully representative of the composition of the specimens under test which, in molecular diagnostic applications, are likely to be obtained from a variety of sources including serum, sputum, synovial fluid, cervical or urethral scrapings, and fecal matter. To avoid the potential for differences in signal detection due to differences in target recovery during extraction and matrix differences between buffers and biological samples, it is typical practice that buffered aqueous solutions of positive control are added to human plasma (after inactivation of endogenous RNAses by a chaotropic agent). This is done at the beginning of the assay protocol so that sample preparation and PCR of the positive control more closely approximates that of a true specimen. In order to serve as an appropriate diluent for the positive control, the human plasma itself is screened to be absent of adventitious agents and processed (e.g. defibrinated and stabilized with preservatives such as sodium azide) to ensure stability. The development of more highly automated real-time PCR instrumentation has also driven the development of positive controls that do not require pre-addition to human plasma as part of the test procedure. They contain the synthetic target pre-set to the desired concentration range in human plasma. Such controls are more exact mimics of the test specimen. These are sometimes referred to as "full process controls" because they can undergo the full processing of an assay protocol just as a true specimen.

Full process controls for use in real-time PCR present an interesting challenge: The synthetic RNA or DNA target must now be stable in human plasma where it normally would be subject to enzymatic degradation. One useful solution to this challenge was the development of viral encapsidated constructs which contain the target sequence of interest. The protein coat surrounding the target nucleic acid affords a suitable level of protection that permits preparation of stable, human plasma based positive controls. One such construct in wide use today is based on the single-stranded RNA bacteriophage MS2 which has been modified to include RNA sequences from infectious agents such as HIV-1 or HCV (Pasloske *et al.*, 1998). This approach has been referred to as Armored® RNA. The MS2 RNA coliphage has a very simple icosahedral structure with a mass of ~3 M Dalton and diameter of 268 Angstrom (Verbraeken and Fiers, 1972; Stonehouse *et al.*, 1996). The packaging system is plasmid-directed with assembly of the particles occurring in an *E. coli* host. The pCP-1 plasmid houses the target sequence and the MS2 package. The packaging system gene organization carried on pCP-1 consists of the coat protein gene of bacteriophage MS2 downstream of an IPTG inducible *lac* promoter and restriction sites into which the target sequence of interest has been inserted (Figure 4.5).

This technology has found useful application for manufacturing stable, plasma-based controls that can be subjected to full sample preparation, extraction and amplification for RNA targets (Hietala., *et al*, 2006).

For DNA controls that can be subjected to the full analytic process of extraction and amplification, a synthetic construct based on the bacteriophage Lambda packaging system may be used (Figure 4.6). The DNA bacteriophage vector is constructed by the insertion of a target sequence into Lambda GT11, which also encodes a temperature sensitive Lambda repressor (C1857) which is inactivated at 42°C. The recombinant Lambda genome is used to infect *E. coli* host cells at 30°C. In the presence of active repressor, the Lambda DNA is integrated into the host cell chromosome and replicates along with the host DNA as a prophage. Upon inactivation of the repressor by temperature shift to 42°C, the Lambda DNA is induced to enter a lytic cycle, and the Lambda packaging system produces Lambda bacteriophage containing the recombinant Lambda DNA encoding the inserted sequence. The Lambda phage particles are released from the host cells by lysis and the phage particles are isolated using a series of conventional purification procedures (Yang *et al.*, 2002). This technology has also found useful application for manufacturing stable, plasma-based full process controls for DNA targets in real-time PCR applications (Sum *et al.*, 2004).

The desire to take advantage of the fully automated capabilities of real-time PCR has been a key motivator for the development and production of full process controls. The increased stability of these controls in solution and a virtually unlimited array of RNA and

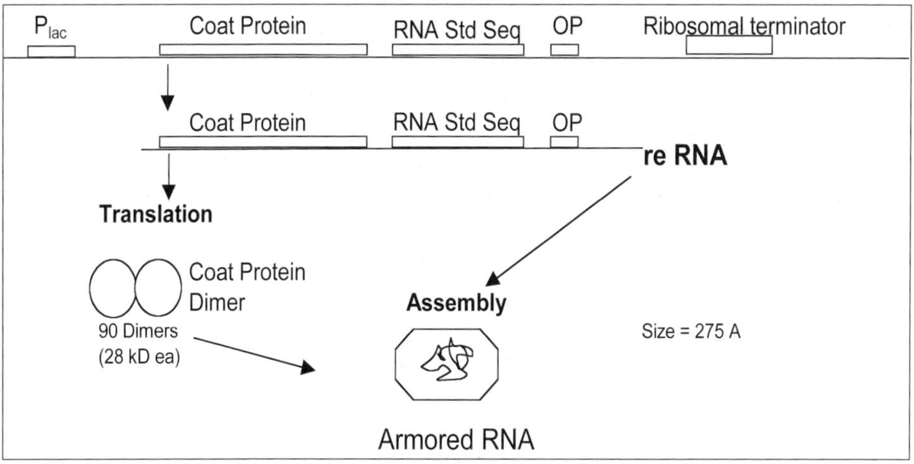

Figure 4.5 Armored® RNAProcessing and Assembly. The RNA bacteriophage MS2 construct, including the cloned RNA target sequence, is under the expression control of the *lacZ* gene promoter. Once induced by IPTG, the RNA is transcribed and, for the coat protein gene, translated. Target sequence RNA packaging occurs by the same mechanism for MS2. In practice, a substantial portion of the Armored® RNA remains in the host, *E. coli* cell and must be released by mechanical or chemical disruption.

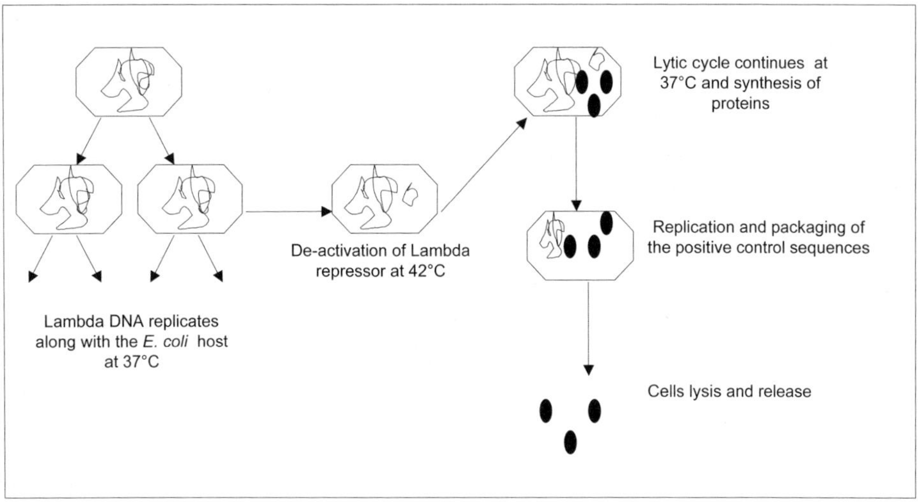

Figure 4.6 Bacteriophage Lambda DNA Packaging. The DNA bacteriophage Lambda construct, including the cloned DNA target sequence, is under the expression control of a temperature sensitive repressor. Once induced by a slight shift in growth temperature (from 37 to 42°C) the bacteriophage engages its lytic cycle and the target sequence DNA packaging occurs. In practice, this process results in substantial lysis of the of host, *E. coli* cell, minimizing the need for mechanical or chemical disruption.

DNA sequence applications, position these reagents as indispensable tools for real-time PCR.

Considerations for preparation of internal and positive controls and calibrators

The laboratory environment

When preparing positive controls, internal and quantitation standards or calibration panels, it is important that the laboratory environment be exceptionally clean and have well-established procedures. Because the synthetic constructs used for preparing the standards and controls are usually highly concentrated (typically $\sim 10^8$ copies/µL or higher) in their stock form, a very real potential exists for cross contamination in the laboratory should mis-handling or spillage occur. Given the robust and stable nature of many of the materials described above (particularly the phage constructs), once contamination occurs, it would be expected to be quite difficult to eradicate fully.

Basic laboratory design considerations should include control of work-flow, ventilation, and air conditioning through filtration devices and proper air pressure balance between adjacent areas. Where it is appropriate, preparation of controls should be conducted within environmentally controlled areas such as Bio-safety cabinets or hoods. If entire room areas are to be used for preparation of larger quantities of real-time PCR standards and controls, it may be useful to consider utilizing an area with HEPA-filtered air to control airborne contamination. Typically, rooms designed to meet ISO Class 8 (Federal Standard 209E Class 100,000) requirements (ISO 14644-1; 1999) will meet most formulation needs. Where high concentrations of RNA/DNA stocks or phage constructs are to be processed or diluted, the handling of such materials within Class II Biological Safety Cabinets (BSC type II) may also provide added isolation from the laboratory environment and minimize the risks of contamination with "positive" synthetic targets into either test samples or internal controls.

Personnel access to areas used for preparation of standards and controls should be limited to minimize the risk of contamination to facilities, equipment and other reagents. Appropriate laboratory procedures should be in place to address personnel attire as well as the flow of staff, materials and waste through these areas. The use of disposable lab coats and gloves should be considered. Only properly attired, authorized personnel who are required for the formulation activity should have access to lab areas dedicated for the preparation of standards and controls. Lab personnel should also receive basic training on cleaning and sanitation concepts, gowning procedures, and equipment and instrumentation operation. Finally, the lab should have a defined cleaning program including sanitizers for floors, walls, doors, and windows. The cleaning agents employed should be shown to be effective in removing or destroying the particular control material in use (e.g. RNA, DNA, Armored® RNA or Lambda targets).

Workflow and separation

Laboratories which prepare their own internal and positive controls need to be very diligent in avoiding cross-contamination between these two materials. Competitive internal controls share the same primer binding regions as their intended target. Therefore, should a positive control, also sharing the intended target binding region accidentally be introduced into its corresponding internal control reagent, it could result in all test specimens appearing as true positive samples. This level of contamination may be obvious. However, if the contamination was to occur at a low level (i.e. approaching the assay limit of detection), such positives may be sporadic and go unnoticed as being unusual unless appropriate negative controls are included in the assay protocols. There also exists the possibility that internal controls, positive controls and calibrators may contaminate other assay reagents such as specimen preparation reagents (diluents and extraction reagents) or amplification reagents (such as "master mix"). Laboratories engaged in the development and use of their own "home brew" real-time PCR reagents need to be particularly alert to the possibilities of cross-contamination of this type. As a preventive measure against contamination, different strategies may be employed. Table 4.1 and Table 4.2 summarize the risk accompanying each type of standard or control for contamination during preparation and potential counter-measures.

Test strategies

Laboratories engaged in preparing standards and controls for their own use must also consider the means by which they will characterize the concentration or copy number of the solution and validate the procedures for routine use. Early experiences with single-target quantitation standards or control formulations containing "naked" transcript RNA

Table 4.1 Control preparation risks and contamination mitigating measures

Control	Contaminated reagent	Production risk	Production risk mitigation
Positive controls	Sample prep reagent (i.e. PCR reaction mix) Internal controls and quantitation standards (QS/IC)	False positive reaction False positive reaction (since QS/IC is added specimen)	Positive control production should be in areas separate from other reagent preparation activities and should have defined flow procedures. Cleaning procedures should include substances known to degrade or neutralize nucleic acids, such as 10% bleach (Prince and Andrus, 1992). Subsequent wiping of areas and equipment with 70% isopropyl alcohol will remove bleach residue and helps control microbial growth. Additional disinfectant solutions should be considered as well for microbial control. Laboratory attire should be used only in positive control preparation areas and access to these areas should be limited/restricted.
Quantitation standards/internal control	Sample prep reagent (PCR reaction mix)	Integrity of assay validation	Same as above.
Full process positive controls (prepared in plasma)	sample prep reagent (PCR reaction mix) QS/IC	False positive reaction	Same as above.
		False positive reaction (QS/IC is added specimen)	Same as above.
		RNAse degradation	Same as above, with additional staff awareness training for RNAse in plasma.
		Significant potential for carry over contamination due to resistance of Armored RNA or Lambda to most cleaning agents	Encapsidated full process positive control production should be distinctly separate. Use of substances known to degrade target (such as 10% bleach) should be used frequently, on all equipment and surfaces. Strict laboratory control procedures.

Table 4.2 Control testing risks and contamination mitigating measures

Control	Contaminated reagent	Test risk	Risk mitigation
Positive control	Sample prep reagent (i.e. PCR reaction mix)	False positive reaction	During testing, positive controls should be handled and added to the sample prep reaction in separate, dedicated hoods or laboratory areas.
	QS/IC	False positive reaction (QS/IC is added specimen)	Dedicated areas should be labeled as pre amplification (only for preparation of PCR reaction mix and specimen), template addition (controls are added/handled), and amplification (area where PCR occurs).
			Dedicated pipettes, disposables, and equipment should also be used. Equipment and supplies used for reagent and specimen prep must not be used for pipetting or processing of amplified DNA or other sources of target.
			Particular attention and caution should be applied when working with highly concentrated materials.
			In the testing laboratory workflow, Positive control handling should occur only when all other reaction tubes are closed or complete.
			Use of 10% bleach or UV lighting (Sakar *et al*, 1991) to decontaminate hood surfaces and pipettes exposed to PC should be considered.
Quantitation standards and internal controls	Sample prep reagent (i.e. PCR reaction mix)	Integrity of assay validation	Same as above

Table 4.2 continued

Control	Contaminated reagent	Test risk	Risk mitigation
Full process positive controls (prepared in plasma)	Sample prep reagent (i.e. PCR reaction mix)	False positive reaction	Same as above
		False positive reaction (QS/IC is added specimen)	
		RNAse Degradation	
		Significant carry over contamination due to resistance of encapsidated RNA or DNA	

or DNA target sequences in buffer have shown that the use of Poisson analysis is an approach that can be quite useful to assign copy number to any dilute solution (Wang and Spadoro,1998). In this approach, a sample is subjected to amplification by PCR and the presence or absence of the target is scored as a simple positive or negative PCR reaction. This method assumes that a solution contains an average number of target molecules per unit volume that are present at small number, such as 1, 2, 3, etc. The method is limited by the sensitivity of the PCR reaction employed. According to the Poisson statistics, the probability of a given unit containing N target copies (P_N) can be calculated as, $P_N = C^N/(N!\ e^C)$ (Wang and Spadoro, 1998). Where N represents the actual number of molecules in a given unit volume. Using this equation, the probability that a given volume can contain 1, 2, 3, etc. copies can be calculated. When diluted to a single copy and repeatedly tested, the probability that a given volume contains 0 copies is equal to the negativity rate. That is, the number of negative results is divided by the number of total tests. The average copy number of this solution can therefore be calculated as:

$$C = -\ ln(N_0/N_T) \qquad 4.5$$

Where, N_0 = the number of negative results and N_T = the total number of tests (Wang and Spadoro,1998). This approach is a *reference-independent* method and does not require the laboratory to pre-define a quantitative relationship between the dilute test sample and a known reference standard. Because of this important attribute, the Poisson method is quite versatile and may be applied to a wide range of situations, particularly where pre-defined reference standards have not been established.

To use this approach in a practical manner, standard or control solutions are diluted to a level yielding approximately 1 copy when added to the final PCR reaction volume (e.g. to 0.02 copies/μL based on a sample volume of 50 μL sample input into a 100 μL PCR reaction). Following the PCR, the rate of positive and negative samples is analyzed as described above and the resulting copy number per PCR is then corrected for the dilution factor to express a final copies/μL result. In order to establish the proper confidence in the resultant copy number, a specific number of replicates is required to obtain the desired accuracy. The number of replicates driving accuracy and confidence should be within the set specifications for each control. Table 4.3 summarizes the confidence level of a particular result at several levels of accuracy based on the Poisson distribution. In a 96-well format (using higher throughput real-time systems based on "wells" in the thermocycler), it is most efficient to include 80 replicates per Poisson test with the remaining wells dedicated to appropriate positive and negative run controls. In this type of set-up, 1 Poisson "test" is equal to 80 replicates (wells or reactions), 2 Poisson tests are equal to 160 replicates and so on. If less replicates are run per test, the confidence in accuracy is proportionally decreased. As a result, one may balance accuracy and confidence (as well as cost and time) by selecting a defined number of Poisson tests. Table 4.3 shows the Confidence Level at a desired Accuracy for a defined number of Poisson tests each with n = 80 replicate wells.

Laboratories who may be considering the flexibility of preparing their own standards and controls, need also consider the time and potential reagent costs required for accurate assignment of concentration by the Poisson method.

Table 4.3 Calculated Poisson confidence levels. Each test incorporates a minimum of n = 80 replicate wells, typically in microwell-plate format. The number of replicate plates used determines the true accuracy at different levels of desired accuracy.

Number of Poisson tests	Desired accuracy (true value ±%)					
	5.0%	7.5%	10.0%	15.0%	20.0%	30.0%
1	26.7%	39.1%	50.5%	69.4%	82.6%	95.3%
2	37.1%	53.1%	66.6%	85.1%	94.4%	99.3%
3	44.6%	62.5%	76.3%	92.2%	97.9%	99.9%
4	50.5%	69.4%	82.7%	95.8%	99.2%	100.0%
5	55.5%	74.8%	87.2%	97.6%	99.7%	
6	59.7%	79.0%	90.5%	98.7%	99.9%	
7	63.3%	82.4%	92.8%	99.2%	99.9%	
8	66.5%	85.2%	94.6%	99.6%	100.0%	
9	69.4%	87.5%	95.9%	99.7%		
10	71.9%	89.4%	96.8%	99.8%		
12	76.3%	92.3%	98.1%	99.9%		
15	81.4%	95.2%	99.1%	100.0%		
18	85.2%	97.0%	99.6%			
21	88.2%	98.1%	99.8%			
24	90.5%	98.8%	99.9%			
27	92.4%	99.2%	100.0%			
30	93.8%	99.5%				
36	95.9%	99.8%				
57	99.0%	100.0%				
105	100.0%					

Two-point bracket method

An alternative approach to the Poisson analysis for concentration determination for samples, standards and controls is the "Bracketing" method. This is also known as a two-point interpolation method. This approach is a *reference-dependent* method that uses well-characterized and quantified secondary or tertiary standards that are placed at concentrations which are above and below the anticipated concentration of the test sample. Conceptually, the approach is quite similar to the two-point calibrations performed with basic laboratory pH meters and two pH reference solutions. The test sample concentration is determined by interpolation from the two known concentration points. As applied to real-time PCR reactions, for one to determine the concentration in this manner, the slope and y-intercept of the line between calibrators is calculated and the concentration of the unknown sample is determined by interpolation.

When utilizing this method in real-time PCR, increased precision is obtained by running several replicates for both the standard and the two points that bracket the standard. Often, visual inspection of the average of the growth curves, with staggered critical thresholds (Ct) at the AFI (average fluorescence intensity level), gives confidence in the expected parallel AFI/RFI (absolute or relative fluorescence increase) response. Since most real-time PCR analyzers allow for runs of fairly large numbers of samples, it is convenient to run n = 24 for both the standard and the two bracket points (Figure 4.7). Because the two-point bracket methods also measures the slope of the line between the two bracket points, this slope value can be used to indicate unexpected Ct delays in both the sample and the bracket (standard) points which could result in an incorrect assessment of standard titer.

To take advantage of this attribute, the linear regression of the two-point bracket line should be defined in advance and used to establish test validity criteria. Factors affecting the reaction growth curve would be expected to produce varying sample Ct values and, as a result, subsequent titer value determination as presented in Figure 4.8. In this example, an unexpected bracket slope value is driven by a delay in the Ct value for the low reference point. This has a subsequent effect on sample Ct determination.

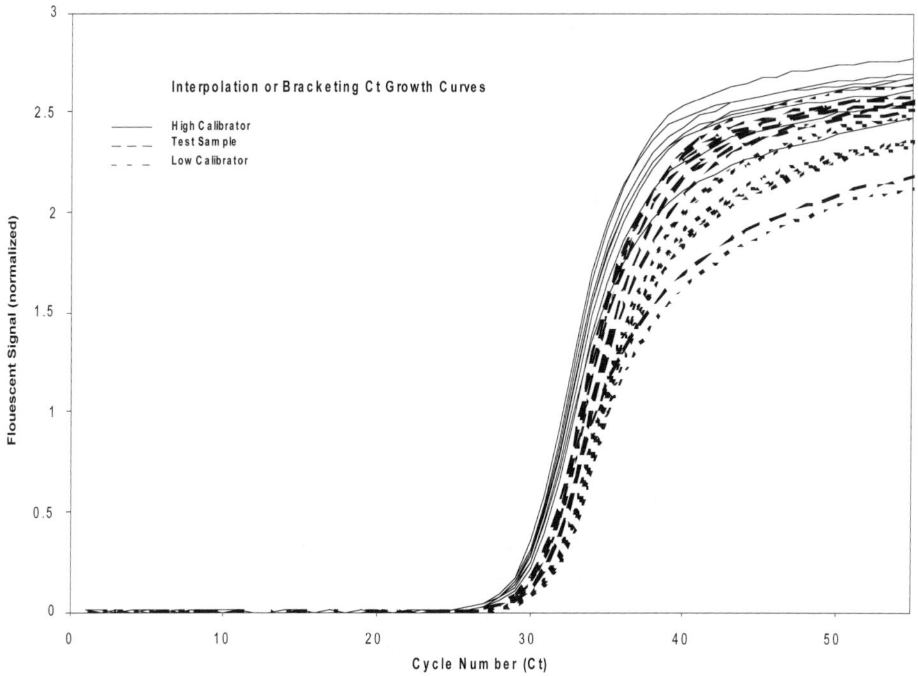

Figure 4.7 Raw Data for Ct Values Measured for Test Sample and Bracketing Calibrators. The mean values of replicate measurements of high and low calibrators (high —, low ···) which "bracket" the replicate measurements of the test sample (– –) are used to determine the unknown test sample concentration from the resulting two-point line (Figure 4.8).

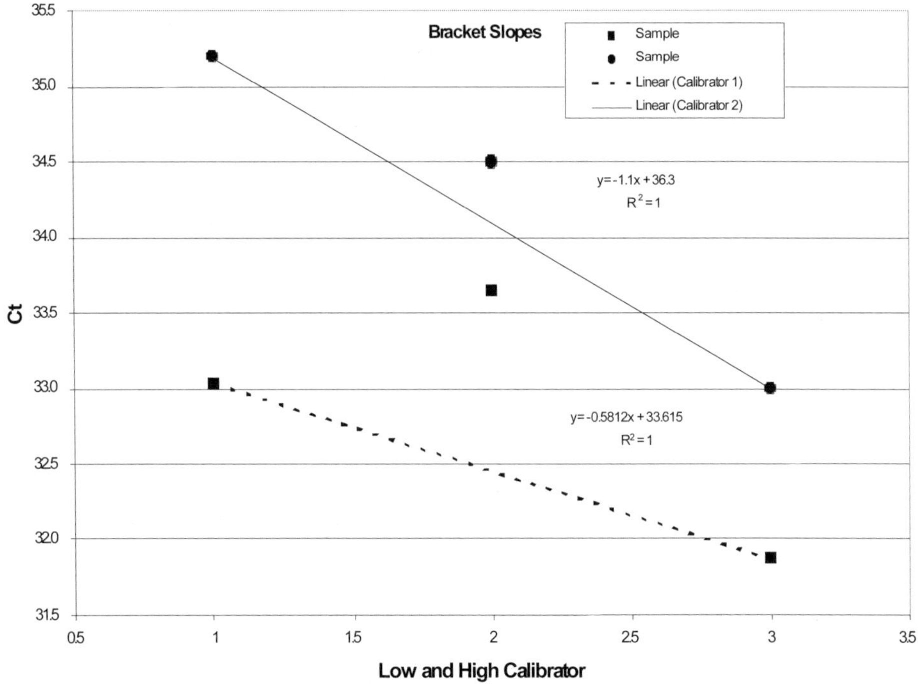

Figure 4.8 Relation of Slope Value and Sample Ct Determination. The selection of two reference points from which to interpolate an unknown sample can influence the accuracy of the measure. The greater slope for Calibrator Set #2 introduces greater variation in the interpolated value than Calibrator Set #1 which has the bracketing calibrator concentrations set closer together. In this example, both test samples (• and ■) would be read off the Calibrator Set #2 (•).

Future trends

The rapid advances in the use of real-time PCR applications that have taken place in the past few years have raised expectation as to what the future may bring for extending even further the capabilities of the technology, especially in the field of molecular diagnostics. As in other fields driven by innovation and technology, it is not unexpected that one will see smaller, faster, and more accurate/more sensitive products. Additionally, the quantity of information that is yielded from real-time PCR reactions will likely increase significantly. It is anticipated that the cost of these materials and products will also be reduced. Continued advances to be seen in the formulation, stabilization and quantitation of assay standards and controls will further contribute to their ability to perform even more reliably. As seen with the recent enhancements seen with encapsidated standards and controls such as Armored® RNA or Lambda constructs, there will be continuing trends towards reagents that are more stable at room temperature for extended periods of time while preserving a high degree of accuracy in the assignment of concentration.

The use of reference standards, calibrators and controls are critical to assuring that valid results are achieved when developing or running real-time PCR assays. Well-characterized reference standards can provide a basis for accurate calibration. With these materials, the analytic result may be relied upon to reflect the true concentration of the target nucleic acid. While up until this point in time, the availability of internationally recognized reference standards has been somewhat limited to a small number of major human viruses (e.g. HCV, HIV and HBV), one should also expect (and demand) a wider diversity of such characterized reference standards will become more readily available.

At present, many large laboratories and manufactures of diagnostic products produce their own reference standards, while the smaller labs tend to purchase reference materials from independent control manufactures (Madej, 2001). This is largely due to the technical complexity inherent in these activities. Many of the technical challenges that arise in producing standards, calibrators and controls which meet the requirements of real-time PCR applications will be addressed and resolved. One such challenge is to produce standard and control materials that perform in the real-time PCR reaction without "matrix effects." These effects, defined as the characteristics of manufactured control or calibrator materials that deviate significantly from the way patient specimens behave in specific procedures, with whatever response characteristics are used for measurement because the source of the difference has not been identified (Clinical and Laboratory Standards Institute, 2005), remain a common technical issue for assay developers. As human plasma is often used as the diluent for the formulation of standards, calibrators and controls, it should be expected to react in the test system in the same manner as the patient specimen reacts. This will pose continuing technical challenges due to the inherent problem of matrix effects often arising from unknown differences between the sample and the standard. In contrast, differences attributable to identifiable sources of interference from endogenous components (e.g. bilirubin, protein etc.) will be continue to be addressed by diligent reagent development and novel formulations. Other aspects to be addressed in the further development of human plasma for use as a standard or control matrix is the extensive pre-screening required to demonstrate the absence of infectious agents required prior to use. The potential for occupational exposure and infection while working with these materials will continue to pose a concern and should continue to be minimized through the use of standard precautions (Centers for Disease Control and Prevention, 1996) and guidelines for prevention of infections in the work place (CLSI, 2005). Also, enhanced procedures for inactivating endogenous agents in human plasma will be beneficial. This will be particularly important with the development of new international reference materials which may contain novel infectious agents.

A clear future trend that will be driven by the technical challenges and issues of using human-sourced materials as the dilution matrix for controls and calibrators will be the development of fully synthetic formulations. While such diluents would still be required to mimic the properties of normal human plasma in terms of nucleic acid target recovery efficiency, in concept, they would obviate the need for precautionary screening and would have the desirable benefit of being predictably consistent from batch to batch. This last attribute is presently lacking with the use of human plasma due to complexity (Anderson *et al*, 2002). With such synthetic diluents, it becomes a more viable option for smaller laboratories to produce their own standards and controls safely and accurately.

As real-time PCR applications continue to expand in scope and complexity, a broader range of laboratory guidelines and procedures will be needed to meet the demands of increasingly more complex test applications and molecular diagnostic systems. Consensus guidelines for the use and evaluation of standards and controls have been previously published (CLSI, 2005), but these will require continuous updating to keep pace with the technology. Guidelines for applications relating to detection and analysis of genotypes, subtypes or serovars for viral or bacterial agents will be beneficial. Such guidelines, in combination with the availability of sequence-matched standards and controls for such applications may help reduce variability or differences in analytic results. As the choice of fluorescent dyes and real-time detection hardware capabilities move towards increasing the number of available signal channels, one can envision that the use of multiple controls with closely related target sequences (perhaps even to the level of single nucleotide polymorphisms or SNPs) for single assays may become the new technology benchmark.

Conclusions

Recent advances in reagent development, hardware and software with multichannel discrimination capabilities have driven the wide adoption of real-time PCR in life-science and diagnostic laboratories. The ability to readily multiplex PCR applications affords the opportunity to incorporate a variety of positive controls and internal controls into the design of an assay for simultaneous detection—enhancing the confidence in both the integrity of specimen transport and preparation as well as confirming the efficiency of the amplification reaction itself. By relating the internal control or quantitation standard fluorescent signal (Ct) to the fluorescent signal of a known primary or secondary reference standard, one can achieve high confidence in the accuracy of measured concentration.

For life science research, diagnostic and microbiological applications, the ability to standardize real-time PCR applications between laboratories can facilitate broader collaborative studies and limit data variation due simply to minor differences in assay performance between labs. This can facilitate data analysis between laboratories and potentially enhance the understanding of the system under study. In molecular diagnostic applications, the same holds true, but it has the added benefit of providing an analytic result that may be suitable for medical decision-making; therapeutic choices and, as a result, have a direct impact on a patient's well-being.

Once developed, a real-time PCR assay should not be assumed to be a static tool. Changes in reagent lots, particularly for the fluorescent oligonucleotide probes, primers and enzymes may result in subtle changes in performance, which, if not corrected, may yield precise but somewhat inaccurate results. Assay calibration and, where available or applicable, participation in external quality assurance programs can monitor and correct for drifts in accuracy. In addition, attention to potential sources of contamination in the standards and controls themselves (e.g. cross-contamination) or contamination of the assay working reagents should be monitored. Adherence to established laboratory practices for segregation, work-flow and cleaning can minimize the potential impact. This is particularly necessary when utilizing some of the more recently developed encapsidated standards or controls (e.g. Armored® RNA or cloned bacteriophage Lambda sequences) due to their greater resistance to RNAse or DNAse degradation.

The future development of more complex multicontrols and standards will be facilitated by a wider bandwidth afforded by future advances in hardware, software and dye chemistry. Synthetic plasma-like formulations for diagnostic applications can be foreseen to improve lot-to-lot consistency as well as reduce the inherent risks associated with human-sourced materials. The resulting increase in information content for single assays, validated by reliable and accurate standards and controls will drive changes in how real-time PCR itself is applied to basic questions in life science and molecular diagnostics.

Acknowledgments

The authors wish to acknowledge the valuable contributions of our colleagues at Roche Molecular Systems, Inc. These dedicated and skilled scientists have developed many of the concepts presented here for the purposes of establishing novel molecular diagnostic tools that are highly reliable and accurate. Our presentation of this information is on their behalf. We are indebted to Zhaung Wang for his advice and input during the preparation of this work.

Web resources

http://www.niaid.nih.gov
Reference sites for funding of research. Subscribe to email alerts. Health and science topics and publications.

http://dorakmt.tripod.com/genetics/realtime.html
The real-time PCR system overview with multiple references and web links.

http://www.ncbi.nlm.nih.gov/entrez/query.fcgi?DB = pubmed
PubMed is a service of the US National Library of Medicine that includes over 16 million citations from MEDLINE and other life science journals. PubMed includes links to full text articles and other related resources.

http://cyclers.gene-quantification.info/
A comparison of the most prominent real-time PCR cyclers is described. The specifications and advantages of the various systems are shown and described.

http://www.eurekalert.org/
EurekAlert is a global news service operated by AAAS. Universities, corporations, and other organizations engaged in research can bring their news to the media with this service.

http://biowww.net
Biowww.net is a site with resources focusing on lab bench work related troubleshooting. Subjects covered include: methods and reagents, discussions in molecular biology, cell biology, genetics, immunology, neuroscience, bioinformatics, and proteomics research.

http://www.biocompare.com/
Biocompare, Inc. connects buyers and sellers of life science products on the internet. This is a media-based marketplace for life science information.

http://www.clsi.org
The Clinical and Laboratory Standards Institute's (CLSI) which was formally know as NCCLS, developments globally applicable voluntary consensus documents for healthcare testing. Numerous documents (standards and guidelines) are available at this site. To search for documents of interest, from the home page, click on the "shop" button.

http://www.who.int/bloodproducts/ref_materials/en/
WHO provides International Biological Reference Preparations which serve as reference sources of defined biological activity expressed in an internationally agreed standard unit. These preparations are the basis for a uniform reporting system, helping physicians and scientists involved in patient care, regulatory authorities and manufacturers to communicate in a common language.

http://www.drugdisc.com/section.asp
This is the news service of the Drug Discovery Technology® & Development World Congress. Coverage on vital topics in drug discovery and development are presented.

http://www.cap.org
This is a resource provided by the College of American Pathologists (CAP). Educational programs and a publication search program are available. Laboratory accreditation and improvement programs are addressed.

http://www.phppo.cdc.gov/CLIA/
This is a link to CDC's division of lab services. There are many hyperlinks to resources about lab quality and proficiency testing. CLIA regulations and links to the Federal Code are listed.

http://www.nibsc.ac.uk/
The National Biological Standards Board (NBSB) run by the UK government is responsible for safeguarding and advancing public health by assuring the quality and safety of biologicals, through its management of the National Institute for Biological Standards and Control (NIBSC).

http://www.virology.net/
This web site has a collection of other web sites all related to information on viruses and subjects of interest to virologists, for example, AIDS and HIV, tutorials, groups/organizations, and emerging viruses.

http://www.fda.gov
The US food and drug administration web site contains a search engine and reference documents on widespread topics of interest to people in health care related professions.

References

Achenbach, J.E., Topliff, C.L., Vassilev, V.B., Donis, R.O., Eskridge, K.M., and Kelling, C.L. (2004). Detection and quantitation of bovine respiratory syncytial virus using real-time quantitative RT-PCR and quantitative competitive RT-PCR assays. J. Virol. Meth. *121*, 1–6.

Anderson, N.H. and Anderson, N.G. (2002). The Human Plasma Proteome. Mol. and cell. proteo. *1*, 845–867.

Beld, M., Minnaar, R., Weel, J., Sol, C., Damen, M., van der Avoort, H. Wertheim-van Dillen, P., van Breda, A., and Boom, R. (2004). Highly sensitive assay for detection of enterovirus in clinical specimens by reverse transcription-PCR with an armored RNA internal control. J. Clin. Microbiol. *42*, 3059–3064.

Brey, B.J., Radcliff, R.P., Clark, D.L., and Ellinson, J.L.E. (2006). Design and Developemnt of an internal control plasmid for the detection of Mycobacterium avium subsp. Paratuberculosis using real-time PCR. Mol. and Cell. Probes *20*, 51–59.

Brightwell, G. Pearce, M., and Leslie, D. (1998). Development of internal controls for PCR detection of Bacillus anthracis. Mol. Cell. Probes. *12*, 367–377.

CAP—College of American Pathologists. (2006). *Surveys and anatomical pathology educational programs catalog.* CAP. Northfield, Ill. USA.

CDC—U.S. Centers for Disease Control and Prevention (1996). Guideline for isolation precautions in hospitals. Infection Control and Hospital Epidemiology. CDC. 17; *1*, 53–80

CLSI—Clinical and Laboratory Standards Institute (2005). Protection of Laboratory Workers from Occupationally Acquired Infections; Approved Guideline—Third Edition. CLSI document M29-A3. Wayne, PA. USA. (Formally know as NCCLS)

CLSI—Clinical and Laboratory Standards Institute (2005). Use of External RNA Controls in Gene Expression Assays; Proposed Guideline. CLSI document MM16-P. Wayne, PA. USA. (Formally know as NCCLS)

CLSI—Clinical and Laboratory Standards Institute (2005). Evaluation of Matrix Effects; Approved Guideline—Second Edition. CLSI Document EP14-A2. Wayne, PA. USA. (Formally know as NCCLS)

CLSI—Clinical and Laboratory Standards Institute (2005). Interference Testing in Clinical Chemistry; Approved Guideline—Second Edition. CLSI Document EP7-A2. Wayne, PA. USA. (Formally know as NCCLS)

Dingle, K.E., Crook, D. and Jeffery, K. (2004). Stable and noncompetitive RNA internal control for routine clinical diagnostic reverse transcription-PCR. J. Clin. Microbiol. *42*, 1003–1011.

Donia, D., Divizia, M., and Pana, A. (2005). Use of armored RNA as a standard to construct a calibration curve for real-time RT-PCR. J. Virol. Methods. *116*, 157–163.

Du, Z.Y., Chen, J.S., and Hiruki, C. (2006). Optimization and Application of a Multiplex RT-PCR System for Simultaneous Detection of Five Potato Viruses Using 18s rRNA as an Internal Control. Plant Disease. *90*, 185–189.

Eisler, D.L., McNabb, A., Jorgensen, D.R., and Isaac-Renton, J.L. (2004). Use of an internal positive control in a multiplex reverse transcription-PCR to detect West Nile virus RNA in mosquito pools. J. Clin. Microbiol. *42*, 841–843.

FDA—United States Food and Drug Administration. (2005). Nucleic acid based In Vitro diagnostic devices for detection of microbial pathogens, Draft guidance for industry and FDA staff. FDA Guidance Document, December 8, 2005. Rockville, MD USA.

Haberhausen, G., Pinsl, J., Kuhn, C., and Markert-Hahn, C. (1998). Comparative study of different standardization concepts in quantitative competitive reverse transcription-PCR assays. J. Clin. Microbiol. *36*, 628–633.

Hashimoto, J.G., Beadles-Bohling, A.S., and Wiren, K.M. (2004). Comparison of RiboGreen™ and 18S rRNA quantitation for normalizing real-time RT-PCR expression analysis. BioTechniques. *36*, 54–60.

Henley, W.N., Schuebel, K.E., and Nielsen, D.A. (1996). Limitations imposed by heteroduplex formation on quantitative RT-PCR. Biochem. Biophys. Res. Comm. *226*, 113–117.

Hietala, S.K., and Crossley, B.M. (2006). Armored RNA as virus surrogate in a real-time reverse transcriptase PCR assay proficiency panel. J. Clin. Microbiol. *44*, 67–70.

Higuchi, R., Dollinger, G., Walsh, P.S., and Griffith, R. (1992). Simultaneous amplification and detection of specific DNA sequences. Biotechnology (NY). *10,* 413–417.

Holman, M.S., Caporale, D.A., Goldberg, J., Lacombe, E., Lubelczyk, C., Rand, P.W., and Smith, R.P. (2004). *Anaplasma phagocytophilum, Babesia microti,* and *Borrelia burgdorferi* in *Ixodes scapularis,* Southern Coastal Maine. Emerg. Infect. Dis. *10,* 744–746.

Holmes, H.C., Davis, C.L., and Heath, A.B. (2003). Report of an international collaborative study to evaluate a WHO HIV-1 RNA genotype reference panel. World Health Organization Expert Committee on Biological Standardization. Geneva, 17 to 21 February 2003.

Kim, B.-G., Jeong, H.-S., Baek, S.-Y., Shin, J.-H., Kim, J.-O., Min, K.-L., Ryu, S.-R., Min, B.-S., Kim, D.-K., Jeong, Y.-S., and Park, S.-N. (2005). Validation of one-step real-time RT-PCR assay in combination with automated RNA extraction for rapid detection and quantitation of Hepatitis C virus for routine testing in clinical specimens. J. Microbiol. Biotechnol. *15,* 595–602.

Kim, J.-O., Shin, J.-H., Baek, S.-Y., Min, K.-I., Min, B.-S., Ryu, S.-R., Kim, B.-G., Kim, D.-K., Ahn, M.-J., Park, M.-K., Song, H.-W., Lee, C.-K., Lee, S.-H., and Park, S.-N. (2004). A collaborative study on Korean standard JE vaccine for potency assay. J. Microbiol. Biotechnol. *14,* 745–750.

Kleinman, S.H., Strong, D.M., Tegtmeier, G.G., Holland, P.V., Gorlin. J.B., Cousins, C., Chiacchierini, R.P. and Pietrelli, L.A. (2005). Hepatitis B virus (HBV) DNA screening of blood donations in minipools with the COBAS AmpliScreen HBV test. Transfusion. *45,* 1247–1257.

Lee, M.A., Leslie, D.L., and Squirrell, D.J. (2004). Internal and external controls for reagent validation. Real-Time PCR, An Essential Guide, K. Edwards, J. Logan, and N. Saunders, ed. (Wymondham, U.K., Horizon Scientific Press), pp. 85–102.

Lefevre, J., Hankins, C., Pourreaux, K., Voyer, H., and Coutlee, F. (2003). Real-time PCR assays using internal controls for the quantitation of HPV-16 and B-globin DNA in cervicovaginal lavages. J. Virol. Meth. *114,* 135–144.

Livak, K.J., Flood, S.J.A., Marmaro, J., Giusti, W., and Deetz, K. (1995). Oligonucleotides with fluorescent dyes at opposite ends provide a quenched probe system useful for detecting PCR product and nucleic acid hybridization. PCR Methods Appl. *4,* 357–362.

Madej, R. (2001). Using Standards and Controls in Molecular Assays for Infectious Diseases. Molecular Diagnosis Vol. 6 No. 4.

Malorny, B.J., Hoorfar, J., Bunge, C., and Helmuth, R. (2003). Multicenter validation of the analytical accuracy of Salmonella PCR, toward an international standard. Appl. Environ. Microbiol. *69,* 290–296.

Marubini, E., Verderio, P., Raggi, C.C., Pazzagli, M., and Orlando, C. (2004). Statistical diagnostics emerging from external quality control of real-time PCR. Internat. J. Biol. Mark. *19,* 141–146.

Morrison, T.B., Weis, J.J., and Wittwer, C.T. (1998). Quantification of low copy transcripts by continuous SYBR Green I Dye monitoring during amplification. BioTechniques. *24,* 954–958.

NCCLS—National Committee for Clinical Laboratory Standards. (2003). *Quantitative Molecular Methods for Infectious Diseases; Approved Guideline.* NCCLS Document mm6-A. Wayne, PA. USA.

NCCLS—National Committee for Clinical Laboratory Standards. (2004). *Protocols for Determination of Limits of Detection and Limits of Quantitation; Approved Guideline.* NCCLS Document EP17-A. Wayne, PA. USA.

Nicolas, L., Milon, G., and Prina, E. (2002). Rapid differentiation of old world *Leishmania* species by LightCycler polymerase chain reaction and melting curve analysis. J. Microbiol. Meth. *51,* 295–299.

Orito, E., Mizokami, M., and Suzuki, K. (1995). Loss of serum HCV at week 4 of interferon-a therapy is associated with more favorable long-term response in patients with chronic Hepatitis C. J. Med. Virol. *46,* 109–115.

Palmer, S., Wiegand, A.P., Maldarelli, F., Bazmi, H., Mican, J.M., Polis, M., Dewar, R.L., Planta, A., Liu, S., Metcalf, J.A., Mellors, J.W., and Coffin, J.M. (2003). New real-time reverse transcriptase-initiated PCR assay with single-copy sensitivity for Human Immunodeficiency Virus type 1 RNA in plasma. J. Clin. Microbiol. *41,* 4531–4536.

Pasloke, B.L., Walkerpeach, C.R., Obermoeller, R.D., Winkler, M., and Dubois, D.B. (1998). Armored RNA technology for production of ribonuclease-resistant viral RNA controls and standards. J. Clin. Microbiol. *36,* 3590–3594.

Prince, A.M. and Andrus, L. 1992. PCR, How to kill unwanted DNA. BioTechniques. *12,* 359–360.

Poon, L.L.M., Wong, B.W.Y., Chan, W.H., Leung, C.S.W., Yuen, K.Y., Guan, Y., and Peiris, J.S.M. (2004). A one step quantitative RT-PCR for detection of SARS coronavirus with an internal control for PCR inhibitors. J. Clin. Virol. *30,* 214–217.

Raggi, C.C., Verderio, P., Pazzagli, M., Marubini, E., Simi, L., Pinzani, P., Paradiso, A., and Orlando, C. (2005). An Italian program of external quality control for quantitative assays based on real-time PCR with TaqMan® probes. Clin. Chem. Lab. Med. *43*, 542–548.

Saldanha, J., Heath, A., Aberham, C., Albrecht, J., Gentili, G., Gessner, M. and Pisani, G. (2003). WHO Collaborative Study to Establish a Replacement WHO International Standard for HCV RNA NAT Assays. World Health Organization Expert Committee on Biological Standardization. Geneva, 17 to 21 February 2003.

Saldanha, J., Lelie, N., and Heath, A. (1999). Establishment of the first international standard for nucleic acid amplification technology (NAT) assays for HCV RNA. Vox Sanguinis. *76*, 149–158.

Saldanha, S., Lelie, N., Yu, M.W. and Heath, A. (2000). An International Collaborative Study To Establish A WHO International Standard for B19 DNA NAT Assays. World Health Organization Expert Committee on Biological Standardization. Geneva, 30 October to 3 November 2000.

Sarkar, G., and Sommer, SS. (1991). Parameters affecting susceptibility of PCR contamination to UV inactivation. Biotechniques *16*, 590–594.

Saunders, N.A. (2004). Real Time PCR An Essential Guide. In Real-Time PCR, Quantitative Real-Time PCR (Wymondham, U.K., Horizon Bioscience) pp. 103–123.

Schwaiger, M., and Cassinotti, P. (2003). Development of a quantitative real-time RT-PCR assay with internal control for the laboratory detection of tick borne encephalitis virus (TBEV) RNA. J. Clin. Virol. *27*, 136–145.

Stonehouse, N.J., Valegard, K., Golmohammadi, R., van den Worm, S., Walton, C., Stockley, P.G., and Liljas, L. (1996). Crystal structures of MS2 capsids with mutations in the subunit FG loop. J. Mol. Biol. *256*, 330–339.

Sum, S.S.-M., Wong, D.K.-H., Yuen, M.F., Yuan, H.-J., Yu, J., Lai, C.-L., Ho, D., and Zhang, L. (2004). Real-time PCR assay using molecular beacon for quantitation of Hepatitis B virus DNA. J. Clin. Microbiol. *42*, 3438–3440.

Tettelin, H., Radune, D., Kasif, S., Khouri, H., and Salzberg, S.L. (1999). Optimized Multiplex PCR, Efficiently Closing a Whole-Genome Shotgun Sequencing Project. Genomics. *62*, 500–507.

Ullmannova, V., and Haskovec, C. (2003). The Use of Housekeeping Genes (HKG) as an Internal Control for the Detection of Gene Expression by Quantitative Real-Time RT-PCR. Folia Biologica. *49*, 211–216.

Verbraeken, E., and Fiers, W. (1972). Studies on the bacteriophage MS2 XX. Expansion of the virion in low salt. Virology *50*, 690–700

Wang, Z., and Spadoro, J. (1998). Determination of target copy number of quantitative standards used in PCR based diagnostic assays. In Gene Quantificaction, F. Ferrer, ed. (Birkhauser, Boston) pp 31–43.

Watzinger, F., and Lion, T. (1998). Multiplex PCR for quality control template RNA/cDNA in RT-PCR assays. Leukemia. *12*, 1983–1993.

Wittwer, C.T., Hermann, M.G., Moss, A.A., and Rasmussen, R.P. (1997). Continuous fluorescence monitoring of rapid cycle DNA amplification. BioTechniques. *22*, 130–138.

Wu, R. (1978). Annual Reviews. Ann. Rev.Biochem. *47*, 607–634.

Yang, Q., and Catalano, C.E. (2003). Biochemical characterization of bacteriophage Lambda genome packaging *in vitro*. Virol. *34*, 276–287.

Zimmerman, K., Reiger, M., GroB, P., Turecek, P.L., and Schwarz, H.P. (2000). Sensitive single-stage PCR using custom synthesized internal controls. BioTechniques. *28*, 694–702.

Quantification of Microorganisms: not Human, not Simple, not Quick

5

Ian M. Mackay, Stephen A. Bustin, José Manuel Andrade, Mikael Kubista, and Theo P. Sloots

Abstract

Most real-time PCR applications in microbiology are for qualitative (resulting in a yes or no answer) detection of a microorganism. In terms of disease relevance, the importance of qualitative PCR to microbiology has been proven, however it is less clear just how punctilious the clinical microbiology laboratory must be to produce relevant quantitative PCR (qPCR) results. Despite a decade of qPCR experience, commercial development of applications is limited and many of our approaches remain entrenched among the PCR techniques used to monitor human mRNA levels rather than addressing adequately the diverse needs of the microbiology field. Real-time PCR has permeated every aspect of microbiology, but its applications have particular value in the clinical microbiology laboratory where the speed, sensitivity, reproducibility and accuracy of this tool help to produce robust data in a clinically relevant timeframe. Other areas within microbiology have also gained from the use of real-time PCR; e.g. gene therapy has found benefit from real-time qPCR applications that monitor the production, replication and administration of viral vectors used to transport therapeutic genes into host cells or tissues. Studies of the host's response to microbial replication suggest a vision of the future wherein patient specimens may be used to provide an indication not only of the type of microorganism present and its replicative status, but also the stage of disease and the type of immune response underway. To make such vision reality, we must first discuss and reach consensus on the best, microbiology-specific real-time qPCR approaches to permit the production of comparable microbial load data. This process must include the development of clear definitions associating microbial load with clinical outcome, the production of more reference materials, the development of more quality assessment schemes and of commercial kits. It may be that we find the perfect estimate of microorganism numbers is not as important as reproducible and clinically relevant data. The increased identification of newly emergent or previously unknown endemic pathogens demands that we must strive harder than ever to expand our understanding of infectious diseases, and for that we need reliable results from reliable tools.

Introduction

The majority of real-time PCR applications in microbiology are for qualitative (resulting in a yes or no answer) detection of a virus, bacterium, fungus or parasite. However the technique also permits quantification of target nucleic acids for both pure research purposes and for a myriad of applications in the clinical microbiology laboratory. While much more user-friendly than earlier conventional competitive quantitative PCR (qPCR), quantification by real-time PCR retains many time-consuming requirements that must be addressed in the context of microbiology in order to produce valid data. A direct relationship is assumed to exist between the amount of microbial nucleic acid and the number of microorganisms in a patient's specimen; minimizing amplification variability while correcting for that which remains can strengthen this association (Ferré, 1992).

Real-time PCR has simplified our ability to visualize the impact of each assay variable, thus permitting improved optimization, standardization and normalization of qPCR assays. It is then disappointing that despite a decade of qPCR experience, commercial development of applications is limited and many of our approaches remain entrenched in the techniques used to monitor human mRNA levels, rather than adequately addressing the diverse needs of the microbiology field.

Why is it important to quantify microorganisms?

The most clinically important reasons for quantifying an infecting microorganism are to permit the medical practitioner rapid insight into the efficacy of an ongoing antimicrobial therapy and to identify when treatment should commence. Should a treatment prove ineffective, as identified by stable or increasing microbial concentration or "load," then the clinician has rapid access to objective data which may aid the selection of an improved therapeutic course or the adaptation of the current approach, thus tailoring treatment to best suit the microbiological and clinical picture.

Another application that benefits from quantifying microbial load data is the monitoring of patients undergoing organ transplantation who are subject to extensive immunosuppression. Treatment of this group can be quickly deployed in response to change in the status of low-level latent or persistent infections. Quantification also has the potential to clarify the extent of virus–host interactions and to propose roles for pathogenicity (Clementi, 2000; Limaye *et al.*, 2000). In addition, the microbial load can frequently be related to disease progression in immune-competent subjects for example by detecting an increase in pathogen numbers preceding clinical symptoms. However, early doubts were cast upon the simplicity of the assumptions made about this relationship (Watzinger *et al.*, 2004). Relying solely upon qPCR has again been called into question for the quantification of HIV-1 because of discrepancies between results from nucleic acid and other diagnostic approaches (Rodríguez *et al.*, 2006).

Collection and testing of serial patient samples has an important role for all of these qPCR applications because compared to testing a single sample, it allows the gathering of more robust data and can reveal clearly visible trends caused by changes in microbial load over time. In the research laboratory, qPCR is used to corroborate microarray data and can determine the successful *in vitro* propagation of virus during attempted isolation using cell culture. The latter role is especially important for discriminating between viral replication and residual inoculum when working with molecularly identified viruses that produce little or no cytopathic effect and/or for which serological reagents do not exist (Schildgen *et al.*, 2006).

While the importance of qPCR seems apparent, the conditions required to produce relevant quantitative results in the clinical microbiology laboratory are less well defined. How pedantic do we have to be? Historically, *in vitro* culture has provided a clear-cut qualitative method indicating success or failure. Culture can be adapted to quantify the microbe but it is a time-consuming process that likely under-represents the number of microorganisms in the original specimen. The serological methods introduced the "equivocal" result and made it commonplace, complicating the diagnosis of infectious disease. Serology is generally employed to quantify the host's antibody response to an invading microorganism. DNA techniques produce data that can be either clear-cut or equivocal, depending on the microorganisms and clinical situation. Perhaps the question of pedantry can be addressed by also asking whether the mere presence of an invading microorganism is suitably diagnostic or whether the amount of the putative pathogen is a better indication of clinical outcome. Consensus on the answer to this question for many microorganisms could go a long way toward simplifying the development of real-time qPCR techniques for microbiology.

What's already on the shelf?

Several benefits arise from employing commercial kits for qPCR in the clinical microbiology laboratory; principal among these is doing away with incompatible in-house methods and the resulting improvement in data comparability between laboratories around the world. The introduction of quantitative molecular testing was first commercialized for only a handful of microorganisms by providers including Roche Diagnostics, Bayer Diagnostics and bioMérieux (formerly Organon Teknika; listed in (Schutten and Niesters, 2001; Niesters, 2002; Watzinger *et al.*, 2006)). Targets included *Chlamydia trachomatis*, cytomegalovirus (CMV), hepatitis B and C viruses (HBV and HCV) and the human immunodeficiency virus-1 (HIV-1). These analyte specific reagents (ASRs) have generally been upgraded alongside developments in laboratory systems and PCR detection formats. ASR kits are also available from artus™ (now produced by QIAGEN™) with additional targets including parvovirus B19, influenza A virus, influenza B virus, enteroviruses, West Nile virus, orthopoxviruses, dengue viruses, additional human herpesviruses (herpes simplex viruses (HSV) 1 and 2 and varicella zoster virus), the SARS coronavirus, *Bacillus anthracis* virulence plasmids, *Borrelia burgdorferi sensu lato*, *Salmonella* serovars, *Listeria monocytogenes*, *Mycobacterium tuberculosis* complex and the malaria parasites *Plasmodium falciparum*, *P. vivax*, *P. ovale* and *P. malariae*. However, despite a description to the contrary, not all of these kits include a quantitative capacity; some contain neither a quantification calibrator nor the protocol required to perform true quantification.

Change the paradigm, change the field

There are still some glaring omissions from the list of kits and from qPCR studies in the research literature. Why are many common microbial infectious agents not the subject of comprehensive investigation by real-time qPCR? The answers are manifold and include the real and perceived technical difficulties of developing these assays, the new costs, the lack of commercial support and the ingrained belief that quantification is only for "special" microorganisms. On this last point there is no question that the quantification of some microorganisms is clinically unwarranted (although there are always reasons for research studies); the mere presence of Ebola virus or variola virus infection is sufficient to enact a clinical regimen of support and treatment because persistent or subclinical infections are not common features of infection. However for other targets, interpreting the qPCR data may be extremely difficult because of the often innocuous presence of the organism in the human host, e.g. *Escherichia coli, Streptococcus pneumoniae, Neisseria* spp., HSV or CMV. Our understanding of the remaining microbial targets which have rarely if at all been the subject of qPCR studies could benefit from good quality quantitative research studies to better understand the role each pathogen plays, especially in illnesses not commonly associated with infection and whether they circulate among asymptomatic populations. Several respiratory viruses including the picornaviruses fall into this category since they are sometimes detected in patients lacking overt clinical signs of illness. Microorganisms causing treatable infections are generally acknowledged as the most worthy qPCR targets, leaving many of the most common (and frequent) viral pathogens that are serviced by relatively few therapeutic options, without any commercial alternative for quantification. It will be important for future studies to challenge infectious disease paradigms using real-time qPCR to ensure our beliefs hold up to close scrutiny.

How does real-time qPCR improve upon previous qPCR methods?

Real-time PCR offers some immediate and significant improvements over previous conventional PCR methods for the quantification of microorganisms. Principally the closed tube or "homogeneous" nature of real-time PCR has been a boon to the clinical microbiology laboratory by dramatically reducing the risk that amplicon will contaminate subsequent assays. The technique can also be used to amplify efficiently and representatively an enormous range of template concentrations without affecting assay efficiency, making it a dramatic improvement over earlier qPCR methods. This "dynamic range" extends to at least eight $\log_{10}$ of template molecules (Ryncarz *et al.*, 1999; Abe *et al.*, 1999).

The versatility of real-time qPCR is made possible because its fluorescence data are collected from the linear phase of the exponential PCR. In this region conditions of DNA duplication are optimal, rather than in the end-point or "plateau" phase where the amount of amplicon present has little relationship with the original template concentration. End-point amplicon quantities are especially affected by conditions of saturation by inhibitory PCR by-products and specific double-stranded DNA (amplicon). The fractional cycle at which fluorescence crosses a defined y-axis value (see Chapter 1 for more detail) is known as the threshold cycle (C_T). The C_T values from amplification of an unknown specimen extract can be compared to similar data generated by the amplification of a dilution series of standards or calibrators after such data are presented as a standard or calibration curve.

Collecting data from the linear phase of the reaction also makes possible the detection of an equally broad dynamic range of amplicon; the lack of such scope was a significant cause for additional amplicon handling among methods predating real-time PCR (Figure 5.1). Dilution of an amplicon is no longer required before detecting it nor is repetition of an assay using a diluted template required because a preliminary result fell above the upper limits of the detection assay. Both of these problems were common to conventional end-point qPCR assays because the detection formats could not encompass amplicon

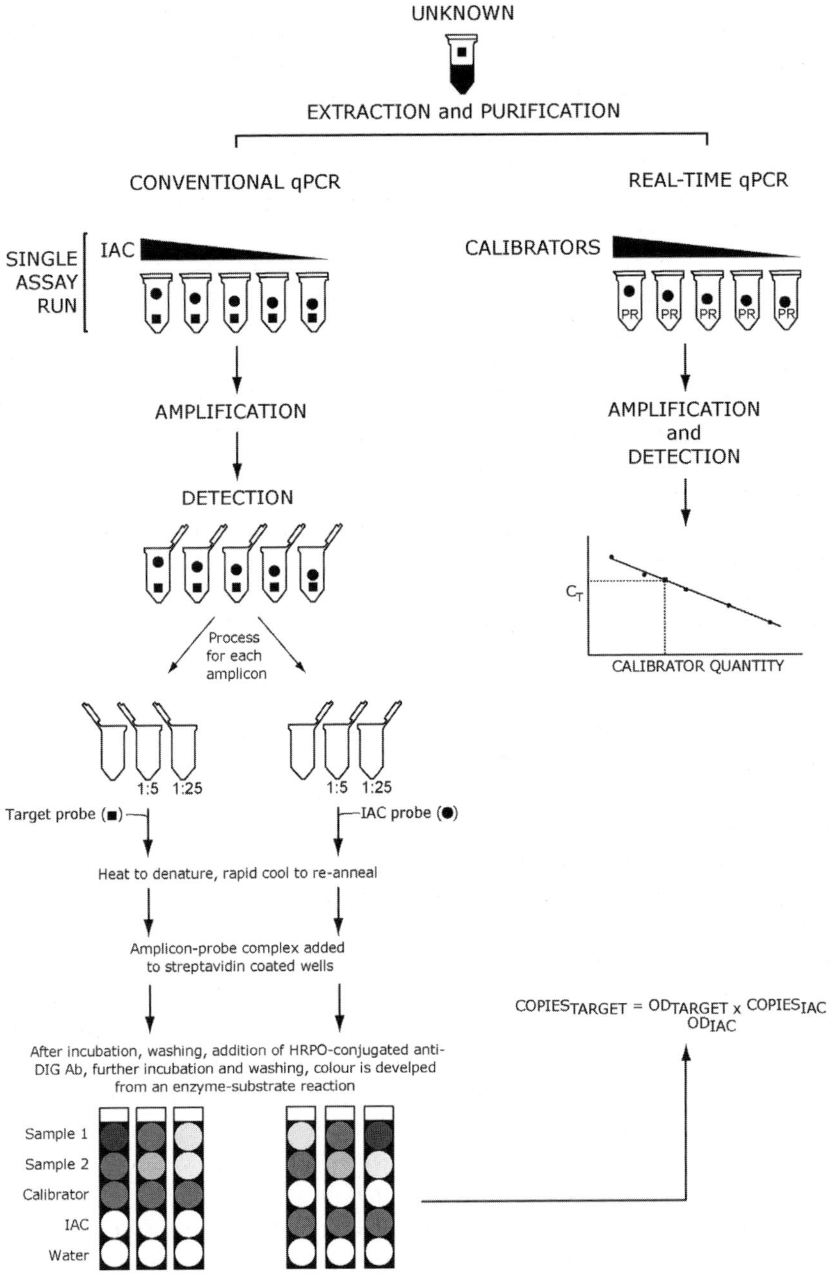

Figure 5.1 A simplified comparison of the complexities of conventional, competitive qPCR versus real-time qPCR methods. Competitive qPCR required the amplification of a fixed amount of target (filled circle) and a dilution series of a characterized competitive internal amplification control (IAC; filled square). The two molecules were then discriminated after the PCR was completed. Using a PCR-ELISA approach, the highest amplicon dilution that gave an optical density (OD) within a pre-established range provided the data for the equation shown. Antibody, Ab; digoxigenin, DIG; horse-radish peroxidise, HRPO; passive reference, PR; quantitative PCR, qPCR

from high template loads whilst maintaining an adequate template sensitivity (Kawai *et al.*, 1999; Schaade *et al.*, 2000; Weinberger *et al.*, 2000; Brechtbuehl *et al.*, 2001). However, the reliance on C_T data evokes one detrimental outcome: the impact from inhibition is enhanced when compared with an end-point quantification approach. Poorly optimized conditions or amplification inhibitors (other than those caused by the amplification process itself) delay amplification and this has a greater impact on the C_T values than on the final amount of amplicon since final amounts often remain unchanged after 30 to 45 cycles of PCR.

Real-time PCR can also exhibit low inter-assay and intra-assay variability and since it is a PCR method, it is also extremely sensitive (Abe *et al.*, 1999; Locatelli *et al.*, 2000; Schutten *et al.*, 2000). However, one could argue that only the latter is an innate feature of real-time PCR and that the remainder requires suitable reference materials, regular quality control and protocols that are well developed and strictly adhered to. Nonetheless real-time PCR permits template quantification, of some sort, from a wide variety of sample types which contain a large yet unpredictable range of template concentrations, as is often the case in samples collected for microbiological investigations. It is a superior tool well suited to the quick decision making required in a clinical situation since time-to-result or "turnaround time" is reduced (Locatelli *et al.*, 2000; Tanaka *et al.*, 2000).

Before real-time PCR, the most objective and accurate approach to quantification was the analysis, following co-amplification (two templates amplified in the same vessel; Figure 5.1), of a competing nucleic acid template of known concentration along with the "wild-type" target nucleic acid of unknown concentration (Gilliland *et al.*, 1990; Siebert and Larrick, 1992; Clementi *et al.*, 1995; Orlando *et al.*, 1998). While the molecule of known concentration is not always employed as a competitor its amplification absolutely depends on the same reagents as the microbial template. Since there are generally more similarities than differences we employ a single title for these molecules which helps simplify further discussion. We refer to these molecules as internal amplification controls (IACs; discussed below); however, they have been used in several roles and known by many names (including internal control, internal standard, competitive standard, internal calibrator or a mimic) indicating the distinguished history of an IAC in the PCR detection of microorganisms (Wang *et al.*, 1989; Niesters, 2002).

The IAC was the driving force behind the development of competitive qPCR and worked most efficiently in this role when it was of almost identical sequence to the assay's main target (Pannetier *et al.*, 1993). When both the IAC and unknown target were amplified together for the purpose of qPCR, the amount of the unknown template was determined from a comparison of signals. This approach commonly required several reactions containing a range of IAC concentrations to accommodate the detrimental effects of excessive competition if one template was more abundant than the other. Each reaction contained a different amount of IAC but a constant volume of the unknown template. Alternatively one could perform a pre-qPCR reaction to obtain a crude estimate of the template load and use only one or two similar IAC concentrations.

Despite similar results when comparing data generated either by absolute calibration curves, relative calibration curves or C_T values (Johnson *et al.*, 2000), competitive PCR has not been extensively examined for quantification using a real-time PCR format (Halford, 1999; Halford *et al.*, 1999). The absence of software with the ability to automatically calculate the concentration of an unknown by comparing the signals generated during co-amplification makes the development of real-time competitive qPCR applications difficult (Kleiber *et al.*, 2000; Pfaffl *et al.*, 2002). In the only application published to date, competitive amplification successfully defined the amount of nodavirus RNA in cell culture samples (Grove *et al.*, 2006). The method was reportedly less sensitive than a non-competitive version (presumably because of competitive inhibition), required the use of four aliquots of the sample extract and did not monitor reverse transcriptase (RT) or amplification efficiencies. An earlier experimental approach employed a wild-type and two co-extracted, co-amplified heterologous plasmid templates employing an algorithm to determine the number of target copies (Gruber *et al.*, 2001). While this study did not include a normalization step, the dual template amplification correlated over four $\log_{10}$ of template. The second IAC could not be simultaneously detected by the software of the time.

While not surprising, the point must be reiterated that PCR data cannot be related to microbial viability unless preceded by a biological isolation or enrichment step (Josefsen *et al.*, 2004; Nitsche *et al.*, 2006). While real-time qPCR is useful to monitor changes in the amount of a microorganism's genome or gene products, it must be used in concert with biological methods, at least during the developmental process, if one wants to assign biological relevance to the results.

Optimization and consistency are critical for obtaining reproducible results yet real-time RT-PCR assays are capable of significantly less variation than conventional RT-PCR protocols which can be subject to significant error (Souazé *et al.*, 1996). In principle, quantification of RNA by real-time RT-qPCR is straightforward: the more copies of an RNA molecule there are at the beginning of the assay, the fewer cycles of amplification are required to generate a specific number of products (Figure 5.2). Therefore, the C_T is inversely correlated to the amount of RNA in the original sample.

The potential to quantify nucleic acids using PCR has proven most popular in the field of gene expression or transcriptome analysis (PhorTech, 2003) but it has also been employed in microbiology since the 1990s (Zhang *et al.*, 1991). Because comprehensive lists of applications in microbiology have recently been published elsewhere (Espy *et al.*, 2006; Watzinger *et al.*, 2006), the remainder of this chapter will instead seek to critically examine the impact of reverse transcription on real-time qPCR assay performance, describe the mathematical approaches employed to obtain quantitative data, discuss the role of quality control and assurance, identify some sources of reference material and review two uncommon uses of real-time PCR quantification in the multidisciplinary field of microbiology.

Calibrating the process of microbial quantification

Although the terminology is often confused, real-time PCR does not inherently imply qPCR. Thus no discussion of nucleic acid quantification can ensue without first mentioning the necessary inclusion of the correct positive and negative controls to determine assay function and the relevant reference standards (ideally of biological origin as per terminology employed by the International Organisation for Standardisation) or calibrators (derived from a standard and by far the more commonly used molecule for in-house qPCR) required to generate accurate data. The externally amplified calibration curve (see Figure 5.1 and 5.2) relies upon titration and amplification of an identical template, performed within

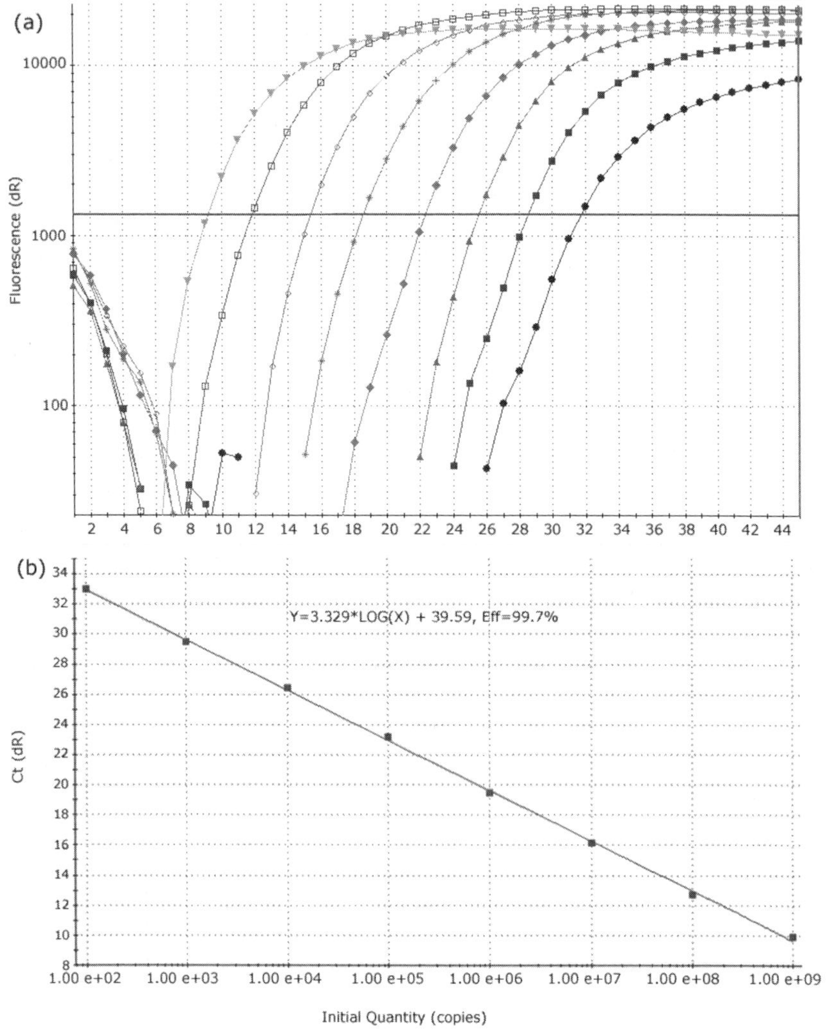

Figure 5.2 (a) Amplification plot for a standard or calibration curve. Sense-strand oligodeoxynucleotides were serially diluted from 1×10^9 to 1×10^2 molecules and subjected to real-time RT-PCR assays on a Stratagene MX3005P. The software calculated the threshold cycle (C_T), which is defined as the number of PCR cycles where the fluorescence generated from the amplification plot crosses a defined fluorescence threshold (bold horizontal line). It depends on the sensitivity of the detection system and can vary significantly depending on assay-specific background levels. This provides several orders of linear dynamic range, in this case seven $\log_{10}$s. The graph of normalized fluorescence vs. cycle number during PCR has three stages. Initially, the fluorescent signal is below the detection limit of the signal detector. In the second stage, the signal can be detected as it continues to increase in direct proportion to the increase in the products of PCR. As PCR product continues to increase, the ratio of polymerase to PCR product decreases, product ceases to grow exponentially and the signal is roughly linear and finally reaches a plateau. (b) A plot of C_T against the log of initial oligonucleotide copy number results in a straight line that is linear over at least seven orders of magnitude and linear regression analysis permits the calculation of the copy number of any unknown target relative to that standard curve. The standard curve also reveals the amplification efficiency of the reaction (slope) and gives some indication of its sensitivity (y-intercept).

the same experimental run but in separate vessels to the unknown target. However, the calibration curve can itself serve as a source of error when inter-tube fluorescence variation is not controlled. This, as with similar omissions throughout the qPCR process, is usually a trade-off between two realities: (i) obtaining some form of quantitative data in a clinically useful timeframe (while under the stresses of a high throughput laboratory environment) and (ii) the need to address multiple sources of error to achieve suitable quantitative data. Suitability is usually guided by the laboratory or its clients and sometimes by participation in quality assessment programs. Clinical relevance may ultimately be a better guide.

No matter what calibrators and controls are used to determine the presence of inhibitors, monitor successful amplification or generate quantitative data, it is imperative to accurately determine their concentration to ensure that they are used at suitable levels. A spectrophotometer alone is inadequate for quantifying these molecules; however in combination with an experimental and statistical analysis, the reliability of the data can be greatly enhanced (Taswell, 1981; Glasel, 1995; Bagnarelli *et al.*, 1995; Rodrigo *et al.*, 1997; Wang and Spadoro, 1998; Sykes *et al.*, 1998). Determination of small *in vitro* synthesized single-stranded DNA or RNA molecule concentrations may require the aid of a fluorescence-based signal amplication system such as PicoGreen® or RiboGreen® (Invitrogen).

When employed, calibrator and control templates should be handled as identically as possible to the test samples. Practically this means that the sample matrix (for example serum, urine or bronchoalveolar lavage) should be used to provide the "background" into which the calibrator is added prior to nucleic acid extraction and purification. In some cases it is not possible to obtain matrix in a standardized form from negative samples e.g. precise collection of a nasal swab or bronchoalveolar lavage is extremely difficult; one may weigh the swab after collection although this cannot discriminate between patient cells and absorbed fluids (Maggi *et al.*, 2003).

Careful development and implementation of calibrators will permit better standardization and ensure that comparison between assay results and testing locations is possible (Niesters, 2001). Without question, the reliability of any qPCR methods is intimately associated with the choice and quality of the assay controls (Alexandre *et al.*, 1998; Celi *et al.*, 2000).

The mathematical mechanics of quantification

Unfortunately for those of us less interested in this aspect of qPCR, mathematics make the method. One cannot escape that useful qPCR data requires some understanding of how the results are obtained. In every PCR cycle the nucleic acid template is copied, ideally giving rise to a doubling of the number of template molecules in the reaction vessel. The doubling process is described by (Kubista *et al.*, 2006):

$$N = N^0 2^x \tag{5.1}$$

N is the number of template molecules after X amplification cycles and N^0 is the initial number of double stranded template molecules. This equation is based on four assumptions discussed below.

Assumption 1: Initial template is double-stranded

Eukaryotic and most prokaryotic genomes are double stranded. But many viral genomes are single-stranded. Also, when analyzing RNA, reverse transcriptase protocols produce single-stranded cDNA for PCR. The single stranded DNA is then converted into double-stranded DNA in the first PCR cycle, thus true amplification starts in cycle two. Hence, when template is single-stranded the doubling process is described by (Kubista *et al.*, 2006):

$$N = N^0 2^{(x-1)} \tag{5.2}$$

As before, N is the number of double-stranded molecules present after × cycles, and N^0 is the initial number of template molecules, but in this case they are single-stranded. The correction for this extra cycle, which has been missed in many publications, often cancels out, as we will see below. However, in some quantification the omission can introduce a two-fold error.

Assumption 2: PCR efficiency is 100%; both strands of all template molecules are copied into full length products each cycle.

Experience shows that PCR efficiency almost never reaches 100%. Even the complete copying of both strands of short, purified amplicon templates rarely exceeds an efficiency of 95%. This reduced efficiency, in the absence of inhibition, is due to inefficient primer hybridization, template folding and probe and dye interference. It can sometimes be improved by redesigning the oligonucleotides to reduce mispriming or formation of folded structures, by increasing primer and/or template concentrations to push equilibrium towards the primed state (although, this can also cause primers to form primer–dimer products; (Lind *et al.*, 2006)), by adjusting annealing temperature, buffer conditions (in particular concentrations of K^+ and Mg^{2+}), thermostable DNA polymerase concentration, by changing or reducing the length of the amplicon or reducing the dye or probe concentration (Kubista *et al.*, 2001; Bengtsson *et al.*, 2003). In biological samples the PCR efficiency is further reduced by concomitants in the sample matrix which have not been removed during the nucleic acid extraction process (Akane *et al.*, 1994; Al-Soud and Rådström, 1998; Al-Soud *et al.*, 2000). In particular, matrices containing high amounts of lipids tend to be inhibitory. The following equation accounts for PCR inhibition:

$$N = N^0 \left(1 + E_{ts}\right)^{(x-1)} \tag{5.3}$$

E_{ts} is the PCR efficiency and it has a value $0 \le E_{ts} \le 1$. E_{ts} has two subscripts indicating that it depends on both the target (t) assay and on the sample (s). E_{ts} can be estimated *in situ* using the methods of serial dilutions of or standard additions to the test sample itself (Ståhlberg *et al.*, 2003). A standard or calibrator curve is produced from the data acquired and the PCR efficiency is estimated. Since the calibrator data are all based on the same single test sample these approaches give the best estimate of its PCR efficiency. The standard additions approach uses the specimen under investigation to provide the matrix for the calibrator or standard. The sample is divided into aliquots that are spiked with increasing amounts of the standard in pure form. One aliquot is left without standard

(Massart *et al.*, 1997). The method has some limitations including a lack of evidence that the added standard will behave the same way as the natural target, increased use of precious sample and the costs of extra reagents and time. Nonetheless, it is presently the best method we have to prepare standards that resemble the true sample matrix. The method of serial dilutions is analogous. The test sample is diluted serially, for example 4-fold, and extracts from the dilutions are analyzed by (RT) real-time PCR. A standard curve is produced based on the data and the PCR efficiency for the particular sample is estimated. Another possibility to estimate E_{ts} is to add purified template to target-free sample matrix (i.e. a negative sample) and use it to construct a calibration curve.

For practical purposes it is helpful to factorize E_{ts} into a contribution that is target-specific, E_t, and one related to sample matrix-specific inhibition, E_s:

$$E_{ts} = E_t \times E_s \tag{5.4}$$

The target specific contribution E_t is obtained from a calibration curve based on purified template in target-free matrix. E_s depends on the sample and can be estimated using an alien spike or endogenous control (see below). Sometimes an average value of $\overline{E}_t$ is assumed for all samples.

Assumption 3: PCR efficiency is constant throughout the amplification process.

The equations above assume that PCR efficiency is constant during the course of the reactions. This may not be the case. If the original template is long e.g. genomic DNA, it may form folded structures that amplify poorly. If the template is supercoiled, which is the case for plasmids, mitochondria, bacterial genomes and circular viruses then PCR efficiency is significantly lower than for the short template molecules that are produced during the course of the PCR (Nogva and Rudi, 2004). The effect of the length and structure of the initial template can be reduced by pre-analytical treatment of the sample with nucleases to remove supercoils or to excise the target region. Another possibility is to compare with calibrators based on the same starting material.

Assumption 4: Fluorescence is proportional to the amount of template

In real-time PCR we are not measuring the number of amplicons directly, but rather we measure the intensity of the emitted fluorescence as it increases with the number of amplicons formed. This fluorescence depends on the dye or probe used, the amplified sequence (Nygren *et al.*, 1998), the length of the amplicon (longer amplicons associate with more intercalating dye), the optical properties of the instrument (filters, lamp etc), the data acquisition temperature and the instrument settings (gain etc). Further, although the fluorescence generally increases along with the number of amplicons formed, it is not strictly a direct relationship. Indeed, during the first cycles we do not detect any fluorescence increase because the signal is indistinguishable against the fluorescent background. But not even in the phase where we can observe increasing fluorescence can we assume that fluorescence intensity is strictly proportional to the number of amplicons present (Nygren *et al.*, 1998). Binding of a probe to template is governed by thermodynamic principles and the fraction of probe bound depends on the concentrations of both the probe and amplicon. When using degradable probes, such as TaqMan® and Qzyme™, the active probe concentration decreases during the course of the reaction, leading to less efficient probing as the reaction proceeds. Dye binding shows cooperative effects at low binding ratios as well as self quenching at high binding ratios. Both phenomena give rise to non-linear behavior.

$$I = k'(N_t) \times N_t = k'(N_t) \times N_t^0 \left(1+E_{ts}\right)^{(x-1)} \quad (5.5)$$

The factor $k'(N_t)$ describes the dependence of fluorescence intensity on the number of amplicons in the sample; it is generally unknown, which complicates analysis of PCR response curves. However, in real-time PCR one typically compares the number of amplification cycles required to reach a certain fluorescence threshold level so the unknown factor is not a problem, since at the point of comparison all samples contain the same number of fluorescent molecules. For ease of reading we have used CT to denote the C_T value described previously. With X = CT, $k'(N_t) = k'$, the intensity is now independent of N, and we obtain a value for every sample A, B, C etc.

$$I = k_t' \times N_t = k_t' \times N_{tA}^0 \left(1+E_{tA}\right)^{(CT_{tA}-1)} = k_t' \times N_{tB}^0 \left(1+E_{tB}\right)^{(CT_{tB}-1)} = \ldots \quad (5.6)$$

This can be rearranged to give the relative initial number of single stranded DNA target molecules in two samples:

$$\frac{N_{tA}^0}{N_{tB}^0} = \frac{\left(1+E_{tB}\right)^{(CT_{tB}-1)}}{\left(1+E_{tA}\right)^{(CT_{tA}-1)}} \quad (5.7)$$

Assuming that the sample matrix is relatively constant among the samples, we can replace the sample specific efficiencies with an average target specific efficiency $\bar{E}_t$:

$$\frac{N_{tA}^0}{N_{tB}^0} = \frac{\left(1+E_{tB}\right)^{(CT_{tB}-1)}}{\left(1+E_{tA}\right)^{(CT_{tA}-1)}} \approx \left(1+\bar{E}_t\right)^{(CT_{tB}-CT_{tA})} \quad (5.8)$$

With this assumption the "–1" in the exponentials cancel out. Hence, expression is valid for both single and double-stranded targets.

Setting $N^0_t = 1$ and $CT_t = CT_t\,(sc)$ (CT for a single copy sample) we obtain:

$$N_t^0 = \left(1+\bar{E}_t\right)^{(CT_t(sc)-CT_t)} \quad (5.9)$$

Taking the logarithm of both sides and rearranging gives:

$$CT_t = CT_t(sc) - \frac{1}{\log\left(1+\bar{E}_s\right)}\log\left(N_t^0\right) \quad (5.10)$$

This equation is the basis for quantification using an external calibration curve otherwise known as absolute quantification. But the name is misleading since concentrations are not determined in an absolute sense. Instead, concentrations of test samples are estimated by comparison to those of calibrator samples. Plotting the threshold values (CT_t) versus the logarithm of the initial number of template molecules in calibrator samples should give a straight line (Figure 5.2.) with:

$$slope_t = -\frac{1}{\log\left(1+\overline{E}_t\right)} \quad (5.11)$$

Inserting $\overline{E}_t = 1$ we find that a 100% efficient PCR gives rise to a slope of −3.32. From the slope the mean PCR efficiency is calculated as:

$$\overline{E}_t = 10^{-\frac{1}{slope_t}} - 1 \quad (5.12)$$

The slope is determined by linear regression. Most software for linear regression will also calculate the standard error of the slope:

$$SE(slope) = \sqrt{\frac{\sum_{i=1}^{n}\left(CT_{measured,i} - CT_{predicted,i}\right)^2}{n-2}} \Bigg/ \sqrt{\sum_{i=1}^{n}\left(\log\left(N_i^0\right) - \overline{\log\left(N^0\right)}\right)^2} \quad (5.13)$$

In Microsoft® Excel® the LINEST function in array mode calculates the standard error of slope. The standard error of the PCR efficiency is obtained from the SE(slope) by error propagation based on truncated Taylor expansion:

$$SE\left(\overline{E}_t\right) = SE(slope) \times \frac{\ln 10\left(1+\overline{E}_t\right)}{slope^2} \quad (5.14)$$

The 95% confidence interval for the PCR efficiency is then given by a Student's t-test with n-2 degrees of freedom:

$$\overline{E}_t \pm t_{95\%,2-tails,,n-2} \times SE\left(\overline{E}_t\right) \quad (5.15)$$

The standard error and confidence interval for CT values and log concentrations can be calculated by standard error propagation or by the Fieller's theorem (Fieller, 1940; Verderio *et al.*, 2004). If the concentrations of the calibrators are available in copy numbers, the intercept determines the CT of a single copy sample:

$$intercept = CT(sc) \quad (5.16)$$

Calibrator concentrations can also be given in relative values, such as dilution factors of a stock. The concentrations of the test samples will then be obtained in the same relative measure. Using relative concentrations for calibrators only affects the intercept. The PCR efficiency is still obtained from the slope by equation 5.12.

Microbial quantification: relatively speaking

Determining the amount of template by PCR can be achieved in two ways: relatively or "absolutely". "Absolute" quantification strategies claim to determine the exact number of nucleic acid targets present in the sample and can express this in relation to a specific unit that is easily interpreted by both scientists and medical practitioners and comparable across different real-time PCR instruments and between laboratories. Such denominators include unit volume of serum, plasma, whole blood, urine or cerebrospinal fluid, the number of cells in blood or the mass of tissue or genomic nucleic acid (Freeman *et al.*, 1999; Pfaffl *et al.*, 2002). The decision about which denominator to use should preferably be made in consultation with one's peers to improve the chance of producing compatible results. Absolute quantification is a valuable tool when there is only a single sample available from a patient or when viral load data are to be compared to a predetermined, clinically relevant threshold value or range. Absolute quantification is the more commonly reported approach utilized by the microbiology laboratory and is easier for DNA analysis because calibrator samples are readily prepared based on purified target DNA diluted is suitable sample matrix. RNA quantification is more complex in an absolute sense because proper usage requires RNA calibrators, which are more cumbersome to generate. Calibration curves based on DNA do not account for any variations in cDNA synthesis yields. An alternative strategy is relative quantification. In relative quantification the amount of target RNA in a sample is measured relative to a second RNA, which serves as the reference. It can be an endogenous reference gene (ERG) transcript either from the host or from the same microorganism, or an RNA spike. Both references would control for variation in reverse transcription efficiency, while the former also controls for variations in the overall expression level. The relative expression of the target RNA to the reference RNA reflects the degree of microbial transcription and is the measure compared among samples.

Equation 8 showed the relative expression of one RNA target in two samples. The relative expression of two RNA targets in one sample a sample is given instead by (Kubista *et al.*, 2006):

$$\frac{N_{aA}^{0}}{N_{bA}^{0}} = K_{RS}\frac{\eta_{bA}\left(1+E_{bA}\right)^{(CT_{bA}-1)}}{\eta_{aA}\left(1+E_{aA}\right)^{(CT_{aA}-1)}} \tag{5.17}$$

Where indices *a* and *b* refer to RNA molecules and *A* indicates the sample, η is the reverse transcription efficiency, which depends on the RNA and may also vary from sample to sample. It also depends on the reverse transcription protocol used, including the primers and the reverse transcriptase. K_{RS} is the relative sensitivity of the two qPCR assays (Ståhlberg *et al.*, 2004a). Clearly, measuring the relative expression of two targets in a sample requires determining a number of parameters. Usually these methods are employed for the expression of two human genes to identify how much the target gene's expression varies from the reference gene. The same equations apply to the measurement of a viral RNA target relative to an RNA spike used as control, or relative to an endogenous RNA reference such as ribosomal RNA or a validated ERG.

$$\frac{\dfrac{N^0_{aA}}{N^0_{bA}}}{\dfrac{N^0_{aB}}{N^0_{bB}}} = \frac{\dfrac{\eta_{bA}\left(1+E_{bA}\right)^{(CT_{bA}-1)}}{\eta_{aA}\left(1+E_{aA}\right)^{(CT_{aA}-1)}}}{\dfrac{\eta_{bB}\left(1+E_{bB}\right)^{(CT_{bB}-1)}}{\eta_{aB}\left(1+E_{aB}\right)^{(CT_{aB}-1)}}} \tag{5.18}$$

In equation 17 the expression K_{RS} cancels. If the two samples are reverse transcribed using the same protocols, and the sample to sample variation is small, the relative reverse transcription yields for the two RNA molecules in the two samples should be the same (Ståhlberg *et al.*, 2004a).

$$\frac{\eta_{bA}}{\eta_{aA}} = \frac{\eta_{bB}}{\eta_{aB}} \tag{5.19}$$

Further, we assume that the degree of PCR inhibition in the two samples is the same. Then the sample specific PCR efficiencies can be replaced with averages values.

$$\frac{\dfrac{N^0_{aA}}{N^0_{bA}}}{\dfrac{N^0_{aB}}{N^0_{bB}}} = \frac{\left(1+\bar{E}_b\right)^{(CT_{bA}-CT_{bB})}}{\left(1+\bar{E}_a\right)^{(CT_{aA}-CT_{bB})}} = \frac{\left(1+\bar{E}_a\right)^{(CT_{aB}-CT_{aA})}}{\left(1+\bar{E}_b\right)^{(CT_{bB}-CT_{bA})}} \tag{5.20}$$

This simplifies to:

$$\frac{\dfrac{N^0_{tar,test}}{N^0_{ref,test}}}{\dfrac{N^0_{tar,control}}{N^0_{ref,control}}} = \frac{\left(1+\bar{E}_{tar}\right)^{\Delta CT_{tar}(control-test)}}{\left(1+\bar{E}_{ref}\right)^{\Delta CT_{ref}(control-test)}} = R \tag{5.21}$$

This is the expression usually recommended to calculate the relative amounts of two RNA molecules in two samples. It assumes that:

- for every RNA assay the PCR efficiency is the same in all samples
- the ratio of the reverse transcriptase yields of the two RNAs is the same in all samples.

Assigning "a" to the microbial RNA target (tar), b to the reference (ref), and A to the test sample and B to the control sample, we obtain:

$$R=\left(1+\bar{E}\right)^{\Delta CT_{GOI}\left(control-test\right)-\Delta CT_{RG}\left(control-test\right)}=\left(1+\bar{E}\right)^{-\Delta\Delta CT} \qquad (5.22)$$

This expression takes into account different efficiencies between test and control samples (Sagner *et al.*, 2004), with efficiencies defined as $0 \leq E \leq 1$ (Pfaffl, 2001). The four CT values are measured for the two RNA molecules in the two samples and the average PCR efficiencies for the two assays are estimated from calibration curves. The expression ratio controls for reverse transcription so DNA calibrator curves can be used. The ratio "R" is the relative expression of target RNA in the two samples using the RNA reference as normalizer.

If the PCR efficiencies for the two assays are equal, we obtain:

$$R=\left(1+\bar{E}\right)^{-\Delta\Delta CT}=2^{-\Delta\Delta CT} \qquad (5.23)$$

This is the ΔΔCT method. Finally, if only very rough estimates of relative expression are needed, 100% PCR efficiency can be assumed. This gives:

$$R=\left(1+\bar{E}\right)^{-\Delta\Delta CT}=2^{-\Delta\Delta CT} \qquad (5.24)$$

Earlier we learnt that a calibration curve with an appropriately added IAC (sometimes called a "spike") is a good approach to assay for microbial and cellular DNA, while a calibration curve combined with an exogenous control is a good approach to assay for RNA targets. The underlying assumption in these approaches is that PCR efficiency does not vary appreciably among samples such that we can assume a fixed PCR efficiency for the target nucleic acids. If this assumption is wrong we may obtain quite erroneous data. Indeed, a study of lymphoma based on serial dilutions showed that several samples would be misclassified if a constant PCR efficiency was assumed (Ståhlberg *et al.*, 2003). How can we recognize if there is a problem with PCR efficiencies?

The development of a real-time PCR signal over time (Figure 5.2), also called the real-time PCR response curve, contains information about PCR efficiency (equation 5.5). Although the efficiency is difficult to estimate from the response curve, its shape can be used for quality control. The method of kinetic outlier detection (KOD) compares the response curves of samples and identifies those with anomalous responses (Bar *et al.*, 2003). These will have PCR efficiencies that deviate significantly from normal and their C_T values are likely to be erroneous. These samples should be rerun and if they retain a deviant PCR efficiency they must either be discarded or re-analyzed by serial dilutions or standard additions to account for their anomalous efficiencies. A web-based solution for KOD will soon be available at LabonNet (kttp://www.labonnet.com).

Microorganisms are not human transcripts; defining the differences

For a number of reasons qPCR data can be difficult to replicate. Real-time qPCR assays must include the capacity to correct for variations caused by a range of factors; without this capacity qPCR data are at best more appropriately described as being semi-quantitative and at worst they produce biologically irrelevant results (Tricarico *et al.*, 2002; Dheda *et al.*, 2005). Whether such a claim is overly meticulous is debatable. However the continued production of data which cannot be compared between laboratories except in the broadest sense should give pause for thought; poor reproducibility will stymie attempts to collaborate in research studies and diagnostic quality assurance programs and confound attempts to improve our understanding of the interaction between microorganism and host (Hoorfar *et al.*, 2003). To this day it is rare to find well-controlled real-time qPCR assays among those in the clinical microbiology literature. That could quickly change if pressure was brought to bear by the editorial staff and peer reviewers of scientific journals (Mikovits *et al.*, 1998). In the interim it would be useful to identify what is actually required to perform robust real-time qPCR in the clinical microbiology laboratory, under realistic conditions.

Some real-time PCR systems have already overcome the simplest technical variable, fluorescence changes caused by reaction mix composition, mix volume variation or because of non-specific quenching. One approach has been to include a non-participating, or "passive," internal reference molecule incorporating a carboxy-X-rhodamine (ROX) fluorophore. This doesn't require a separate amplification since the reference is present in the reaction buffer and its signal is used to indicate volume variations and non-specific fluorescence quenching. The corrected or "normalized" values (not to be mistaken for normalization of qPCR data using a reference or calibrating template), obtained from a ratio of the emission intensity of the fluorophore and ROX, are termed RQ+. To further control amplification fluctuations, the fluorescence from a "no-template" control reaction (RQ-) is subtracted from RQ+ for each sample, resulting in the ΔRQ value that indicates the magnitude of the reference signal generated for the given PCR (Gelmini *et al.*, 1997). This approach embodies important criteria for the design of a PCR control to be employed for correcting unwanted variation or error; it can perform its job within the same tube as the template under study, it is relatively simple to design and implement and there is no interference with enzymes or oligonucleotides. The passive reference does not determine template quality, enzyme inhibition nor account for variation in nucleic acid amount so these issues must be addressed during a pre-analytical phase. If the issues are suitably addressed however, a passive reference provides the ideal approach to account for inter-tube variation of the sort that can occur between unknowns and the calibrator dilutions of an external calibration curve.

Important steps to consider before quantifying: the pre-analysis phase

Coming to terms with the complexities of performing high-quality real-time qPCR is a particular and daunting challenge for the diagnostic clinical microbiology community. The largest single adjustment is to understand that these assays cannot be performed as simply as qualitative real-time PCRs. A number of pre-analysis steps are required to ensure that the template is adequately prepared for quantification and this sets quantitative molecular methods apart from the increasingly high-throughput methods that populate the clinical microbiology laboratory today.

In addition to minimizing the usual technical variables including staff competence, equipment performance and upkeep, reagent quality and batch variation and use of sample replicates, molecular quantification of a microorganism relies upon the use of measures to account for:

1 Variation in template amount between samples
2 Poor or variable quality template (due to variable purification, mis-handling or poor storage)
3 The presence of inhibitors (of both fluorescence and amplification)
4 Variation in the performance of RNA dependent DNA polymerases
5 Variation in the performance of DNA dependent DNA polymerases
6 Variation in amplification between sample and calibrators.

In addition to causing acute infections, viruses are the underlying cause of a significant proportion of human cancers. They also pose a major bioterrorism threat despite requiring specific expertise to produce, engineer, and disseminate in weaponized form. In particular, RNA viruses comprise the most abundant group of human, animal and plant pathogens with new, highly pathogenic viruses such as Avian influenza continually emerging. Newly identified endemic RNA viruses also comprise a growing population of human pathogens that are poorly characterized by qPCR. We will therefore dwell, and frequently refer to, RNA virus targets in the following sections. However the approaches we discuss are often compatible with the quantification of DNA templates from viruses, bacteria, fungi and parasites.

Along with technical variation, the first three points in the previous list should be addressed before performing any quantification. We advocate the following considerations for real-time qPCR to ensure optimal assay reproducibility:

- The quantity of RNA should be determined using a single method for all assays and the amount used for the reverse transcription step must be kept as constant as possible. This is often difficult for a microbial target since its genome or gene transcript may be a small and variable fraction of the total nucleic acid present.
- An assessment of the quality of any RNA preparation used for quantification is essential. In particular information should be provided regarding the absence of inhibitors in each sample and the integrity of the RNA.
- The cDNA priming method must be consistent, with specific priming the preferred option, together with as short an incubation time as possible and as high a reaction temperature as practical.
- Calibration curve data should be analyzed so the amplification efficiency and sensitivity of every PCR assay can be reported and to confirm that the quantification of any unknowns has been carried out within the dynamic range of that assay.
- A common IAC should be used in every reaction of the qPCR to permit correction for inter-tube amplification variation.
- When reporting data as a relative change, the C_T range of target detection should be quoted, since this allows the reader to gauge the likely accuracy and reliability of the quantitative data, particularly if the mRNA levels of the targets are widely different.

- Any signal detected in the negative controls should be reported and qualified by melt curve (if possible) or other analysis.
- Normalization of qPCR (see below) must be transparent and reference genes, if used for comparison, must be validated for each experimental set-up, clinical condition and matrix type and appropriate data should be included in every publication.

A more detailed discussion of these and other considerations can be found in print (Bustin, 2004) and online (http://www.gene-quantification.info/).

The amount and quality of nucleic acid templates

Many viruses and other microorganisms contain a DNA genome and these pathogens are increasingly detected alongside RNA viruses within the same sample where they may collaborate in contributing to disease (Jennings *et al.*, 2004; Arden *et al.*, 2006). Thus, rather than preparing separate DNA and RNA extracts, it is preferable for the clinical microbiology laboratory to use a purification method capable of efficiently preparing both classes of nucleic acids in order to maximize the usefulness of the precious patient specimen e.g. when detecting enterovirus and *Neisseria meningitidis* from a small volume pediatric sample.

The importance of preparing intact microbial RNA for generating reproducible, quantitative data cannot be overstated. Since most of the RNA isolation procedures take place in a strong denaturant which protects the template, appropriate tissue handling including collection, transport and storage prior to its extraction is essential to avoid template loss at these earlier stages.

Similarly, the importance of quantification and quality assessment of the RNA template for any qPCR applications cannot be overstated yet these are usually overlooked by applications in microbiology. The most commonly used method, absorbance measurement at OD_{260} and analysis of $OD_{260/280}$ in a spectrophotometer, is insufficiently accurate. In addition, this method can waste significant amounts of precious RNA. The most accurate and preferred means of quantification uses an RNA-binding dye, RiboGreen; however, this does not provide qualitative information. There are two other options: first, LabChip® technology allows the rapid analysis of very small quantities of RNA. It is based on microfluidics technology, and a LabChip cassette consists of a series of interconnected microchannels and reservoirs embedded in a palm-sized chip architecture. Migrations through the channels are monitored and controlled using a bench top instrument such as the Agilent 2100 Bioanalyser or the Bio-Rad Experion. The RNA LabChip accurately quantifies as little as 50 pg of RNA and calculates the ratio of 28S:18S rRNA in the sample to provide a simultaneous assessment for each sample, and by inference for the quality of the microbial nucleic acids. Results can be viewed as gel-like images, electropherogrammes or in tabular formats. However, this analysis relates to rRNA and provides no information with respect to mRNA integrity. Integrity can be determined using the second option, the 3′:5′ assay. This assay is modeled on the standard approach adopted by microarray users (Auer *et al.*, 2003) and measures the integrity of GAPDH mRNA, which is taken as representative of the integrity of all mRNAs in a given RNA sample. It

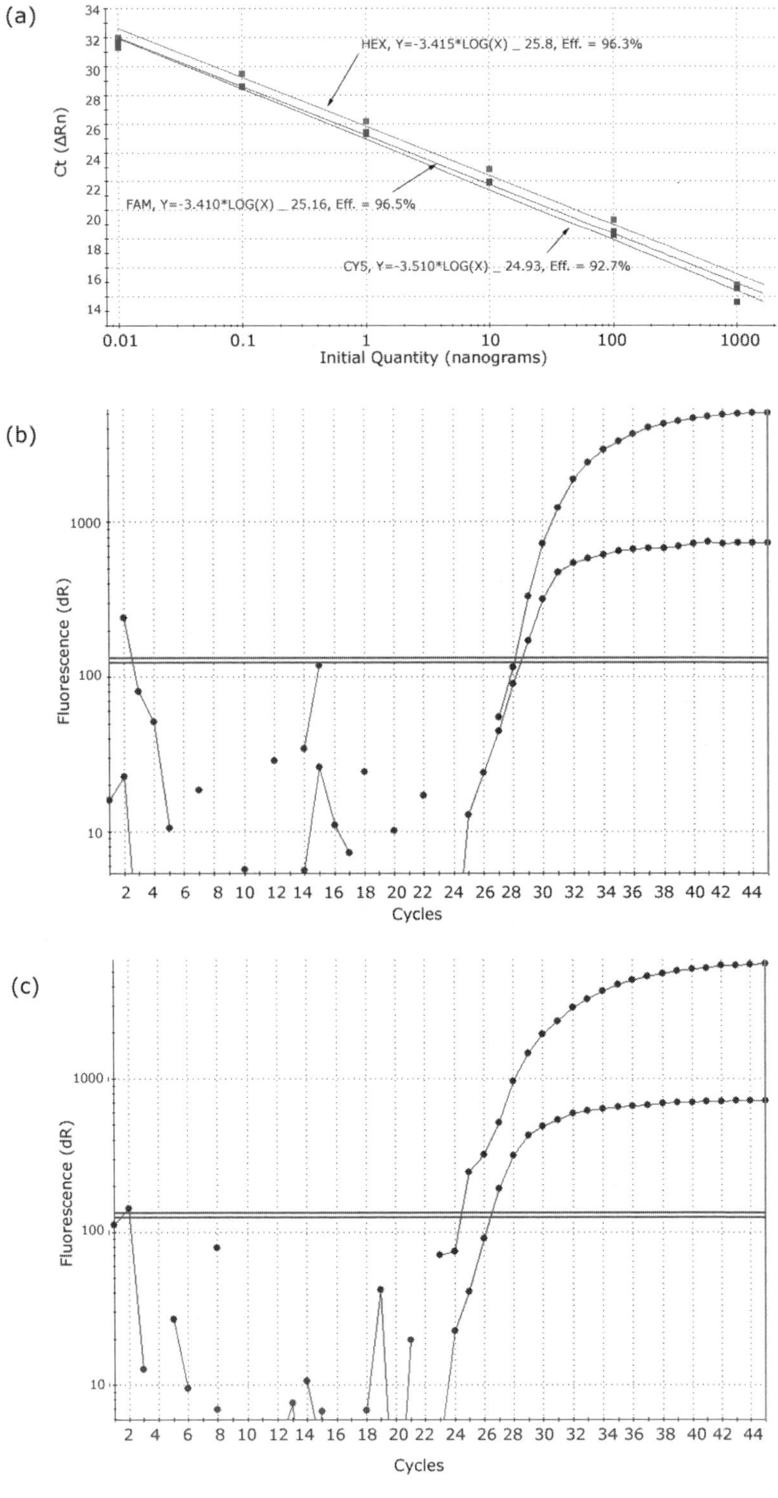
(a)
HEX, Y=-3.415*LOG(X) _ 25.8, Eff. = 96.3%
FAM, Y=-3.410*LOG(X) _ 25.16, Eff. = 96.5%
CY5, Y=-3.510*LOG(X) _ 24.93, Eff. = 92.7%
Ct (ΔRn)
Initial Quantity (nanograms)
(b)
Fluorescence (dR)
Cycles
(c)
Fluorescence (dR)
Cycles

synthesizes cDNA from the 3′ end of the RNA and then calculates the ratio of abundance of amplicons located at the 3′ and 5′ ends, respectively (Figure 5.3). A ratio of > 5 suggests RNA degradation (Hands *et al.*, 2007)

Inhibition, inefficiency and the internal amplification control

As introduced earlier, the IAC is a molecule that is co-amplified with the unknown target. IACs are today most frequently used to indicate the occurrence of false negative reactions or any reduction in amplification efficiency due to inhibition. We will refer to an IAC that has been engineered to bind the same primers but is discriminated by an oligoprobe that hybridizes to a unique sequence within the amplicon, as an homologous IAC (Mulder *et al.*, 1994; Tarnuzzer *et al.*, 1996; Alexandre *et al.*, 1998; Aberham *et al.*, 2001; Stöcher *et al.*, 2002) whereas one that consists of a completely different sequence from that of the microbial target we call an heterologous IAC; the latter is our preferred approach for microbiology (Templeton *et al.*, 2003; Templeton *et al.*, 2005; Scheltinga *et al.*, 2005). When an IAC is added into the sample prior to template extraction and purification (termed an extraction or full-process IAC) or just prior to amplification (amplification IAC), it is described as an exogenous IAC, since it does not occur naturally within the sample matrix but is still co-amplified within the same reaction vessel. An endogenous, heterologous IAC occurs naturally within the sample; these are not engineered molecules and thus require separate primers for amplification. Endogenous IACs are most commonly known as "housekeeping" or endogenous reference genes (ERGs) and they can also be used to measure purification efficiency and/or identify inhibition. Of particular concern is the possibility that, since its concentration cannot be controlled from patient sample to patient sample, an overwhelming amount may be co-purified and detrimentally compete with the microbial template for amplification.

It is also possible to control for template loss during purification, which otherwise goes unnoticed. In this application it is preferable to add the IAC to the lysis buffer rather than the crude sample itself to avoid degradation of the IAC molecule, unless the IAC has been protected or "armoured" with a cell coat (e.g. MS-2 RNA; (Pasloske *et al.*, 1998)). Since the purification efficiency can differ between different tissue types it is also best to limit a quantitative study, or the use of a kit, to the same type of tissue and employ the same method of extraction.

If the exogenous IAC is only used to monitor inhibition, it should be added to the master mix, minimizing pipetting error incurred by addition to each separate vessel. In general, inhibition does not cause problems for amplification of calibrators or reference

Figure 5.3 (opposite) Integrity of RNA. Following reverse transcription with oligo-dT, three sets of primers and probe target amplicons at different positions (5′, centre and 3′) on the RNA, here GAPDH mRNA. The ratio of amplicons reflects the relative success of the oligo-dT primed reverse transcription to proceed along the entire length of the transcript. Clearly, the progress of the reverse transcriptase enzyme past the 5′ amplicon is dependent on the intactness of the mRNA, with the enzyme unable to reach it if the mRNA is degraded. (a) When run in a multiplex reaction, it is apparent from the slopes that the amplification efficiencies of the three reactions are approximately the same, with similar y-intercepts indicating roughly similar sensitivity. (b) For clarity, results from only two amplicons are shown. The amplification plots suggest that this RNA preparation is of very high quality and fully intact. (c) The amplification plots of a moderately degraded RNA, showing a ΔCt of 4.

standards since they are more carefully prepared. The unidentified presence of inhibitors in some biological samples will result in an underestimation of the nucleic acid levels in the test samples (Ståhlberg *et al.*, 2005). Inhibition unrelated to the amplification process itself can have an impact at four main points along the laboratory process of obtaining qPCR data (adapted from Wilson (1997)): inhibitors can interfere with cell lysis, degrade nucleic acids, act directly and detrimentally on the reverse transcriptase and partially or completely interfere with amplification or fluorescent detection. Consequently, inhibitors, such as reagents used during nucleic acid extraction or co-purified components from the biological sample, can therefore significantly reduce the sensitivity and alter the kinetics of real-time qPCR assays, rendering quantitative data invalid (Cone *et al.*, 1992; Wilson, 1997; Perch-Nielsen *et al.*, 2003; Guy *et al.*, 2003; Suñen *et al.*, 2004; Lefevre *et al.*, 2004; Rådström *et al.*, 2004; Jiang *et al.*, 2005). Obviously, inhibitors are likely to distort any comparative quantitative data. However, a survey of practices at this time has revealed that only 6% of researchers test their nucleic acid samples for the presence of inhibitors (Bustin, 2005).

There are various methods that can assess the presence of inhibitors within biological samples but they all generally require a dedicated amplification step; for real-time qPCR this occurs in the pre-analysis phase. As discussed earlier, the simplest method is to carry out a serial dilution of the sample spanning a large concentration range although the effect of some inhibitors can be difficult to remove this way (Kubista *et al.*, 2006). Studies have shown that the degree of PCR inhibition can vary substantially among similar biological samples and that inhibition may affect different assays in different ways (Ståhlberg *et al.*, 2003). Different polymerases also react differently to inhibitors and it is worth checking different matrix extracts for inhibition using several different thermostable DNA polymerases to identify the one most efficient at amplifying the template. This then raises the question of how many patients from each type of disease, and from normal populations, should be tested before we can be satisfied that we have exposed our reagents and protocols to every possible form of inhibition? Comprehensive studies are needed to address this question for the benefit of all clinical microbiology laboratories using real-time qPCR. When using a full-process IAC, signal due to amplification of the IAC must be reproducible and reach a predetermined C_T value or, preferably, range when co-amplified with the template under investigation (Pasloske *et al.*, 1998; Reiss and Rutz, 1999; Smith *et al.*, 2003; Burggraf and Olgemöller, 2004). Delay in the C_T value indicates the presence of inhibition, poor quality template or low recovery of template (Levy, 1999). By comparing that value to the value obtained from a similar matrix sample containing only the IAC one can determine template loss or the presence and extent of inhibition (Niesters, 2004). Of course the pedant could also question the validity of this approach since there is no mention of a calibrator to control for inter-tube amplification variability although replicates are often used to address this aspect. The presence of partial inhibition is not usually a significant problem for uniplex qualitative real-time PCR but becomes a much more important factor for multiplex assays and for quantification. Any reduction in signal during quantification will result in an underestimation of the amount of microorganism present. The IAC used during pre-analysis must not significantly interfere with the detection of small quantities of the microbial template if it is to determine the extent, not just the presence, of inhibition therefore its use must be carefully evaluated. Such an IAC must also play no interfering role during the analysis phase of real-time qPCR testing.

When the IAC was employed as a competitive template for conventional qPCR (Chehab *et al.*, 1987) it was carefully added to ensure it would not overwhelmingly and detrimentally compete for reagents with the wild-type template (Brightwell *et al.*, 1998; Rosenstraus *et al.*, 1998). A major caveat for the use of an IAC as a competitor was that it had, or should have had, identical amplification efficiency to the template under investigation (Zimmermann and Manhalter, 1996). To achieve this, extensive sequence similarity was required between the microbial and IAC templates which subsequently led to the formation of heteroduplex amplicons (Zimmermann and Manhalter, 1996). These mixed DNA hybrids consisted of one strand from the IAC amplicon and one from the intended target's amplicon; they reduced the sensitivity of detection and the efficiency of amplification for the intended target and made discrimination from the unknown sequence impossible by classical methods such as restriction enzyme digestion and agarose gel electrophoresis (Freeman *et al.*, 1999; Pierce *et al.*, 2005). An ideal competitive IAC should be of the same length and GC content as the intended target but differ from any microbial target sequence in all other regions apart from the oligonucleotide hybridization sites. Unfortunately the current use of very small amplicons may leave practically no unique sequence to reduce heteroduplex formation. Thus it is more practical to employ an heterologous IAC, ensuring that amplification efficiency is as close to the assay's microbial target template as possible.

It is useful to add the pre-analysis IAC to samples that will be stored or transported at, or just after, specimen collection to monitor any degradation of the sampled material. This is much more important for RNA than for DNA, since RNA is the less stable nucleic acid. Practically such a plan is a very difficult proposition for clinical microbiology samples since the spike would usually have to be added by non-laboratory personnel in the hospital, clinic, collecting centre or general practice. Pre-addition of a stabilized IAC to specimen tubes could be the solution.

Biological material derived from cultured microorganisms should be avoided as a calibrator or an IAC without completely understanding and accounting for technical issues due to this source of template. For example, *in vitro* virus culture suffers from the production of defective interfering particles that may contain template (genome) but be non-infectious, or alternatively may lack template but retain the capacity for receptor attachment and fusion and thus contribute to a gold standard biological detection system such as a plaque-forming or indirect immunofluorescent assay, falsely elevating the viral particle count (discussed for respiratory viruses in Chapter 8).

It is also important that the type of IAC nucleic acid is compatible with the microbial target. For analysis of DNA, a DNA-based calibration curve and IAC is a reliable combination. For analysis of RNA, an RNA-based calibration curve and IAC is the best choice. It is also possible to use a DNA-based calibration curve, when quantifying the amount of cDNA produced by the reverse transcription (Ståhlberg *et al.*, 2004a), in combination with an RNA IAC that controls for extraction and reverse transcriptase yields. A combination of an RNA-based calibration curve with a DNA IAC is not very useful, since DNA and RNA purifies differently. An RNA IAC should share any important sequence features with the microbial RNA target that may affect stability. A polyA-tail and a 5′ cap that protect from degradation should be present for mRNA targets while other features may be more important in IACs used alongside viral genomic RNA targets.

The IAC should have the same secondary (double vs. single strand) and tertiary (supercoiled) structures as the intended target. A naturally supercoiled plasmid, for example, is not a good extraction control for genomic or viral DNA. Most recently a simple qPCR-based assay has been described that consists of a plant-derived universal exogenous IAC template sequence, primers and a detecting probe (Nolan *et al.*, 2006) and there are now heterologous IAC systems commercially available e.g. from Applied Biosystems.

Reverse transcription: the Achilles heel of quantification

A real-time reverse transcription qPCR (RT-qPCR) assay consists of three steps: (1) the conversion of RNA into cDNA using an RNA-dependent DNA polymerase, (2) the amplification of the cDNA using the PCR and (3) the detection and quantification of amplification products in real-time (Gibson *et al.*, 1996).

Although seemingly a small step, the conversion of RNA into an amplifiable cDNA template is, in practice, an important contributor to the variability observed in any RT-qPCR experiments (Kolls *et al.*, 1993; Bustin *et al.*, 2005). Quite apart from biological variability, there are significant technical issues. One major drawback of the reverse transcription step is its lack of standardization, which means its variable efficiency can continue uncontrolled, resulting in inaccurate quantification. Reverse transcriptase yields can vary up to 100-fold with the choice of enzyme and this variation is target-dependent (Ståhlberg *et al.*, 2004b). Furthermore, results also depend on the priming strategy and the concentration of RNA (Bustin and Nolan, 2004; Ståhlberg *et al.*, 2004a). Hence it is essential to at least use the same enzyme, priming strategy, and experimental conditions, if quantitative (or qualitative) RNA measurements are to be comparable between laboratories. Furthermore, the choice of smaller amplicons is a preferred workaround for degradation of longer RNA templates (Fleige and Pfaffl, 2006).

Real-time RT-qPCR assays can be performed either as a one tube, single reverse transcriptase and PCR enzyme method or a separate reverse transcriptase and PCR enzyme technique using one or two tubes (Figure 5.4). A single enzyme such as *Tth* polymerase is able to function both as an RNA- and DNA-dependent DNA polymerase and can be used for the one tube method without secondary additions to the reaction mix (Cusi *et al.*, 1994; Juhasz *et al.*, 1996). Its main advantages are the reduced hands-on time and reduced potential for contamination. Its main disadvantage is that since all reagents are added to the reaction tube at the beginning of the reaction, it is not possible to optimize the two reactions separately. Furthermore, this reaction is characterized by extensive accumulation of primer dimers, which may obscure the true results in quantitative assays (Vandesompele *et al.*, 2002a).

In the separate techniques reverse transcription and PCR enzymes can be used either in "coupled" (single tube) or "uncoupled" (two tube) reactions. In the coupled method the reverse transcriptase synthesizes cDNA in the presence of either gene-specific or oligo-dT primers. Following the reverse transcription reaction, PCR buffer, a thermostable DNA polymerase, and gene-specific primers are added and the PCR is performed in the same

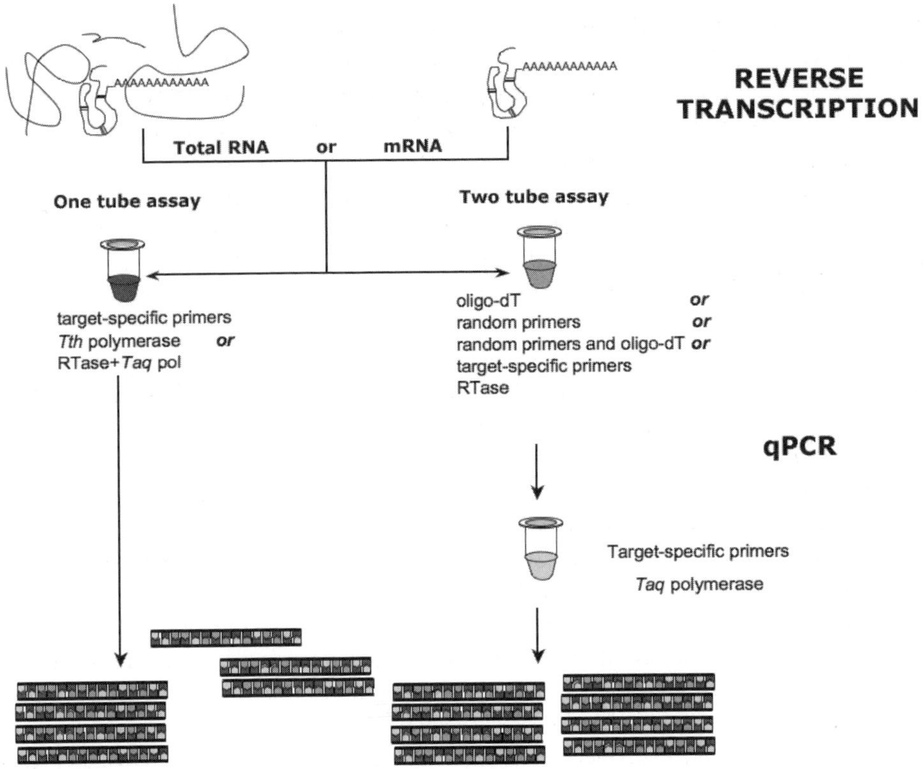

Figure 5.4 RT-qPCR protocols. Most assays targeting viruses will use total RNA and target-specific priming. In contrast, most assays targeting non-viral targets tend to use random priming or a combination of random and oligo-dT priming.

tube. In the uncoupled alternative the reverse transcriptase synthesizes cDNA in a first tube, under optimal conditions, using random, oligo-dT or gene-specific primers. An aliquot from the first reaction is then transferred to another tube containing the thermostable DNA polymerase, DNA polymerase buffer, and PCR primers and the PCR is carried out under conditions that are optimal for the DNA polymerase. Inter- assay variation of two enzyme protocols can be very small when carried out properly, with correlation coefficients ranging between 0.974 and 0.988 (Vandesompele *et al.*, 2002a). The disadvantages of this approach, particularly for the high throughput clinical microbiology laboratory are that it takes longer, there are additional opportunities for contamination and the reverse transcriptase can inhibit subsequent PCR assays even after inactivation, resulting in an overestimation of amplification efficiency and target quantification, particularly for rarely expressed transcripts, with the yield of amplified RNA increased up to 20-fold (Suslov and Steindler, 2005).

Random priming or oligo-dT priming both allow a pool of cDNA to be produced during a single reaction. However, it has been shown that priming using random hexamer primers does not result in equal efficiencies of reverse transcription for all targets in the sample and that there is not a linear correlation between input target amount and cDNA yield when specific targets are measured (Bustin and Nolan, 2004; Lacey *et al.*, 2005). A recent comparison of the efficiency of reverse transcription priming by random primers of varying lengths showed that 15-nucleotide-long random oligonucleotides consistently yielded at least twice the amount of cDNA as random hexamers (Stangegaard *et al.*, 2006). The 15-mers were more efficient at priming, resulting in reverse transcription of > 80% of the template, while random hexamers induced reverse transcription of only 40%. Not surprisingly, this resulted in the detection of one order of magnitude more genes in whole transcriptome DNA microarray experiments.

Oligo-dT primers should only be used with intact RNA and generally only with those microorganisms known to have a polyA tract associated with the genome or transcript. Even then the cDNA molecules may be truncated, since the reverse transcriptase cannot proceed efficiently through highly structured regions. Accordingly, oligo-dT-primed assays should be targeted towards the 3′ end of the transcript. Oligo-dT priming is not recommended when using RNA extracted from paraffin tissue sections, since formalin fixation results in the loss of the polyA tails on mRNA (Lewis and Maughan, 2004).

Gene-specific primers are the most specific and most sensitive method for converting mRNA into cDNA (Lekanne Deprez *et al.*, 2002) and are the recommended choice when RNA quantity is not a limiting factor. Even if the amount of RNA is a limiting factor, a recent report demonstrates the use of specific primers for the reliable and specific amplification of 72 genes from limiting amounts of RNA using a multiplexed tandem PCR approach (Stanley and Szewczuk, 2005). Nevertheless, as with random priming there may be differences in the efficiencies at which individual reverse transcription reactions occur. These variations must be controlled by reference of the PCR data generated from an unknown sample to that from a calibrator sample (when using $\Delta\Delta C_t$ analysis) or to a calibration curve. Specific priming of RNA dilutions results in a linear response of target cDNA yield (Bustin and Nolan, 2004) and so a further advantage of using this priming method is that the efficiency of the combined real-time qPCR reaction can be confirmed by analysis of the slope of the calibration curve. The inclusion of a calibrator sample or calibration curve in every assay is an important control for measuring inter-assay variability that may occur when multiple samples are run on different assays. However the extent of the linearity is entirely dependent on the level of conservation between the oligonucleotide and its target site.

Nucleotide target sequence variation

Sequence variation has always been and will most likely always remain an issue of significant concern when developing PCR assays but it is particularly troublesome when it plays a role in quantitative assays. Even quantitative kits targeting DNA viruses can be subject to problems caused by sequence variation highlighting the importance of kit comparisons (Ruiz *et al.*, 2005).

In particular, the low efficiency of proofreading and post-replicative repair activities of viral RNA polymerases results in relatively high mutation rates, causing potential problems for an assay that relies on exquisite sequence specificity. Furthermore, populations of RNA viruses are extremely heterogeneous existing as quasispecies in which RNA viral genomes are statistically defined but individually indeterminate (see Chapter 8 for more detail). This has raised some concerns about the detrimental effect of primer-template mismatches on the polymerization efficiency and problems associated with geographically disparate reference sequences (Mullan *et al.*, 2004) but is of equal concern from season to season in the same location. In our experience, amplifying the same nucleic acid extract using subtlety different primer pairs quite usefully demonstrates that sequence variation is common. Each primer pair may preferentially amplify different quasispecies demonstrating just how problematic the design of useful oligoprobes can be.

While some degree of sequence mismatch can be tolerated by primers and even by some fluorogenic oligoprobe chemistries during qualitative real-time PCR, any mismatch is considered undesirable for quantitative purposes. Depending on its position, it can lead to destabilization of primer or oligoprobe binding to template and may result in reduced amplification efficiency. This reduction is not reflected by the perfect match between oligonucleotides and the templates used to produce calibration curves, and creates disparity. A practical outcome for the clinical laboratory is that some virus lineages and sublineages will appear to have higher or lower viral loads than others due to better or worse hybridization efficiency, respectively. One approach to combat the impact of microbial variation is to narrow down the genetic identity of the microorganism using qualitative assays, during the pre-analysis phase of the real-time qPCR process (Watzinger *et al.*, 2006). This approach also avoids the loss of specificity and sensitivity that can result from using degenerate primers (which are essentially a mixture of primers which vary at one position) or pan-microorganism primers designed to detect all members of a species, serovar or genus.

For example, the extreme sequence heterogeneity of HIV-1 poses a significant challenge to its efficient detection with nucleic acid-based assays. Not only must the primers bind to regions that are conserved and not subject to rapid mutation, but also the oligoprobes used for a particular specific chemistry assay must not hybridize to variable sequences. This has led to the development of assays using an intercalating dye, e.g. SYBR®-Green, that rely on the use of melting curves to ensure that the obtained signal is specific (Ruelle *et al.*, 2004). These assays are better than branched-chain DNA assays in determining viral load (Leutenegger *et al.*, 2001), especially during the monitoring of HAART treatment (Gibellini *et al.*, 2004) and have a superior linearity over a wide concentration range (Klein *et al.*, 2003) however we do not recommend their general use for quantification.

Normalization of real-time qPCR data: the analysis phase

All PCR assay results are subject to variability caused by technical as well as biological variation. Once technical variability (e.g. pipetting, instrumental performance and reagent quality) is recognized and kept to a minimum the chances of identifying clinically relevant changes in nucleic acid levels are improved. Apart from sequence variation, normalization of amplification data encompasses perhaps the most important biological variable in need of control. However, it is difficult to identify how best to address this variation for microbial real-time qPCR applications. Perhaps this is why normalization is commonly overlooked by the clinical microbiology laboratory or why approaches are often selected from the human gene expression literature. It is essential that microbiologically relevant approaches to normalization are identified in order to produce robust real-time qPCR data for this field.

Data normalization is essentially the process of correcting deviations from expected performance caused by the amount and quality of template amplified. The results are more meaningful if they have not been influenced by inhibitors or degraded or lost template so it is essential to account for these factors during the pre-analytical phase. Normalization is also required to correct for changes in enzyme function principally indicated by amplification performance. Further contributing to the poor uptake of normalization is the often confusing use of the word. Normalization is sometimes used to describe the adjustment of real-time PCR fluorescence data to account for amplification-independent fluctuations, or to define a component of the quality control process (Watzinger *et al.*, 2006).

Variable sample collection is another significant cause of error in many aspects of diagnostic microbiology, as is variation within the sample itself exemplified by changes in the cell content or the amount of extracellular fluid between samples from the same patient during the course of an infectious process or during an immunosuppressive or cytotoxic treatment regimen. Sampling and sample variation are particular obstacles to the design of a successful real-time PCR normalization strategy for the clinical microbiology laboratory.

Normalization to an ERG is the most common solution for real-time qPCR in human gene transcription studies and is currently the preferred option so long as it is carried out correctly (Goossens *et al.*, 2005). This strategy can express the target gene's quantity relative to that of one or, preferably, more ERGs. For human transcriptome studies this approach benefits from the fact that all the steps of the final qPCR measurement are controlled. The procedure is simplified as both the gene of interest and the ERGs are co-amplified using real-time RT-PCR; this is a more reliable approach than adding material from an external source. However, it is essential that reference gene expression in the target tissue is carefully analyzed (Perez-Novo *et al.*, 2005) so that only targets exhibiting a minimum of variability are employed (Vandesompele *et al.*, 2002b). Ideally transcription of an ERG target should be minimally regulated and exhibit a constant and cell cycle-independent basal level of transcription (Selvey *et al.*, 2001). The current gold standard combines the evaluation of a panel of several reference genes (e.g. available commercially http://www.tataa.com or http://www.primerdesign.co.uk) together with a method for selecting reference genes with the most stable expression, e.g. GeNorm (Vandesompele *et al.*, 2002b), Bestkeeper (Pfaffl *et al.*, 2004), Normfinder (Andersen *et al.*, 2004) or GenEx (http://www.multid.se). In this way dramatically up and down regulated genes can be identified under the conditions expected within the host tissues and their detrimental variability avoided (Radonic *et al.*, 2004; Radonic *et al.*, 2005). ERGs have been successfully used to quantify gene expression using RT-PCR and to monitor template integrity after its purification since their preparation is also identical to that of the microbial template (Chehab *et al.*, 1987). Studies have shown that an 18S rRNA target meets these criteria for some applications (Thellin *et al.*, 1999; Selvey *et al.*, 2001). This is not the case for other genes including β-actin; unfortunately their use remains widespread, especially for human gene transcript analyses (Härtel *et al.*, 1999).

A very important question raised by the assumptions above, and for which there is currently no definitive answer, is how can we be certain that "ideal" ERG targets remain minimally regulated throughout the course of an often lengthy and complicated infectious event; let alone between hosts? No single gene expresses consistently among all tissues despite promising candidate targets (Radonic *et al.*, 2004). The RNA polymerase gene displays low variation among many tissues, even during mitogenic stimulation of a T-cell line. RNA polymerase mRNA encodes the principle enzyme used in mRNA transcription and is therefore part of a self regulating cycle. It may be assumed that this gene is expressed steadily and independently of the cell state. However authentication is still required for each situation, a daunting task for applications in microbiology.

Specific problems normalizing data to quantify microorganisms

Correcting the data is particularly useful when conducting studies to compare microbial levels in serial specimens collected from the same subject over time. By comparing normalized data, one can better identify a change in ΔCt values and then correlate that with a change in the level of a microorganism irrespective of sample or subject. Similarly, real-time qPCR studies of latent or persistent viral infection can better demonstrate reactivation of the microorganism.

As mentioned, most of the work on real-time qPCR normalization has been conducted for human gene expression studies and consequently, on RNA templates. Nonetheless similar approaches to the ones discussed below should function just as well for DNA templates although using samples previously extracted with a method that claims specificity for RNA may result in lower DNA yields. However, it remains unclear how useful currently popular gene transcription normalization strategies are for the quantification of microorganisms.

Normalization against input RNA amount is a feasible and, if validated, acceptable option (Bustin, 2002). However, as for gene expression studies there are problems comparing a microbial target to a marker that proliferates at a different rate, since this will affect the mRNA/rRNA ratio. Making this more difficult is the presence of microorganisms in many different environments, including those with little background RNA (virus alone contributes little nucleic acid mass; a single human cell contains approximately 7pg of nucleic acid, matched in mass by approximately 10^6 rhinovirus virions) or among patients with underlying disorders that may affect the expression levels of ERGs. There may also be one or more pathogen-specific effects on host cell expression all of which create problems for the production of reliable quantitative data.

Comparing the target signal against genomic DNA is particularly useful when performing real-time RT-qPCR analyses on RNA obtained from minute samples such as those obtained from laser capture microdissection. The main problems here are (1) that the differential stability of DNA and RNA may distort quantification, (2) that the sample cannot be DNase-treated, (3) that, whilst being an internal control, there is no equivalent reverse transcription step and (4) that reverse transcription genomic DNA may be a source of variation if the gene of interest is disproportionately represented compared to other genes because of target unrelated reasons such as tumor growth.

Normalizing directly to a sample's cellular content is too uncertain for broad application. For example in patients with HIV-AIDS or those with malignant hematopoietic disorders, peripheral blood cell counts may be lowered or raised respectively, compared to otherwise healthy subjects infected by the microorganism under investigation. In these instances the number of extracellular target microbes per cell would be falsely raised or lowered (Huggett *et al.*, 2005). Another example is the detection of intracellular viruses or malarial parasites; in our previous example their numbers may appear falsely lowered or raised respectively. A differential cell count is advisable to avoid, or at least indicate, the possibility of such problems. Similarly, certain tissue conditions such as fibrosis or malignancy may increase the number of cells, but perhaps not in favor of those supporting the growth of the microorganism under investigation therefore resulting in falsely lowered normalized microbial load values (Huggett *et al.*, 2006).

A useful compromise for normalization of real-time qPCR data obtained from the clinical microbiology laboratory is to employ a second distinguishable, accurately quantified, synthetic, heterologous, IAC added to each extract and to an aliquot of extract obtained from the same matrix type for treatment in parallel (Figure 5.5). If poor extraction ef-

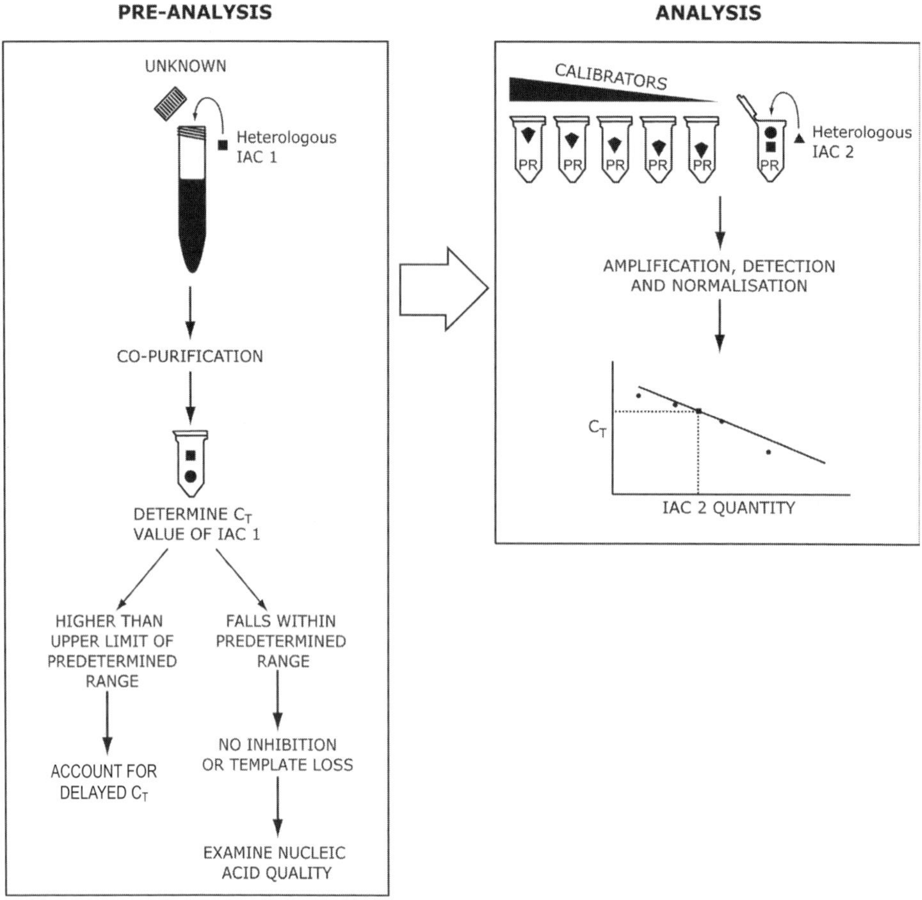

Figure 5.5 A schematic representation of the recommended steps required during the pre-analysis and analysis phase of real-time qPCR. The addition of an heterologous IAC (IAC 1;filled square) along with the chaotropic extraction reagent is recommended to identify extraction efficiency and the presence of inhibiting substances after purification. This requires a "pre-analysis" real-time PCR assay to examine the yield of IAC 1. Once inhibitor free nucleic acids have been efficiently extracted, the quality of the potential microbial template (filled circle) should be examined as described in the text. Only then should one consider the qPCR assay or "analysis" phase. A 2nd heterologous IAC (IAC 2; filled triangle) is added to a pre-determined amount of each unknown template preparation and the signal from the co-amplified IAC is used to normalize the real-time qPCR data. A reference standard or calibrator (filled diamond) dilution series is prepared and amplified alongside the unknown(s). A passive reference (PR) is also included in every tube to account for inter-tube variation caused by fluorescence inhibition or pipetting error. The corrected data then permits one to determine the amount of unknown in the extract by interpolation from the calibration curve.

ficiency or inhibition was identified during pre-analysis, this would need to be addressed before real-time qPCR could be undertaken (discussed earlier). Of particular benefit is the ability to quality control the large scale production of synthetic IACs and distribute them among clinical microbiology and research laboratories, perhaps one day leading to the development of commercial versions (Smith *et al.*, 2003). An IAC used in this way could not account for differences between the relative amounts of intra- and extracellular virus or nucleic acid but then in practical terms, no approach can.

If one agrees that no existing pre-analytical method can accurately identify the DNA or RNA content attributable to the microorganism and that values determined using current strategies more likely reflect the host's inconsistent cellular content and variable transcriptional status, then one must also agree that there is no perfect way to normalize real-time qPCR data in microbiology. However, employing IACs that are unaffected by host-related variation is much closer to an ideal approach. Agreement on these points would go some way to aligning our expectations with the reality of diagnostic clinical microbiology, rather than the unachievable approaches handed down from human transcription studies.

A disadvantage often attached to the use of synthetic, heterologous IACs is the need to carefully choose and laboriously clone the molecule and possibly transcribe RNA from the DNA versions *in vitro* (Huggett *et al.*, 2005). Also the claim is often made that such a molecule will not perform the same way during extraction when compared to deriving the intended target from an intracellular source but there seems to be little or no definitive evidence to support the existence of such a difference. The first reason is false economy; it does take time to choose and clone a synthetic target, but a heterologous target has the potential for use by all assays whereas a panel of suitable ERGs must be identified for every tissue, disease, target and experimental condition; this workload far surpasses that for a cloning and *in vitro* transcription protocol.

Quantification occurs close to home

The diagnostic needs of medical practitioners and researchers must often be met by the local research or diagnostic microbiology laboratories because there are relatively few kits available for quantifying microorganisms by PCR and few internationally validated reference standards or calibrators. However these sources are commonly perceived to produce assays which are less well developed than commercial versions making in-house assay accreditation extremely difficult. In practice this perception may be far from the truth and anecdotal evidence provides many tales of poorly performing commercial kits. In the absence of adequate commercial support or a relevant quality assurance program, a need exists for different laboratories who quantify the same targets to come together to construct suitable reagents for standardizing the performance of their assays. It should be possible to employ a cloning strategy to incorporate suitably conserved microorganism sequences of interest into simple, encapsidated particles (Chapter 4 and (Pasloske *et al.*, 1998)) for use as co-amplified DNA or RNA controls. Surely, in this age of rapid communication and comprehensive protocols for the transport of biological materials, some key reference microbiology laboratories could develop and disseminate this material among their peers with a sharing of costs, in order to at least develop proof-of-concept quality assurance programs for more targets. Success in such endeavors may be the lure that attracts commercial interests to become more actively involved in producing the reagents on a larger scale and under best practice methods.

Emerging roles for real-time PCR quantification

Among the many real-time qPCR applications in microbiology, some are less frequently considered than others. These also suffer from the issues we have discussed above and because they may have a larger role in clinical microbiology in the future, now is a good time to consider these less common applications as part of an overall process of improving real-time qPCR for microbiology.

Viral gene therapy

Gene therapy aims to introduce genetic material to an ill patient for therapeutic purposes with the goal of improving the quality of life; ideally providing a permanent cure for the target illness (Pfeifer and Verma, 2001). Genes are most commonly delivered to patient tissues in modified virus vectors (http://www.wiley.co.uk/genetherapy/clinical/). While still a young science burdened by many significant hurdles, one could envisage that the successful development of viral gene therapy protocols could result in their routine therapeutic use which would require the services of the high-throughput clinical microbiology laboratory. In addition to the use of real-time qPCR to quantify vector concentration in the research laboratory, diagnostic screening applications could be routinely used to determine how much virus is present in treated patients over time, to screen for the emergence of replication competent viruses due to recombination or mutation and to monitor the presence, integrity and activity of the transferred gene (transgene) or its regulatory elements. Traditional methods to detect the efficiency of viral vector-mediated gene transfer into a target cell or tissue (transduction) and to monitor the gene's integrity and transcriptional activity include histochemistry (Blömer *et al.*, 1997), immunofluorescent microscopy (Chen *et al.*, 2002) and Southern or Northern blot hybridization (Woo *et al.*, 2006). However, these techniques are lengthy, often insensitive and they have a low sample throughput. Therefore rapid and reliable methods to quantify targets are essential (Gautheret *et al.*, 1996; Tang *et al.*, 1998; Andreadis *et al.*, 2000; Delenda and Gaillard, 2005).

Studies have recently identified that analysis of DNA, specifically by real-time PCR using TaqMan probes, is as much as 10^3 to 10^4-fold more sensitive than monitoring a green fluorescent protein (GFP) reporter transgene by FACS analysis or determining RNA concentration from culture supernatants, when assessing the titer of a lentiviral vector (Scherr *et al.*, 2001; Sastry *et al.*, 2002). However the question then arises: how does one determine whether particles are infectious (Delenda and Gaillard, 2005)? Because this answer remains elusive, the detection of proviral DNA has become a useful surrogate measure for retrovirus-mediated gene transfer since these vectors have then at least transduced the target cell. Real-time PCR has also been used to detect integration of retroviral vectors (Martin-Rendon *et al.*, 2002). This approach is reliant upon a correlation between a positive signal and vector integration following serial passage of the host cells and identifies the past presence of infectious vector particles. Passaging the cells of a human host is not feasible but more importantly, this approach cannot confirm that the signal supposedly indicating the presence of proviral DNA is not derived from DNA retained within infected cells in some other non-integrated form (Gerard *et al.*, 1996). While PCR detection does not prove integration of the retrovector, identifying transduction is useful to show that a packaged vector preparation was infectious.

The predominantly TaqMan-based real-time PCR studies used for quantification in gene therapy applications, as well as our own earlier conventional competitive quantitative PCR approaches either chose target sequences that were common to the vector class, e.g. the common packaging or regulatory signal of a retrovirus vector (Gerard *et al.*, 1996; Sastry *et al.*, 2002; Lizeé *et al.*, 2003), strong stop retroviral cDNA (Scherr *et al.*, 2001) or the transgene, e.g. *gfp* or neomycin phosphotransferase (Sanburn and Cornetta, 1999; Klein *et al.*, 2000; Mackay *et al.*, 2001) or its promoter sequence (Martin-Rendon *et al.*, 2002)). In this way, variants of the vector backbone could be monitored using the same quantitative assay without re-development for different disease targets. Virus-specific sequences have also been used to monitor the biodistribution of viral vectors within the subject's organs (Hackett *et al.*, 2000). This approach identified that the majority of an intravenous preparation of adenovector was deposited within the lung. Whilst others have reported success using viral RNA templates derived from culture supernatants, the use of a DNA target avoids introducing additional handling and standardization requirements. However one must be very careful to avoid, or at least account for, contamination with DNA from the original plasmid components left over from the preparation phase of a viral vector; DNase treatments are suggested to at least remove any extracellular contaminants (Scherr *et al.*, 2001).

Conventional methods which analyze transgene function such as FACS rely on a fluorescent reporter; therefore they have limited use for testing therapeutically treated patients. On these occasions ELISA methods may indicate the presence of an expression product if a suitably specific antiserum is available. Such reagents are often unavailable commercially or prone to problems with background and cross-reaction. Furthermore, promoter strength and choice of biological host system can significantly influence the apparent number of vector particles (Bowers *et al.*, 2000). Real-time qPCR is a more objective method and it has proven itself capable of reliable vector detection (Deffrasnes *et al.*, 2005).

Microbial load and the host response to infection: the future of pathogenicity?

The application of real-time PCR is beginning to provide evidence to support paradigm shifts in the way we conceptualize the interplay between microbe and host as well as in the way we describe these interactions (Casadevall and Pirofski, 2003). Pathogen-centric views have dogmatically defined virulence as the result of microbial factors, when in fact the role of the host can be clearly shown to exert a significant influence (Mitchell, 1998; Casadevall and Pirofski, 2001). Increasingly, the old view is being challenged by our growing understanding that changing host immunity is as integral to the course of an infectious disease as the microbe present in the infected host. This poses the question "what is a pathogen"? (Casadevall and Pirofski, 2002). Microarray technologies present answers from the host's perspective providing a comprehensive snap-shot of the state of the host or invading microbe's transcriptome however arrays do not permit reliable quantification of the target change over a broad dynamic range (Lockhart and Winzeler, 2000; Lucchini *et al.*, 2001; Clewley, 2004). Real-time PCR is the method of choice for validating and further characterizing experimental data generated by microarrays and it will be at this level that we may be able to answer the pathogenicity question. Considerably more has been published on the role of real-time PCR in the basic detection of cytokine transcripts from stored or *in vitro* stimulated blood cells than as an indicator of microbial impact (Kruse *et al.*, 1997; Härtel *et al.*, 1999; Blaschke *et al.*, 2000; Stordeur *et al.*, 2002). This is set to be a growth area for future real-time PCR applications and perhaps an addition to the repertoire of the clinical microbiology laboratory of the future.

Cytokines play an important role in the innate mammalian immune response, where they mediate the communication with and recruitment of leukocytes and the removal of invading microorganisms. However, it is the host's molecular pattern recognition systems operating via specialized receptors that trigger many of the events leading to cytokine induction post-infection (Strieter *et al.*, 2003). Real-time PCR has identified critical mRNA from these receptors and from cytokine genes as a *de facto* indicator of protein production (Overbergh *et al.*, 1999; Giulietti *et al.*, 2001; Hein *et al.*, 2001; Overbergh *et al.*, 2003). While this is not a perfect relationship insofar as protein levels are modified by more than simply transcript abundance, studies have also shown a good correlation between the transcriptome and the proteome for many genes (Balnaves *et al.*, 1995). Additionally, a molecular approach is valuable because it is often impossible to detect the expressed product in tissue samples due to the low amounts present and short half-lives of cytokine proteins (Stordeur *et al.*, 2002; Broberg *et al.*, 2003). Nonetheless, transcriptional studies using real-time PCR are enhanced by accompanying protein expression data.

Well-controlled quantitative real-time PCR experiments have permitted the differentiation and quantification of IFN-α subclasses as a measure of the innate immune response to infectious or inactivated virus in humans (Löseke *et al.*, 2003). The probe or SYBR Green I-based real-time determination of expression of human IL-4, IFN-γ or mouse IL-12p35, 12p40, 23p19 and IFN-γ cytokines has also been successfully monitored in murine models of microbial immunopathogenesis (Deng *et al.*, 2003; Broberg *et al.*, 2003). In both studies mice were used to provide a laboratory model of infection and immunopathogenesis for human viral pathogens or to study the use of quantitative real-time PCR on practical samples. β-actin was used to normalize the data. Relative quantification by TaqMan-based real-time PCR and glyceraldehyde 3-phosphate dehydrogenase normalization was employed by Song *et al.* to examine the direct impact of cytokines on viral gene expression and the effect of a human herpesvirus-8 (HHV-8) cytokine homologue (vIL-6) on co-infecting viruses (Song *et al.*, 2002). This study demonstrated both interplay between pathogens and the host, indicating the complex nature of microbial infections and identified a potential therapeutic strategy to combat HIV and HHV-8 co-infection. In another study, a SYBR-Green I-based RT-qPCR approach was employed to demonstrate that IL-8 production was upregulated after *Chlamydia trachomatis* protein production commenced within infected HeLa cells (Buchholz and Stephens, 2006). This study aimed to discern the mechanisms of IL-8 production since this cytokine is chronically produced in *C. trachomatis* infections and it is also associated with inflammation-based tissue damage, a hallmark of *Chlamydia*-based disease.

Zhao and colleagues described the use of quantitative TaqMan-based real-time PCR to confirm the results of microarray analysis of *in vitro* adenovirus-infected HeLa cultures

(Zhao *et al.*, 2003). Adenovirus infection downregulated several cytokine genes and genes associated with growth arrest. Some stress response genes were upregulated as was an apoptosis inhibitor. Interestingly, β-actin was not differentially expressed and so was used to normalize the data. SYBR-Green I or TaqMan applications have validated microarray data indicating that parvovirus H-1 NS1 gene transcript accumulation could be correlated with the non-apoptotic death of an infected cell line and that a deletion mutant of African swine fever virus was unable to replicate effectively in infected macrophages, possibly due to a relatively enhanced type I interferon (IFN) response compared to that induced by a parental virus (Afonso *et al.*, 2004; Li *et al.*, 2005). These data have also correlated with protein or Northern blot analyses, respectively. TaqMan-based real-time relative RT-qPCR and ELISA were used to show that an impaired type I interferon response resulted from HRV-16 infection of *in vitro* cultured primary bronchial epithelial cells obtained from patients with asthma (Wark *et al.*, 2005). Since a type I IFN deficiency in asthmatics facilitates virus replication, there may be a therapeutic role for IFN-β in moderating virus-induced exacerbations of asthma. SYBR Green I based real-time PCR once again confirmed microarray data which suggested the human polyomavirus JCV induces a regulatory immune response in transfected and infected cells (Verma *et al.*, 2006). In turn, Western blot data corroborated the PCR results for two of the upregulated genes. Among immunocompetent hosts this cytokine response may constrain the progression of more serious disease and perhaps permit a long term latency and persistent infection. Another respiratory pathogen, *Bordetella pertussis* was investigated by Ishibashi and colleagues using SYBR-Green I-based real-time PCR and they showed that a fimbrial structure on the surface of the bacterium upregulated intercellular adhesion molecule-1 expression whilst the presence of pertussis toxin downregulated its expression thus inhibiting a putative antibacterial mechanism (Ishibashi and Nishikawa, 2002). Once again β-actin was used to normalize these data.

Using a SYBR Green I real-time RT-qPCR, Nelson and colleagues were able to propose that by upregulating expression of siderocalin, a potent bacteriostatic effector, *Streptococcus pneumoniae* and *Haemophilus influenzae* could limit the growth of susceptible competitor species of bacteria (Nelson *et al.*, 2005). On a larger scale, SYBR Green I-based real-time RT-qPCR has been used to validate microarray data for 17 genes expressed in response to SARS-CoV infection of Vero E6 cells, providing insight into the pathophysiology of SARS (Leong *et al.*, 2005).

Despite an increase in the number of published studies employing real-time qPCR to investigate the impact of a microorganism on its host, there are still few studies looking at microbial determinants, or virulence factors, at the same time as the host's response. Furthermore these approaches also fail to account for multiple virulence factors (McClelland *et al.*, 2005; McClelland *et al.*, 2006) which could become an area of research for future development.

Quality and real-time qPCR

Not only does the lack of commercially available kits remain a major problem for quantification of microorganisms, but the absence of standardization among existing in-house assays results in the production of incomparable data (Kao *et al.*, 2005). These weaknesses and others we have outlined throughout the entire chapter mean that many qPCR approaches are not fit to deliver upon the their stated intent, in particular true absolute quantification of the microorganism in a patient specimen.

The implementation of quantitative molecular techniques in the 1990s (Piatak Jr. *et al.*, 1993) confronted us then as it does now with questions related to the accurate and precise detection of different microorganisms, the best approach to standardize these quantitative test systems, how to monitor performance of the calibrators, the relevance of extreme sensitivity and whether a test system developed in-house is less reliable than one available commercially. Questions aside, the quality of data firstly relies upon our continual vigilance over routine technical PCR variables (listed previously). Quality also depends on variables addressed during the pre-analysis phase which include the amount of nucleic acid tested, the quality of the template and the presence of inhibitors. Together with the variability of the reverse transcription step and the subjectivity of data analysis there are clearly many ways to confuse quantitative results (Bustin, 2002). Consequently, it is essential to minimize variability and maximize reproducibility by quality controlling every component of the real-time qPCR assay and adhering to common guidelines for data analysis (Hands *et al.*, 2007). Additional complexity is added to the development of robust real-time qPCR protocols by the variety of different approaches to developing and implementing standards or calibrators. Comparative studies should be performed to clearly identify the advantages and disadvantages of using intact microbes versus naked nucleic acid versus encapsidated pseudoviral targets.

Commercial sources of certified reference materials and calibrators include the National Institute for Biological Standards and Control (NIBSC, UK; http://www.nibsc.ac.uk/), BBI diagnostics (http://www.seracare.com/index.htm, USA), BioRad laboratories and ZeptoMetrix corporation (http://www.zeptometrix.com/nattrol.htm) and include parvovirus B19, hepatitis A virus (HAV), HBV, HCV, HIV, West Nile virus, cytomegalovirus (CMV), human papillomavirus (HPV), *Chlamydia trachomatis* or *Neisseria gonorrhoeae* materials for qualitative, quantitative or genotyping nucleic acid testing assays.

Quality control (QC) systems measure data integrity, correctness and completeness, address errors and document the entire process. To be fit for its stated purpose laboratory data must enable a user to make technically and administratively correct decisions (Kubista *et al.*, 2006). Though simple in concept, QC is not an easy regimen to apply in daily work because most clients of the clinical microbiology laboratory do not specify what they expect to find and to what level of accuracy they require the laboratory to operate. Clients often only ask for "best measurement," which is more a concept than an absolute since the accuracy of a measurement is improved by replicates that average out technical variability. Of course extra testing comes at a cost. It follows then that obtaining data of the highest possible accuracy and precision may not always be required to satisfy a client, but may be essential to perform well in proficiency studies. In proficiency studies the technical competence of the laboratory is tested and measures of accuracy and precision can be derived. Despite this fundamental ambiguity, one very basic idea is clear: a good result should be correct. Correctness relies upon both precision and accuracy.

The importance of analytical precision is not in question although most discussions are concerned with how the replicates should be performed; in the same session, on different days, by different operators, etc. In most cases this is solved by specifying whether the precision calculations were made in a repeatable (repetitions were made in one working session using the same reagents, instruments, by the same staff, etc.) or reproducible (repetitions were made by changing many experimental conditions) manner, or even using an intermediate level of precision (repetitions were made changing only some experimental conditions).

Accuracy is much more difficult to calculate, if possible at all, since it is the degree to which the measured value and the "true" value agree. The main problem here is that "true" values cannot be determined. So, how can we address this fundamental problem? There are three possibilities: using certified reference materials, comparing results with Reference Methods, and participating in inter-laboratory exercises or quality assurance (QA) programs. The two first options are not yet available and we can presently only rely on inter-laboratory exercises.

Quality assurance aids the production of comparable data by formally assessing quality controlled laboratory methods, the suitability of the standard operating procedures, organization and management of the laboratory and result auditing. Within Europe an energetic culture of activities exist to review PCR procedures. The largest provider of European nucleic acid based quality control schemes groups is QCMD (Quality Control for Molecular Diagnostics, http://www.qcmd.org/). Established in 2001, QCMD is endorsed by the European Society for Clinical Microbiology and Infectious Disease (ESCMID). Other groups conducting quality assurance programs include the European Network for Diagnostics of Imported Viral Diseases (ENIVD; http://www.enivd.org), the United Kingdom National External Quality Assessment Service (UKNEQAS; http://www.ukneqas.org.uk), the Institute for Standardisation and Documentation in the Medical Laboratory (INSTAND; http://www.instand-ev.de) the Royal College of Pathologists Australasia (RCPA; http://www.rcpaqap.com.au) and the National Serology Reference Laboratory, Australia which also provides nucleic acid testing quality assurance programs (NRL; http://www.nrl.gov.au).

There are six principles that a quality assurance program should address (EURACHEM working group, 1998):

1. Analytical measurements should be made to satisfy an agreed requirement, i.e. the client should specify what is required and this should be on par with the quality of peer laboratories.
2. Analytical measurements should be made using methods and equipment which are themselves certified as fit for the purpose.
3. Staff making analytical measurements should be both qualified and competent to undertake the task.
4. There should be a regular independent quality assessment of the technical performance of the laboratory, based on predetermined requirements.
5. Analytical measurements made in one location should be consistent with those made elsewhere.
6. Organizations making analytical measurements should have well defined QC and QA procedures.

Most of these principles relate to method validation, which is beyond the scope of this brief discussion. Below we will further consider inter-laboratory exercises.

There are three main types of inter-laboratory exercise (Lawn and Thompson, 1997):

1 Collaborative trials intended to study an analytical procedure in depth. The participating laboratories must strictly follow the guidelines and written procedures provided by the organizing body. There should be minimum freedom to introduce variability, because the aim is to identify any bias, figures of merit, etc. for a particular method, or perhaps even to certify it.
2 Certification exercises aiming to verify one or more targets in a given sample matrix. Here, freedom is exercised regarding the choice of analytical method, procedure, instrument, etc. The main requirement is that the participating laboratories have experience working with the target and/or the matrix.
3 Proficiency testing is a collaborative exercise to evaluate how laboratories perform, i.e. to evaluate their technical competence and compare that with others. Participating laboratories can select the analytical procedure for the purpose. They receive an unknown sample to be analyzed and they report results back to the organization that is in charge of the study.

The goal of proficiency testing is to obtain a regular, objective, and independent assessment of the accuracy of an analytical laboratory's performance using routine test samples and to promote improvements in their operating procedures (Lawn and Thompson, 1997).

Two issues should be stressed. Proficiency testing is not a one-off event. To be useful several exercises are required. In fact, the first exercise is often a disaster for some of the participating laboratories. Three exercises per year are usually recommended (Lawn and Thompson, 1997). Routine is the magic word in these exercises. It is well known that many managers and analysts pamper inter-laboratory test samples. This produces results that do not represent routine operations, but rather how well a laboratory can perform under unrealistically ideal conditions. If a laboratory really is interested in finding out how good their routine is, these samples must be handled as would any other sample.

Simple statistical evaluations of the results include removing outlier samples and outlier laboratories and autoscaling the reported data (so-called z-scores). Autoscaling consists on subtracting the mean of the overall set of values and dividing by the standard deviation (SD). Based on the Gaussian distribution laboratories reporting data outside mean ± 3SD (99% confidence interval) have performed poorly, while laboratories reporting within ± 2SD (95% confidence intervals) performed well. The final general meeting between the laboratory managers should look for and explain major differences and identify sources of error. By the way, never throw away any valuable leftover test samples; although it is not a certified reference material, it is a valuable calibrator for the continued control of one's own assays and laboratory performance.

In 2006 a report of the first European proficiency inter-laboratory test on nucleic acid analysis and real-time PCR was presented from the EU framework 6 program (Ramsden *et al.*, 2006). 137 laboratories from 29 member states participated. The laboratories were provided plasmid calibrators containing the ABL proto-oncogene, and primers and TaqMan probes for the detection of ABL. They were also provided three cloned cDNA samples; two samples contained cells resuspended in RNAlater®. They were asked to:

1 Construct a calibration curve based on the ABL plasmid calibrators provided
2 Estimate ABL cDNA copy numbers in the three cloned samples by real-time PCR
3 Optionally perform RNA extraction and RT and measure cDNA levels by real-time PCR to estimate the amounts of ABL RNA in the resuspended cells.

For the pre-extracted samples more than 95% of the participants showed valid performance, while for the quantification of cells, valid performance was limited to 40 to 50% of the participants. This very large discrepancy can most likely be explained by the large variations in performance of reverse transcription protocols. While all laboratories used the same real-time PCR reagents and protocol (provided by the organizers) to quantify cDNA levels, they used in-house methods for RNA extraction and purification and reverse transcription to analyze the cells. The very large gene specific variation among reverse transcriptases and priming methods (Ståhlberg *et al.*, 2004b) cannot be controlled using reference genes, IACs and DNA based calibrators. The only way to obtain comparable inter-laboratory results when using different reverse transcriptase protocols is to use exogenous calibrators based on RNA.

The above discussion and the European EQUAL project calls for guidelines and standard operating procedures for real-time PCR based diagnostics. In US the FDA is preparing guidance for both industry and FDA staff on the best use of nucleic acid based *in vitro* diagnostic devices for detection of microbial pathogens (http://www.fda.gov/dockets/ecomments). In Europe directives for CE marking for *in vitro* diagnostic medical devices were laid in 1998 (2006). Nonetheless the need for better quality real-time PCR assays encompassing more target microorganisms in clinical microbiology is real but remains poorly addressed. Published QA studies, commonly focus on the same targets as those developed for commercial kits and describe up to a third of participants performing inadequate human enterovirus, HSV, HCV or respiratory virus PCR detection with particular problems detecting low template loads (van Vliet *et al.*, 2001; Schirm *et al.*, 2002; Schloss *et al.*, 2003; Templeton *et al.*, 2006). A common feature of these publications is the use of a wide range of commercial and in-house PCR formats making determination of the best assay type for each microbial target, or in general, difficult to assess although in-house methods have been flagged as performing with less reliability than commercial approaches (Verkooyen *et al.*, 2003). While the main purpose of a quality assurance program is to examine the capability of participants to provide reliable results, method selection is an important factor for improved performance that can only be interpreted when such a diverse range of users come together as occurs during QA programs. It is apparent that regular participation in QA schemes improves the proficiency of PCR testing in microbiology (Mancini *et al.*, 2004) so it is up to the end-users to create the environment for improvement to all aspects of assay quality so that real-time qPCR can reach its true potential.

Future trends

A natural progression of real-time qPCR technology and its role in clinical microbiology would seem to be quantification of microorganisms in parallel with the immune responses they elicit; yet this area remains largely unexplored. Additional areas that will hopefully receive more attention in the future include improved collaboration among those using real-time qPCR for under-characterized microorganisms, the production of more reference materials and calibrators, the production of more commercial real-time qPCR kits for the diagnostic laboratory and the overall need to identify simple, relevant and realistic goals for real-time qPCR in the clinical microbiology laboratory.

Conclusions

Real-time technology has revolutionized the use of PCR assays and permitted the simple, speedy, sensitive and specific generation of quantitative data. Furthermore, the amount of data generated lends itself to analysis using statistics, which increases the confidence in reporting quantitative results. The development of standardized experimental protocols and designs that are rigorously controlled still remains an important challenge. Overcoming this challenge will permit meaningful global comparisons of microbial load data.

While real-time PCR can detect two-fold differences in the amount of DNA (Bubner *et al.*, 2004; Pal *et al.*, 2006), the technique struggles to produce equivalent results from an RNA template due to issues relating to our current inability to directly detect small quantities of RNA. Practically, two-fold differences in the amount of a microorganism may not prove to be clinically relevant. If realistic thresholds could be identified for each microorganism then we could use these to add value to our results. If the amount of target microorganism is below the threshold then the laboratory could report with confidence that infection is not associated with illness. Similarly, a sub-threshold microbial load may indicate successful antimicrobial therapy whereas a rise above threshold may signal reactivation of latent or persistent infection, heralding a poor clinical outcome. Validated thresholds may be one way we can relax some of the technical constraints we are currently bound by while we unproductively attempt to develop near-perfect quantitative assays.

Currently we apply real-time qPCR methods better suited to the study of human transcription and we usually expect them to provide robust data for what is an entirely different set of problems; those of complex infectious disease processes. If we could modify our approach and openly discuss the variables that set clinical microbiology apart, then perhaps we may produce real-time qPCR applications providing more realistic and robust results that benefit a greater number of patient outcomes around the world. While such applications may not produce the perfect estimate of microorganism numbers, they will lessen the likelihood of irreproducible and clinically irrelevant data. The major implication of inaccurate qPCR assays is a slowing of our understanding of infectious disease etiology. Since we live in a time when the discovery of newly emergent or previously unknown endemic pathogens is increasing in frequency, we must strive harder than ever before to expand our understanding of infectious diseases, and for that we need reliable results from trustworthy tools.

Acknowledgments

Part of this work was supported by the Spanish Ministry of Science and Technology (grant SAB2005–0162) to Mikael Kubista. Thanks must go to Katherine Arden and John (F) Mackay for critical review of this document.

Web sites

European Network for Diagnostics of Imported Viral Diseases (ENIVD)
http://www.enivd.org/
A collaboration of scientists from university medical centers, country health departments, and hospitals all over Europe working to improve diagnosis of "imported" viral disease.

Tataa Biocenter
http://www.tataa.com
The Center conducts commissioned research and training within molecular diagnostics and gene expression analysis using real-time PCR technology to specifically quantify nucleic acids.

Quality Control for Molecular Diagnostics (QCMD)
http://www.qcmd.org/
QCMD specializes in the standardization and quality control for molecular diagnostics and genomic technologies.

References

Directive 98/79/EC of the European Parliament and of the Council of 27 October 1998 on *in vitro* diagnostic medical devices. (2006).

Abe, A., Inoue, K., Tanaka, T., Kato, J., Kajiyama, N., Kawaguchi, R., Tanaka, S., Yoshiba, M., and Kohara, M. (1999). Quantitation of hepatitis B virus genomic DNA by real-time detection PCR. J. Clin. Microbiol. 37, 2899–2903.

Aberham, C., Pendl, C., Gross, P., Zerlauth, G., and Gessner, M. (2001). A quantitative, internally controlled real-time PCR assay for the detection of parvovirus B19 DNA. J. Virol. Methods 92, 183–191.

Afonso, C.L., Piccone, M.E., Zaffuto, K.M., Neilan, J., Kutish, G.F., L.u., Z., Balinsky, C.A., Gibb, T.R., Bean, T.J., Zsak, L., and Rock, D.L. (2004). African swine fever virus multigene family 360 and 350 genes affect host interferon response. J. Virol. 78[4], 1858–1864.

Akane, A., Matsubara, K., Nakamura, H., Takahashi, S., and Kimura, K. (1994). Identification of the heme compound copurified with deoxyribonucleic acid (DNA) from bloodstains, a major inhibitor of polymerase chain reaction (PCR) amplification. J. For. Sci. 39[2], 362–372.

Al-Soud, A. and Rådström, P. (1998). Capacity of nine thermostable DNA polymerases to mediate DNA amplification in the presence of PCR-inhibiting substances. Appl. Environ. Microbiol. 64[10], 3748–3753.

Al-Soud, W.A., Jönsson, L.J., and Rådström, P. (2000). Identification and characterization of immunoglobulin G in blood as a major inhibitor of diagnostic PCR. J. Clin. Microbiol. 38[1], 345–350.

Alexandre, I., Zammatteo, N., Ernest, I., Ladriere, J.-M., Hamels,Le.S., Chandelier, N., Vipond, B., and Remacle, J. (1998). Quantitative determination of CMV DNA using a combination of competitive PCR amplification and sandwich hybridization. Biotechniques 25, 676–683.

Andersen, C.L., Jensen, J.L., and Ørntoft, T.F. (2004). Normalization of real-time quantitative reverse transcription-PCR data: A model-based variance estimation approach to identify genes suited for normalization, applied to bladder and colon cancer data sets. Cancer Res. 64, 5245–5250.

Andreadis, S., Lavery, T., Davis, H.E., Le Doux, J.M., Yarmush, M.L., and Morgan, J.R. (2000). Toward a more accurate quantitation of the activity of recombinant retroviruses: Alternatives to titer and multiplicity of infection. J. Virol. 74, 1258–1266.

Arden, K.E., McErlean, P., Nissen, M.D., Sloots, T.P., and Mackay, I.M. (2006). Frequent detection of human rhinoviruses, paramyxoviruses, coronaviruses, and bocavirus during acute respiratory tract infections. J. Med. Virol. 78, 1232–1240.

Auer, H., Lyianarachchi, S., Newsom, D., Klisovic, M.I., Marcucci, U., and Kornacker, K. (2003). Chipping away at the chip bias: RNA degradation in microarray analysis. Nat Gen. 35, 292–293.

Bagnarelli, P., Menzo, S., Valenza, A., Paolucci, S., Petroni, S., Scalise, G., Sampaolesi, R., Manzin, A., and Clementi, M. (1995). Quantitative molecular monitoring of human immunodeficiency virus type 1 activity during therapy with specific antiretroviral compounds. J. Clin. Microbiol. 33, 16–23.

Balnaves, M.E., Bonacquisto, L., Francis, I., Glazner, J., and Forrest, S. (1995). The impact of newborn screening on cystic fibrosis testing in Victoria, Australia. J. Med. Genet. 32, 537–542.

Bar, T., Ståhlberg, A., Muszta, A., and Kubista, M. (2003). Kinetic outlier detection (KOD) in real-time PCR. Nucleic Acids Res. *31*, e105.

Bengtsson, M., Karlsson, H.J., Westman, G., and Kubista, M. (2003). A new minor groove binding asymmetric cyanine reporter dye for real-time PCR. Nucleic Acids Res. *31*, e45.

Blaschke, V., Reich, K., Blaschke, S., Zipprich, S., and Neumann, C. (2000). Rapid quantitation of proinflammatory and chemoattractant cytokine expression in small tissue samples and monocyte-derived dendritic cells: validation of a new real-time RT-PCR technology. J. Immunol. Methods *246*, 79–90.

Blömer, U., Naldini, L., Kafri, T., Trono, D., Verma, I.M., and Gage, F.H. (1997). Highly efficient and sustained gene transfer in adult neurons with a lentivirus vector. J. Virol. *71*, 6641–6649.

Bowers, W.J., Howard, D.F., and Federoff, H.J. (2000). Discordance between expression and genome transfer titering of HSV amplicon vectors: Recommendation for standardized enumeration. Mol. Ther. *1*, 294–299.

Brechtbuehl, K., Whalley, S.A., Dusheiko, G.M., and Saunders, N.A. (2001). A rapid real-time quantitative polymerase chain reaction for hepatitis B virus. J. Virol. Methods 93, 105–113.

Brightwell, G., Pearce, M., and Leslie, D. (1998). Development of internal controls for PCR detection of *Bacillus anthracis*. Mol. Cell. Probes 12, 367–377.

Broberg, E.K., Nygårdas, M., Salmi, A.A., and Hukkanen, V. (2003). Low copy number detection of herpes simplex virus type 1 mRNA and mouse Th1 type cytokine mRNAs by Light Cycler quantitative real-time PCR. J. Virol. Methods *112*, 53–65.

Bubner, B., Gase, K., and Baldwain, I.T. (2004). Two-fold differences are the detection limit for determining transgene copy numbers in plants by real-time PCR. BMC Biotechnology 4, 14.

Buchholz, K.R. and Stephens, R.S. (2006). Activation of the host cell proinflammatory interleukin-8 response by *Chlamydia trachomatis*. Cellular microbiology 8[11], 1768–1779.

Burggraf, S. and Olgemöller, B. (2004). Simple technique for internal control of real-time amplification assays. Clin. Chem. *50*, 819–825.

Bustin, S.A. (2002). Quantification of mRNA using real-time reverse transcription PCR (RT-PCR): trends and problems. J. Mol. Endocrinol. *29*, 23–39.

Bustin, S.A. (2004). A-Z of quantitative PCR. In A-Z of quantitative PCR, S.A.Bustin, ed. (California: IUL Press).

Bustin, S.A. (2005). Real-time, fluorescence-based quantitative PCR: a snapshot of current procedures and preferences. Expert Rev. Mol. Diagn. 5, 493–498.

Bustin, S.A., Benes, V., Nolan, T., and Pfaffl, M.W. (2005). Quantitative real-time RT-PCR—a perspective. J. Mol. Endocrinol. *34*, 597–6001.

Bustin, S.A. and Nolan, T. (2004). Pitfalls of quantitative real-time reverse-transcription polymerase chain reaction. J. Biomol. Tech. *15*, 155–166.

Casadevall, A. and Pirofski, L. (2001). Host pathogen interactions: the attributes of virulence. J. Infect. Dis. *184*, 337–344.

Casadevall, A. and Pirofski, L. (2002). What is a pathogen? Ann. Med. *34*, 2–4.

Casadevall, A. and Pirofski, L. (2003). Microbial virulence results from the interaction between host and microorganism. Trends Microbiol. *11*, 157–158.

Celi, F.S., Mentuccia, D., Proietti-Pannunzi, L., di Gioia, C.R.T., and Andreoli, M. (2000). Preparing poly-A-containing RNA internal standards for multiplex competitive RT-PCR. Biotechniques *29*, 454–458.

Chehab, F.F., Doherty, M., Cai, S., Kan, Y.W., Cooper, S., and Rubin, E.M. (1987). Detection of sickle cell anaemia and thalassaemias. Nature *329*, 293–294.

Chen, H., Wurm, T., Britton, P., Brooks, G., and Hiscox, J.A. (2002). Interaction of the coronavirus nucleoprotein with the nucleolar antigens and the host cell. J. Virol. *76*, 5233–5250.

Clementi, M. (2000). Quantitative molecular analysis of virus expression and replication. J. Clin. Microbiol. 38, 2030–2036.

Clementi, M., Menzo, S., Manzin, A., and Bagnarelli, P. (1995). Quantitative molecular methods in virology. Arch. Virol. *140*, 1523–1539.

Clewley, J.P. (2004). A role for arrays in clinical virology: fact or fiction? J. Clin. Virol. *29*, 2–12.

Cone, R.W., Hobson, A.C., and Huang, M.-L.W. (1992). Coamplified positive control detects inhibition of polymerase chain reactions. J. Clin. Microbiol. *30*, 3185–3189.

Cusi, M.G., Valassina, M., and Valensin, P.E. (1994). Comparison of M-MLV reverse transcriptase and Tth polymerase activity in RT-PCR of samples with low virus burden. Biotechniques *17*[6], 1034–1036.

Deffrasnes, C., Côté, S., and Boivin, G. (2005). Analysis of replication kinetics of the human metapneumovirus in different cell lines by real-time PCR. J. Clin. Microbiol. *43*, 488–490.

Delenda, C. and Gaillard, C. (2005). Real-time quantitative PCR for the design of lentiviral vector analytical assays. Gene Ther. 12, S36-S50.

Deng, X., L.i., H., and Tang, Y.-W. (2003). Cytokine expression in respiratory syncytial virus-infected mice as measured by quantitative reverse-transcriptase PCR. J. Virol. Methods *107*, 141–146.

Dheda, K., Huggett, J.F., Chang, J.S., Kim, L.U., Bustin, S.A., Johnson, M.A., Rook, G.A.W., and Zumla, A. (2005). The implications of using an inappropriate reference gene for real-time reverse transcription PCR data normalization. Anal. Biochem. *344*, 141–143.

Espy, M.J., Uhl, J.R., Sloan, S.P., Buckwalter, S.P., Jones, M.F., Vetter, E.A., Yao, J.D.C., Wengenack, N.L., Rosenblatt, J.E., Cockerill III, F.R., and Smith, T.F. (2006). Real-time PCR in clinical microbiology: Applications for routine laboratory testing. Clin. Microbiol. Rev. *19*, 165–256.

EURACHEM working group. (1998). The fitness for purpose of analytical methods. *A Laboratory Guide to Method Validation and Related Topics*, Eurachem Guide 1998. Available at: www.eurachem.ul.pt/guides/valid.pdf

Ferré, F. (1992). Quantitative or semi-quantitative PCR: reality versus myth. PCR Meth. Appl. 1–9.

Fieller, E.C. (1940). The biological standardization of insulin. J. R. Stat. Soc. *7*[1], 1–64.

Fleige, S. and Pfaffl, M.W. (2006). RNA integrity and the effect on the real-time qRT-PCR performance. Mol. Aspects Med. *27*, 126–139.

Freeman, W.M., Walker, S.J., and Vrana, K.E. (1999). Quantitative RT-PCR: Pitfalls and potential. Biotechniques *26*, 112–125.

Gautheret, A., Aubin, J.-T., Poirel, L., Chopineau, S., Agut, H., and Dussaix, E. (1996). Detection and variant identification of HHV-6 by a non-radioactive hybridization microplate assay for amplimers detection. J. Virol. Methods *58*, 33–40.

Gelmini, S., Orlando, C., Sestini, R., Vona, G., Pinzani, P., Ruocco, L., and Pazzagli, M. (1997). Quantitative polymerase chain reaction-based homogeneous assay with fluorogenic probes to measure c-*cer*B-2 oncogene amplification. Clin. Chem. *43*, 752–758.

Gerard, C.J., Arboleda, M.J., Solar, G., Mule, J.J., and Kerr, W.G. (1996). A rapid and quantitative assay to estimate gene transfer into retrovirally transduced hematopoietic stem/progenitor cells using a 96-well format PCR and fluorescent detection system universal for mmLV-based proviruses. Hum. Gen. Ther. *7*, 343–354.

Gibellini, D., Vitone, F., Gori, E., La Placa, M., and, R.e., M.C. (2004). Quantitative detection of human immunodeficiency virus type 1 (HIV-1) viral load by SYBR green real-time RT-PCR technique in HIV-1 seropositive patients. J. Virol. Methods *115*, 183–189.

Gibson, U.E.M., Heid, C.A., and Williams, P.M. (1996). A novel method for real time quantitative RT-PCR. Genome Res. *6*, 995–1001.

Gilliland, G., Perrin, S., and Bunn, H.F. (1990). Competitive PCR for quantitation of mRNA. In PCR protocols: A guide to methods and applications., Academic Press), pp. 60–69.

Giulietti, A., Overbergh, L., Valckx, D., Decallonne, B., Bouillon, R., and Mathieu, C. (2001). An overview of real-time quantitative PCR: applications to quantify cytokine gene expression. Methods *25*, 386–401.

Glasel, J.A. (1995). Validity of nucleic acid purities monitored by 260nm/280nm absorbance ratios. Biotechniques *18*, 62–63.

Goossens, K., van Poucke, M., van Soom, A., Vandesompele, J., van Zeveren, A., and Peelman, L.J. (2005). Selection of reference genes for quantitative real-time PCR in bovine preimplantation embryos. BMC Dev. Biol. 5, 27.

Grove, S., Faller, R., Soleim, K.B., and Dannevig, B.H. (2006). Absolute quantitation of RNA by a competitive real-time RT-PCR method using piscine nodavirus as a model. J. Virol. Methods *132*, 104–112.

Gruber, F., Falkner, F.G., Dorner, F., and Hämmerle, T. (2001). Quantitation of viral DNA by real-time PCR applying duplex amplification, internal standardization, and two-colour fluorescence detection. Appl. Environ. Micro. *67*, 2837–2839.

Guy, R.A., Payment, P., Krull, U.J., and Horgen, P.A. (2003). Real-time PCR for quantification of *Giardia* and *Cryptosporidium* in environmental water samples and sewage. Appl. Environ. Microbiol. 69[9], 5178–5185.

Hackett, N.R., El Sawy, T., Lee, L.Y., Silva, I., O'Leary, J., Rosengart, T.K., and Crystal, R.G. (2000). Use of quantitative TaqMan real-time PCR to track the time-dependent distribution of gene transfer vectors *in vivo*. Mol. Ther. *2*, 649–656.

Halford, W.P. (1999). The essential prerequisites for quantitative RT-PCR. Nat. Biotechnol. *17*, 835.

Halford, W.P., Falco, V.C., Gebhardt, B.M., and Carr, D.J.J. (1999). The inherent quantitative capacity of the reverse transcription-polymerase chain reaction. Anal. Biochem. *266*, 181–191.

Hands, R.E., Nolan, T., and Bustin, S.A. (2007). Analysis of mRNA expression by real-time PCR. Nat. Prot. (in press).

Härtel, C., Bein, G., Kirchner, H., and Klüter, H. (1999). A human whole-blood assay for analysis of T-cell function by quantification of cytokine mRNA. Scand. J. Immunol. *49*, 649–654.

Hein, J., Schellenberg, U., Bein, G., and Hackstein, H. (2001). Quantification of murine IFN-γ mRNA and protein expression: Impact of real-time kinetic RT-PCR using SYBR Green I dye. Scand. J. Immunol. *54*, 285–291.

Hoorfar, J., Cook, N., Malorny, B., Wagner, M., De Medici, D., Abdulmawjood, A., and Fach, P. (2003). Making an internal amplification control mandatory for diagnostic PCR. J. Clin. Microbiol. *41*, 5835.

Huggett, J., Dheda, K., Bustin, S., and Zumla, A. (2005). Real-time RT-PCR normalisation; strategies and considerations. Genes Immun. *6*, 279–284.

Huggett, J., Dheda, K., and Bustin, S.A. (2006). Normalization. In Real-time PCR, M.T.Dorak, ed. (New York: Taylor & Francis), pp. 83–91.

Ishibashi, Y. and Nishikawa, A. (2002). *Bordetella pertussis* infection of human respiratory epithelial cells up-regulates intercellular adhesion molecule-1 expression: role of filamentous hemagglutinin and pertussis toxin. Microb. Pathogen. *33*, 115–125.

Jennings, L.C., Anderson, T.P., Werno, A.M., Beynon, K.A., and Murdoch, D.R. (2004). Viral etiology of acute respiratory tract infections in children presenting to hospital. Pediatr. Infect. Dis. J. *23*, 1003–1007.

Jiang, J., Alderisio, K.A., Singh, A., and Xiao, L. (2005). Development of procedures for direct extraction of *Cryptosporidium* DNA from water concentrates and for relief of PCR inhibitors. Appl. Environ. Microbiol. *71*[3], 1135–1141.

Johnson, M.R., Wang, K., Smith, J.B., Heslin, M.J., and Diasio, R.B. (2000). Quantitation of dihydropyrimidine dehydrogenase expression by real-time reverse transcription polymerase chain reaction. Anal. Biochem. *278*, 175–184.

Josefsen, M.H., Jacobsen, N.R., and Hoorfar, J. (2004). Enrichment followed by quantitative PCR both for rapid detection and as a tool for quantitative risk assessment of food-borne thermotolerant Campylobacters. Appl. Environ. Micro. *70*, 2588–2592.

Juhasz, A., Ravi, A., and O'Connell, C.D. (1996). Sensitivity of tyrosinase mRNA detection by RT-PCR: rTth DNA polymerase vs. mmLV-RT and AmpliTaq polymerase. Biotechniques *20*[4], 592–600.

Kao, C.L., King, C.C., Chao, D.Y., W.u., H.L., and Chang, G.J. (2005). Laboratory diagnosis of dengue virus infection: current and future perspectives in clinical diagnosis and public health. J. Microbiol. Immunol. Infect. *38*, 5–16.

Kawai, S., Yokosuka, O., Kanda, T., Imazeki, F., Maru, Y., and Saisho, H. (1999). Quantification of hepatitis C virus by TaqMan PCR: comparison with HCV amplicor monitor assay. J. Med. Virol. *58*, 121–126.

Kleiber, J., Walter, T., Haberhausen, G., Tsang, S., Babiel, R., and Rosenstraus, M. (2000). Performance characteristics of a quantitative, homogeneous TaqMan RT-PCR test for HCV RNA. J. Mol. Diagn. *2*, 158–166.

Klein, D., Bugl, B., Günzburg, W.H., and Salmons, B. (2000). Accurate estimation of transduction efficiency necessitates a multiplex real-time PCR. Gene Ther. *7*, 458–463.

Klein, S.A., Karsten, S., Rüster, B., Klebba, C., Pape, M., Ottmann, O.G., Hoelzer, D., and Roth, W.K. (2003). Comparison of TaqMan™ real-time PCR and p24 Elisa for quantification of *in vitro* HIV-1 replication. J. Virol. Methods *107*, 169–175.

Kolls, J., Deininger, P., Cohen, C., and Larson, J. (1993). cDNA equalization for reverse transcription-polymerase chain reaction quantitation. Anal. Biochem. *208*, 264–269.

Kruse, N., Pette, M., Toyka, K., and Rieckmann, P. (1997). Quantification of cytokine mRNA expression by RT PCR in samples of previously frozen blood. J. Immunol. Methods *210*, 195–203.

Kubista, M., Andrade, J.M., Bengtsson, M., Forootan, A., Jonák, J., Lind, K., Sindelka, R., Sjöback, R., Sjögreen, B., Strömbom, L., Ståhlberg, A., and Zoric, N. (2006). The real-time polymerase chain reaction. Mol. Aspects Med. *27*, 95–125.

Kubista, M., Ståhlberg, A., and Bar, T. (2001). Light-up probe based real-time Q-PCR. Proceedings of SPIE 4264, 53–58.

Lacey, H.A., Nolan, T., Greenwood, S.L., Glazier, J.D., and Sibley, C.P. (2005). Gestational profile of Na+/H+ exchanger and Cl-/HCO3- anion exchanger mRNA expression in placenta using real-time QPCR. Placenta *26*, 93–98.

Lawn, R.E. and Thompson, M. (1997). Proficiency Testing in Analytical Chemistry. R.S.o.Chemistry, ed.

Lefevre, J., Hankins, C., Pourreaux, K., Voyer, H., and Coutlée, F. (2004). Prevalence of selective inhibition of HPV-16 DNA amplification in cervicovaginal lavages. J. Med. Virol. 72, 132–137.

Lekanne Deprez, R.H., Fijnvandraat, A.C., Ruijter, J.M., and Moorman, A.F. (2002). Sensitivity and accuracy of quantitative real-time polymerase chain reaction using SYBR green I depends on cDNA synthesis conditions. Anal. Biochem. *307*, 63–69.

Leong, W.F., Tan, H.C., Ooi, E.E., Koh, D.R., and Chow, V.T.K. (2005). Microarray and real-time RT-PCR analyses of differential human gene expression patterns induced by severe acute respiratory syndrome (SARS) coronavirus infection of Vero cells. Microbes Infect. *7*, 248–259.

Leutenegger, C.M., Higgins, J., Matthews, T.B., Tarantal, A.F., Luciw, P.A., Pedersen, N.C., and North, T.W. (2001). Real-time TaqMan PCR as a specific and more sensitive alterntaive to the branched-chain DNA assay for quantitation of simian immunodeficiency virus RNA. AIDS. Res. Hum. Retro. *17*, 243–251.

Levy, J.P. (1999). Xenotropism: the elusive viral receptor finally uncovered. Proc. Natl. Acad. Sci. USA *96*, 802–804.

Lewis, F. and Maughan, N.J. (2004). Extraction of Total RNA from Formalin-Fixed Paraffin-Embedded Tissue. In A-Z of quantitative PCR, S.A.Bustin, ed. (La Jolla, California: IUL).

Li, J., Werner, E., Hergenhahn, M., Poirey, R., Luo, Z., Rommelaere, J., and Jauniaux, J-C. (2005). Expression profiling of human hepatoma cells reveals global repression of genes involved in cell proliferation, growth, and apoptosis upon infection with parvovirus H-1. J. Virol. *79*[4], 2274–2286.

Limaye, A.P., Jerome, K.R., Kuhr, C.S., Ferrenberg, J., Huang, M.-L., Davis, C.L., Corey, L., and Marsh, C.L. (2000). Quantitation of BK virus load in serum for the diagnosis of BK virus-associated nephropathy in renal transplant recipients. J. Infect. Dis. *183*, 1669–1672.

Lind, K., Ståhlberg, A., Zoric, N., and Kubista, M. (2006). Combining sequence-specific probes and DNA binding dyes in real-time in real-time PCR for specific nucleic acid quantification and melting analysis. Biotechniques *40*, 315–319.

Lizeé, G., Aerts, J.L., Gonzales, M.I., Chinnasamy, N., Morgan, R.A., and Topalian, S.L. (2003). Real-time quantitative reverse transcriptase-polymerase chain reaction as a method for determining lentiviral vector titres and measuring transgene expression. Hum. Gene Ther. *14*, 497–507.

Locatelli, G., Santoro, F., Veglia, F., Gobbi, A., Lusso, P., and Malnati, M.S. (2000). Real-time quantitative PCR for human herpesvirus 6 DNA. J. Clin. Microbiol. *38*, 4042–4048.

Lockhart, D.J. and Winzeler, E.A. (2000). Genomics, gene expression and DNA arrays. Nature *405*, 827–836.

Löseke, S., Grage-Griebenow, E., Wagner, A., Gehlhar, K., and Bufe, A. (2003). Differential expression of IFN-α subtypes in human PBMC: evaluation of novel real-time PCR assays. J. Immunol. Methods *276*, 207–222.

Lucchini, S., Thompson, A., and Hinton, J.C.D. (2001). Microarrays for microbiologists. Microbiology *147*, 1403–1414.

Mackay, I.M., Metharom, P., Sloots, T.P., and Wei, M.Q. (2001). Quantitative PCR-ELAHA for the determination of retroviral vector transduction efficiency. Mol. Ther. *3*, 801–808.

Maggi, F., Pifferi, M., Vatteroni, M., Fornai, C., Tempestini, E., Anzilotti, S., Lanini, L., Andreoli, E., Ragazzo, V., Pistello, M., Specter, S., and Bendinelli, M. (2003). Human metapneumovirus associated with respiratory tract infections in a 3-year study of nasal swabs from infants in Italy. J. Clin. Microbiol. *41*, 2987–2991.

Mancini, C., Pisani, G., Azzi, A., Zerbini, M.L., Gentili, G., and Bisso, G.M. (2004). Inter-laboratory comparison of qualitative and quantitative detection of hepatitis C. (HCV) virus RNA in diagnostic virology: a multicentre study (MS) in Italy. J. Clin. Virol. *30*, 313–319.
Martin-Rendon, E., White, L.J., Olsen, A., Mitrophanous, K.A., and Mazarakis, N.D. (2002). New methods to titrate EIAV-based lentiviral vectors. Molecular Therapy 5[5], 566–570.
Massart, D.L., Vandeginste, B., Buydens, L., De Jong, S., Lewi, P., and Smeyers-Verbeke, J. (1997). In Handbook of Chemometrics and Qualimetrics.
McClelland, E.E., Bernhardt, P., and Casadevall, A. (2005). Coping with multiple virulence factors: Which is most important? PLoS Pathog *1*, e40.
McClelland, E.E., Bernhardt, P., and Casadevall, A. (2006). Estimating the relative contributions of virulence factors for pathogenic microbes. Infect. Immun. *74*, 1500–1504.
Mikovits, J.A., Young, H.A., Vertino, P., Issa, J.-P.J., Pitha, P.M., Turcoski-Corrales, S., Taub, D.D., Petrow, C.L., Baylin, S.B., and Ruscetti, F.W. (1998). Infection with human immunodeficiency virus type 1 upregulates DNA methyltransferase, resulting in de novo methylation of the gamma interferon (IFN-) promoter and subsequent downregulation of IFN- production. Mol. Cell. Biol. *18*, 5166–5177.
Mitchell, T.J. (1998). Molecular basis of virulence. Arch. Dis. Child. *78*, 197–209.
Mulder, J., McKinney, N., Christopherson, C., Sninsky, J., Greenfield, L., and Kwok, S. (1994). Rapid and simple PCR assay for quantitation of human immunodeficiency virus type 1 RNA in plasma: application to acute retroviral infection. J. Clin. Microbiol. *32*, 292–300.
Mullan, B., Sheehy, P., Shanahan, F., and Fanning, L. (2004). Do *Taq*-generated RT-PCR products from RNA viruses accurately reflect viral genetic heterogeneity? J. Viral Hepatitis *11*, 108–114.
Nelson, A.L., Barasch, J.M., Bunte, R.M., and Weiser, J.N. (2005). Bacterial colonization of nasal mucosa induces expression of siderocalin, an iron-sequestering component of innate immunity. Cell Microbiol. *7*, 1404–1417.
Niesters, H.G. (2004). Molecular and diagnostic clinical virology in real time. Clin. Microbiol. Infect. *10*, 5–11.
Niesters, H.G.M. (2001). Quantitation of viral load using real-time amplification techniques. Methods *25*, 419–429.
Niesters, H.G.M. (2002). Clinical virology in real time. J. Clin. Virol. *25*, S3-S12.
Nitsche, A., Stern, D., Ellerbrok, H., and Pauli, G. (2006). Detection of infectious poxvirus particles. Emerg. Infect. Dis. *12*, 1139–1141.
Nogva, H.K. and Rudi, K. (2004). Potential influence of the first PCR cycles in real-time comparative gene quantifications. Biotechniques 37[2], 246–248.
Nolan, T., Hands, R.E., Ogunkolade, W., and Bustin, S.A. (2006). SPUD: A quantitative PCR assay for the detection of inhibitors in nucleic acid preparations. Anal. Biochem. *351*, 308–310.
Nygren, J., Svanvik, N., and Kubista, M. (1998). The interactions between the fluorescent dye thiazole orange and DNA. Biopolymers. 46, 39–51.
Orlando, C., Pinzani, P., and Pazzagli, M. (1998). Developments in quantitative PCR. Clin. Chem. Lab. Med. *36*, 255–269.
Overbergh, L., Giulietti, A., Valckx, D., Decallonne, B., Bouillon, R., and Mathieu, C. (2003). The use of real-time reverse transcriptase PCR for the quantification of cytokine gene expression. J. Biomol. Tech. *14*, 33–43.
Overbergh, L., Valckx, D., Waer, M., and Mathieu, C. (1999). Quantification of murine cytokine mRNAs using real-time quantitative reverse transcriptase PCR. Cytokine. *11*, 305–312.
Pal, A., Sirota, L., Maudru, T., Peden, K., and Lewis, Jr., A.M. (2006). Real-time, quantitative PCR assays for the detection of virus-specific DNA in samples with mixed populations of polyomaviruses. J. Virol. Methods 135, 32–42.
Pannetier, C., Delassus, S., Darche, S., Saucier, C., and Kourilsky, P. (1993). Quantitative titration of nucleic acids by enzymatic amplification reactions run to saturation. Nucleic Acids Res. *21*, 577–583.
Pasloske, B.L., Walkerpeach, C.R., Obermoeller, R.D., Winkler, M., and DuBois, D.B. (1998). Armored RNA technology for production of rbonuclease-resistant viral RNA controls and standards. J. Clin. Microbiol. 36[12], 3590–3594.
Perch-Nielsen, I.R., Band, D.D., Poulsen, C.R., El-Ali, J., and Wolff, A. (2003). Removal of PCR inhibitors using dielectrophoresis as a selective filter in a microsystem. Lab on a chip 3, 212–216.
Perez-Novo, C.A., Claeys, C., Speleman, F., van Cauwenberge, P., Bachert, C., and Vandesompele, J. (2005). Impact of RNA quality on reference gene expression stability. Biotechniques 39, 52, 54, 56.

Pfaffl, M. (2001). A new mathematical model for relative quantification in real-time RT-PCR. Nucl. Acids Res. 29[9], e45.

Pfaffl, M.W., Horgan, G.W., and Dempfle, L. (2002). Relative expression software tool (REST©) for group-wise comparison and statistical analysis of relative expression results in real-time PCR. Nucleic Acids Res. *30*, e36.

Pfaffl, M.W., Tichopad, A., Prgomet, C., and Neuvians, T.P. (2004). Determination of stable housekeeping genes, differentially regulated target genes and sample integrity: BestKeeper—Excel-based tool using pair-wise correlations. Biotechnol Lett *26*, 509–515.

Pfeifer, A. and Verma, I.M. (2001). Gene therapy: Promises and problems. Annu. Rev. Gen. Human Gen. 2, 177–211.

PhorTech. (2003). Rapid rise in platforms for realtime PCR transforms DNA amplification instrumentation market. http://www.phortech.com/02ampi.htm.

Piatak Jr., M., Luk, K.-C., Williams, B., and Lifson, J.D. (1993). Quantiative competitive polymerase chain reaction for accurate quantitation of HIV DNA and RNA species. Biotechniques *14*, 70–80.

Pierce, K.E., Sanchez, J.A., Rice, J.E., and Wangh, L.J. (2005). Linear-after-the-exponential (LATE)-PCR: Primer design criteria for high yields of specific single-stranded DNA and improved real-time detection. Proc. Natl. Acad. Sci. USA *102*, 8609–8614.

Radonic, A., Thulke, S., Bae, H.G., Müller, M.A., Sieget, W., and Nitsche, A. (2005). Reference gene selection for quantitative real-time PCR analysis in virus infected cells: SARS coronavirus, Yellow fever virus, Human Herpesvirus-6, Camelpox virus and Cytomegalovirus infections. Virol. J. *2*.

Radonic, A., Thulke, S., Mackay, I.M., Landt, O., Siegert, W., and Nitsche, A. (2004). Guideline to reference gene selection for quantitative real-time PCR. Biochem. Biophys. Res. Commun. *313*, 856–862.

Ramsden, S.C., Daly, S., Geilenkeuser, W.J., Duncan, G., Hermitte, F., Marubini, E., Neumaier, M., Orlando, C., Palicka, V., Paradiso, A., Pazzagli, M., Pizzamiglio, S., and Verderio, P. (2006). EQUAL-quant: an international external quality assessment scheme for real-time PCR.

Rådström, P., Knutsson, R., Wolffs, P., Lövenklev, M., and Löfström, C. (2004). Pre-PCR processing Strategies to generate PCR-compatible samples. Mol. Biotech. *26*, 133–146.

Reiss, R.A. and Rutz, B. (1999). Quality control PCR: A method for detecting inhibitors of *Taq* DNA polymerase. Biotechniques *27*, 920–926.

Rodrigo, A.G., Goracke, P.C., Rowhanian, K., and Mullins, J.I. (1997). Quantitation of target molecules from polymerase chain reaction-based limiting dilution assays. AIDS. Res. Hum. Retro. *13*, 737–742.

Rodríguez, B., Sethi, A.K., Cheruvu, V.K., Mackay, W., Bosch, R.J., Kitahata, M., Boswell, S.L., Mathews, W.C., Bangsberg, D.R., Martin, J., Whalen, C.C., Sieg, S., Yadavalli, S., Deeks, S.G., and Lederman, M.M. (2006). Predictive value of plasma HIV RNA level on rate of CD4 T-cell decline in untreated HIV infection. JAMA *296*[12], 1498–1506.

Rosenstraus, M., Wang, Z., Chang, S.-Y., DeBonville, D., and Spadoro, J.P. (1998). An internal control for routine diagnostic PCR: design, properties, and effect on clinical performance. J. Clin. Microbiol. *36*, 191–197.

Ruelle, J., Mukadi, B.K., Schutten, M., and Goubau, P. (2004). Quantitative real-time PCR on LightCycler® or the detection of human immunodeficiency virus type 2 (HIV-2). J. Virol. Methods *117*, 67–74.

Ruiz, G., Peña, P., de Ory, F., and Echevarría, J.E. (2005). Comparison of commercial real-time PCR assays for quantifaction of Epstein-Barr virus DNA. J. Clin. Microbiol. *43*[5], 2053–2057.

Ryncarz, A.J., Goddard, J., Wald, A., Huang, M.-L., Roizman, B., and Corey, L. (1999). Development of a high-throughput quantitative assay for detecting herpes simplex virus DNA in clinical samples. J. Clin. Microbiol. *37*, 1941–1947.

Sagner, G., Tabiti, K., Gutekunst, M., and Soong, R. (2004). Method for the efficiency-corrected real-time quantification of nucleic acids. [US Patent 6,691,041].

Sanburn, N. and Cornetta, K. (1999). Rapid titer determination using quantitative real-time PCR. Gene Ther. *6*, 1340–1345.

Sastry, L., Johnson, T., Hobson, M.J., Smucker, B., and Cornetta, K. (2002). Titering lentiviral vectors: comparison of DNA, RNA and marker expression methods. Gene Ther. 9, 1155–1162.

Schaade, L., Kockelkorn, P., Ritter, K., and Kleines, M. (2000). Detection of cytomegalovirus DNA in human specimens by LightCycler PCR. J. Clin. Microbiol. *38*, 4006–4009.

Scheltinga, S.A., Templeton, K.E., Beersma, M.F.C., and Claas, E.C.J. (2005). Diagnosis of human metapneumovirus and rhinovirus in patients with respiratory tract infections by an internally controlled multiplex real-time RNA PCR. J. Clin. Virol. 33, 306–311.
Scherr, M., Battmer, K., Blömer, U., Ganser, A., and Grez, M. (2001). Quantitative determination of lentiviral vector particle numbers by real-time PCR. Biotechniques *31*, 520–526.
Schildgen, O., Jebbink, M.F., de Vries, M., Pyrc, K., Dijkman, R., Simon, A., Müller, A., Kupfer, B., and van der Hoek, L. (2006). Identification of cell lines permissive for human coronavirus NL63. J. Virol. Methods.
Schirm, J., van Loon, A.M., Valentine-Thon, E., Klapper, P.E., Reid, J., and Cleator, G.M. (2002). External quality assessment program for qualitative and quantitative detection of hepatitis C virus RNA in diagnostic virology. J. Clin. Microbiol. *40*[8], 2973–2980.
Schloss, L., van Loon, A.M., Cinque, P., Cleator, G., Echevarria, J-M., Falk, K.I., Klapper, P., Schirm, J., Vestergaard, B.F., Niesters, H., Popow-Kraupp, T., Quint, W., and Linde, A. (2003). An international external quality assessment of nucleic acid amplification of herpes simplex virus. J. Clin. Virol. 28, 175–185.
Schutten, M. and Niesters, H.G.M. (2001). Clinical utility of viral quantification as a tool for disease monitoring. Expert review of molecular diagnostics. *1*[2], 153–162.
Schutten, M., van den Hoogen, B., van der Ende, M.E., Gruters, R.A., Osterhaus, A.D.M.E., and Niesters, H.G.M. (2000). Development of a real-time quantitative RT-PCR for the detection of HIV-2 RNA in plasma. J. Virol. Methods *88*, 81–87.
Selvey, S., Thompson, E.W., Matthaei, K., Lea, R.A., Irving, M.G., and Griffiths, L.R. (2001). β-actin-an unsuitable internal control for RT-PCR. Mol. Cell. Probes *15*, 307–311.
Siebert, P.D. and Larrick, J.W. (1992). Competitive PCR. Nature *359*, 557–558.
Smith, R.D., Brown, B., Ikonomi, P., and Schechter, A.N. (2003). Exogenous reference RNA fpor normalizartion of real-time quantitative PCR. Biotechniques *34*[1], 88–91.
Song, J., Ohkura, T., Sugimoto, M., Mori, Y., Inagi, R., Yamanishi, K., Yoshizaki, K., and Nishimoto, N. (2002). Human interleukin-6 induces human herpesvirus-8 replication in a body cavity-based lymphoma cell line. J. Med. Virol. *68*, 404–411.
Souazé, F., Ntodou-Thomé, A., Tran, C.Y., Rostène, W., and Forgez, P. (1996). Quantitative RT-PCR: Limits and accuracy. Biotechniques *21*, 280–285.
Stangegaard, M., Dufva, I.H., and Dufva, M. (2006). Reverse transcription using random pentadecamer primers increases yield and quality of resulting cDNA. Biotechniques *40*, 649–657.
Stanley, K.K. and Szewczuk, E. (2005). Multiplexed tandem PCR: gene profiling from small amounts of RNA using SYBR Green detection. Nucleic Acids Res. 33, e180.
Ståhlberg, A., Åman, P., Ridell, B., Mostad, P., and Kubista, M. (2003). Quantitative real-time PCR method for the detection of B-lymphocyte monoclonality by comparison of κ and λ immunoglobulin light chain expression. Clin. Chem. *49*[1], 51–59.
Ståhlberg, A., Håkansson, J., Xian, X., Semb, H., and Kubista, M. (2004a). Properties of the reverse transcription reaction in mRNA quantification. Clin. Chem. *50*, 509–515.
Ståhlberg, A., Kubista, M., and Pfaffl, M. (2004b). Comparison of reverse transcriptases in gene expression analysis. Clin. Chem. *50*, 1678–1680.
Ståhlberg, A., Zoric, N., Aman, P., and Kubista, M. (2005). Quantitative real-time PCR for cancer detection: the lymphoma case. Expert review of molecular diagnostics. 5[2], 221–230.
Stordeur, P., Poulin, L.F., Craciun, L., Zhou, L., Schandené, L., de Lavareille, A., Goriely, S., and Goldman, M. (2002). Cytokine mRNA quantification by real-time PCR. J. Immunol. Methods *259*, 55–64.
Stöcher, M., Leb, V., Hölzl, G., and Berg, J. (2002). A simple approach to the generation of heterologous competitive internal controls for real-time PCR assays on the LightCycler. J. Clin. Virol. *25*, S47-S53.
Strieter, R.M., Belperio, J.A., and Keane, M.P. (2003). Host innate defenses in the lung: the role of cytokines. Curr. Opin. Infect. Dis. *16*, 193–198.
Suñen, E., Casas, N., Moreno, B., and Zigorraga, C. (2004). Comparison of two methods for the detection of hepatitis A virus in clam samples (*Tapes* spp) by reverse transcription-nested PCR. Int. J. Food Microbiol. *91*, 147–154.
Suslov, O. and Steindler, D.A. (2005). PCR inhibition by reverse transcriptase leads to an overestimation of amplification efficiency. Nucl. Acids Res. 33, e81.

Sykes, P.J., Brisco, M.J., Snell, L.E., Dolman, G., Neoh, S.-H., Peng, L.-M., and Toogood, I. (1998). Minimal residual disease in childhood acute lymphoblastic leukaemia quantified by aspirate and trephine: is the disease multifocal? Br. J. Haematol. *103*, 60–65.

Tanaka, N., Kimura, H., Iida, K., Saito, Y., Tsuge, I., Yoshimi, A., Matsuyama, T., and Morishima, T. (2000). Quantitative analysis of cytomegalovirus load using a real-time PCR assay. J. Med. Virol. *60*, 455–462.

Tang, Y.-W., Rys, P.N., Rutledge, B.J., Mitchell, P.S., Smith, T.F., and Persing, D.H. (1998). Comparative evaluation of colorimetric microtiter plate systems for detection of herpes simplex virus in cerebrospinal fluid. J. Clin. Microbiol. *36*, 2714–2717.

Tarnuzzer, R.W., Macauley, S.P., Farmerie, W.G., Caballero, S., Ghassemifar, M.R., Anderson, J.T., Robinson, C.P., Grant, M.B., Humphreys-Beher, M.G., Franzen, L., Peck, A.B., and Schultz, G.S. (1996). Competitive RNA templates for detection and quantitation of growth factors, cytokines, extracellular matrix components and matrix metalloproteinases by RT-PCR. Biotechniques *20*, 670–674.

Taswell, C. (1981). Limiting dilution assays for the determination of immunocompetent cell frequencies. J. Immunol. *126*, 1614–1619.

Templeton, K.E., Forde, C.B., van Loon, A.M., Claas, E.C.J., Niesters, H.G.M., Wallace, P., and Carman, W.F. (2006). A multi-centre pilot proficiency programme to assess the quality of molecular detection of respiratory viruses. J. Clin. Virol. *35*, 51–58.

Templeton, K.E., Scheltinga, S.A., Graffelman, A.W., van Schie, J.M., Crielaard, J.W., Sillekens, P., van den Broek, P.J., Goossens, H., Beersma, M.F.C., and Claas, E.C.J. (2003). Comparison and evaluation of real-time PCR, real-time nucleic acid sequence-based amplification, conventional PCR, and serology for diagnosis of *Mycoplasma pneumoniae*. J. Clin. Microbiol. 41[9], 4366–4371.

Templeton, K.E., Scheltinga, S.A., van den Eeden, W.C.J.F.M., Graffelman, A.W., van den Broek, P.J., and Claas, E.C.J. (2005). Improved diagnosis of the etiology of community-acquired pneumonia with real-time polymerase chain reaction. Clin. Inf. Dis. *41*, 345–351.

Thellin, O., Zorzi, W., Lakaye, B., De Borman, B., Coumans, B., Hennen, G., Grisar, T., Igout, A., and Heinen, E. (1999). Housekeeping genes as internal standards: use and limits. J. Biotech. *75*, 291–295.

Tricarico, C., Pinzani, P., Bianchi, S., Paglierani, M., Distante, V., Pazzagli, M., Bustin, S.A., and Orlando, C. (2002). Quantitative real-time reverse transcription polymerase chain reaction: normalization to rRNA or single housekeeping genes is inappropriate for human tissue biopsies. Anal. Biochem. 309, 293–300.

van Vliet, K.E., Muir, P., Echevarria, J.M., Klapper, P.E., Cleator, G.M., and van Loon, A.M. (2001). Multicenter proficiency testing of nucleic acid amplification methods for the detection of enteroviruses. J. Clin. Microbiol. *39*, 3390–3392.

Vandesompele, J., de Paepe, A., and Speleman, F. (2002a). Elimination of primer–dimer artifacts and genomic coamplification using a two-step SYBR green I real-time RT-PCR. Anal. Biochem. *303*, 95–98.

Vandesompele, J., de Preter, K., Pattyn, F., Poppe, B., van Roy, N., de Paepe, A., and Speleman, F. (2002b). Accurate normalization of real-time quantitative RT-PCR data by geometric averaging of multiple internal control genes. Genome Biol 3, 1–12.

Verderio, P., Orlando, C., Casini Raggi, C., and Marubini, E. (2004). Confidence interval estimation for DNA and mRNA concentration by real-time PCR: A new environment for an old theorem. Int. J. Biol. Mark. *19*[1], 76–79.

Verkooyen, R.P., Noorhoek, G.T., Klapper, P.E., Reid, J., Schirm, J., Cleator, G.M., Ieven, M., and Hoddevik, G. (2003). Reliability of nucleic acid amplification methods for detection of *Chlamydia trachomatis* in urine: Results of the first international collaborative quality control study among 96 laboratories. J. Clin. Microbiol. *41*[7], 3013–3016.

Verma, S., Ziegler, K., Ananthula, P., C.o., J.K.G., Frisque, R.J., Yanagihara, R., and Nerurkar, V.R. (2006). JC virus induces altered patterns of cellular gene expression: Interferon-inducible genes as major transcriptional targets. Virology *345*, 457–467.

Wang, A.M., Doyle, M.V., and Mark, D.F. (1989). Quantitation of mRNA by the polymerase chain reaction. Proc. Natl. Acad. Sci. USA *86*, 9717–9721.

Wang, Z. and Spadoro, J. (1998). Determination of target copy number of quantitative standards used in PCR based diagnostic assays. In Gene quantification, F.Ferré, ed. (Boston: Birkhauser).

Wark, P.A.B., Johnston, S.L., Bucchieri, F., Powell, R., Puddicombe, S., Laza-Stanca, V., Holgate, S.T., and Davies, D.E. (2005). Asthmatic bronchial epithelial cells have a deficient innate immune response to infection with rhinovirus. J. Exp. Med. *201*[6], 937–947.

Watzinger, F., Ebner, K., and Lion, T. (2006). Detection and monitoring of virus infections by real-time PCR. Mol. Asp. Med. 27, 254–298.

Watzinger, F., Suda, M., Preuner, S., Baumgartinger, R., Ebner, K., Baskova, L., Niesters, H.G.M., Lawitschka, A., and Lion, T. (2004). Real-time quantitative PCR assays for detection and monitoring of pathogenic human viruses in immunosuppressed pediatric patients. J. Clin. Microbiol. *42*, 5189–5198.

Weinberger, K.M., Wiedenmann, E., Böhm, S., and Jilg, W. (2000). Sensitive and accurate detection of hepatitis B virus DNA using a kinetic fluorescence detection system (TaqMan PCR). J. Virol. Methods *85*, 75–82.

Wilson, I.G. (1997). Inhibition and facilitation of nucleic acid amplification. Appl. Environ. Micro. *63*, 3741–3751.

Woo, P.C.Y., Lau, S.K.P., L.i., K.S.M., Poon, R.W.S., Wong, B.H.L., Tsoi, H.-W., Yip, B.C.K., Huang, Y., Chan, K.-H., and Yuen, K.-Y. (2006). Molecular diversity of coronaviruses in bats. Virology *351*, 180–187.

Zhang, L.Q., Simmonds, P., Ludlam, C.A., and Brown, A.J.L. (1991). Detection, quantification and sequencing of HIV-1 from the plasma of seropositive individuals and from factor VIII concentrates. AIDS *5*, 675–681.

Zhao, H., Granberg, F., Elfineh, L., Pettersson, U., and Svensson, C. (2003). Strategic attack on host cell gene expression during adenovirus infection. J. Virol. *77*[20], 11006–11015.

Zimmermann, K. and Manhalter, J.W. (1996). Technical aspects of quantitative competitive PCR. Biotechniques *21*, 268–279.

Multiplex rtPCR in Microbiology

6

Nick M. Cirino, Norma P. Tavakoli, Susan Madison-Antenucci, and Christina Egan

The value of an idea lies in the using of it.

Thomas A. Edison

Abstract

Real-time PCR, or rtPCR, with its ability to detect and identify microorganisms is contributing to improvements in public health, and is facilitating more rapid attribution of disease-causing agents, whether they are old foes or newly emerging pathogens. Capitalizing on the strengths of rtPCR, we can further expand the capabilities of the methodology, so as to detect multiple target nucleic acid sequences in a single reaction. We refer to this type of an assay as a multiplex diagnostic assay, capable of detecting two (i.e. duplex) or more target signatures simultaneously. While multiplexing of rtPCR assays is achieved on a limited basis, typically two to four target sequences, there are examples of highly multiplexed assays such as DNA arrays or gene chips that can interrogate > 10 000 oligonucleotide sequences in a single sample. This chapter will focus on the multiplex ability of rtPCR assays. Through integration of multiple assays into a single reaction, information about assay quality (e.g. internal positive or inhibition controls) can be simultaneously generated, and additional target pathogen sequences queried. Multiplexing therefore reduces analytical costs, improves turnaround time, expands testing capability and capacity, and adds data richness to analyses. In this chapter, we will describe the current applications of multiplex rtPCR to clinical diagnostics and public health, and we will review current applications of rtPCR to various classes of pathogenic microorganisms including viruses, bacteria, fungi, and parasites. Given a basic understanding of multiplexing concepts and critical parameters, it is fairly simple to convert conventional or rtPCR assays to multiplex formats.

Introduction and overview

Previous chapters have described the principles of rtPCR, the various fluorescence resonance energy transfer (FRET) chemistries employed to monitor amplicon accumulation, how to design and verify primer and probe oligonucleotide specificity, and subsequent chapters will describe how this technology is being integrated into research, public health, and bioterrorism preparedness. In this chapter, we will describe additional advances in rtPCR detection technology that permit monitoring of multiple dyes, and hence different target amplicon accumulation, in a single reaction vessel. A simple example of this multiplex rtPCR approach would be if three individual rtPCR assays were being analyzed for *Bacillus anthracis*: one which targets a chromosomal sequence, one which targets the anthrax toxin encoding plasmid (i.e. pXO1), and one which targets the capsule producing plasmid (i.e. pXO2). Obviously, all three of these assays are essential in determining the virulence of *B. anthracis*, since absence of either plasmid renders the bacterium non-lethal. Run individually against a known pathogenic *B. anthracis*, the assays would provide a composite, high-confidence, positive result, assuming that all other assay controls (e.g. inhibition controls, positive controls and negative controls) also gave predicted results. Clearly, however, there would be significant hands-on time in preparing the various reactions and analyzing the various target assays. Now imagine having all the target assays in a single tube: all three *B. anthracis* assays (each with a unique dye on the probes), a PCR positive control (e.g. *E. coli* 16S with a unique dye on the probe), and a PCR inhibition control (e.g. a rare template assay and a known quantity of the template). This multiplex assay would require simultaneous detection of at least five dyes, the maximum number of dyes that can currently be discriminated by the optics of rtPCR systems.

In conventional PCR, multiple templates can be amplified in a single reaction and can then be distinguished on the basis of size via gel electrophoresis. However, in rtPCR, amplicon discrimination is achieved by use of multiple fluorogenic probes, or alternatively by acquisition of melting curves from non-specific DNA intercalating fluorophores. Fluorogenic probes emit fluorescence at specific wavelengths. If multiple probes are selected that fluoresce at different wavelengths, then positive signals from the different primer/probe sets can be distinguished. Multiplex rtPCR has the same advantages of singleplex rtPCR. These advantages include:

- ability to monitor the increasing amounts of amplicon as they accumulate
- elimination of the need for post-amplification processing
- fast turn-around time, in particular if rapid cycling conditions are used
- increased automation
- decreased risk of cross-contamination, since the PCR product remains contained throughout
- excellent sensitivity
- the addition of a probe to a PCR reaction adds a layer of specificity to the detection of amplicon
- reproducibility of assays
- capability to be made quantitative.

In addition, since the assay is multiplexed:

- Lower specimen volume is required to run the assay. This is an important issue when the volume of patient specimen received by a diagnostic laboratory is limited. An example is the detection of multiple viral agents in cerebrospinal fluid (CSF), especially in pediatric patients from whom only small sample volumes are available.
- Costs are reduced, since less reagents and fewer consumables are needed and the requirement for hands-on time is reduced.

Much of the current generation rtPCR detector equipment can detect four or five dyes with good resolution and little spectral emission overlap, or bleed-through, of emitted light.

To maximize the potential number of targets interrogated in a single multiplex assay, many assay developers have switched from fluorescent quenchers, like the tetra-methylcarboxy-rhodamine (TAMRA) dye, to dark quenchers. Dark quenchers absorb light analogously to fluorescent quenchers, but they do not emit the absorbed light; instead, it is released as heat. The amount of heat produced by the quenchers does not affect the reaction. All of the dyes used in the reaction can be employed as FRET donors/reporters to monitor amplicon accumulation. As more sophisticated optical filters and dyes become available, the number of assays that can be integrated will likely increase to seven or eight. Nevertheless, because optical filtration and novel dyes are extremely expensive, decaplex rtPCR detectors will likely be cost-prohibitive for most applications and will likely never reach the commercial marketplace.

The presence of the single dark quencher in the reaction allows for quenching of all the reporter dyes. Since FRET is limited by distance, minimization or limitation of probe concentration will be a consideration. Similarly, since it is possible to have more than one target sequence being amplified in the reaction, it will be essential to limit primer concentrations in order to prevent nucleotide depletion and *Taq* sequestration. Primer limiting, which is a common practice in many PCR applications, is absolutely necessary when multiplexing assays for targets that may be present at concentrations differing by 100 or 1000 fold. For example, if a multiplex assay had been developed for an *E. coli* chromosomal gene and a cloning vector plasmid like BlueScript™, which is present in hundreds to thousands of copies per bacterium, the PCR reaction would favor the plasmid target amplification, since the plasmid constitutes 99.9% of the targets in solution. Table 6.1 illustrates how probes and nucleotide depletion can affect analytical results. By multiplexing, the amplification efficiency of the low-concentration target is artificially reduced. Decreasing the amplification efficiency by as little as 5%, can significantly affect the cycle at which the detection threshold is reached (the C_T). In our experience, most FRET-based detectors "see" amplicon accumulation at about 10^{11} molecules (i.e. they detect single molecules after 37–40 cycles of amplification), so limiting primers and probes to the amount needed to generate 10^{12}–10^{13} amplicons is sufficient and should not affect first hit kinetics or amplification efficiency. For dilute solutions of template DNA, there is a statistical probability that the template, primer, and *Taq* polymerase will not meet and have sufficient time to amplify the template before the denaturation phase, so the first amplification, or "first hit" may not occur on the first PCR cycle. A 5% reduction in amplification efficiency translates to a 10-fold decrease in product by cycle 40 and hence a potential false negative result if the cycling were terminated at cycle 40. Many multiplex assays are run for additional cycles (e.g. 45) to ensure late or delayed detection of lower concentrations of second targets in a single reaction. Typically, for a well-developed and controlled assay, the analyst will see detection of positive targets before cycle 40 and nothing between cycles 40 and 45. This nuance of multiplex assays should be considered when you plan to combine assays designed to detect sequences in a single sample that may be of very different concentrations.

Careful planning of the multiplex process is essential when the assays are run simultaneously in order to minimize the chances of reduced efficiency and cross-reactivity,. Combining only those rtPCR assays that have been developed with consistent software, sensitivity parameters, mastermix concentrations, cycling conditions, and detection platforms will significantly improve a developer's success in achieving a robust and predictive

Table 6.1 Effect of disparate template concentrations on multiplex rtPCR assay detection.

Cycle number	*E. coli* singleplex assay Chromosomal target (ε=100%) Predicted copies/amplicons	*E. coli* singleplex assay Plasmid target (ε=100%) Predicted copies/amplicons	Multiplex assay Plasmid (ε=100%) Predicted amplicons	Multiplex assay Chromosomal (ε=95%) Predicted amplicons
0	1	1000	1000	1
1	2	2000	2000	1.9
10	1024	1.0×10^{6}	1.0×10^{6}	613
25	3.4×10^{7}	3.4×10^{10}	3.4×10^{10}	9.3×10^{6}
40	$\mathbf{1.1 \times 10^{12}}$	1.1×10^{15}	1.1×10^{15}	$\mathbf{1.4 \times 10^{11}}$
45	3.5×10^{13}	3.5×10^{16}	*3.5×10^{16}*	3.5×10^{12}

ε is amplification efficiency. On most current detection platforms, 1 copy is detected (i.e., crosses the cycle threshold) between cycles 38 and 40. In this example, we will say cycle 40, which means we need to have approximately 1.1×10^{12} amplicons present before the FRET change indicates a positive result. If the lower copy number target in a multiplex assay, chromosomal DNA in this example, is amplified just 5% less efficiently throughout the assay, this will lead to 10 fold fewer copies by cycle 40 and hence a false negative result. By adding 5 additional cycles of amplification, we can detect even the lower copy number target (i.e., the number of amplicons exceeds the detection threshold of 1.1×10^{12}). * Primer limitation in the multiplex assay will preclude amplification to this actual level.

multiplex rtPCR assay in the long run. Indeed, forethought and upstream time commitment translates to significantly reduced time, cost, and resources needed to validate the final multiplex assay. This is true whether developing multiplex conventional PCR assays or rtPCR assays, and is especially critical when developing assays for RNA targets. To amplify RNA, the template must be converted to DNA using a reverse transcriptase (RT) enzymatic step prior to PCR. The additional RT step can significantly affect the outcome of the reaction, and incorporation of additional RT primers can be problematic in the integration of a multiplex real-time RT-PCR for viral hemorrhagic fever viruses, for example. Priming of the reverse transcription reaction is generally achieved through use of a primer with sequence that is specific to the viral agent (this primer can be one of the PCR primers), or, through use of random primers (generally hexanucleotides). Our experience in developing rtPCR assays for two RNA viruses (enterovirus and parechovirus) has been that the combination of a separate RT reaction using random primers, followed by an rtPCR assay, is 10–100 fold more sensitive than a two-step, single-tube real-time RT-PCR assay. Other assays that we have developed, including assays for the detection of eastern equine encephalitis (EEE), Powassan virus and West Nile virus (WNV), have shown similar sensitivities when we have compared the use of random primers in a separate RT reaction followed by rtPCR, and the use of specific primers in a two-step real-time RT-PCR assay. In the subsequent sections, we will describe a range of state-of-the-art multiplex diagnostic assays and their utility in detecting and identifying organisms that commonly cause illness, or that cause significant outbreaks due to either (re)emergence or intentional release (i.e. bioterrorism).

Rapid and sensitive detection of viral agents is of the utmost importance in preventing the spread of diseases and particularly emerging infectious diseases such as the severe acute respiratory syndrome (SARS) coronavirus and the highly pathogenic avian influenza virus. Early diagnosis is also key to patient management. The sooner a patient is diagnosed, the sooner the appropriate drug therapy can be administered. Equally important, the sooner the diagnosis, the sooner inappropriate therapy can be ceased. These are critical issues for appropriate laboratory diagnosis, since many antiviral agents have unwanted side effects and are costly, and since over-use can lead to the development of resistant strains.

Limitations of multiplex rtPCR

Multiplex rtPCR is limited by the number of fluorescent dyes that can be used in a reaction. Currently, the maximum number of reactions that can be multiplexed is four to five. This limit is established by the number of spectrally discriminable probes available, and by the constraints of the instrument that is being used to perform the assay. Certain instruments have a variable excitation capability that allows access to a broader range of fluorophores. Instruments with a monochromatic light source for the excitation of the fluorophores are limited to employing fluorophores that can be efficiently excited by a blue laser (for example, the ABI 7700™ and ABI 7900™). Instruments with a variable wavelength light source and a high resolution detector can accommodate a wider range of fluorophores, thus increasing multiplex capability. For example, the BioRad iCycler™, the Stratagene Mx4000™, and the newer ABI instruments such as the ABI 7300™ and ABI 7500™ have greater sensitivity for longer wavelength (red) dyes thus allowing five color detection.

Multiplex assays require additional optimization in order for them to maintain the sensitivity of the singleplex assays. In general, rtPCR reactions are performed at very high primer concentrations, so that the reaction will proceed quickly. However, at excessively high concentrations, the primers form dimers or higher multimers, and the efficiency of the reaction is decreased. In multiplex PCR the problem is exacerbated, since there are additional primer sets. Therefore, great care must be taken when a multiplex rtPCR is designed. In addition primers should be evaluated so that they do not form hairpins or interact with the other pairs or with one another.

The choice of fluorophore is important. In general, FAM is used as the reporting fluorophore for singleplex assays because it gives a strong fluorescence. However, one school of thought holds that FAM should not be used in multiplex assays for this very reason: the strong fluorescence of FAM can mask the fluorescence of weaker dyes. In our experience this has not been a problem. As long as other dyes are used which are spectrally separated from the FAM dye layer, there should be no issue of masking the signal of weaker emitting dyes.

It is, nevertheless, important to multiplex with appropriate reporters that are spectrally well separated. For example, it is not possible to multiplex with hexachlorofluorescein (HEX) and VIC™ (Applied Biosystems, Foster City, CA), since their emission wavelength maxima are similar (553 and 554 nm respectively). In addition, bleed-through into other dye layers can be problematic and can lead to false positive results. The concentration of the primers and probes should be kept to a minimum. In optimization of the reaction the concentrations of primer and probe selected should be those that give the lowest threshold cycle (C_T) while maintaining a relatively high ΔRn (i.e. change in relative fluorescent signal upon FRET). The ΔRn will be lower than the singleplex assay due to limiting primer and/or probe concentrations.

Primer interactions

Intercalating dyes such as SYBR™ Green (FMC Bioproducts, Rockland, ME, USA) interact with double stranded (ds) DNA molecules in a non-specific manner; when excited by a light source, they emit fluorescence. The design of primers for use with SYBR Green is extremely important since primers that produce non-specific amplification or form primer–dimers can lead to spurious fluorescence emission which affects the interpretation of the results. Multiplexing is not recommended for assays that use intercalating dyes. This is because different targets cannot be distinguished through determination of the fluorescence emission; also, the presence of multiple primers in the reaction will favor primer–dimer formation. Some groups have tried optimizing multiplex PCR reactions employing SYBR Green by various methods, including use of different polymerases in order to improve amplification efficiency and acquisition of data at high temperature (> 75°C) in order to eliminate background signals resulting from primer–dimer formation (Jothikumar *et al.*, 2003; Varga and James, 2005). However, hot start modifications to the polymerase do not greatly influence amplification efficiency (Arezi *et al.*, 2003), and in general performance of a multiplex assay using SYBR Green results in an increase in background fluorescence and a decrease in the amplification of desired products.

The design of primer/probe sets is a crucial factor in obtaining a specific and sensitive rtPCR assay. This is especially important for the detection of viruses, because there are many different strains/variations of most viral agents. For example, there are over 100 strains of rhinovirus, 70 strains of enterovirus, and 50 strains of adenovirus. In addition, because the complete genetic sequence of many of the different strains has not been determined, comparative sequence information is not available for use in performing multiple alignments and selecting conserved regions for designing primers and probes.

Mutation rates are higher in RNA viruses than in DNA viruses (Drake and Holland, 1999; Elena and Sanjuan, 2005; Holland and Domingo, 1998). For DNA viruses there is approximately one mutation per 100–1000 viral genomes, whereas for RNA viruses there is approximately one mutation per viral genome. Nucleotide base misincorporations result in the generation of quasispecies and the creation of a mixture of virus mutants that differ from one another in single, or multiple nucleotide sequences. This variability allows viruses greater adaptability and is important in virus evolution. However, it is a major problem for the design of a molecular method for a diagnostic assay because even if the test is carefully developed and works well to detect a specific viral agent, it may not detect even a subtle variant of the viral agent that also causes infections. An example is the influenza virus, which mutates by shift and drift (Wright and Webster, 2001). The high mutation rate, while ensuring that the virus escapes the host's immunological defenses, renders detection of the virus problematic by molecular methods. There is therefore a need for routine sequencing of the complete genome of influenza viruses, and other RNA viruses that are in circulation, to perform surveillance and to update molecular assays so that the current viruses can be reliably detected. For example, the current real-time assays that detect influenza A do not detect the H5N1 variant (the cause of the current Asian bird flu), and specific assays have to be designed for this variant.

One of the advantages of conventional PCR over rtPCR is that degenerate primers can be designed that can detect multiple different sequences. For example, there are specific assays using degenerate primers that can detect flaviviruses [including Saint Louis encephalitis virus (SLE), WNV, Powassan virus, Yellow fever virus, dengue virus serotypes 1–4, Japanese encephalitis virus (JE)] and alphaviruses [including Sindbis virus (SIN), EEE, western equine encephalitis virus (WEE), Venezuelan equine encephalitis virus (VEE)] (Pfeffer *et al.*, 1997; Scaramozzino *et al.*, 2001). Such detection at the family level is not yet possible with rtPCR. In fact, the design of an assay to detect all strains of SLE has been very problematic in our laboratory. Most of the sequence information available for SLE strains is in the non-conserved regions of the genome. Therefore, after unsuccessfully searching for conserved regions in these sequences we decided to undertake our own sequencing of conserved regions of 25 strains in order to obtain suitable information for designing primers and probes. This would obviously be a difficult undertaking for a routine diagnostic laboratory without the facilities to perform this type of research. One option when sequence conservation is limited is to design shorter probes. For example, MGB™ probes (Applied Biosystems, Foster City, CA) can be as short as 13 bp, and locked nucleic acid (LNA) probes (Roche Molecular Diagnostics, Indianapolis, IN) are only nine bp. The numbers of reporter fluorophores on MGB probes is limited. In addition, shorter probes are extremely vulnerable to sequence polymorphisms.

The use of conventional PCR in clinical diagnostic laboratories has been problematic because the methodology is labor-intensive, readily susceptible to contamination, and technically demanding. False-positive and also false-negative results are a major concern in a clinical setting, and unfortunately they are common. Advances in technology have simplified and automated PCR procedures, thus improving the reliability and reproducibility of the assays and transforming rtPCR into a valuable tool for diagnostics. As can be seen in Table 6.2, multiplex rtPCR assays have been developed for many viral agents, since these assays first appeared in the literature in 1999.

Multiplex rtPCR in the detection of viral agents

The diagnosis of viral respiratory infections has generally been by virus isolation in culture and immunofluorescent assays (IFAs). These methods are labor-intensive, slow, and insensitive, and they cannot detect fastidious pathogens. Their utility is therefore limited. In addition, serological tests do not detect early infection, at the stage when the immune response is still developing (Pantaleo *et al.*, 1993), and they require commercial antibodies. Rapid differential diagnosis of pathogens causing respiratory infections can be achieved by multiple multiplex rtPCR assays. Templeton *et al.* (2004) designed two multiplex assays, one for the detection of influenza A virus (IFAV), influenza B virus (IFBV) and human respiratory syncytial virus (HRSV) and the other for the detection of human parainfluenza virus 1, 2, 3 and 4 (HPIV). This group of viruses causes the majority of lower respiratory infections in children and in the elderly and immuno-compromised individuals (Greenberg, 2002; Hall, 2001). It should be noted, however, that even though seven viral agents were included in these two multiplex assays, the paired assays obviously do not cover the spectrum of viral agents, including rhinoviruses, enteroviruses, human coronaviruses, adenovirus and human metapneumovirus, that also cause respiratory infections. The assays do aid in detecting co-infections which could be overlooked if testing for multiple viral agents was not performed. The molecular beacons used for these assays were labeled with FAM, Texas Red™, and hexachlorofluorescein (HEX) for the IFV and HRSV assay, and with the same three reporters with the addition of indodicarbocyanine (Cy5) for the HPIV assay. After multiplexing, the sensitivities of all assays were similar to the sensitivities of the singleplex assays, except for the HPIV4 assay, which had sensitivity reduced by one $\log_{10}$.

Multiple strains of each virus were tested using the two multiplex assays and all were detected (Templeton *et al.*, 2004). The necessary specificity of the primers and probes for an rtPCR reaction can often mean that emergent strains having mutations in the primer/probe region no longer hybridize to the primer/probe set, or they do so at a lower efficiency, thus decreasing the sensitivity of the assay. It is therefore important, in the development of a molecular assay, to first design primers and probes based on alignment of available sequences from multiple strains of a particular virus, and then to test the assay on all viral types and subtypes characterized to date. For example, until recently, influenza virus type A (IFAV) subtypes H1N1, H2N2, and H3N2 have been associated with widespread epidemics in humans. However, other reassortants, including H1N2 and the highly pathogenic avian influenza virus H5N1, have also emerged. In order to decrease the possibility of obtaining false negatives, the specificity of the assays for different strains and subtypes should be determined. The duplex PCR assay for the detection of IFAV and IFBV reported by (Hindiyeh *et al.*, 2005) gave a positive signal for the commonly detected subtypes that infect humans, as well as for the avian viruses H7N7 and H9N2. However,

Table 6.2 Multiplex rtPCR assays for the detection of viral agents

Multiplexed targets	Reference
Adenovirus subgroups A, B and C	Claas *et al.,* 2005
Adenovirus subgroups A, B, C, E and 8 serotypes of subgroup D	Gu *et al.*, 2003
Adenovirus (all currently known serotypes)	Ebner *et al.*, 2005
Influenza A virus and Influenza A virus subtype H5N1	Payungporn *et al.*, 2006
Influenza A and B viruses and human metapneumovirus	Gunson *et al.*, 2005
Respiratory syncytial virus A and B and rhinovirus	Gunson *et al.*, 2005
Parainfluenza 1, 2 and 3	Gunson *et al.*, 2005
Coronavirus 229E, OC43 and NL63	Gunson *et al.*, 2005
Respiratory syncytial virus A and B	van Elden *et al.*, 2003
Coronavirus 229E and OC43	van Elden *et al.*, 2004
Human metapneumovirus, rhinovirus and equine arteritis virus (IC)	Scheltinga *et al.*, 2005
Influenza A and B viruses	van Elden *et al.*, 2001
Influenza A and B viruses	Hindiyeh *et al.*, 2005
Influenza A and B viruses and Respiratory syncytial virus	Boivin *et al.*, 2004
Influenza A and B viruses and Respiratory syncytial virus	Templeton *et al.*, 2004
Parainfluenza viruses 1, 2, 3 and 4	Templeton *et al.*, 2004
Herpes simplex virus 1 and 2	Ryncarz *et al.*, 1999
Herpes simplex virus 1 and 2	Espy *et al.*, 2000
Herpes simplex virus 1 and 2	Schalasta *et al.*, 2000
Herpes simplex virus 1 and 2	O'Neill *et al.*, 2003
Herpes simplex virus 1 and 2	Filen *et al.*, 2004
Herpes simplex virus 1 and 2	Haas *et al.*, 2004
Herpes simplex virus 1 and 2	Ramaswamy *et al.*, 2004
Herpes simplex virus 1 and 2	Corey *et al.*, 2005
Herpes simplex virus 1 and 2	Issa *et al.*, 2005
Herpes simplex virus 1 and 2	Aryee *et al.*, 2005
Herpes simplex virus 1, 2 and human endogenous retrovirus (IC)	Whiley *et al.*, 2004
Herpes simplex virus 1 and 2 and varicella zoster virus	Stranska *et al.*, 2004
Orthopoxvirus, herpes simplex virus 1 and 2	Carletti *et al.*, 2005
Orthopoxvirus and Variola virus	Aitichou *et al.*, 2005
Cytomegalovirus and human apoprotein B gene	Sanchez and Storch, 2002
Human cytomegalovirus and Beta-actin DNA (IC)	Hanfler *et al.*, 2003
Cytomegalovirus polymerase (*pol*) and glycoprotein genes (*gB*)	Herrmann *et al.*, 2004
Cytomegalovirus and human herpes virus 6	Pradeau *et al.*, 2006
Human herpes virus 6 A and B	Safronetz *et al.*, 2003

Table 6.2 continued

Multiplexed targets	Reference
Feline herpes virus, *Chlamydophila felis* and feline 28S ribosomal DNA	Helps *et al.*, 2003
Hepatitis B and murine cytomegalovirus (IC)	Garson *et al.*, 2005
Hepatitis B virus lamivudine resistance markers	Geng *et al.*, 2006
Hepatitis B virus genotypes B and C	Payungporn *et al.*, 2004
Hepatitis B virus and Hepatitis C virus	Mercier *et al.*, 1999
Hepatitis B virus, Hepatitis C virus and HIV type 1	Meng *et al.*, 2001
Hepatitis B virus, Hepatitis C virus and HIV type 1	Candotti *et al.*, 2004
Hepatitis C and glyceraldehyde-3-phosphate-dehydrogenase (IC)	Pugnale *et al.*, 2005
Hepatitis C virus genotypes 1–4	Rolfe *et al.*, 2005
Enterovirus and parechovirus	Corless *et al.*, 2002
Human parvovirus B19 and plasmid standard (IC)	Gruber *et al.*, 2001
Canine parvovirus 2 and antigenic variants (types 2a, 2b, 2c)	Decaro *et al.*, 2006
Ebola virus subtypes Zaire and Sudan	Gibb *et al.*, 2001
Human Immunodeficiency virus type 1 and Rnase P (IC)	Luo *et al.*, 2005
Human Immunodeficiency virus type 1 and human albumin gene (IC)	Eriksson *et al.*, 2003
HIV types 1 and 2 and human T-lymphotrophic virus types I and II	Vet *et al.*, 1999
Human T-lymphotrophic virus types 1 and 2	Estes and Sevall, 2003
Norwalk-like viruses	Kegeyama *et al.*, 2003
Norovirus genogroups I and II	Pang *et al.*, 2005
Norovirus genogroups I and II	Hoehne and Schreier, 2006
Equine infectious anemia virus and green fluorecsent protein RNA (IC)	Cook *et al.*, 2002
Tobacco rattle virus and Potato mop top virus	Mumford *et al.*, 2000
Barley yellow mosaic virus and Barley mild mosaic virus	Mumford *et al.*, 2004
Bluetongue virus types 2 Italian field and South African vaccine strains	Orru *et al.*, 2004
Plum pox virus strains D and M	Varga and James, 2005
Lyssavirus genotypes 1, 5 and 6	Wakeley *et al.*, 2005
Vesicular stomatitis virus (Indiana and New Jersey serotypes)	Rasmussen *et al.*, 2005
Dengue virus 1, 2, 3 and 4	Johnson *et al.*, 2005

the avian H5N3 and the highly pathogenic H5N1 viruses were not detected, so a separate assay will be required in areas where these strains are circulating.

A multiplex rtPCR for the detection of H5N1 strains of IFAV was developed that uses TaqMan®-minor groove binding (MGB) probes labeled with FAM, VIC, and NED™

(Applied Biosystems, Foster City, CA) to detect sequences corresponding to the matrix (M), hemagglutinin (H5), and neuraminidase (N1) genes of IFAV (Payungporn *et al.*, 2006). TaqMan-MGB probes are generally shorter than regular TaqMan® probes; however since the probe is coupled with a minor groove binding moiety the T_M of the probe is enhanced. In addition TaqMan-MGB probes have a non-fluorescent quencher attached at the 3′ end that does not interfere with fluorescent signal detection. A positive signal for the M gene indicates IFAV infection, whereas a triple signal indicates IFAV H5N1 infection, making this a useful assay for IFAV outbreaks. Van Elden *et al.* (2001) also developed a duplex assay for the detection of IFAV and IFBV. This TaqMan assay had both probes for the IFAV and IFBV targets labeled with the FAM reporter dye, and any specimen that was found to be positive had to be typed further. The authors also suggested the use of subtype-specific primers in sequential TaqMan PCRs; this is conceivably a more rapid approach to typing in detail.

Two common causes of acute respiratory infections in children are human rhinovirus (HRV) and human metapneumovirus (HMPV). Because viral culture for HRV is insensitive and because HMPV grows very slowly, PCR is the preferred alternative for diagnosis of these viral agents. A triplex assay was designed to detect both viruses and an internal control (Scheltinga *et al.*, 2005). Two of the probes were molecular beacons labeled with HEX-BHQ1 and CY5-BHQ2 and the HRV probe was a TaqMan probe labeled with FAM-BHQ1. Individual assays were first optimized by Mg^{2+}, primer and probe titration and then by varying the annealing temperature. The thermal profile in this assay included an extension step as well as an annealing step (initial cDNA synthesis step of 30 min at 50°C, followed by 15 min at 95°C, and 50 cycles of 30 s at 95°C, 30 s at 55°C, and 30 s at 72°C). The PCR parameters for the LightCycler® instrument (Roche Applied Science, Indianapolis, IN) is similar to those for conventional PCR instruments in having denaturation, annealing, and extension steps in each cycle, because the molecular beacon probes are not hydrolysis probes; they are displaced instead of being hydrolyzed. During the denaturation step (95°C) the nucleotide pairs are dissociated, and the molecular beacons fluoresce. At the annealing temperature (55°C) the molecular beacons hybridize to specific targets that are exact complements, and the amount of fluorescence is proportional to the amount of product. At the extension temperature (72°C) the molecular beacons dissociate from the target to ensure that polymerization can occur without interference from the molecular beacon. An extension step is also required in TaqMan assays that have a larger amplicon size to allow time for the polymerization to proceed to completion. In addition, some groups use primers from a traditional PCR assay in an rtPCR assay and retain the PCR profile of the traditional PCR assay. The following caveats should be noted: if the annealing temperature is decreased from 60°C to 55°C, the specificity of the assay is reduced; if an extension step is added, the time required to perform the assay is increased; if the profile is different from that of other assays that are performed in the laboratory, performance of the different assays requiring different cycling times on the same plate is precluded and requires a dedicated instrument. The assay reported by Scheltinga *et al.* (2005) detected a $TCID_{50}$ of 0.01 for both viruses and was shown to be specific, except for the weak amplification of high titer stock of enterovirus RNA. Enteroviruses and HRV both belong to the family *Picornaviridae* and have sequence similarity in the 5′ untranslated region (5′ UTR), which is the region of the genome used as a target for the molecular detection of all known subtypes of these viruses. Due to the sequence conservation, this region cannot be used for accurate differentiation of enteroviruses from HRVs, or for serotype and genotype analysis.

The ideal assay to be developed for use in a diagnostic laboratory would one that can detect all pathogens using one assay. Gunson *et al.* (2005) came one step closer to achieving this goal by designing four triplex rtPCR reactions to detect 12 respiratory viral agents, namely IFAV and IFBV, HRSV types A and B, HRV, HPIV 1, 2, and 3, HMPV and coronaviruses 229E, OC43, and NL63. Some of the primers and probes were previously reported and were adapted, or used as reported (Bredius *et al.*, 2004; Mackay *et al.*, 2003; Templeton *et al.*, 2004; van Elden *et al.*, 2003; van Elden *et al.*, 2004); others were newly designed. The probes were labeled with FAM-BHQ, VIC-TAMRA, and CY5-BHQ and were used at a concentration of 300 nm. The concentrations of the primers ranged from 125 nm to 1000 nm. The C_T values of the positive controls were similar in the singleplex and triplex assays, and the endpoint dilutions were detected in pooled specimens containing 10 different viral targets, showing that mixed infections would not reduce the sensitivity of the assays.

Human adenovirus infections can cause a variety of diseases, including enteritis, upper respiratory tract infection, encephalitis, and cystitis (La Rosa *et al.*, 2001). There are at least 50 serotypes, belonging to six major subgroups (Xu *et al.*, 2000). Since the disparity in DNA sequence among the adenovirus serotypes precludes the design of a single primer-probe set to detect all serotypes, Ebner *et al.* (2005) designed a two-reaction rtPCR assay. This group found that mismatches between the primers and probes and the target sequence led to decreased sensitivity of detection; if they used low- stringency conditions for the PCR reaction to try to overcome this limitation, the sensitivity of the assay was compromised. Their solution was to design multiple primers and probes. One reaction contained three sets of primers and three probes, all labeled with the same fluorophore (FAM-TAMRA), and the other reaction contained one set of primers and two probes. The latter probes were labeled with FAM-MGB and differed in sequence by one base pair. The resultant two-reaction pan-adenovirus assay detects all currently known human serotypes of adenovirus and allows quantitation of viral load in a range of clinical specimens.

An alternative assay for the detection of adenovirus was comprised of five degenerate primers and seven probes, which were all labeled with FAM-TAMRA (Gu *et al.*, 2003). This assay was reported as being able to detect all serotypes from adenovirus subgroups A, B, C, and E as well as eight serotypes from subgroup D. Absolute quantitation was equivalent across all viral types tested, and sensitivity was shown to be < 10 viral genomes. Claas *et al.* (2005) also developed a rtPCR assay for the detection of subgroups A, B and C adenoviruses. They used three adenovirus-specific primers, two of which had five mismatches in order to detect all of the strains tested, as well as a set of primers to detect the internal control. Three molecular beacon probes were used in the reaction. The FAM-labeled probe detected the species A and C viruses, the HEX- labeled probe detected the subgroup B viruses, and the Cy5-labeled probe detected the internal control.

PCR assays are a major improvement over virus culture and antigen assays for the diagnosis of herpes simplex virus (HSV) and varicella zoster virus (VZV) especially in infections of the central nervous system (Gu *et al.*, 2003; Madhaven *et al.*, 1999; Mitchell *et al.*, 1997). Stranska *et al.* (2004) reported a multiplex assay to detect HSV-1 and HSV-2 and VZV based on a combination of the MagnaPure™ extractor and a TaqMan rtPCR; this assay was shown to be more sensitive and reproducible than the traditional shell vial culture method of diagnostics. The reporter dyes were FAM, VIC and 6-carboxy-4,7,2′,7′-tetrachloro-fluorescein (TET), and their fluorescence was normalized against that of internal reference dye [6-carboxy-X-rhodamine (ROX)]. A number of real-time multiplex assays that detect and distinguish HSV-1, HSV-2 and in some cases VZV have been developed and published by various groups (Corey *et al.*, 2005; Filen *et al.*, 2004; Gu *et al.*, 2003; Madhaven *et al.*, 1999; Mitchell *et al.*, 1997; Ryncarz *et al.*, 1999; Whiley *et al.*, 2004). Some of these assays use a single primer set but different probes to distinguish HSV-1 and HSV-2 (Corey *et al.*, 2005; Whiley *et al.*, 2004).

A duplex assay using LUX™ (Light Upon eXtension) primers has been developed to detect orthopox virus DNA and variola DNA (Aitichou *et al.*, 2005; Corey *et al.*, 2005; Whiley *et al.*, 2004). The LUX is a self-quenched fluorogenic primer labeled with a fluorophore close to the 3′ end, in a hairpin structure. When the primer is incorporated and extended during PCR, the fluorophore is no longer quenched, and it emits fluorescence. In this assay, one LUX primer was labeled with FAM and the other with carboxy 4′,5′ dichloro-2′,7′-dimethoxyfluorescein (JOE). The advantage of using LUX primers is that the PCR products retain the fluorescent label after extension, and therefore melting curve analysis of PCR products can be performed to help resolve problems of primer–dimer formation and false positive results.

Gibb *et al.* (2001) designed a TaqMan assay that can detect and differentiate Ebola virus subtypes Zaire and Sudan. This was a duplex assay in which one set of degenerate primers amplifies both viral targets, whereas two differentially labeled probes discriminate between the two subtypes. The two probes differed in seven nucleotides and were therefore specific for each strain.

Development of a molecular assay for the detection of noroviruses, like many other viruses, is difficult due to the genetic diversity of the genus *Norovirus*. Even the highly conserved regions within a genogroup only show 64% sequence identity at the nucleotide level (Vinje *et al.*, 2000). Pang *et al.* (2005) designed a duplex assay to target genogroups I and II. A mixture of two probes was used to detect genogroup I; both probes were labeled with VIC, and one of them had degeneracies at two positions. In general, degenerate probes are not recommended for rtPCR, since the population of probe having a 100% match to a particular sequence is reduced, thus reducing the probe concentration. Further, each individual probe sequence will have a different melting temperature (T_M). This is inadvisable; the T_M of the probe should be 8–10°C higher than the T_M of the primers, to ensure that the probe is fully hybridized during primer extension. Furthermore, the use of universal thermal cycling parameters is only possible if primers and probes with recommended T_M values are selected. Degeneracies in the probe therefore affect probe concentration and T_M values and they reduce PCR efficiency. In comparing random and specific primers, Pang *et al.* (2005) found that the random primers were preferable; they reported that separate RT and PCR reactions, using random primers in the RT reaction, were 100 times more sensitive than a reaction in which the RT and PCR were performed in a two-step, single tube reaction. By using shorter MGB probes in their duplex assay, Hoehne *et al.* (2006) were able to design a single probe to detect genogroup I; this assay had a greater dynamic range than the assay reported by Pang *et al.* (2005).

A novel method for detecting, differentiating and quantitating the two main vesicular stomatitis virus (VSV) serotypes was developed by Rasmussen *et al.* (2005) and based on a primer-probe energy transfer (PriProET) system (Rasmussen *et al.*, 2003; Vinje *et al.*, 2000), this system combines probe-based real-time monitoring of PCR amplification with confirmation of probe hybridization from the T_M curve. Two primers are used in the reaction: the reverse primer is labeled with a donor fluorophore, and a fluorescent probe is labeled with a reporter fluorophore. During each amplification cycle, the primers are extended. The fluorescent probe anneals to the extended reverse strand, and energy transfer occurs from the donor to the reporter fluorophore. The use of a 5′-exonuclease-deficient DNA polymerase prevents breakdown of the probe. The fluorescence from the reporter quantitatively reflects the amount of amplicon formed. In this particular assay, the VSV–specific forward primer was labeled with FAM at the 5′ end, and the two serotype-specific probes were labeled with Texas Red or Cy5 at the 3′ end. During the hybridization step, the specific probe binds to the extended forward strand which is now labeled with the FAM fluorophore. Although the forward primer and the probe do not overlap, the nucleotide at the 3′ end of the forward primer is adjacent to the complimentary nucleotide at the 3′ end of the probe and this close proximity allows energy transfer from the donor (FAM) to the reporter fluorophore (Texas Red or Cy5). Following the PCR reaction, the T_M profiles of the probes confirmed the specificity of the reactions. For this assay, optimization of the annealing temperature, $MgCl_2$ concentration, and primers by checkerboard titration was performed.

Development of an rtPCR assay for the detection of rabies virus is also difficult, since there are multiple virus strains with significant sequence variation. Wakeley *et al.* (2005) developed a four-probe assay to detect genotypes 1 and 6 as well as an internal control (β-actin mRNA). The assay would be most suitable for use in a limited geographic region where these particular genotypes are prevalent.

Traditionally, the diagnosis of human enterovirus and parechovirus disease has been done by cell culture or serological assays. Corless *et al.* (2002) developed a two-step reverse transcriptase TaqMan rtPCR assay that can detect 22 different enteroviruses and parechovirus types 1 and 2. This assay is 100-fold more sensitive than cell culture, and it detects some of the Coxsackie A serotypes that do not grow in culture. One primer and probe set was directed to the 5′ UTR of enteroviruses, and the other to the 5′ UTR of parechoviruses 1 and 2. This region of the genome is the most conserved for both viruses. The assay is a useful tool for rapid diagnosis of cases of meningitis and can lead to shorter hospitalization periods and minimization of inappropriate treatment. Rapid diagnosis further enables treatment with new drug regimens such as pleconaril, thus reducing fatal outcomes for neonates with enteroviral infections (Rasmussen *et al.*, 2003; Rotbart and Webster, 2001).

PCR is a rapid and reliable method for the detection of dengue virus in patient specimens. Because dengue virus isolation by culture can take up to 3 weeks, viral culture is not a suitable method to aid in patient management. A timely diagnosis can permit supportive treatment. If left untreated, the case fatality rate for dengue hemorrhagic fever is 10% or higher; serological results can be confusing, since cross-reaction is observed with other flaviviruses. The four types of dengue virus, although closely related, do not have sufficient sequence conservation that they can be detected with a single PCR primer-probe set. Johnson *et al.* (2005) developed a tetraplex real-time RT-PCR assay to detect dengue types 1, 2, 3, and 4. A range of reagent concentrations and annealing temperatures were evaluated in order to optimize the tetraplex assay; the assay's sensitivity was found to be comparable to that of the singleplex assays.

Multiplex rtPCR utilizing an internal control for the detection of viral agents

In addition to the positive and negative rtPCR controls, all molecular diagnostic tests with pre-PCR treatment, such as nucleic acid extraction and reverse transcription steps, should have an internal control that takes into account the efficiency of the extraction and reverse transcription and the presence of PCR inhibitors in the reaction. Some assays have an endogenous control such as glyceraldehyde-3-phosphate dehydrogenase (GAPDH), β-actin, or β-globin. The expression of these housekeeping genes is assumed to remain constant and therefore they are thought to be good normalizers. However, studies have shown that their expression is regulated and does not remain constant (Glare *et al.*, 2002; Ke *et al.*, 2000; Wu and Rees, 2000). Therefore, the use of an endogenous control for detecting efficient extraction, reverse transcription, and PCR inhibition is not optimal. An exogenous standard is preferable as an internal control, because it is a more accurate and reliable method of detecting efficient extraction, reverse transcription, and PCR inhibition. Addition of a known amount of an exogenous standard (either RNA or DNA, depending on whether RNA or DNA is being detected) to a specimen at the start of the extraction process and then determination of the efficiency of recovery, allows any irregularities to be detected. An exogenous standard can be any quantified wild-type or synthetic RNA or DNA containing target gene sequences. In our laboratory we have used a short sequence of the green fluorescent protein (GFP) gene inserted into a plasmid DNA or as transcribed RNA as the exogenous standard. The advantage of using GFP transcript or plasmid as an exogenous internal control is that it is a non-human and non-viral nucleic acid and therefore will be absent in patient specimens. This excludes interference with the results of the internal control assay which is a problem if a control was selected that is already present in patient specimens. In the past our laboratory used a plasmid or transcript based on glyceraldehyde 3-phosphate dehydrogenase (GAPDH) as the internal control. However, due to the presence of this gene in human cells, results of the internal control assay varied considerably. In our current clinical testing, during the lysis step of the extraction process, the GFP plasmid or transcript is spiked into the patient specimen. Following extraction (and reverse transcription, in the case of a two-step assay detecting an RNA virus), a rtPCR assay is performed to detect GFP. If the C_T value falls outside of a pre-determined range, it indicates a problem with extraction, reverse transcription, PCR inhibition, or a technical problem such as pipetting error.

A number of multiplex real-time assays have been developed that have the viral target multiplexed with the internal control. Gruber *et al.* (2001) developed a rtPCR assay for the detection of parvovirus B19, and they included an external standard. This standard was co-extracted with the viral genome and both targets were amplified and detected using a two-color fluorescence system. Sanchez and Storch (2002) created a multiplex, quantitative PCR that amplifies CMV and human DNA (apoprotein B gene), thus allowing the CMV viral load to be determined (which is based on the content of human DNA in the specimen and accurately reflects the number of cells in the specimen).

Similarly, Cook *et al.* (2002) used GFP as the internal control for a multiplex assay for the detection of equine infectious anemia virus (EIAV). Primers and probes directed at the *gag* gene of EIAV RNA and a modified form of GFP were designed, and the assay was performed as a duplex. One probe was labeled with FAM at the 5′ end and BHQ 1 at the 3′ end; the other probe had Texas Red at the 5′ end and BHQ 2 at the 3′ end. In order to decrease the competition for common PCR reagents between the two reactions, an 80% reduction in GFP primer concentration relative to viral primers was required; through this reduction, the multiplex assay could be optimized.

In the same vein, a duplex assay was developed for human immunodeficiency virus 1 (HIV-1), with an internal control included as part of the multiplex (Eriksson *et al.*, 2003; Wu and Rees, 2000). The target sequences were a conserved region of the HIV-1 *pol* gene and a section of the human albumin gene (*alb*). This method combines the quantitation of HIV-1 DNA and cellular genomes in the same reaction, and it is used to measure the level of HIV-1 DNA in $CD4^+$ cells. Comparisons of levels of peripheral virion HIV-1 RNA and HIV-1 DNA are useful in monitoring disease progression and evaluating treatment. The rtPCR was designed with FAM-TAMRA and VIC-TAMRA probes, and the primer concentrations for the *alb* target were limited. The expectation is that the number of cellular genomes will be greater than the quantity of HIV-1 DNA; therefore, in order to favor the HIV-1 DNA PCR, the albumin assay was minimized. An assay to detect HIV-1 DNA in dried blood spots was reported in which targets in the HIV-1 long terminal repeat and in the RNase P gene internal control were used (Eriksson *et al.*, 2003; Luo *et al.*, 2005; Wu and Rees, 2000).

Whiley *et al.* (2004) developed a LightCycler™ PCR assay for the detection and typing of HSV based on melting curve analysis, and with the *glycoprotein D* gene used as the target. Since the sample type being tested was generally untreated swab specimens, there was a real concern regarding obtaining false negatives due to the presence of inhibitors. Therefore, an internal control, a human endogenous retrovirus, was incorporated into the assay. Despite the fact that measures were taken to prevent the internal control assay from interfering with the HSV assay (concentrations of HSV primers and probes were higher than those of the internal control primers and probe), the sensitivity of the HSV assay was reduced, leading the authors to suggest that incorporation of the internal control should be optional, depending on the level of inhibition seen in the specimen tested.

Molecular beacons have also been used to design multiplex rtPCR assays. Helps *et al.*(2003) designed a multiplex assay to simultaneously detect feline herpesvirus (FHV) and *Chlamydophila felis* (*C. felis*) together with the feline 28S ribosomal DNA internal control. FHV and *C. felis* are two pathogens commonly associated with respiratory and ocular disease in cats. In this assay, the probe for the 28S rDNA was a molecular beacon labeled with Texas Red at the 5′ end and with DABCYL at the 3′ end. The probes directed at the *C. felis* OmpA gene and the FHV thymidine kinase gene were fluorescent quencher probes labeled with FAM-BHQ1 and CY5-BHQ2 respectively. To optimize the multiplex assay the concentration of the *Taq* DNA polymerase was increased from 0.75 to 2 U, and the $MgCl_2$ concentration was increased from 3 to 6 mM per reaction. In addition, the primer concentration for the FHV assay was reduced. These are all variables that should be explored in the optimization of multiplex PCR reactions. The authors report the sensitivity of the FHV assay to be better than that of culture, but they did not perform *C. felis* isolation. In addition, they believe the assay to be especially well suited for epidemiologic studies, since large numbers of samples can be screened at low cost and with little effort.

Multiplex rtPCR in determining viral load

Human cytomegalovirus (HCMV) is an opportunistic pathogen found in immuno-compromised patients; it is a major cause of morbidity and mortality (Eriksson *et al.*, 2003; Griffiths, 1994; Luo *et al.*, 2005). Reactivation of the virus, acquired in childhood, can be a life threatening complication after solid organ transplantation. Early detection of the virus and close monitoring of the infection are important factors in preventing or treating the reactivation of latent virus. Based on the viral load of the patient and the rate of increase of viral load since the start of the infection, appropriate antiviral therapy can be administered. A number of multiplex rtPCR assays for the detection of CMV have been developed and published. These include assays where the multiplex consisted of PCRs in which two target regions in CMV were selected (Herrmann *et al.*, 2004); PCRs in which one target in CMV was selected and an internal control was included; and PCRs in which there was a combination of viral targets. Pradeau *et al.* (2006) for example, reported a multiplex assay consisting of targets for CMV and human herpes virus 6 (HHV-6), a closely related virus in the family *Herpesviridae*. Once a patient is infected with either of these viruses, the infection remains for life, in a latent state with the potential to reactivate. Reactivation can occur in immunosuppressed patients such as organ recipients or those infected with HIV. With effective treatments available, the diagnosis and quantification of these infections will greatly aid in the management of affected patients. Determination of viral load is also important for patients infected with Epstein Barr virus (EBV). Reactivation of both CMV and EBV occurs in immuno-compromised individuals and can cause post- transplant complications (Eriksson *et al.*, 2003; Griffiths, 1995).

Employing rtPCR has facilitated quantitative measurements and is useful in evaluating the response to antiviral treatment. In the case of hepatitis C virus (HCV), the current therapy for the chronic infection is pegylated interferon-α and ribavirin administered for 24 to 48 weeks. Although these drugs are effective in more than half of the patients treated, they have several side-effects and are expensive. If the viral load of patients undergoing treatment is determined after 12 weeks, the drug responders can be distinguished from the non-responders, and drug administration can cease for the non-responders. A quantitative duplex rtPCR assay was designed for HCV detection, with GAPDH used as the endogenous internal control (Pugnale *et al.*, 2005). Quantitation of HCV in plasma was performed over the course of antiviral treatment of eight chronic hepatitis C patients. Because the C_T values for GAPDH did not vary significantly over the course of antiviral treatment, the values were considered representative of the quantity of input RNA. Relative quantitation of HCV RNA in a single patient at various time points was performed through normalization of the HCV value against the endogenous control and expression of the value relative to a reference. When the assay was performed at four time points, it was possible to monitor the level of HCV in the patient's plasma. Of the eight patients in the study, one did not respond to the antiviral treatment. The assay therefore could aid in establishing a treatment regimen and for follow-up.

The response of HCV infection to therapy is linked to viral genotype (Fried *et al.*, 2002), and the duration of therapy using interferon can also be selected based on the genotype (Hadziyannis *et al.*, 2004). It is therefore important to have a rapid, sensitive, and accurate method for genotyping of HCV. Rolfe *et al.* (2005) designed universal PCR primers targeting highly conserved sites of the 5′ non-coding (NC) region of HCV, and probes for four of the genotypes that targeted type-specific motifs within this region. A second confirmatory PCR panel was also designed, since there was no probe in the first panel to detect genotype 4, and since some cross-reaction was observed among certain genotypes.

Garson *et al.* (2005) describe a duplex rtPCR assay for the detection of hepatitis B virus (HBV) that used a murine CMV internal control. This assay was effective over a wide dynamic range and was reported to be as sensitive as any test yet described. Since HBV DNA concentrations can vary enormously (an approximately nine $\log_{10}$ range was reported) within a single patient, the wide dynamic range is important in such an assay. Assay sensitivity is also essential, since patients on antiviral therapy show low titers of virus that may not be detected by less sensitive assays. The assay was found to accurately quantify all eight genotypes of HBV which uses the World Health Organization International HBV standard as the calibration standard for quantification.

A duplex rtPCR assay has been used for the quantitation of proviral human T-lymphotropic virus (HTLV) types 1 and 2 (Estes and Sevall, 2003). A calibration standard consisting of serial dilutions of HTLV-1 and HTLV-2 DNA was included in each run. In a separate PCR reaction, detection of the human β-globin gene was used to normalize for human DNA input, to determine proviral DNA load, and to act as an amplification control. When the duplex assay was compared with singleplex assays, the sensitivity and quantitation were found to be equivalent.

The sensitive and early detection of drug resistance is a major consideration in therapeutic decision making, and it is helpful in monitoring the viral dynamics associated with treatment. Lamivudine is a nucleoside analog that is administered to patients affected with chronic HBV. A rtPCR assay for the detection of lamivudine resistance mutations in HBV, involving the use of TaqMan-MGB probes, has been reported (Geng *et al.*, 2006). The resistance to lamivudine is mainly associated with the mutations in the reverse transcriptase region of the polymerase gene (Chayama *et al.*, 1998). These mutations result in a protein with amino acid substitutions within the YMDD motif from methionine to valine or isoleucine at codon 204 of the C domain of the polymerase (Chayama *et al.*, 1998). Therefore, two TaqMan-MGB probes were constructed that recognized YIDD and YVDD; the probes were labeled with FAM and HEX dyes respectively to differentiate the amplification of each allele. The assay was sensitive and specific and was also able to distinguish mixed populations.

Multiplex rtPCR for detecting co-infections

Co-infections are common in immunosuppressed patients (Michaelides *et al.*, 2002; Permar *et al.*, 2006; Weinberg *et al.*, 2004). Their detection is important, given that certain organisms can accelerate the progression of disease resulting from infection of another pathogen (Shieh *et al.*, 2003). In addition, there is a poor understanding of the role of co-infections and common acute infections in chronic disease. In many cases of acute infection, further investigations are not performed once the first agent is detected, even though that agent may not be the cause of symptoms. Treatment of such patients may therefore not be appropriate or effective. For these reasons, and also to aid in epidemiological studies of co-infections, the use of assays that can detect multiple agents is strongly recommended.

In theory, co-infections can be detected by multiplex rtPCR. However, multiplex PCR does not generally perform optimally when there are significantly different concentrations of target nucleic acid at the beginning of a reaction (see Table 6.1). The inhibition of the less abundant target is most probably due to several factors, including sequestration or inhibition of the *Taq* enzyme by the accumulating PCR product for the more abundant target, and accumulation of pyrophosphate released when nucleotides are added during DNA synthesis. As discussed earlier, one strategy to improve the dynamic range of multiplex rtPCR is to limit the primer concentration of the assay with the highest amount of target. This is obviously not possible in testing of patient specimens since their viral populations are unknown beforehand. When a high positive result is obtained in a real-time multiplex reaction, it is prudent to test the specimen individually for each assay in the multiplex, to ensure that the high positive did not mask a low positive for a different agent.

Multiplex reactions have been used in screening donated blood for the presence of pathogenic retroviruses. Vet *et al.* (1999) used molecular beacons to develop a multiplex rtPCR assay for the detection of four viruses: HIV-1, HIV-2, HTLV-I and HTLV-II. The molecular beacons were labeled at the 5′ end with FAM, TET, tetramethylrhodamine (TMR), and 5-carboxyrhodamine 6G (RHD). This assay reliably detected as few as 10 target molecules, and was useful for the simultaneous quantitation of all four targets. All six pairwise combinations of the four retroviral DNAs were tested and were detected. However, although clinical samples were tested and correctly identified, no specimens from patients known to be co-infected were tested. While there appeared to be no cross-talk in this assay (i.e. the signal generated from one target was independent of signals generated from other targets), the authors did note an artifact that resulted from a portion of the rhodamine fluorescence being interpreted by the spectrofluorometric thermal cycler as TMR fluorescence. This is a good example of the limitations of the instruments and fluorophores that arise when rtPCR assays are multiplexed. The assay was performed on the Applied Biosystems Prism 7700 (a laser-based instrument, i.e. a monochromatic light source); that was one of the first rtPCR instruments in widespread use, but is no longer manufactured. The improved design of instruments currently in use (both laser-based and tungsten-halogen lamp-based) should overcome the artifact observed in the above mentioned assay.

The optimization of a real-time duplex assay for the detection of HBV and HCV (Mercier *et al.*, 1999; Shieh *et al.*, 2003) involved the evaluation of various concentrations of primer, probe and $MgCl_2$. By a change in the HBV probe concentration, the annealing temperature of the HBV assay could be increased from 55°C to 60°C, with the latter being the optimal temperature for the HCV assay. The assay, which required a separate RT step, was reported as being just as sensitive as the singleplex assay and suitable for screening donated blood. Although simultaneous detection of HCV and HBV was performed using mixed cultures, no co-infections were detected in the donor blood samples that were tested, and no specimens from patients with known co-infections were tested.

Some assays, although designed as multiplex assays, are qualitative and unable to differentiate among targets. An example is a multiplex real-time assay for the detection of HBV, HCV, and HIV-1 reported by Meng *et al.* (2001). This assay has primer and probe sets directed at three targets within the viral genomes, as well as one set directed at an internal control. The detection probes for HBV, HCV, and HIV-1 were labeled with the same fluorogenic reporter and quencher dyes, while the internal control probe was labeled with a different reporter but the same quencher dye. The addition of the internal control allowed the detection of PCR inhibition; this internal control also acted as an amplification and detection control. This assay is a high-throughput, automated multiplex assay, since it uses a fully automated sample preparation station (GT-X; Roche Applied Science, Indianapolis, IN) and can test 90 blood samples or pools in four hours. With a pool size of 500 units, and with the use of a combination of one GT-X and two ABI analyzers, 18 000 units of blood can be screened in eight hours. Obviously, due to the inability of this assay to discriminate among the three viral targets, it cannot distinguish single infections from multiple infections.

Other multiplex PCR tests suitable for large-scale testing of blood donations include a multiplex, quantitative real-time RT-PCR assay for the detection and quantitation of HBV, HCV, and HIV-1 in plasma and serum samples (Candotti *et al.*, 2004). This assay is performed as a two-step reaction with a separate initial RT reaction. Optimum conditions for the multiplex reaction were established using 2X the concentration of reverse transcriptase (20 U per reaction) and 2X the normal concentration of dNTPs (1.2 mm) in the PCR reaction. Primer and probe concentrations were also adjusted and differed from the concentrations optimal for the singleplex assays. The $MgCl_2$ concentration of five mm provided optimal HBV and HCV amplification but reduced HIV-1 amplification efficiency. The sensitivity of the assays remained similar to that of the singleplex assays, although there was evidence of competition between the HCV and HIV-1 RNA amplifications. The detection of HIV-1 was delayed rather than inhibited which can be a problem for quantitation of HIV-1 when both RNAs are present in the specimen. In addition to detecting multiple viral agents in competition experiments, this assay detected co-infections in specimens from known dual-infected patients.

Depending on the sequence similarity between target regions, it is possible in some instances to develop PCR assays that utilize the same set of primers but have different probes. This is the case for duplex assays, developed by Corey *et al.* (2005) and by Whiley *et al.* (2004), designed to detect and differentiate HSV-1 and HSV-2. Corey *et al.* (2005) reported an assay with primers in the glycoprotein B (gB) gene of HSV which were designed to amplify HSV-1 and HSV-2 with equal efficiencies. The two type-specific probes were labeled with different fluorophores (VIC-TAMRA and FAM-TAMRA) and were normalized against that of ROX. This assay was able to detect co-infections in patient specimens, as well as in competition experiments with mixed viral culture. Competition experiments showed that when the fold-difference in quantity between the targets was between 1000- and 10 000-fold, then the target with the lower quantity was likely to be out-competed, thus creating a problem with false negatives in dual infections in which one virus titer is significantly higher than the other.

We have developed a TaqMan rtPCR assay for the detection of HSV-1, HSV-2, and VZV. The assay is used routinely for the detection of these viruses in CSF from encephalitic patients, and also in genital swabs from patients with genital herpes. Dual infections have been detected a number of times, especially in HIV-positive patients.

Use of melting curve analysis for virus multiplexing

In addition to multiplexing by color, multiplexing based on melting temperature (T_M) is in routine use, to discriminate among sequences. This method is based on acquisition of melting curves while varying the temperature and continuously monitoring fluorescence. Product identification and allele discrimination can be performed on the basis of the melting of different duplexes at different temperatures.

Varga *et al.* (2005) described a multiplex rtPCR with melting curve analysis that used SYBR™ Green I to identify Plum pox virus isolates of strains D and M. The amplicon sizes were 114 bp and 380 bp and had distinct melting temperatures, thus allowing their differentiation. T_D values are affected by the concentrations of DMSO, $MgCl_2$, and template DNA, and by the rate of temperature transition (Giglio *et al.*, 2003; Ririe *et al.*, 1997). It should be noted that sequence variation in different isolates of the same strain can lead to T_D variations, as can the GC content (Ririe *et al.*, 1997). Melting curve analysis has also been used for the detection and differentiation of other viral agents (Haas *et al.*, 2004; Issa *et al.*, 2005; Orru *et al.*, 2004; Payungporn *et al.*, 2004; Ramaswamy *et al.*, 2004; Ririe *et al.*, 1997; Safronetz *et al.*, 2003).

A rtPCR assay for the detection of IFAV and HRSV was reported that uses the melting curve analysis feature of the LightCycler™ instrument to distinguish the amplified products (Boivin *et al.*, 2004; Safronetz *et al.*, 2003). The authors stated that this method enables a large number of viral targets to be detected simultaneously, since there is not an issue of the limitation of assays employing probes due to the instrument's ability to spectrally discern only a handful of fluorophores. However, it should be noted that when melting curve analysis is used as a method of multiplexing, the assumption must be made that there is no interaction between or among the primers, and that each viral amplicon has a reproducible and discriminant denaturing temperature (T_D). Moreover, quantitation of the target is not possible. These disadvantages have prevented melting curve analysis from being used routinely in the multiplexing of a large number of targets.

In order to distinguish orthopoxvirus infections from HSV and VZV, an rtPCR assay based on melting temperature analysis was developed by Carletti *et al.* (2005). PCR products of 300 bp, 260 bp, and 360 bp for orthopoxvirus, VZV, and HSV respectively were distinguished after rtPCR, through performance of a melting curve analysis. The T_D values were selected so as to be sufficiently different to allow this analysis. When melting temperatures failed to discriminate specific targets within each virus family, a further characterization was performed using Restriction Fragment Length Polymorphism (RFLP) analysis. Aryee *et al.* (2005) also chose to detect HSV by melting curve analysis; they additionally performed a duplex quantitative rtPCR using TaqMan-MGB probes to quantitate and type HSV. The use of rtPCR for the detection of HSV-1 and HSV-2 by melting curve analysis has been shown to be more sensitive than the use of shell vial or cell culture methods (Espy *et al.*, 2000; Safronetz *et al.*, 2003). However, care should be taken in the selection of probe sequences, even for DNA viruses, since the selection of probe sequences in regions of the genome that are targets of antiviral drugs, or that are subject to variation for other reasons, can compromise the sensitivity of assays.

A slightly different method of melting curve analysis using two probes hybridizing in head to tail configuration has been described by Schalasta *et al.* (2000) for the detection and differentiation of HSV-1 and HSV-2. A rapid cycle rtPCR system was developed based on the LightCycler system. Two specifically designed hybridization probes internal to the amplification primers were labeled with different fluorescent dyes: fluorescein was coupled at the 3′ end of one probe, and LC Red 640 at the 5′ end of the other. The probes hybridized in a head to tail configuration to the target and when excited at 470 nm the FAM fluorophore transferred energy to LC-Red 640, which then emitted light of a longer wavelength. A mismatch within one of the probes decreases the probe/product T_M from 69°C to 64°C, thus allowing discrimination between HSV-1 and HSV-2.

A rather unusual multiplex rtPCR assay was reported by O'Neill *et al.* (2003) for the detection of HSV-1 and HSV-2 and VZV. This group performed a nested multiplex PCR using SYBR Green, and they differentiated the PCR products using the Light Cycler melting curve software. The first round of PCR was performed as a multiplex on a block thermal cycler using three sets of primers. The second round of PCR was performed on an iCycler™ (Bio-Rad Laboratories, Hercules, CA) using three sets of nested primers. A single fluorescent reading was taken at the end of each extension cycle, and a melting curve analysis was performed from 63° to 95°C, with a fluorescent reading taken at each 0.1°C increase in temperature. Each target gave a distinct T_D, and there were no overlaps between the targets that were tested. The assay was just as sensitive as the conventional multiplex assay for HSV-1 and VZV and more sensitive for HSV-2. This assay is unusual because one of the major reasons for using rtPCR is to reduce the possibility of contamination through the closed tube system. However, this assay requires the PCR product from the first round to be added to the second-round mastermix, thereby re-introducing the possibility of contamination. Further, SYBR Green is the fluorescent indicator and as we stated previously, the use of multiple primers in SYBR Green assays is not encouraged, since it can lead to primer–dimer formation and therefore to non-specific fluorescence.

Commercial multiplex kits available for diagnosis of viral agents

Several manufacturers have developed kits that detect multiple viral agents. The RealArt™ HSV1/2 PCR kit (Artus GmbH, Hamburg, Germany) and the LightCycler HSV1/2 detection kit (Roche Molecular Diagnostics, Indianapolis, IN) detect HSV-1 and HSV-2 and distinguish them by melting curve analysis. Vetdetect (Anigen Diagnostics, Hong Kong SAR, China) offers a number of real-time kits, including a multiplex kit for the detection of avian influenza virus (AIV) and Newcastle disease virus (NDV) for use on veterinary and environmental samples. This is a TaqMan assay, and the mastermix contains probes labeled with FAM and VIC. The ProFlu-1™ assay is a research use only (RUO) rtPCR assay for the detection of IFAV, IFBV and HRSV (Prodesse, Waukesha, WI). The development of rtPCR assays by commercial vendors is widespread and will undoubtedly continue to increase rapidly. However, it is essential that every diagnostic laboratory validates any such assay in-house, to ensure optimal performance.

Multiplex rtPCR in the detection of parasites

Diarrheal diseases

Diarrheal diseases are exceedingly common; their etiological agents can be viral, bacterial, or parasitic. Even if we limit consideration only to parasites, several different organisms can cause disease in humans, and symptoms of infection are similar. Given the non-specific clinical presentation, it is particularly helpful to be able to simultaneously screen for several parasites known to most commonly cause diarrheal illness. The traditional method of diagnosis has been microscopic examination of stool samples or water concentrates. However, microscopic detection is time-consuming, often insufficiently sensitive to detect the low numbers of parasites capable of causing disease, and requires extensive training and experience for accurate identification. As a means to increase sensitivity and reduced turnaround time, several multiplex assays have been developed. As will be seen, different methods of detection have been employed, and each assay has been designed to detect a particular subset of the suspected agents.

Entamoeba, *Giardia*, and *Cryptosporidium*

A multiplex rtPCR assay has been developed for the simultaneous detection of *Entamoeba histolytica, Giardia lamblia* and *Cryptosporidium parvum* (Verweij *et al*., 2004). Detection is based on primers and detection probes complementary to small ribosomal subunit rRNA. Exogenous DNA from a herpesvirus that infects seals was added to all samples as a control for inhibition of the amplification reaction. For each parasite, 20 stool samples that had been confirmed positive by other means, such as microscopy, PCR or positive antigen test, were evaluated in the multiplex assay. In each case there was 100% concordance with the results of the original method. In addition, the specificity of the primers was tested against DNA derived from four parasites; *E. dispar, Enterocytozoon bieneusi, Encephalitozoon intestinalis* and *Cyclospora cayetanensis* and DNA from 12 different bacterial and yeast cultures; *Bacillus cereus, Enterococcus faecalis, Staphylococcus aureus*, coagulase-negative *Staphylococcus, Escherichia coli, Klebsiella pneumoniae, Proteus mirabilis, Pseudomonas aeruginosa, Salmonella enterica* serovar Typhimurium, *Shigella flexneri, Yersinia enterocolitica*, and *Candida albicans*. None of these organisms gave detectable amplification with any of the parasite-specific primers or the control primers. Also tested were stool samples from patients confirmed positive for *E. dispar* by PCR and samples from healthy patients with no known history of parasitic infection. None of these *E. dispar* or healthy patient samples gave positive results for *E. histolytica, G. lamblia* or *C. parvum*, again confirming the specificity of the multiplex reaction.

Sensitivity was not determined either in terms of the minimum numbers of oocysts or cysts or the minimum amount of DNA detected, although all DNA from patient samples was purified from a small volume (200 μl, ~0.5 g/ml) of stool sample. Sensitivity of the rtPCR assay was compared to the sensitivity of microscopic examination of acid-fast-stained fecal smears for an immunocompromised patient diagnosed with a *C. parvum* infection. Eight samples were taken over a one-year period. Samples collected at four time points gave positive results by both microscopy and rtPCR, three samples were positive by rtPCR only, and one was negative by both methods. The results are in agreement with previous reports indicating that detection of parasite-derived DNA by conventional or rtPCR is often more sensitive than microscopy.

A microarray-based detection method has been developed for the simultaneous detection of *E. histolytica, E. dispar*, two assemblages of *G. lamblia, C. parvum* (type 2) and *C. hominis* (type1) (Wang *et al*., 2004b). DNA isolated from cultures of *Entamoeba*, trophozoites of *Giardia*, or oocysts of *Cryptosporidium* was amplified by multiplex conventional PCR, using six to eight primer sets designed to be genus-specific, but not species-specific.

Amplified DNA served as a template for primer extension, to generate fluorescently labeled ssDNA. The fluorescently labeled DNA was used as hybridization probes on microarrays. The microarrays were generated from oligonucleotides (23–30 nt) specific for sequences common to two species, or assemblages, of parasites, and they were also spotted with primers designed to distinguish between the related organisms.

When assessed for cross-reactivity and specificity, hybridization patterns were sufficiently different to allow species identification. However, both false positive and false negative results were obtained. For some microarray primers, the sequence differences between the two strains were minimal, which contributed to cross hybridization. Of the three pairs of parasites tested, *C. parvum* and *C. hominis* were the least distinguishable. Only seven species-specific oligonucleotides were used for *Cryptosporidium*, compared with 13 each for the other two parasites. In addition, fluorescent probes derived from *C. hominis* cross-hybridized strongly to one oligonucleotide designed to be *C. parvum* specific. The *C. hominis* probe also hybridized weakly to four other *C. parvum* oligonucleotide spots. Similarly, hybridization with *C. parvum* resulted in one strongly positive and four weakly positive reactions with *C. hominis* specific oligonucleotides. In addition, *C. parvum* did not hybridize strongly to one of the spots intended to be common to the two species. In spite of the cross-hybridization observed for the oligonucleotide set that was queried, distinguishable patterns were observed for the two species.

The *Cryptosporidium* arrays were also tested against two non-human-pathogenic species; *C. meleagridis*, primarily found in birds, and *C. muris*, which infects rodents. DNA amplified from both of these species resulted in profiles that were distinct. The *C. meleagridis* profile contained 11 strong, four intermediate, and 10 weak hybridization signals, which were distinct from the pattern obtained with the three other *Cryptosporidium* probes. The profile observed for *C. muris* was consistent with information from previous phylogenetic studies indicating divergence from the human pathogens, as only four array primers resulted in a hybridization signal. Not surprisingly, the microarray probes detected corresponded to genes encoding heat shock protein 70 and rRNA.

Sensitivity of the microarray detection method was tested using DNA isolated from the two *G. lamblia* isolates. The ability to detect DNA corresponding to 200 trophozoites and to five trophozoites was tested, for six different genes. For three genes, hsp70, triose phosphate isomerase, and ORF c4, the number of hybridization positive spots was equal for both the 200-trophozoite and five-trophozoite samples, and all of the oligonucleotides corresponding to the gene were detected. For glutamate dehydrogenase, only four out of five spots were detected at low DNA concentration for one *G. lamblia* isolate, while all five were detected when more DNA was present. The second *G. lamblia* isolate produced hybridization to all spots at both DNA concentrations. Only for one gene tested, giardin, and only for one of the isolates, did the DNA corresponding to five trophozoites fail to hybridize to any of the gene-specific oligonucleotides, while all of the oligonucleotide spots gave positive signals when a larger quantity of DNA was present. Again, positive signals were obtained at both DNA concentrations for the second isolate. The data indicate that this method is likely to be sufficiently sensitive to detect the small numbers of parasites that are capable of resulting in infection. Clearly, the oligonucleotides included require testing for sensitivity and specificity. However, once optimized, the degree of multiplexing possible via microarray analysis enables multiple organisms and multiple gene targets to be queried simultaneously.

Malaria

Malaria is an extremely widespread disease with 300–500 million cases worldwide, resulting in over one million deaths, annually. Transmitted by the bite of an infected anopheline mosquito, malaria is caused by four different species of *Plasmodium* parasites, *P. falciparum*, *P. malariae*, *P. ovale*, and *P. vivax*. Traditionally, diagnosis has been by microscopy of peripheral blood smears. Differentiation of the four species is possible by microscopy when conducted by trained and experienced personnel. However, low parasitemia levels, mixed infections, and alterations in morphology of infected red blood cells following drug treatment challenge the abilities of even the most skilled microscopist. As a result, a number of molecular methods of detection have been investigated.

Multiplex rtPCR for the detection of malaria parasites

An rtPCR assay has been developed that is capable of distinguishing the four species of *Plasmodium* species known to cause malaria in humans. The assay utilizes the intercalating dye SYBR Green to detect amplicons and melting curve analysis to distinguish products from each of the four species (Mangold *et al.*, 2005). As with other malaria assays, the gene targeted for identification encodes the 18S rRNA of the small ribosomal subunit; due to the multiple copies found in the genome, this gene confers an advantage in terms of sensitivity (Gardner *et al.*, 2002). Patient samples used for detection were in the form of blood spotted onto treated filter paper, and the limit of detection was calculated at approximately one parasite/μl. Identification of each of the four species was made based on denaturation curves: the T_D for each of the products differed from the others by at least 2°C. A consistent difference in the average T_D was observed between plasmid-derived DNA and patient samples. The T_D was also dependent upon the method used to extract DNA from the filter paper, possibly due to the salt concentration in the extraction buffers. Thus, when this method is used, melting curves may need to be determined for the specific application. Beyond that limitation, a particular utility of the study is that detection and species identification are possible from approximately 10 μl of a dried blood spot.

A multiplex TaqMan rtPCR with the capability to identify three of the four *Plasmodium* species has been described (Perandin *et al.*, 2004). The assay uses probes labeled with FAM, TET, and VIC to detect *P. falciparum*, *P. vivax* and *P. ovale*, respectively. The limit of detection ranged from one to five parasites/μl of infected blood, depending upon the species tested, and average C_T values ranged from 29 to 39, based on at least four replicates. The C_T values were linear over a 10 000-fold range of input DNA. Specificity of the reaction was evaluated using *Toxoplasma gondii*, *Leishmania infantum* and human DNA, all of which gave no detectable fluorescent signal. Accurate species identification by rtPCR was evaluated by comparison to microscopy, nested conventional PCR, and DNA sequencing. DNA template for conventional and rtPCR was isolated from 200 μl of blood drawn from patients, and thin blood smears were prepared from the same samples. Of 122 samples tested, 61 were found negative by microscopy. Real-time PCR, nested PCR, and DNA sequencing confirmed negative results for all but one sample, which was found positive for

P. falciparum. All samples found positive by microscopic evaluation were confirmed positive by rtPCR, with the exception of *two P. malariae* samples. The two samples were negative by the triplex rtPCR assay, since a *P. malariae* primer/probe set was not included. Most discrepancies that occurred were in identification of species, particularly for those samples identified by microscopy as non-*P. falciparum* or mixed infection. With the exception of the two *P. malariae*-containing samples, species identification by rtPCR, nested conventional PCR, and DNA sequencing gave complete agreement for all positive samples.

A two-step rtPCR assay has been developed for the diagnosis of malaria (Rougemont *et al.*, 2004) and is capable of identifying all four *Plasmodium* species that infect humans. Detection of PCR product is by dual-label fluorogenic probe, and the target of the assay is again the small ribosomal subunit 18S rRNA gene. In the first step, a primer/probe combination is used to screen for the presence of any of the four *Plasmodium* species. The specificity of this reaction was tested against human genomic DNA, *Leishmania, Aspergillus, Toxoplasma, Pneumocystis,* and *Neurospora;* due to sequence conservation in the region of the gene targeted, false-positive results were obtained for all but the human and *Leishmania* DNA. In the second step of the assay, two parallel reactions were performed to detect *P. falciparum* or *P. vivax* and *P. ovale* or *P. malariae*. TaqMan probes used in the first multiplexed assay were labeled with 5′ FAM and 3′ TAMRA, for hybridization to *P. falciparum* amplicons, combined with 5′ VIC and 3′ TAMRA for detection of *P. vivax*. In the second multiplexed assay TaqMan-MGB probes were labeled with 5′ FAM for P. *malariae* detection and 5′ VIC for *P. ovale*. The four species-specific reactions did not show cross-reactivity for the four plasmodium species nor was cross-reactivity observed for the other eukaryotic organisms tested.

Sensitivity was evaluated with plasmids carrying cloned 18S DNA from each of the four species as well as a *P. ovale* variant. Individual assays had a detection limit of between one and 10 copies. Comparison of rtPCR performed with EDTA-treated blood samples with species identification by microscopy gave a concordance rate of 86%. Six samples found positive by rtPCR but negative by microscopy were subsequently confirmed positive by immunochromatographic or nested PCR methods. Three of the patients had a previous history of malaria infection, and three had a recent history of prophylactic chemotherapy. The results highlight the difficulty of microscopic diagnosis for low parasitemias or following drug treatment. Three samples were initially deemed positive by microscopy but gave negative results by rtPCR. All three of the samples were subsequently found negative, both by a second microscopist and by nested PCR. Thus, for the 14% of the samples in which initial results disagreed, the rtPCR result was confirmed over the findings based on microscopic evaluation.

Ligase detection fluorescent microsphere assay

A multiplex post-PCR amplification method has been developed that uses fluorescent microspheres coupled to oligonucleotides to simultaneously detect the four species of Plasmodium (McNamara *et al.*, 2006). The PCR target is, as in assays described above, the 18S rRNA. Following amplification with a single set of genus-specific primers (McNamara *et al.*, 2004) the PCR product is hybridized to species-specific primers bearing 5′ oligonucleotide tagged (TAG) sequences and an adjacent common primer with a 3′ biotin label. In the presence of the PCR-amplified product from the corresponding species, 5′ TAG labeled and 3′ biotin labeled oligonucleotides are ligated. The ligation products are hybridized to oligonucleotides complementary to the TAG sequence and are associated with a fluorescent microsphere. A streptavidin–phycoerythrin conjugate is then bound to the 3′ biotin, and doubly labeled fluorescent ligation products are detected in a fluorescent microsphere assay.

In the assay described, amplification was achieved by 27 cycles of PCR with a detection limit of approximately one to 10 infected red blood cells per microliter of blood sample. Specificity of the ligation reaction was determined using cloned DNA fragments to generate all pair-wise and three-way combinations, and the four-species combination. In each case the only ligation products observed were representative of the input template(s) (McNamara *et al.*, 2004). Comparison of microscopic evaluation, rtPCR and the ligase detection assays was performed with *P. falciparum* cultured in human red blood cells and with *P. vivax* infected monkeys. DNA was extracted either from 200 μl of cultured infected cells or whole blood drawn from primates. Concordance of results for all three assays was 76%. Pair-wise concordance was highest for microscopy and ligase detection (90.3%) followed by rtPCR and ligase detection (84.5%). The lowest agreement between assays was observed for microscopy results and rtPCR (76.6%). Typically, non-concordance of results was observed at low parasitemias, although one sample with 7000 infected cells/μl as judged by microscopy and with high fluorescence intensity in the ligase detection assay was negative by rtPCR.

One distinct advantage of the ligase detection assay over rtPCR is the ability to expand the assay beyond the handful of fluorescent dyes that can be detected simultaneously with current rtPCR detectors. Up to 100 uniquely labeled fluorescently labeled microspheres are available, making multiple allele detection along with anti-malarial drug resistance testing an attainable goal.

Multiplex rtPCR in the detection of bacteria

Real-time PCR has been an invaluable diagnostic tool, especially for detection of certain bacteria that are unculturable or fastidious. This technology has been particularly useful in epidemiologic investigations, to determine the etiology of certain bacterial outbreaks, as well as to identify drug-resistant characteristics of microorganisms. As the technology has evolved, the ability to multiplex has increased the utility of this type of analysis. Multiplex rtPCR has additional benefits over singleplex rtPCR reactions, including the conservation of clinical specimens, reduced reagent cost, and overall reduced analysis time, since multiple assays are run in one tube. This section of the chapter will discuss the use of multiplex rtPCR for bacterial pathogen identification and will highlight some specific applications of the method to the identification of bacteria.

One of the most important advantages of the use of rtPCR is the great sensitivity of the assay. Samples submitted for analysis may, as a result of mishandling, be exposed to extended transport or temperature, or they may be improperly collected, resulting in non-viable organisms that cannot be isolated by culture. Melo and co-workers found multiplex rtPCR to be more sensitive than culture for the diagnosis of *Yersinia pestis*, for both retrospective and more recent samples (Melo *et al.*, 2003). The diagnostic potential of

PCR for the direct detection of *B. anthracis* in human specimens was confirmed during the anthrax mailing events of 2001 (Hoffmaster *et al.*, 2002; Melo *et al.*, 2003). In a subset of these patients, cultures were negative because of the initiation of early antibiotic treatment; nevertheless, PCR yielded positive results, revealing the causative agent.

Several pathogenic bacterial organisms are fastidious or require specialized growth medium or environment. Thus, another important use of multiplex rtPCR is as a screening technique for these difficult-to-culture organisms. For example, *Campylobacter* species must be grown anaerobically and can take days to weeks to isolate by culture. Clinical specimens and associated food samples are often sent for testing to a public health or food laboratory to identify a contaminated food source to link with infected individuals. In many instances, a large number of samples must be tested, to identify a contaminated food product. If rtPCR is used as a screening tool it can assist in an epidemiological investigation. These assays can also assist the laboratory to focus its attention on particular food sources, during efforts to identify a causative bacterial agent. Valuable information can be provided to epidemiologists before culture results can become available for organisms such as *Campylobacter*. Another example of a fastidious organism is *Mycobacterium*. Multiplex rtPCR has also been utilized for genotyping and forensic analysis of this genus of bacteria in recent years. Hillemann *et al.* have distinguished strains of *M. tuberculosis* in order to track the spread of a new drug-resistant strain of *M. tuberculosis* (Hillemann *et al.*, 2006).

Multiplex rtPCR assays are ideally suited to the analysis of clinical specimens because sample is often limited and turnaround time is an important consideration for patient treatment. Much of the research applying multiplex rtPCR to the detection of bacterial pathogens has focused on the development of syndromic panels, such as respiratory, enteric, genital tract, and encephalitis-causing pathogens, as well as on the development of biothreat-related multiplex assays. The application of rtPCR to the detection of biothreat agents, constituting one of the most significant advances of this technology, will be extensively described in an upcoming chapter. The use of syndromic panels enables screening for numerous infectious agents from minimal patient sample. For an encephalitic assay panel run on a limited-quantity CSF sample, multiplex rtPCR is an ideal tool for a rapid diagnosis. Table 6.3 lists the multiplex real-time assays that have been developed for bacterial pathogens.

Enteric bacterial pathogens

Real-time PCR has great value for the detection of enteric pathogens in clinical specimens and in the associated environmental samples that are needed to determine the source of human infections, such as contaminated food or water. Several multiplex rtPCR assays designed to detect Shiga-toxin producing *E. coli* strains have been reported (Table 6.3). These assays were some of the earliest documented uses of multiplex rtPCR in the field of bacteriology. At least 10 real-time multiplex assays have been developed that use multiple probe chemistries and rtPCR platforms. The use of assorted probe technologies and various platforms gives diagnostic laboratories more flexibility in which assays they will adopt. Assays have been developed to be used by high-throughput laboratories, or by laboratories that test very few samples and need reagents that are inexpensive such as SYBR Green.

Yoshitomi *et al.* (2006b) have developed a multiplex assay targeting three markers of *E. coli* O157:H7 and other Shiga toxin producing strains of *E. coli* using SYBR Green.

Table 6.3 Multiplex rtPCR detection of bacterial pathogens

Bacterial pathogen(s) detected	Gene targets	Specimen type	Platform	Probe chemistry	Reference
Anaplasma phagocytophilum and *Borrelia burgdorferi*	16S RNA, *fla*	Blood, ticks	BioRad iCycler®	TaqMan®	Courtney *et al.*, 2004
Anaplasma spp. and *Ehrlichia* spp.	16S RNA	Blood	Cepheid SmartCycler®	TaqMan®	Sirigireddy and Ganta, 2005
B. pertussis, *B. parapertussis*	IS481- *B. pertussis*, IS1001-*B. parapertussis*	Nasopharyngeal swabs	Roche LightCycler®	HybProbe	Sloan *et al.*, 2002
B. pertussis, *B. parapertussis*	IS481, IS1001	Various respiratory specimens including dried slides	Roche LightCycler®	HybProbes	Cloud *et al.*, 2003
Borrelia spp.	*ospA*, *glpQ*, *rrs-rrlA* IGS	Ticks	Not given	TaqMan®	Ullman *et al.*, 2005
Campylobacter jejuni and, *C. coli*	*mapA*, *ceuE*	Bacterial isolates	ABI Prism® 7700	TaqMan®	Best *et al.*,, 2003
Campylobacter spp.	16s rDNA, 23s rDNA	Chicken rinse samples	ABI Prism® 7700, Rotor-Gene 3000	TaqMan®	Josefsen *et al.*, 2004
Chlamydia trachomatis, Neisseria gonorrhoeae	*porA*, Ct- major outer membrane protein	Urogenital specimens	ABI Prism® 7500, Roche LightCycler®	TaqMan®, HybProbe	Whiley and Sloots, 2005
C. trachomatis LGV	Cryptic plasmid pLGV440, *ompL*	Rectal swabs	ABI Prism® 7000, ABI Prism® 7500	TaqMan® MGB	Halse *et al.*, 2006
C. pneumoniae, *M. pneumoniae*	P1- *M. pneum*, PST-1 fragment-CT	Various respiratory specimens	ABI Prism® 7700	TaqMan®	Welti *et al.*, 2003

Table 6.3 continued

Bacterial pathogen(s) detected	Gene targets	Specimen type	Platform	Probe chemistry	Reference
C. pneumoniae, *L. pneumophila*, *L. micdadei*, *M. pneumoniae*	Targets not given	Bacterial isolates	Not given	Not given	Khanna *et al.*, 2005
Clostridium difficile	*tcdA* and *tcdB*	Feces	Cepheid SmartCycler®	Molecular beacon	Belanger *et al.*, 2003
Enterohemorrhagic *E. coli*	*stx* 1 and 2	Bacterial Isolates	Roche LightCycler®	SYBR® Green	Bellin *et al.*, 2001
Shiga toxin producing *E. coli*	*stx* 1 and 2	Stool	Cepheid SmartCycler®	Molecular beacon	Belanger *et al.*, 2002
Enterohemorrhagic *E. coli*	*stx* 1 and 2 and *uidA*	Bacterial isolates	Cepheid SmartCycler®	SYBR® Green	Yoshitomi *et al.*, 2006
E. coli O157:H7	*stx* 1 and 2 and *eae*	Stool	ABI Prism® 7700	TaqMan®	Sharma, 2006
Enterohemorrhagic *E. coli*	*rfbE* and *eae*	Bacterial isolates and Feces (bovine)	ABI Prism® 7700	TaqMan® **Reverse Transcription rtPCR multiplex	Sharma, 2006
E. coli O157:H7	*Stx* 1 and 2	Various environmental matrices	BioRad iCycler	5′nuclease	Ibekwe, 2002, Ibekwe and Grieve, 2003
E. coli O157:H7, Shiga toxin producing *E. coli*	*stx* 1 and 2, *eaeA*	Beef, feces	ABI Prism® LS 7200 Sequence Detection System	TaqMan®	Sharma *et al.*, 1999
E. coli O157:H7/H-	*stx* 1 and 2, *uidA*	Bacterial Isolates	Cepheid SmartCycler®	TaqMan® MGB	Jinneman *et al.*, 2003

Enterotoxigenic *E. coli*	Heat lABI Prism®le and Heat stABI Prism®le toxin	Various produce and foods	Cepheid SmartCycler®	TaqMan®	Grant *et al.*, 2006
E. coli O157:H7	*stx* 1 and 2	Bacterial isolates	Roche LightCycler®	SYBR® Green	Jothikumar and Griffiths, 2002
Haemophilus influenzae, *N. meningitidis*, and *S. pneumoniae*	*bexA*, *ctrA*, and *ply*, respectively	CSF, serum, plasma, whole blood	ABI Prism® 7700	5′ nuclease	Corless *et al.*, 2001
H. influenza, N. meningitidis, *S. Pneumoniae*	*ctrA*, *ply*, *bexA*	Various clinical specimens	ABI Prism® 7700	5′ nuclease	Smith *et al.*, 2004
Listeria monocytogenes, Salmonella serovars	*fimI* and *hlyA*	Bacterial culture	Roche LightCycler®	SYBR® Green	Jothikumar *et al.*, 2003
Listeria monocytogenes, Salmonella serovars	*invA, hlyA*	Meat	Roche LightCycler®	SYBR® Green	Wang *et al.*, 2004
Legionella pneumophila, *Legionella* spp.	16S, *mip*	Sputum, throat swab, BAL	BioRad iCycler®	Molecular beacon	Templeton *et al.*, 2003
Methicillin- resistant *S. aureus*	*mecA, nuc*	Bacterial isolates	Stratagene Mx4000®	Molecular beacon	Elsayed *et al.*, 2003
Methicillin- resistant *S. aureus*	3 Staph cassette chromosome *mec, orfX*	Nasal swabs	Cepheid SmartCycler®	Molecular beacon	Huletsky *et al.*, 2004
Methicillin- and vancomycin-resistant *S. aureus*	*mecA, vanA, lukF*	Bacterial Isolates	Stratagene Mx4000®	Molecular beacon	Sinsimer *et al.*, 2005
Methicillin- resistant *S. aureus*	*mecA, nuc*	Bacterial isolates	Roche LightCycler®	HybProbe	Costa *et al.*, 2005
Methicillin- resistant *S. aureus*	*mecA, nuc*	Swabs	Roche LightCycler®	HybProbe	Hope *et al.*, 2004
Methicillin- resistant *S. aureus*	Type I,II,III, and IV *ccrB*	Bacterial isolates	ABI Prism® 7700	TaqMan®	Francois *et al.*, 2004

Table 6.3 continued

Bacterial pathogen(s) detected	Gene targets	Specimen type	Platform	Probe chemistry	Reference
Mycoplasma genitalium	*gyrA*, human RNaseP	Vaginal samples	ABI Prism® 7900HT	TaqMan®	Blaylock *et al.*, 2004
Mycobacterium tuberculosis	N.A.	TB isolates	RotorGene™	TaqMan®	Hillemann *et al.*, 2006
Staphylococcus aureus, *Streptococcus uberis*, *S. agalactiae*	*cfb, pla*	Milk	Roche LightCycler®	HybProbe	Gillespie and Oliver, 2005
Vancomycin-resistant *S. aureus*	*vanA, vanB*	Rectal swabs	Roche LightCycler®	HybProbe	Palladino *et al.*, 2004
Vancomycin-resistant *S. aureus*	*vanA, vanB*	Bacterial isolates	Roche LightCycler®	HybProbe	Palladino *et al.*, 2003
Vibrio vulnificus	*vvh, viuB*	Shellfish	Cepheid SmartCycler®	SYBR® Green	Panicker *et al.*, 2004

This method uses rtPCR and melting curve analysis for a triplex assay. *E. coli* O157:H7 can be distinguished from other Shiga toxin-producing enteric pathogens via the combination of targets for two Shiga toxin genes as well as a highly conserved base substitution of the beta-glucoronidase gene. In 2001, Bellin *et al.* developed a multiplex rtPCR based on both Shiga toxin genes, using a LightCycler platform (Bellin *et al.*, 2001; Hillemann *et al.*, 2006). This was a duplex rtPCR assay in which results could be obtained in as little as 45 minutes.

Relatively few of the *E. coli* multiplex rtPCR assays were validated for using human stool samples, the most common clinical specimen type submitted to laboratories for testing (Belanger *et al.*, 2002). This is an important parameter to consider when developing multiplex rtPCR assays. The clinical specimen type or matrix most frequently tested should always be part of the validation process, to ensure that the multiplex assay is optimized for it and to ensure that there is no interference by inhibitors common to the sample type. Belanger *et al.* (2002) developed a duplex rtPCR assay that detected Shiga toxin-producing bacteria, using molecular beacons on the Cepheid SmartCycler platform. In a complicated matrix such as stool, the authors found that the assay had a positive predictive value of 100%. Multiplex rtPCR assays for *E. coli* O157:H7 were also useful in establishing an epidemiological link between a patient and environmental source. Several multiplex assays have been developed for the detection of *E. coli* O157:H7 from dairy wastewater, soil, and manure (Ibekwe *et al.*, 2002; Ibekwe and Grieve, 2003). These assays were shown to be highly sensitive as well as specific; these characteristics are important for sample types that may contain a large number of related strains.

Another enteric pathogen that has been the target of multiplex rtPCR assay development is *Clostridium difficile*. Belanger (2003) utilized a duplex assay that could detect *C. difficile* toxins A and B. The assay was able to detect as few as 10 genome copies of this pathogen, and it could detect the most common strains of *C. difficile*. This rtPCR assay was validated using stool samples, and the targeted genes were amplified in 28 of 29 positive samples. The gold standard for *C. difficile* toxin identification is a cytotoxin, assay but these assays are often time-consuming and their results can be problematic, due to interfering agents present in stool samples. As for other *Clostridium* species such as *C. botulinum* toxins, the use of multiplex rtPCR has been extremely beneficial as a screening mechanism (Shone *et al.*, 2006). Often the specimen is limiting when diagnosing infant botulism cases, and there may be insufficient for the mouse bioassay. By utilizing results from the screening PCR assay, physicians can get results more rapidly and they can undertake an investigation into the source of the infection. Our laboratory has observed an excellent correlation of results, between multiplex rtPCR assay and the gold-standard method, the mouse bioassay.

Grant *et al.* (2006) developed a multiplex rtPCR assay for the detection of *E. coli* toxins in various food sample types. The authors reported a detection limit of 10 colony-forming units (CFUs) per reaction in produce samples; sufficient to detect an infectious dose of enterotoxigenic *E. coli*. In another application, Yoshitomi *et al.* (2006) developed a multiplex assay that identified the presence of Shiga toxin-producing *E. coli* and the extremely pathogenic *E. coli* strain, *E. coli* O157:H7.

While multiplex rtPCR has been successfully utilized for detection of *E. coli*, Wang *et al.* (2004a) reported a multiplex assay to detect two enteric pathogens, *Salmonella* serovars and *Listeria monocytogenes* from raw meat samples. Several groups have developed multiplex assays to detect *Vibrio vulnificans* from shellfish homogenates (Harwood *et al.*, 2004; Panicker *et al.*, 2004). These assays were reported to be highly sensitive, detecting one CFU per gram of oyster tissue homogenate (Panicker *et al.*, 2004). For high-throughput testing of food such as milk samples, multiplex rtPCR can be invaluable. An assay developed to detect *Staphylococcus aureus*, *Streptococcus uberis*, and *S. agalactiae* minimized expenditures of reagents and laboratory time by testing for three agents in milk samples simultaneously (Gillespie and Oliver, 2005).

Other enteric bacterial pathogen assays have been developed. Best (2003) reported a duplex assay to detect *Campylobacter jejuni* and *Ca. coli*. While this assay was aimed at detection of *Campylobacter* spp. laboratory isolates, another group utilized rtPCR to detect this organism in food samples (Josefsen *et al.*, 2004). Josefsen *et al.* (2004) used a multiplex rtPCR assay to enhance detection of this pathogen, by incorporating an enrichment protocol to detect *Campylobacter* species from direct rinse samples of chicken. They found that the use of screening enrichment broths by rtPCR was beneficial due to the presence of other competing bacteria; enrichment enhanced the detection of this organism.

Commercial assays have recently been made available for the detection of enteric pathogens. ProGastro from Prodesse is a triplex rtPCR assay that is used to detect *Salmonella*, *Shigella*, and *Campylobacter*. To promote utilization of this assay for a range of applications, in clinical laboratories, public health laboratories, and large commercial laboratories, the company has made the assay available for a wide array of commonly used rtPCR instruments, including the Cepheid SmartCycler, ABI 7500, BioRad iCycler, and Corbett Research RotorGene.

Genital tract infections

Diagnosis of bacterial infections of the genitourinary tract has relied heavily on molecular methodologies due to the difficulty of culturing many of the organisms responsible for clinical illness. For example, *M. genitalium* and *C. trachomatis* are highly fastidious organisms. Often, swab specimens are submitted for testing, so the sample volume is limited. This can prohibit repeat testing or testing for other infectious agents, if equivocal results have been obtained from original testing. Multiplex rtPCR assays have been used to detect several of these pathogens simultaneously. Whiley and Sloots (2005) developed a highly sensitive multiplex assay for the detection *C. trachomatis* and *Neisseria gonorrhoeae* from urogenital specimens. Two probe technologies were compared; 5′ nuclease and hybridization probes, but no appreciable difference was noted between them. A two target assay was been developed for the detection of *M. genitalium* from urine and vaginal swab specimens (Hardick *et al.*, 2006). This two target assay added a confirmatory assay to the single rt-PCR, thus reducing the need to run a second confirmatory PCR to verify original results.

A recent publication by Halse *et al.* (2006) described the development of a duplex real-time assay to detect a rare disease, *Lymphogranuloma venereum* (LGV) caused by the bacterium *Chlamydia trachomatis*. This pathogen was detected using a duplex assay; the first target identified *C. trachomatis*, and the second target was specific for serovar L-2. This pathogen has recently emerged as an important public health concern, and multiplex rtPCR has been a critical tool for the rapid diagnosis of several outbreaks in many HIV-positive males. Use of the assay permitted rapid diagnosis and early institution of antibiotic treatment (Fenton and Imrie, 2005).

Real-time PCR can be used to quantitatively to determine the concentration of organisms, and thus a relative measure of the degree of infectivity can be obtained. A high or low bacterial load can provide useful information for clinicians and assist in treatment given to the patient. Blaylock *et al.* (2004) utilized multiplex rtPCR to amplify regions of the gyraseA gene of *Mycoplasma genitalium* and the RNase P gene found in human cells. Their assay exploits rtPCR's ability to detect genes in both bacteria and humans. This assay is not only extremely valuable for providing a physician with rapid results but it also assists epidemiologists in their efforts to trace contacts and prevent sexually transmitted disease.

Encephalitis

Bacterial pathogens that induce encephalitis are of great concern; diagnosis must be obtained as quickly as possible, due to the rapidity with which illness can progress, especially in the elderly and young. Three pathogens cause 90% of all bacterial meningitis infections (Tzanakaki *et al.*, 2005). One assay developed to detect these agents was a triplex rtPCR designed to test as many serotypes of *Neisseria meningitis*, *Haemophilus influenzae*, and *Streptococcus pneumoniae* as possible (Corless *et al.*, 2001). The main seven *N. meningiditis* serotypes, 23 serotypes of *S. pneuomoniae*, and two serotypes of *H. influenzae* that cause bacterial encephalitis could be detected by this assay in three specimen matrices, CSF, plasma, and blood (Corless *et al.*, 2001). In a retrospective study, examining a large number of culture-negative samples, specimens were identified in which these three pathogens could be detected by the real-time assay but not by using the older conventional PCR assay (Blaylock *et al.*, 2004). The investigators determined that the rtPCR assay was more sensitive than the older PCR assay that they were using at that time. Smith *et al.* (2004) also published a triplex rtPCR assay for these three pathogens and validated the assay utilizing an automated extraction system, resulting in more rapid turnaround time and increased throughput.

Respiratory pathogens

A large number of assays have been designed for identification of bacterial respiratory pathogens. There is a growing body of literature on the numerous respiratory pathogens and on the validation of these assays with various specimen types. For bacterial agents *Bordetella pertussis* and *B. parapertussis*, duplex assays were among the first multiplex rtPCR assays developed and they had great utility for testing large numbers of nasopharyngeal swabs (Sloan *et al.*, 2002). These organisms can be difficult to culture but in recent years they have been seen with increasing frequency by laboratories. They take days to grow, and they have become a remerging public health issue in recent years (Storsaeter and Wolter, 2006). The development of a duplex rtPCR assay for these agents helped to decrease the turn-around time for reporting of tests, assisted in diagnosis and in earlier institution of antibiotic treatment (an important consideration for the population that these respiratory agents often infect: the young, the immunocompromised and the elderly in which rapid treatment is necessary), aided in the identification and tracking of outbreaks. The majority of the assays reported in the literature exhibit excellent sensitivity, detecting an average of less than one organism per reaction (Sloan *et al.*, 2002).

As the previous example highlighted, detection of multiple organisms within the same genus can be accomplished in one reaction and this approach can also be of great value for the many members of the genus *Legionella*. Instead of having to run a large number of assays or wait for sequencing results (which can take days, depending on access to sequencing equipment), the analyst can use multiplex rtPCR to produce results in a fraction of the time needed for sequencing. Templeton *et al.* (2003) developed a multiplex rtPCR assay that can differentiate *L. pneumophila*, a leading cause of respiratory illness, from other *Legionella* spp. The same group reported a multiplex rtPCR assay that combines detection of *Legionella* spp. with detection of viral agents of pneumonia (Templeton *et al.*, 2005).

One group has developed an rtPCR assay that can detect three respiratory pathogens, *Chlamydia pneumoniae*, *Legionella pneumophila*, and *Mycoplasma pneumoniae* (Welti *et al.*, 2003). This multiplex assay was performed using two tubes, one for detection of *M. pneumoniae* and *C. pneumoniae*, and the other for amplification of *L. pneumophila* DNA. This was an important development for multiple bacterial agent detection, in that the rtPCR assay was quantitative. By generating standard curves of positive plasmid control DNA with each specimen, the assay produced a quantitative result for every specimen tested. It is arguable whether a quantitative assay is particularly useful or cost-efficient, for causative agents of respiratory illnesses, since collection efficiencies and transport time can often invalidate the result, especially for quantitative respiratory pathogen rtPCR assays. Most of the real-time assays designed for detection of bacterial pathogens are qualitative in nature. Nevertheless the assay constitutes proof of concept for bacterial diagnostics, and it should be applicable to other specimen matrices such as whole blood, urine, or stool.

Many of the above mentioned-assays are designed to detect two targets per reaction. Very few assays have been developed that can detect more than four pathogens. Recently, a multiplex assay was designed to detect five respiratory agents. This multiplex assay, referred to as the "pneumoplex," is now commercially available (Khanna *et al.*, 2005). While this multiplex assay is not real-time, the same group has also developed an rtPCR assay that detects four pathogens, *Mycoplasma pneumoniae*, *Chlamydia pneumoniae*, *Legionella pneumophilia* and *Legionella micdadei*. This assay incorporates an internal control as well and is thus a pentaplex real-time assay. The problem encountered most frequently in development of assays with more than three targets has been the lack of availability of multicolor detection for most real-time instruments. Many instruments have been claimed to have four or more channels that can independently capture emissions from dyes, but our laboratory has found significant bleed-through issues associated with the development of assays that are based on such capabilities. Several newer instruments have five or more channels to capture data and these should perform adequately for pentaplex assays.

Tick-borne bacterial infections

Because a bite from a tick can result in either a single infection or a co-infection by multiple bacterial pathogens, multiplex rtPCR can be of great diagnostic value. Several assays have been developed to detect multiple tick-borne infections. Two of the most common tick-borne infectious agents are due to *Anaplasma phagocytophilum* and *Borrelia burgdorferi*. Courtney *et al.* (2004) have developed a multiplex assay that could detect less than one infected *A. phaocytophilum* cell, or 50 *B. burgdoferi* bacteria. The assays were developed to detect these bacteria from ticks, but blood samples were also examined as a potential specimen matrix.

Sirigireddy and Ganta (2005) added to the diagnostic tools available for tick-borne infections, by developing an assay that can detect five separate tick-borne pathogens, belonging to multiple *Anaplasma* and *Ehrlichia* species, in blood. This multiplex assay is of note because it uses reverse transcriptase PCR to detect RNA from singly infected or co-infected ticks. By identifying tick-borne single infection or co-infection, the assay enables rapid treatment of the patient, which can be crucial. This study was carried out using a dog model but with additional validation the multiplex real-time RT PCR should be applicable to ticks that bite humans and other vertebrate hosts.

Ullmann *et al.* (2005) developed a multiplex rtPCR assay for species determination of *Borrelia*. This assay was developed to differentiate various *Borrelia* species in ticks. One assay targets the rrs-rrlA IGS region of the genome of *Borrelia* spp., to detect *Borrelia burgdorferi* sensu lato and *Borrelia miyamotoi*. The second assay targets the ospA gene of *B. burgdorferi* sensu lato to detect *B. burgdorferi sensu stricto, Borrelia bissettii*, and *Borrelia andersonii*. The final assay targets the glpQ gene of *B. miyamotoi* sensu lato to differentiate *B. miyamotoi* LB-2001 and *Borrelia lonestari*. This type of identification assay is very useful for epidemiological purposes, since it can enable tracking of the spread of infection by *Borrelia* spp. into different vectors, throughout a region, or world-wide.

Doyle *et al.* (2005) provided a similar scheme for the identification of *Ehrlichia* species that cause human clinical infection. *Ehrlichia chaffeensis, E. ewingii*, and *E. canis* were detected through amplification of the genus-specific disulfide bond formation protein gene dsb. *Ehrlichia* genus-specific dsb primers amplify DNA from all known *Ehrlichia* species but not DNA from other Rickettsial organisms such as *Anaplasma platys, Anaplasma phagocytophilum, Rickettsia conorii*, or *Rickettsia typhi*. This triplex rtPCR assay has the ability to detect as few as 50 gene copies.

Drug-resistant bacteria

Not only is multiplex PCR used to detect certain pathogens in a sample, but it has also been useful for determining mutations in certain bacterial genes that are involved in the development of drug resistance. Rapid detection of drug resistance in bacterial infections is critical for effective patient treatment. Drug resistance and multiplex detection of genes that are responsible for it have been well defined for methicillin- and vancomycin-resistant *Staphylococcus aureus* (Costa *et al.*, 2005; Eisner *et al.*, 2005; Elsayed *et al.*, 2003; Francois *et al.*, 2003; Francois *et al.*, 2004; Hope *et al.*, 2004; Huletsky *et al.*, 2004; Palladino *et al.*, 2003b; Palladino *et al.*, 2003a; Sinsimer *et al.*, 2005). In 2003, a duplex real-time assay was developed to detect the mecA and nuc genes in bacterial cells (Elsayed *et al.*, 2003; Sinsimer *et al.*, 2005). Another multiplex rtPCR assay was then developed for the detection of methicillin- resistant *S. aureus* from nasal swabs. This assay could detect 25 CFUs of MRSA and utilized five primers that were specific to the staphylococcal cassette chromosome and the orfX gene. The assay employed molecular beacons and could be completed in less than one hour. This was an important development for the detection of drug resistant *S. aureus*, since samples are often contaminated with other *S. aureus* or with coagulase-negative *Staphylococcus* that can carry the mecA gene.

A quantitative multiplex rtPCR assay was developed that targeted *S. aureus* regulatory genes involved in virulence (Elsayed *et al.*, 2003; Sinsimer *et al.*, 2005; Francois *et al.*, 2006). Use of this assay allowed simultaneous detection and typing of *S. aureus* isolates, without the need to wait for DNA sequencing.

Assays for two vancomycin resistance genes were duplexed in a rtPCR reaction (Palladino *et al.*, 2003b; Palladino *et al.*, 2003a). This multiplex assay, designed for use on rectal swabs, combined enrichment with the real-time assay. It was found to give superior results compared to direct detection from rectal swabs. In 2005, Sinismer *et al.* developed a multiplex real-time assay that could identify both methicillin- and vancomycin-resistant *S. aureus*, using molecular beacon probe technology (Palladino *et al.*, 2003b; Palladino *et al.*, 2003a; Sinsimer *et al.*, 2005). This assay is based on the mecA gene, the vanA resistance gene, and the virulence factor Panton-Valentine leucocidin (lukF) in bacterial isolates. If specific virulence information is readily available to the physician, patients can be more effectively treated.

Commercial multiplex kits available for diagnosis of bacterial agents

The importance of multiplex rtPCR for the detection of bacterial pathogens has, not surprisingly, been recognized by commercial vendors. Several multiplex rtPCR kits that have been developed for bacteria are in the testing phase or are currently available (see Web Resources section below). These kits are used for the identification of a range of bacteria, in various specimen types including human blood and stool, as well as environmental and food samples. The numbers of FDA-approved assays or RUO kits for bacteria lag well behind the numbers of assays and kits available for virus detection by multiplex rtPCR. However, we can expect in the next few years to see a great upturn in the number of commercially available rtPCR kits for bacteria.

Future trends

As the costs associated with nucleic-acid amplification continue to decrease, the diagnostic capabilities of clinical laboratories will increase. Emerging technologies and new analytical platforms for high-throughput DNA sequencing are becoming mature and should soon be available for routine use in diagnostic laboratories. With concomitant advances in the fields of microfluidics, nucleotide and fluorescent dye chemistries, and information processing, highly multiplexed nucleic acid detection and identification technologies will gradually come to be applied in a vast range of situations in which sensitivity, specificity, and speed are priorities.

Discussion

Clearly, many researchers and diagnosticians are working to bring multiplex rtPCR assays into mainstream analytical testing. While the acceptance and integration of PCR-based methods into clinical diagnostic labs have not proceeded rapidly, the accumulated experience in molecular methodologies is serving to bring rtPCR and multiplex analyses into laboratories at this level. The closed-tube nature of rtPCR, combined with the reduced expenditures of time and money required for multiplex analyses, makes this a very attractive platform, for a market in which there is very little margin for cost increase, and low tolerance for false results. The integrated analytical controls that are possible with multiplex rtPCR should make the methodology very attractive to the regulatory bodies that monitor diagnostic labs so as to ensure quality control.

While it is clear that multiplex rtPCR is a very powerful approach to microbiological diagnostics, there are some specific limitations that must be considered when deciding on the best approach for microbial detection. One basic limitation is the fact that all PCR-based diagnostic methods are directed, i.e. they can only detect what they are designed to detect, as opposed, to, for example, bacterial culture methods, which will grow bacteria non-specifically; i.e. whether or not one is looking for *Staphylococcus* or *Bacillus* in a culture, both will grow. This consideration is important if the pathogen that you wish to detect has a high degree of genetic diversity or genetic drift; in that case, it may rapidly become non-detectable by a previously validated PCR-based assay. In addition, it is essentially impossible to develop a robust assay for a newly emerging pathogen because any nucleic acid sequence information available will be limited and because the range of genetic diversity in the new pathogen will be unknown.

Another limitation for rtPCR is assay development costs. Beyond the initial expense of developing a single-target assay, there are costs associated with evaluation and verification of multiplex assays against all of the target sequences that could potentially be present in a sample, either individually and combined, as well as confirmation of the compatibility of all the primer and probe sets. Validation guidelines are available (see Website Resources section below) which describe measures to minimize the costs of validation while creating statistically robust assays to satisfy regulatory agencies.

As surveyed in this chapter, many FRET chemistries, detection platforms, and mastermixes are available for rtPCR and multiplex assays. Unfortunately, not all chemistries work on each platform, nor do all mastermixes perform equivalently. Thus, platform, reagent, and chemistry compatibilities are still an issue. In addition, the results of any nucleic acid-based analysis will be only as high in quality as the nucleic acids put into the reaction vessel, so sample extraction and processing are critical for multiplex rtPCR, as they are for all microbial rtPCRs. Many assay developers do not address clinical sample processing; too often, the assays are developed and validated using synthetic, pristine nucleic acids. Diagnosticians must be certain that the sample processing utilized for controls is compatible with their particular matrix type; they must also determine that the assays being employed yield results comparable to those reported by assay vendors/developers if a commercial kit exists for the same targets. As indicated previously, inclusion of process control samples in the testing process is essential for this purpose.

Web resources

Topic	**Site**
WHO: Infectious Diseases	http://www.who.org
Validation of Analytical Procedures: Methodology	http://www.fda.gov/cder/guidance/1320fnl.pdf
Multiplex rtPCR Probes	http://biosearchtech.com/hot/amplify.asp
Multiplex Probe Compatibility	http://www.epibio.com/
Real-time PCR Resources	http://www.gene-quantification.info/
	http://www.horizonpress.com/gateway/rt-pcr.html
Bacterial Multiplex rtPCR kits	www.prodesse.com
	www.cepheid.com
Molecular Beacon Probes	http://www.molecular-beacons.com
Hybridization Probes	http://www.roche-applied-science.com/index.jsp
TaqMan Probes	http://www.appliedbiosystems.com/
LUX Primers	https://www.invitrogen.com/

References

Aitichou, M., Javorschi, S., and Ibrahim, M.S. (2005). Two-color multiplex assay for the identification of orthopox viruses with real-time LUX- PCR. Molecular and Cellular Probes *19*, 323–328.

Arezi, B., Xing, W., Sorge, J.A., and Hogrefe, HH. (2003). Amplification efficiency of thermostable DNA polymerases. Anal. Biochem. *321*, 226–235.

Aryee, E.A.N., Bailey, R.L., Natividad-Sancho, A., Kaye, S., and Holland, M.J. (2005). Detection, quantification and genotyping of Herpes Simplex Virus in cervicovaginal secretions by real-time PCR: a cross sectional survey. Virol. J. *2*, 61.

Belanger, S.D., Boissinot, M., Clairoux, N., Picard, F.J., and Bergeron, M.G. (2003). Rapid detection of Clostridium difficile in feces by real-time PCR. J. Clin. Microbiol. *41*, 730–734.

Belanger, S.D., Boissinot, M., Menard, C., Picard, F.J., and Bergeron, M.G. (2002). Rapid detection of Shiga toxin-producing bacteria in feces by multiplex PCR with molecular beacons on the smart cycler. J. Clin. Microbiol. *40*, 1436–1440.

Bellin, T., Pulz, M., Matussek, A., Hempen, H.G., and Gunzer, F. (2001). Rapid detection of enterohemorrhagic Escherichia coli by real-time PCR with fluorescent hybridization probes. J. Clin. Microbiol. *39*, 370–374.

Best, E.L., Powell, E.J., Swift, C., Grant, K.A., and Frost, J.A. (2003). Applicability of a rapid duplex real-time PCR assay for speciation of Campylobacter jejuni and Campylobacter coli directly from culture plates. FEMS Microbiol. Lett. *229*, 237–241.

Blaylock, M.W., Musatovova, O., Baseman, J.G., and Baseman, J.B. (2004). Determination of infectious load of Mycoplasma genitalium in clinical samples of human vaginal cells. J. Clin. Microbiol. *42*, 746–752.

Boivin, G., Cote, S., Dery, P., De Serres, G., and Bergeron, M.G. (2004). Multiplex Real-Time PCR Assay for Detection of Influenza and Human Respiratory Syncytial Viruses. J. Clin. Microbiol. *42*, 45–51.

Bredius, R.G., Templeton, K.E., Scheltinga, S.A., Class, E.C., Kroes, A.C., and Vossen, J.M. (2004). Prospective study of respiratory viral infections in pediatric hemopoietic stem cell transplantation patients. Pediatric Infectious Disease Journal 23, 518–522.

Candotti, D., Temple, J., Owusu-Ofori, S., and Allain, J-P. (2004). Multiplex real-time quantitative RT-PCR assay for hepatitis B virus, hepatitis C virus, and human immunodeficiency virus type 1. J. Virol. Methods*118*, 39–47.

Carletti, F., Di Caro, A., Calcaterra, S., Grolla, A., Czub, M., Ippolito, G., Capobianchi, M.R., and Horejsh, D. (2005). Rapid, differential diagnosis of orthopox- and herpesviruses based upon real-time PCR product melting temperature and restriction enzyme analysis of amplicons. J. Virol. Methods*129*, 97–100.

Chayama, K., Suzuki, Y., Kobayashi, M., Tsubota, M., Hashimoto, M., Miyano, Y., Koike, H., Kobayashi, M., Koida, I., Arase, Y., Saitoh, S., Murashima, N., Ikeda, K., and Kumada, H. (1998). Emegence and takeover of YMDD motif mutant hepatitis B virus during long-term lamivudine therapy and re-takeover by wild type after cessation of therapy. Hepatology *27*, 1711–1716.

Claas, E.C.J., Schilham, M.W., deBrouwer, C.S., Hubacek, P., Echavarria, M., Lankester, A.C., van Tol, M.J.D., and Kroes, A.C.M. (2005). Internally controlled real-time PCR monitoring of adenovirus DNA load in serum or plasma of transplant recipients. J. Clin. Microbiol. *43*, 1738–1744.

Cloud, J.L., Hymas, W.C., Turlak, A., Croft, A., Reischl, U., Daly, J.A., and Carroll, K.C. (2003). Description of a multiplex Bordetella pertussis and Bordetella parapertussis LightCycler PCR assay with inhibition control. Diagn. Microbiol. Infect. Dis. *46*, 189–195.

Cook, R.F., Cook, S.J., L.i., F., Montelaro, R.C., and Issel, C.J. Development of a multiplex real-time reverse transcriptase-polymerase chain reaction for equine infectious anemia virus (EIAV). J. Virol. Methods*105*, 171–179. 2002. Ref Type: Generic

Corey, L., Huang, M.L., Selke, S., and Wald, A. (2005). Differentiation of herpes simplex virus types 1 and 2 in clinical samples by a real-time Taqman PCR assay. J. Med. Virol. 76, 350–355.

Corless, C.E., Guiver, M., Borrow, R., Edwards-Jones, V., Fox, A.J., Kaczmarski, E.B., and Mutton, K.J. (2002). Development and evaluation of a "real-time" RT-PCR for the detection of Enterovirus and Parechovirus RNA in CSF and throat swab samples. J. Med. Virol. 67, 555–562.

Corless, C.E., Guiver, M., Borrow, R., Edwards-Jones, V., Fox, A.J., and Kaczmarski, E.B. (2001). Simultaneous detection of Neisseria meningitidis, Haemophilus influenzae, and Streptococcus pneumoniae in suspected cases of meningitis and septicemia using real-time PCR. J. Clin. Microbiol. *39*, 1553–1558.

Courtney, J.W.; Kostelnik, L.M.; Zeidner, N.S.; and Massung, R.F. (2004). Multiplex real-time PCR for detection of anaplasma phagocytophilum and Borrelia burgdorferi. J. Clin. Microbiol. *42*, 3164–3167.

Costa, A.M., Kay, I., and Palladino, S. (2005). Rapid detection of mecA and nuc genes in staphylococci by real-time multiplex polymerase chain reaction. Diagn. Microbiol. Infect. Dis. *51*, 13–17.

Doyle, C.K., Labruna, M.B., Breitschwerdt, E.B., Tang, Y.W., Corstvet, R.E., Hegarty, B.C., Bloch, K.C., L.i., P., Walker, D.H., and McBride, J.W. (2005). Detection of medically important Ehrlichia by quantitative multicolor TaqMan real-time polymerase chain reaction of the dsb gene. J. Mol. Diagn. *7*, 504–510.

Drake, J. W and Holland, J.J. (1999). Mutation rates among RNA viruses. Proceedings of the National Academy of Sciences USA *96*, 13910–13913.

Ebner, K., Suda, M., Watzinger, F., and Lion, T. (2005). Molecular Detection and Quantitative Analysis of the Entire Spectrum of Human Adenoviruses by a Two-Reaction Real-Time PCR Assay. J. Clin. Microbiol. *43*, 3049–3053.

Eisner, A., Gorkiewicz, G., Feierl, G., Leitner, E., Kofer, J., Kessler, H.H., and Marth, E. (2005). Identification of glycopeptide-resistant enterococci by VITEK 2 system and conventional and real-time polymerase chain reaction. Diagn. Microbiol. Infect. Dis. *53*, 17–21.

Elena, S. F and Sanjuan, R. (2005). Adaptive Value of High Mutation Rates of RNA Viruses: Separating Causes from Consequences. J. Virol. *79*, 11555–11558.

Elsayed, S., Chow, B.L., Hamilton, N.L., Gregson, D.B., Pitout, J.D., and Church, D.L. (2003). Development and validation of a molecular beacon probe-based real-time polymerase chain reaction assay for rapid detection of methicillin resistance in Staphylococcus aureus. Arch. Pathol. Lab Med. *127*, 845–849.

Eriksson, L.E., Leitner, T., Wahren, B., Bostrom, A-C, and Falk, K.I. (2003). A multiplex real-time PCR for quantification of HIV-1 DNA and human albumin gene CD4+ cells. Acta Pathologica, Microbiologica et Immunologica Scandinavica *111*, 625–633.

Espy, M.J., Uhl, J.R., Mitchell, P.S., Thorvilson, J.N., Svien, K.A., Wold, A.D., and Smith, T.F. (2000). Diagnosis of Herpes Simplex Virus Infections in the Clinical Laboratory by LightCycler PCR. J. Clin. Microbiol. *38*, 795–799.

Estes, M and Sevall, J.S. (2003). Multiplex PCR using real time DNA amplification for the rapid detection and quantitation of HTLV I or II. Molecular and Cellular Probes *17*, 59–68.

Fenton, K.A. and Imrie, J. (2005). Increasing rates of sexually transmitted diseases in homosexual men in Western Europe and the United States: why? Infect. Dis. Clin. North Am. *19*, 311–331.

Filen, F., Strand, A., Allard, A., Blomberg, J., and Herrmann, B. (2004). Duplex real-time polymerase chain reaction assay for detection and quantification of herpes simplex virus type 1 and herpes simplex virus type 2 in genital and cutaneous lesions. Sexually Transmitted Diseases *31*, 331–336.

Francois, P., Koessler, T., Huyghe, A., Harbarth, S., Bento, M., Lew, D., Etienne, J., Pittet, D., and Schrenzel, J. (2006). Rapid Staphylococcus aureus agr type determination by a novel multiplex real-time quantitative PCR assay. J. Clin. Microbiol. *44*, 1892–1895.

Francois, P., Pittet, D., Bento, M., Pepey, B., Vaudaux, P., Lew, D., and Schrenzel, J. (2003). Rapid detection of methicillin-resistant Staphylococcus aureus directly from sterile or nonsterile clinical samples by a new molecular assay. J. Clin. Microbiol. *41*, 254–260.

Francois, P., Renzi, G., Pittet, D., Bento, M., Lew, D., Harbarth, S., Vaudaux, P., and Schrenzel, J. (2004). A novel multiplex real-time PCR assay for rapid typing of major staphylococcal cassette chromosome mec elements. J. Clin. Microbiol. *42*, 3309–3312.

Fried, M.W., Shiffman, N.L., Reddy, K.R., Smith, C., Marinos, G., Goncales, J.R., F.L., Haussinger, D., Diago, M., Carosi, G., Dhumeaux, D., Craxi, A., Lin, A., Hoffman, J., and, Y.u., J. (2002). Combination of peginterferon alfa-2a plus ribavirin in patients with chronic hepatitis C virus infection. N. Engl. J. Med. *347*, 975–982.

Gardner, M.J., Hall, N., Fung, E., White, O., Berriman, M., Hyman, R.W., Carlton, J.M., Pain, A., Nelson, K.E., Bowman, S., Paulsen, I.T., James, K., Eisen, J.A., Rutherford, K., Salzberg, S.L., Craig, A., Kyes, S., Chan, M.S., Nene, V., Shallom, S.J., Suh, B., Peterson, J., Angiuoli, S., Pertea, M., Allen, J., Selengut, J., Haft, D., Mather, M.W., Vaidya, A.B., Martin, D.M., Fairlamb, A.H., Fraunholz, M.J., Roos, D.S., Ralph, S.A., McFadden, G.I., Cummings, L.M., Subramanian, G.M., Mungall, C., Venter, J.C., Carucci, D.J., Hoffman, S.L., Newbold, C., Davis, R.W., Fraser, C.M., and Barrell, B. (2002). Genome sequence of the human malaria parasite Plasmodium falciparum. Nature *419*, 498–511.

Garson, J.A., Grant, P.R., Ayliffe, U., Ferns, R.B., and Tedder, R.S. (2005). Real-time PCR quantitation of hepatitis B virus DNA using automated sample preparation and murine cytomegalovirus internal control. J. Virol. Methods *126*, 207–213.

Geng, H., Hua, B., Wang, H., Cao, Y., Sun, Y., and, Y.u., A. (2006). Dual-probe assay for detection of lamivudine-resistance hepatitis B virus by real-time PCR. J. Virol. Methods *132*, 25–31.

Gibb, T.R., Norwood, J.R., D.A., Woollen, N., and Henchal, E.A. (2001). Development and Evaluation of a Fluorogenic 5′ Nuclease Assay To Detect and Differentiate between Ebola Virus Subtypes Zaire and Sudan. J. Clin. Microbiol. 39, 4125–4130.

Giglio, S., Monis, P.T., and Saint, C.P. (2003). Demonstration of preferential binding of SYBR green I to specific DNA fragments in real-time multiplex PCR. Nucl. Acids Res. *31*, e136.

Gillespie, B.E. and Oliver, S.P. (2005). Simultaneous detection of mastitis pathogens, Staphylococcus aureus, Streptococcus uberis, and Streptococcus agalactiae by multiplex real-time polymerase chain reaction. J. Dairy Sci. *88*, 3510–3518.

Glare, E.M., Divjak, M., Bailey, M.J., and Walters, E.H. (2002). Beta-Actin and GAPDH housekeeping gene expression in asthmatic airways is variable and not suitable for normalising mRNA levels. Thorax *57*, 765–770.

Grant, M.A., H.u., J., and Jinneman, K.C. (2006). Multiplex real-time PCR detection of heat-labile and heat-stable toxin genes in enterotoxigenic Escherichia coli. J. Food Prot. *69*, 412–416.

Greenberg, S.B. (2002). Respiratory viral infections in adults. Curr. Opin. Pulmonary Med. *8*, 201–208.

Griffiths, P.D. Cytomegalovirus. Zuckerman, A.J., Banatvala, J.E., and Pattison, J.R. (1994). Principles and Practice of Clinical Virology, 69. Chichester, Wiley.

Griffiths, P.D. (1995). Viral complications after transplantation. Journal of Antimicrobial Chemotherapy *36* (Suppl. B), 91–106.

Gruber, F., Falkner, F.G., Dorner, F., and Hammerle, T. (2001). Quantitation of viral DNA by real-time PCR applying duplex amplification, internal standardization, and two-color fluorescence detection. Appl. Environ. Microbiol. *67*, 2837–2839.

Gu, Z., Belzer, W., Gibson, C.S., Bankowski, M.J., and Hayden, R.T. (2003). Multiplexed, real-time PCR for quantitative detection of human adenovirus. J. Clin. Microbiol. *41*, 4636–4641.

Gunson, R.N., Collins, T.C., and Carman, W.F. (2005). Real-time RT-PCR detection of 12 respiratory viral infections in four triplex reactions. J. Clin. Virol. *33*, 341–344.

Haas, I., Muhlbauer, G., Bozic, M., Stelzl, E., Koidl, C., Berger, A., Berg, J., Vince, A., Rabenau, H., and Kessler, H.H. (2004). Evaluation of a new assay for detection of herpes simplex virus type 1 and type 2 DNA by real-time PCR. J. Lab. Med. *28*, 361–367.

Hadziyannis, S.J., Sette, J.r., H., Morgan, T.R., Balan, V., Diago, M., Marcellin, P., Ramadori, G., Bodenheimer, J.r., H., Bernstein, D., Rizzetto, M., Zeuzem, S., Pockros, P.J., Lin, A., and Ackrill, A.M. (2004). Peginterferon-alpha2a and ribavirin combination therapy in chronic hepatitis C: a randomised study of treatment duration and ribavirin dose. Ann. Intern. Med. *140*, 346–355.

Hall, C.B. (2001). Respiratory syncytial virus and parainfluenza virus. N. Engl. J. Med. *344*, 1917–1928.

Halse, T.A., Musser, K.A., and Limberger, R.J. (2006). A multiplexed real-time PCR assay for rapid detection of Chlamydia trachomatis and identification of serovar L-2, the major cause of Lymphogranuloma venereum in New York. Mol. Cell Probes *20*, 290–297.

Hardick, J., Giles, J., Hardick, A., Hsieh, Y.H., Quinn, T., and Gaydos, C. (2006). Performance of the gen-probe transmission-mediated amplification research assay compared to that of a multitarget real-time PCR for Mycoplasma genitalium detection. J. Clin. Microbiol. *44*, 1236–1240.

Harwood, V.J., Gandhi, J.P., and Wright, A.C. (2004). Methods for isolation and confirmation of Vibrio vulnificus from oysters and environmental sources: a review. J. Microbiol. Methods *59*, 301–316.

Helps, C., Reeves, N., Egan, K., Howard, P., and Harbour, D. (2003). Detection of *Chlamydophila felis* and feline Herpesvirus by Multiplex Real-Time PCR Analysis. J. Clin. Microbiol. *41*, 2734–2736.

Hillemann, D., Warren, R., Kubica, T., Rusch-Gerdes, S., and Niemann, S. (2006). Rapid detection of Mycobacterium tuberculosis Beijing genotype strains by real-time PCR. J. Clin. Microbiol. *44*, 302–306.

Hindiyeh, M., Levy, V., Azar, R., Varsano, N., Regev, L., Shalev, Y., Grossman, Z., and Mendelson, E. (2005). Evaluation of a Multiplex Real-Time Reverse Transcriptase PCR Assay for Detection and Differentiation of Influenza Viruses A and B during the 2001–2002 Influenza Season in Israel. J. Clin. Microbiol. *43*, 589–595.

Hoehne, M and Schreier, E. (2006). Detection of Norovirus genogroup I and II by multiplex real-time RT-PCR using a 3′-minor groove binder-DNA probe. BMC Infectious Diseases *6*, 69.

Hoffmaster, A.R., Meyer, R.F., Bowen, M.D., Marston, C.K., Weyant, R.S., Thurman, K., Messenger, S.L., Minor, E.E., Winchell, J.M., Rassmussen, M.V., Newton, B.R., Parker, J.T., Morrill, W.E., McKinney, N., Barnett, G.A., Sejvar, J.J., Jernigan, J.A., Perkins, B.A., and Popovic, T. (2002). Evaluation and validation of a real-time polymerase chain reaction assay for rapid identification of Bacillus anthracis. Emerg. Infect. Dis. *8*, 1178–1182.

Holland, J and Domingo, E. (1998). Origin of Evolution of Viruses. Virus Genes *16*, 13–21.

Hope, W.W., Morton, A.P., Looke, D.F., Schooneveldt, J.M., and Nimmo, G.R. (2004). A PCR method for the identification of methicillin-resistant Staphylococcus aureus (MRSA) from screening swabs. Pathology *36*, 265–268.

Huletsky, A., Giroux, R., Rossbach, V., Gagnon, M., Vaillancourt, M., Bernier, M., Gagnon, F., Truchon, K., Bastien, M., Picard, F.J., van, B.A., Ouellette, M., Roy, P.H., and Bergeron, M.G. (2004). New real-time PCR assay for rapid detection of methicillin-resistant Staphylococcus aureus directly from specimens containing a mixture of staphylococci. J. Clin. Microbiol. *42*, 1875–1884.

Ibekwe, A.M. and Grieve, C.M. (2003). Detection and quantification of Escherichia coli O157:H7 in environmental samples by real-time PCR. J. Appl. Microbiol. *94*, 421–431.

Ibekwe, A.M., Watt, P.M., Grieve, C.M., Sharma, V.K., and Lyons, S.R. (2002). Multiplex fluorogenic real-time PCR for detection and quantification of Escherichia coli O157:H7 in dairy wastewater wetlands. Appl. Environ. Microbiol. *68*, 4853–4862.

Issa, N.C., Espy, M.J., Uhl, J.R., and Smith, T.F. (2005). Sequencing and resolution of amplified herpes simplex virus DNA with intermediate melting curves as genotype 1 or 2 by LightCycler PCR assay. J. Clin. Microbiol. *43*, 1843–1845.

Jinneman, K.C., Yoshitomi, K.J., and Weagant, S.D. (2003). Multiplex real-time PCR method to identify Shiga toxin genes stx1 and stx2 and Escherichia coli O157:H7/H- serotype. Appl. Environ. Microbiol. *69*, 6327–6333.

Johnson, B.W., Russell, B.J., and Lanciotti, R.S. (2005). Serotype-specific detection of dengue viruses in a fourplex real-time reverse transcriptase PCR assay. J. Clin. Microbiol. *43*, 4977–4983.

Josefsen, M.H., Jacobsen, N.R., and Hoorfar, J. (2004). Enrichment followed by quantitative PCR both for rapid detection and as a tool for quantitative risk assessment of food-borne thermotolerant campylobacters. Appl. Environ. Microbiol. *70*, 3588–3592.

Jothikumar, N. and Griffiths, M.W. (2002). Rapid detection of Escherichia coli O157:H7 with multiplex real-time PCR assays. Appl. Environ. Microbiol. *68*, 3169–3171.

Jothikumar, N., Wang, X., and Griffiths, M.W. (2003). Real-time multiplex SYBR green I-based PCR assay for simultaneous detection of Salmonella serovars and Listeria monocytogenes. J. Food Prot. *66*, 2141–2145.

Ke, L.D., Chen, Z., and Yung, W.K.A. (2000). A reliability test of standard-based quantitative PCR: exogenous vs endogenous standards. Molecular and Cellular Probes *14*, 127–135.

Khanna, M., Fan, J., Pehler-Harrington, K., Waters, C., Douglass, P., Stallock, J., Kehl, S., and Henrickson, K.J. (2005). The pneumoplex assays, a multiplex PCR-enzyme hybridization assay that allows simultaneous detection of five organisms, Mycoplasma pneumoniae, Chlamydia (Chlamydophila) pneumoniae, Legionella pneumophila, Legionella micdadei, and Bordetella pertussis, and its real-time counterpart. J. Clin. Microbiol. *43*, 565–571.

La Rosa, A.M., Champlin, R.E., Mirza, N., Gajewski, S., Giralt, S., Rolston, K.V., Raad, I., Jacobson, K., Kontoyiannis, D., Elting, L., and Whimbey, E. (2001). Adenovirus infections in adult recipients of blood and marrow transplants. Clinical Infectious Diseases 32, 871–876.

Luo, W., Yang, H., Rathbun, K., Pau, C-P, and, O.u., C-Y. (2005). Detection of Human Immunodeficiency Virus Type 1 DNA in Dried Blood Spots by a Duplex Real-Time PCR Assay. J. Clin. Microbiol. *43*, 1851–1857.

Mackay, I.M., Jacob, K.C., Woolhouse, D., Waller, K., Syrmis, M.W., Whiley, D.M., Siebert, D.J., Nissen, M., and Sloots, T.P. (2003). Molecular assays for detection of human metapneumovirus. J. Clin. Microbiol. *41*, 100–105.

Madhaven, H.N., Priya, K., Anand, A.R., and Therese, K.L. (1999). Detection of Herpes simplex virus (HSV) genome using polymerase chain reaction (PCR) in clinical samples. Comparison of PCR with standard laboratory methods for the detection of HSV. J. Clin. Virol. *14*, 145–151.

Mangold, K.A., Manson, R.U., Koay, E.S., Stephens, L., Regner, M., Thomson, R.B., Jr., Peterson, L.R., and Kaul, K.L. (2005). Real-time PCR for detection and identification of Plasmodium spp. J. Clin. Microbiol. *43*, 2435–2440.

McNamara, D.T., Kasehagen, L.J., Grimberg, B.T., Cole-Tobian, J., Collins, W.E., and Zimmerman, P.A. (2006). Diagnosing infection levels of four human malaria parasite species by a polymerase chain reaction/ligase detection reaction fluorescent microsphere-based assay. Am. J. Trop. Med. Hyg. *74*, 413–421.

McNamara, D.T., Thomson, J.M., Kasehagen, L.J., and Zimmerman, P.A. (2004). Development of a multiplex PCR-ligase detection reaction assay for diagnosis of infection by the four parasite species causing malaria in humans. J. Clin. Microbiol. *42*, 2403–2410.

Melo, A.C., Almeida, A.M., and Leal, N.C. (2003). Retrospective study of a plague outbreak by multiplex-PCR. Lett. Appl. Microbiol. *37*, 361–364.

Meng, Q., Wong, C., Rangachari, A., Tamatsukuri, S., Sasaki, M., Fiss, E., Cheng, L., Ramankutty, T., Clarke, D., Yawata, H., Sakakura, Y., Hirose, T., and Impraim, C. (2001). Automated multiplex assay system for simultaneous detection of hepatitis B virus DNA, hepatitis C virus RNA, and human immunodeficiency virus type 1 RNA. J. Clin. Microbiol. *39*, 2937–2945.

Mercier, B., Burlot, L., and Ferec, C. (1999). Simultaneous screening for HBV DNA and HCV RNA genomes in blood donations using a novel TaqMan PCR assay. J. Virol. Methods 1–9.

Michaelides, A., Glare, E.M., Spelman, D.W., Wesselingh, S.L., Hoy, F.J., Mijch, M.A., and Kotsimbos, T.C. (2002). Beta-Herpesvirus (Human Cytomegalovirus and Human Herpesvirus 6) Reactivation in At-Risk Lung Transplant Recipients and in Human Immunodeficiency Virus-infected Patients. J. Infect. Dis. *186*, 173–180.

Mitchell, P.S., Espy, M.J., Smith, T.F., Toal, D.R., Rys, P.N., Berbari, E.F., Osmon, D.R., and Persing, D.H. (1997). Laboratory diagnosis of central nervous system infections with herpes simplex virus by PCR performed with cerebral spinal fluid specimens. J. Clin. Microbiol. *35*, 2873–2877.

O'Neill, H.J., Wyatt, D.E., Coyle, P.V., McCaughey, C., and Mitchell, F. (2003). Real-Time Nested Multiplex PCR for the Detection of Herpes Simplex Virus Types 1 and 2 and *Varicella zoster* Virus. J. Med. Virol. *71*, 557–560.

Orru, G., De Santis, P., Solinas, F., Savini, G., Piras, V., and Caporale, V. (2004). Differentiation of Italian field and South African vaccine strains of bluetongue virus serotype 2 using real-time PCR. J. Virol. Methods *122*, 37–43.

Palladino, S., Kay, I.D., Costa, A.M., Lambert, E.J., and Flexman, J.P. (2003a). Real-time PCR for the rapid detection of vanA and vanB genes. Diagn. Microbiol. Infect. Dis. *45*, 81–84.

Palladino, S., Kay, I.D., Flexman, J.P., Boehm, I., Costa, A.M., Lambert, E.J., and Christiansen, K.J. (2003b). Rapid detection of vanA and vanB genes directly from clinical specimens and enrichment broths by real-time multiplex PCR assay. J. Clin. Microbiol. *41*, 2483–2486.

Pang, X.L., Preiksaitis, J.K., and Lee, B. (2005). Multiplex real time RT-PCR for the detection and quantitation of norovirus genogroups I and II in patients with acute gastroenteritis. J. Clin. Virol. *33*, 168–171.

Panicker, G., Call, D.R., Krug, M.J., and Bej, A.K. (2004). Detection of pathogenic Vibrio spp. in shellfish by using multiplex PCR and DNA microarrays. Appl. Environ. Microbiol. *70*, 7436–7444.

Pantaleo, G., Graziosi, C., and Fauci, A.S. (1993). The Immunopathogenesis of Human Immunodeficiency Virus Infection. N. Engl. J. Med. *328*, 327–335.

Payungporn, S., Chutinimitkul, S., Chaisingh, A., Damrongwantanapokin, S., Buranathai, C., Amonsin, A., Theamboonlers, A., and Poovorawan, Y. (2006). Single step multiplex real-time RT-PCR for H5N1 influenza A virus detection. J. Virol. Methods *131*, 143–147.

Payungporn, S., Tangkijvanich, P., Jantaradsamee, P., Theamboonlers, A., and Poovorawan, Y. (2004). Simultaneous quantitation and genotyping of hepatitis B virus by real-time PCR and melting curve analysis. J. Virol. Methods *120*, 131–140.

Perandin, F., Manca, N., Calderaro, A., Piccolo, G., Galati, L., Ricci, L., Medici, M.C., Arcangeletti, M.C., Snounou, G., Dettori, G., and Chezzi, C. (2004). Development of a real-time PCR assay for detection of Plasmodium falciparum, Plasmodium vivax, and Plasmodium ovale for routine clinical diagnosis. J. Clin. Microbiol. *42*, 1214–1219.

Permar, S.R., Griffin, D.E., and Letvin, N.L. (2006). Immune Containment and Consequences of Measles Virus Infection in Healthy and Immunocompromised Individuals. Clin. Vaccine Immunol. *13*, 437–443.

Pfeffer, M., Proebster, B., Kinney, R.M., and Kaaden, O-R. (1997). Genus-specific detection of alphaviruses by semi-nested reverse transcription-polymerase chain reaction. Am. J. Trop. Med. Hygiene 57, 709–718.

Pradeau, K., Couty, L., Szelag, J-C, Turlure, P., Rolle, F., Ferrat, P., Bordessoule, D., Meur, Y.L., Denis, F., and Ranger-Rogez, S. (2006). Multiplex real-time PCR assay for simultaneous quantitation of human cytomegalovirus and herpesvirus-6 in polymorphonuclear and mononuclear cells of transplant recipients. J. Virol. Methods *132*, 77–84.

Pugnale, P., Latorre, P., Rossi, C., Crovatto, K., Pazienza, V., De Gottardi, A., and Negro, F. (2005). Real-time multiplex PCR assay to quantify hepatitis C virus RNA in peripheral blood mononuclear cells. J. Virol. Methods *133*, 195–204.

Ramaswamy, M., McDonald, C., Smith, M., Thomas, D., Maxwell, S., Tenant-Flowers, M., and Geretti, A.M. (2004). Diagnosis of genital herpes by real time PCR in routine clinical practice. Sexually Transmitted Infections *80*, 406–410.

Rasmussen, T.B., Uttenthal, A., de Stricker, K., Belak, S., and Storgaard, T. (2003). Development of a novel quantitative real-time RT-PCR assay for the simultaneous detection of all serotypes of *Foot-and-mouth disease virus*. Arch. Virol. *148*, 2005–2021.

Rasmussen, T.B., Uttenthal, A., Fernandez, J., and Storgaard, T. (2005). Quantitative Multiplex Assay for Simultaneous Detection and Identification of Indiana and New Jersey Serotypes of Vesicular Stomatitis Virus. J. Clin. Microbiol. *43*, 356–362.

Ririe, K.M., Rasmussen, R.P., and Witter, C.T. (1997). Product differentiation by analysis of DNA melting curves during the polymerase chain reaction. Anal. Biochem. *245*, 154–160.

Rolfe, K.J., Alexander, G.J.M., Wreghitt, T.G., Parmar, S., Jalal, H., and Curran, M.D. (2005). A real-time Taqman method for hepatitis C virus genotyping. J. Clin. Virol. *34*, 115–121.

Rotbart, H. A and Webster, A.D. (2001). Treatment of potentially life threatening enteroviral infections with pleconaril. Clinical Infectious Diseases 32, 228–235.

Rougemont, M., Van, S.M., Sahli, R., Hinrikson, H.P., Bille, J., and Jaton, K. (2004). Detection of four Plasmodium species in blood from humans by 18S rRNA gene subunit-based and species-specific real-time PCR assays. J. Clin. Microbiol. *42*, 5636–5643.

Ryncarz, A.J., Goddard, J., Wald, A., Huang, M-I, Roizman, B., and Corey, L. (1999). Development of High-Throughput Quantitative Assay for Detecting Herpes Simplex Virus DNA in Clinical Samples. J. Clin. Microbiol. *37*, 1941–1947.

Safronetz, D., Humar, A., and Tipples, G.A. (2003). Differentiation and quantitation of human herpesviruses 6A, 6B and 7 by real-time PCR. J. Virol. Methods *112*, 99–105.

Sanchez, J. L and Storch, G.A. (2002). Multiplex, Quantitative, Real-Time PCR Assay for Cytomegalovirus and Human DNA. J. Clin. Microbiol. *40*, 2381–2386.

Scaramozzino, N., Crance, J-M, Jouan, A., DeBriel, D.A., Stoll, F., and Garin, D. (2001). Comparison of *Flavivirus* Universal Primer Pairs and Development of a Rapid, Highly Sensitive Heminested Reverse Transcription-PCR Assay for Detection of Flavivirus Targeted to a Conserved Region of the NS5 Gene Sequences. J. Clin. Microbiol. 39, 1922–1927.

Schalasta, G., Arents, A., Schmid, M., Braun, R.W., and Enders, G. (2000). Fast and Type-Specific Analysis of Herpes Simplex Virus Types 1 and 2 by Rapid PCR and Fluorescence Melting-Curve-Analysis. Infection 28, 85–91.

Scheltinga, S.A., Templeton, K.E., Beersma, M.F.C., and Claas, E.C.J. (2005). Diagnosis of human metapneumovirus and rhinovirus in patients with respiratory tract infections by an internally controlled multiplex real-time RNA PCR. J. Clin. Virol. 33, 306–311.

Sharma, V.K. (2006). Real-time reverse transcription-multiplex PCR for simultaneous and specific detection of rfbE and eae genes of Escherichia coli O157:H7. Mol. Cell Probes *20*, 298–306.

Sharma, V.K., an-Nystrom, E.A., and Casey, T.A. (1999). Semi-automated fluorogenic PCR assays (TaqMan) forrapid detection of Escherichia coli O157:H7 and other shiga toxigenic, *E. coli*. Mol. Cell Probes *13*, 291–302.

Shieh, B., Chang, M-J, K.o., W-C, Chen, E-J, W.u., J-C, Lee, C-F, Chang, T-T, and, L.i., C. (2003). Effects of Multiple Virus Coinfections on Disease Progression in HIV-Positive Patients. Intervirology 46, 105–113.

Shone, C., Ferreira, J., Boyer, A., Cirino, nm., Egan, C., Evans, E., Kools, J., and Sharma, S. (2006). The 5th International Conference on Basic and Therapeutic Aspects of Botulinum and Tetanus Neurotoxins. Workshop review: assays and detection. Neurotox. Res. 9, 205–216.

Sinsimer, D., Leekha, S., Park, S., Marras, S.A., Koreen, L., Willey, B., Naidich, S., Musser, K.A., and Kreiswirth, B.N. (2005). Use of a multiplex molecular beacon platform for rapid detection of methicillin and vancomycin resistance in Staphylococcus aureus. J. Clin. Microbiol. *43*, 4585–4591.

Sirigireddy, K.R. and Ganta, R.R. (2005). Multiplex detection of Ehrlichia and Anaplasma species pathogens in peripheral blood by real-time reverse transcriptase-polymerase chain reaction. J. Mol. Diagn. *7*, 308–316.

Sloan, L.M., Hopkins, M.K., Mitchell, P.S., Vetter, E.A., Rosenblatt, J.E., Harmsen, W.S., Cockerill, F.R., and Patel, R. (2002). Multiplex LightCycler PCR assay for detection and differentiation of Bordetella pertussis and Bordetella parapertussis in nasopharyngeal specimens. J. Clin. Microbiol. *40*, 96–100.

Smith, K., Diggle, M.A., and Clarke, S.C. (2004). Automation of a fluorescence-based multiplex PCR for the laboratory confirmation of common bacterial pathogens. J. Med. Microbiol. 53, 115–117.

Storsaeter, J. and Wolter, J. (2006). Is there a need for a new generation of vaccines against pertussis? Expert. Opin. Emerg. Drugs *11*, 195–205.

Stranska, R., Schuurman, R., de Vos, M., and van Loon, A.M. (2004). Routine use of a highly automated and internally controlled real-time PCR assay for the diagnosis of herpes simplex and varicella-zoster virus infections. J. Virol. Methods30, 39–44.

Templeton, K.E., Scheltinga, S.A., Beersma, M.F.C., Kroes, A.C.M., and Claas, E.C.J. (2004). Rapid and Sensitive Method Using Multiplex Real-Time PCR for Diagnosis of Infections by Influenza A and Influenza B Viruses, Respiratory Syncytial Virus, and Parainfluenza Viruses 1, 2, 3, and 4. J. Clin. Microbiol. 42, 1564–1569.

Templeton, K.E., Scheltinga, S.A., Sillekens, P., Crielaard, J.W., van Dam, A.P., Goossens, H., and Claas, E.C. (2003). Development and clinical evaluation of an internally controlled, single-tube multiplex real-time PCR assay for detection of Legionella pneumophila and other Legionella species. J. Clin. Microbiol. *41*, 4016–4021.

Templeton, K.E., Scheltinga, S.A., van den Eeden, W.C., Graffelman, A.W., van den Broek, P.J., and Claas, E.C. (2005). Improved diagnosis of the etiology of community-acquired pneumonia with real-time polymerase chain reaction. Clin. Infect. Dis. *41*, 345–351.

Tzanakaki, G., Tsopanomichalou, M., Kesanopoulos, K., Matzourani, R., Sioumala, M., Tabaki, A., and Kremastinou, J. (2005). Simultaneous single-tube PCR assay for the detection of Neisseria meningitidis, Haemophilus influenzae type b and Streptococcus pneumoniae. Clin. Microbiol. Infect. *11*, 386–390.

Ullmann, A.J., Gabitzsch, E.S., Schulze, T.L., Zeidner, N.S., and Piesman, J. (2005). Three multiplex assays for detection of Borrelia burgdorferi sensu lato and Borrelia miyamotoi sensu lato in field-collected Ixodes nymphs in North Ameerica. J. Med. Entomol. *42*, 1057–1062.

van Elden, L.J., van Loon, A.M., van Alphen, F., Hendriksen, K.A., Hoepelman, A.I., van Kraaij, M.G., Oosterheert, J-J, Schipper, P., Schuurman, R., and Nijhuis, M. (2004). Frequent detection of human coronaviruses in clinical specimens from patients with respiratory tract infection by use of a novel real-time reverse-transcriptase polymerase chain reaction. J. Infect. Dis. *189*, 652–657.

van Elden, L.J., van Loon, A.M., van der Beek, A., Hendriksen, K.A., Hoepelman, A.I., van Kraaij, M.G., Schipper, P., and Nijhuis, M. (2003). Applicability of a real-time quantitative PCR assay for diagnosis of respiratory syncytial virus infection in immunocompromised adults. J. Clin. Microbiol. *41*, 4378–4381.

van Elden, L.J.R., Nijhuis, M., Schipper, P., Schuurman, R., and van Loon, A.M. (2001). Simultaneous Detection of Influenza Viruses A and B Using Real-Time Quantitative PCR. J. Clin. Microbiol. 39, 196–200.

Varga, A and James, D. (2005). Detection and differentiation of Plum pox virus using real-time multiplex PCR with SYBR Green and melting curve analysis: a rapid method of strain typing. J. Virol. Methods *123*, 213–220.

Verweij, J.J., Blange, R.A., Templeton, K., Schinkel, J., Brienen, E.A., van Rooyen, M.A., van, L.L., and Polderman, A.M. (2004). Simultaneous detection of Entamoeba histolytica, Giardia lamblia, and Cryptosporidium parvum in fecal samples by using multiplex real-time PCR. J. Clin. Microbiol. *42*, 1220–1223.

Vet, J.A.M., Majithia, A.R., Marras, S.A.E., Tyagi, S., Dube, S., Poiesz, B.J., and Kramer, F.R. (1999). Multiplex detection of four pathogenic retroviruses using molecular beacons. Proc. Natl. Acad. Sci. USA *96*, 6394–6399.

Vinje, J., Green, J., Lewis, D.C., Gallimore, C.I., Brown, D.W.G., and Koopmans, M.P.G. (2000). Genetic polymorphism across regions of the three open reading frames of "Norwalk-like viruses". Arch. Virol. *145*, 223–241.

Wakeley, P.R., Johnson, N., McElhinney, L.M., Marston, D., Sawyer, J., and Fooks, A.R. (2005). Development of a real-time, TaqMan reverse transcription-PCR assay for detection and differentiation of Lyssavirus genotypes 1, 5, and 6. J. Clin. Microbiol. *43*, 2786–2792.

Wang, X., Jothikumar, N., and Griffiths, M.W. (2004a). Enrichment and DNA extraction protocols for the simultaneous detection of Salmonella and Listeria monocytogenes in raw sausage meat with multiplex real-time PCR. J. Food Prot. *67*, 189–192.

Wang, Z., Vora, G.J., and Stenger, D.A. (2004b). Detection and genotyping of Entamoeba histolytica, Entamoeba dispar, Giardia lamblia, and Cryptosporidium parvum by oligonucleotide microarray. J. Clin. Microbiol. *42*, 3262–3271.

Weinberg, A., Bloch, K.C., Tang, Y.W., Palmer, M., and Tyler, K.L. (2004). Dual infections of the central nervous system with Epstein-Barr virus. J. Infect. Dis. *191*, 234–237.

Welti, M., Jaton, K., Altwegg, M., Sahli, R., Wenger, A., and Bille, J. (2003). Development of a multiplex real-time quantitative PCR assay to detect Chlamydia pneumoniae, Legionella pneumophila and Mycoplasma pneumoniae in respiratory tract secretions. Diagn. Microbiol. Infect. Dis. *45*, 85–95.

Whiley, D.M., Mackay, I.M., Syrmis, M.W., Witt, M.J., and Sloots, T.P. (2004). Detection and differentiation of herpes simplex virus types 1 and 2 by a duplex LightCycler PCR that incorporates an internal control PCR reaction. J. Clin. Virol. *30*, 32–38.

Whiley, D.M. and Sloots, T.P. (2005). Comparison of three in-house multiplex PCR assays for the detection of Neisseria gonorrhoeae and Chlamydia trachomatis using real-time and conventional detection methodologies. Pathology *37*, 364–370.

Wright, P. F and Webster, R.G. (2001). Orthomyxoviruses. Knipe, D. M and Howley, P.M. Fields Virology. [Fourth Edition], 1533–1579. Philadelphia, PA.

Wu, Y-Y and Rees, J.L. (2000). Variation in Epidermal Housekeeping Gene Expression in Different Pathological States. Acta Dermato Venereologica *80*, 2–3.

Xu, W., McDonough, M.C., and Erdman, D.D. (2000). Species-specific identification of human adenoviruses by a multiplex PCR assay. J. Clin. Microbiol. *38*, 4114–4120.

Yoshitomi, K.J., Jinneman, K.C., and Weagant, S.D. (2006). Detection of Shiga toxin genes stx1, stx2, and the +93 uidA mutation of, *E. coli* O157:H7/H-using SYBR Green I in a real-time multiplex PCR. Mol. Cell Probes *20*, 31–41.

The Role of Real-time PCR in Routine Microbial Diagnostics

7

Eric C.J. Claas, Willem. J.G. Melchers, and Adriaan J.C. van den Brule

Abstract

Routine microbial diagnostics have changed significantly over the last two decades. Initially, implementation of the polymerase chain reaction (PCR) resulted in sensitive detection of microbes that were hard or impossible to diagnose by conventional diagnostic procedures such as culture and serology. A further improved version is now rapidly replacing end-point detection PCR in many diagnostic laboratories. In real-time PCR, amplified products are detected by fluorescence at the moment that they are generated and directly related to the input target quantity, so that quantitation is possible. For research purposes this has been shown to be a valuable tool in for example quantifying expression levels of genes, but also routine clinical laboratories have implemented numerous applications of this powerful method. This chapter provides an overview of different routine microbial applications in clinical diagnostics and shows the increasing importance of real-time PCR in the different fields of microbiology.

Introduction

Medical microbiological diagnosis finds its origin in the scientific activities of Van Leeuwenhoek in the late middle ages and of Pasteur and Koch in the middle of the nineteenth century. With microscopy and culture, infectious agents could be diagnosed and preventive and curative procedures could subsequently be developed. In the first half of the twentieth century many human pathogens were characterized by means of these methods. Contemporary methods made it possible to specifically isolate and characterize antigens. These could subsequently be used to identify the same pathogens by means of a direct or indirect immunological route without further culturing.

During the past decades a shift in the microbiological diagnostic potential has been observed and DNA and RNA detection are now generally accepted as important and universal diagnostic targets (Peruski, Jr. and Peruski, 2003). Molecular diagnostics have become an irreplaceable tool in the diagnosis of infectious diseases (Yang and Rothman, 2004). Moreover, the ongoing search for new nucleic acid based techniques that can be applied to the diagnosis of infectious diseases has driven technology forward. Besides qualitative applications of real-time PCR in routine microbiology resulting in identification of a pathogen, the methodology also enables quantitation. Quantifying the pathogen has been playing an increasingly important role in clinical microbiology and therefore should be discussed in a chapter on applications of real-time PCR in clinical microbiology as well.

Initially, an important criterion for implementation of DNA or RNA detection was that conventional microbiological diagnostics had failed because: (i) no adequate serologic tests were available (e.g. Human papillomavirus), (ii) the microorganisms were difficult or impossible to culture (e.g. *Aspergillus* spp., *Borrelia burgdorferi* and many virus types) or, (iii) the culture was time-consuming which limited rapid diagnosis (e.g. *Mycobacterium tuberculosis* and *Toxoplasma gondii*). To exemplify this, several clinical diagnostic problems related to the items described above will be addressed briefly.

Poor serological tests

Cancer of the cervix ranks number two worldwide in cancers that affect women, accounting for 6% of all malignancies in women. This cancer is a major public health problem with an estimated 450 000 new cases per year. It causes high mortality in the developing world and, if precancerous conditions are included, a high morbidity worldwide (Parkin *et al.*, 1993; Bekkers *et al.*, 2004). Epidemiological and molecular studies over the past two decades have convincingly demonstrated that certain types of human papillomaviruses (HPVs) are etiologically related to the development of cervical cancer (Walboomers *et al.*, 1999). To date, approximately 100 HPV genotypes have been identified by sequence analysis, at least 30 of which have been found to infect the genital mucosa. Several HPV types, such as HPV 16, 18, 31, 33, and 35 have been implicated in cervical carcinogenesis while other HPV types such as HPV 6 and 11 are frequently detected in benign lesions such as condylomata acuminate (Munoz *et al.*, 2003). Women infected with these HPV types are considered to be at a higher risk of developing cervical cancer than women who are not infected with HPV or are infected with low-risk HPV types (Cuzick *et al.*, 1995). Since HPV still cannot be cultured efficiently, and the clinical performance of serological assays is very poor, reliable diagnosis of HPV infection is based entirely on molecular tools (Melchers *et al.*, 1991). Identification of high-risk HPV types could make it possible to identify women with an increased risk of cervical disease, and could, therefore, provide additional clinical value. In the United States, both the American College of OBGYN (ACOG) and the American Cancer Society (ACS) recommend that women over 30 yrs old should be screened with an HPV DNA test combined with a cytological examination. Many countries are expected to follow this procedure in the near future.

Inability to culture

Invasive aspergillosis is a major cause of mortality and morbidity in patients with hematological malignancies and stem cell transplant recipients (SCT) (Segal and Walsh, 2006). To improve prognosis in these patients early diagnosis and antifungal therapy is essential. Difficulties in establishing an early diagnosis have prompted investigations towards alternative diagnostic methods. During the last decade, molecular-based assays have emerged as an important tool for improved diagnostic workup, monitoring of antifungal therapy and clinical management of patients with suspected or proven invasive aspergillosis (Florent *et al.*, 2006; Jordanides *et al.*, 2005; Lass-Florl *et al.*, 2004).

Lengthy isolation process

Tuberculosis is a communicable disease with significant morbidity and mortality (Frieden *et al.*, 2003). In the last decade, tuberculosis has re-emerged as one of the leading causes of death. In most developing countries its severity has increased due to the global Human

Immunodeficiency Virus (HIV) pandemic and extensive social restructuring. Moreover, multidrug-resistant strains of *Mycobacterium tuberculosis* seriously threaten tuberculosis control and prevention efforts (Ginsburg *et al.*, 2003). The diagnosis of tuberculosis is largely based on conventional approaches, which rely on clinical features and the results of microscopy and culture. Clinical features are not generally diagnostic and microscopy, while rapid, lacks sensitivity and specificity. Culture methods are sensitive and specific, but they are slow. Several alternative approaches have been attempted for the rapid and specific diagnosis of tuberculosis, but the molecular methods are the most promising (Cheng *et al.*, 2005).

The solution: molecular diagnostics?

Nowadays molecular methods have created new possibilities for diagnosis and treatment in the full area of medical microbiology (Yang and Rothman, 2004). The importance of molecular diagnosis of infectious diseases is obvious. A fast diagnosis can be made in a clinically relevant time-frame resulting in optimized therapy for the individual patient.

The application of molecular techniques such as Southern blot, dot-spot, and *in situ* hybridization for the diagnosis of microorganisms, were impeded by the relatively restricted sensitivity and specificity of these research laboratory techniques. The development of the PCR resolved these restrictions, and, consequently, the clinical use of molecular diagnostics became routine (Saiki *et al.*, 1986). A breakthrough in the implementation of these new technologies in routine medical microbiological laboratories was the development and availability of commercial tests. Following PCR, a number of other amplification techniques have been developed, such as Nucleic Acid Sequence Based Amplification (NASBA), ligase chain reaction (LCR), transcription-mediated amplification (TMA), self-sustaining-sequence replication (SSR), strand displacement amplification (SDA), and Q-beta amplification, which are all based on the principle of amplification of nucleic acids (Deiman *et al.*, 2002; Lisby, 1999; Abramson and Myers, 1993). DNA amplification methods can determine the absence or presence of very small quantities of nucleic acids for certain infectious agents. Exclusive sensitivity and specificity of the tests can be guaranteed subject to a correct laboratory organization. Disadvantageous aspects of the extreme sensitivity are matters such as contamination and the clinical relevance of positive results (Kwok and Higuchi, 1989). Conventional PCR often makes use of agarose gel electrophoresis or post-PCR processes such as capture ELISA or blot hybridization for the detection of amplified products. These probe hybridization approaches also enhanced the sensitivity of detection of amplified products (amplicons). Another approach to lower the detection limit was the use of nested PCR, in which the region amplified by a first PCR reaction was subjected to a second round of PCR using a separate primer set. However, a major argument against nested PCR is the risk of contamination due to the spread of amplicon. Post-PCR processing makes conventional PCR time-consuming and labor-intensive (Aslanzadeh, 2004; Borst *et al.*, 2004).

In real-time PCR, amplification and detection of the target DNA using fluorescent probes are performed in the same, closed, reaction vessel and can be applied to rapid and dynamic assessment of clinical samples for the presence of a specific infectious agent (Bustin *et al.*, 2005). The possibility of contamination by previously amplified PCR products is greatly reduced (Zhang and Fang, 2006). With real-time PCR, DNA copy numbers are determined based on the threshold cycle (C_T values), which is proportional to the copy number of the infecting agent present in the sample. In contrast to conventional PCR, real-time PCR does not rely on the final amount of amplicon. This indication of the amount of pathogen present can be used to determine the clinical relevance of the infection. A breakthrough due to real-time PCR analysis and the development of improved platforms is the potential to perform multiplex PCR tests, allowing detection of multiple targets simultaneously in a single test tube. Another development for improving the reproducibility of PCR assays is the use of general "master-mixes," i.e. premade reaction mixtures containing nucleotides, Mg^{2+} buffer, DNA polymerase and all components for an optimized amplification. As these can be produced in large batches, variation is reduced.

In many clinical molecular diagnostic settings a shift from conventional PCR to real-time PCR is now apparent and it is likely that real-time PCR will change our perspective on how to identify and define the quantity of infectious agents, as well as on the determination of virulence factors and resistance mutations. Whether these assays will be widely applied in routine microbiology remains to be seen.

An extensive and comprehensive review describing the different technology platforms and applications of real-time PCR in clinical microbiology was published recently (Espy *et al.*, 2006). In the remainder of this chapter we will, therefore, focus on the role of real-time PCR in routine microbial diagnostic laboratories, and on important aspects of its implementation.

General aspects of real-time PCR in routine diagnostics

Choice of real-time PCR

For the development of real-time PCR assays both the target choice and assay design are of importance. The selection of targets will not be addressed in this review and is discussed extensively in the numerous papers describing PCR assays for the detection and typing of microorganisms (Gunson *et al.*, 2006; Espy *et al.*; 2006). Assay design in real-time PCR is dependent on the amplification technique (e.g. PCR, NASBA, TMA), probe technology (e.g. hybridization probes, TaqMan® probes, TaqMan-MGB probes or molecular beacons) and the PCR platform used (Chui *et al.*, 2005; Cockerill, III, 2003). The most frequently used PCR devices for home-brewed diagnostic real-time PCR assays are the ABI 7000 and 7500 (Applied Biosystems), LightCycler® (Roche), iCycler iQ™ (BioRad), Rotor-Gene™ (Corbett Research) and the SmartCycler® (Cepheid). Commercially available diagnostic assays most commonly make use of specially developed real-time PCR equipment from the same diagnostic company; e.g. HIV diagnostic assays are provided by different diagnostic companies that have their own real-time PCR platform such as the Cobas TaqMan® (Roche), NucliSense® (bioMerieux) and the mt2000 (Abbott). Due to Food and Drug Administration (FDA) regulations of *in vitro* diagnostic (IVD) assays and for commercial reasons, these IVD real-time PCR devices can only handle IVD approved assays from the same company. This greatly influences the flexibility of molecular diagnostic laboratories and will force laboratories to use home-brewed assays. Fortunately, diagnostic companies have recognized this problem and promised that new generations real-time PCR devices will contain an open channel, which will enable other assays to be used. For example, FDA approved LightCycler assays are now available for mycobacterial diagnosis.

With respect to home-brewed real-time PCR assays, the choice of platform is determined by the throughput of molecular diagnostic assays and the assay speed needed. An example of a low throughput device is the conventional LightCycler, which has a maximum capacity of 32 samples per run that can be analyzed within 30 minutes. This is especially suited for immediate testing in case of medical urgency (so-called STAT testing derived from the Latin *statim* meaning immediately). Widely used high throughput devices, the ABI 7700 or 7000, handle a maximum of 96 samples per run in approximately 2 hours. To date, three real-time PCR strategies are typically seen in molecular diagnostic laboratories: (1) different assays performed in a single run on a high throughput device with all the assays standardized using an identical amplification program; (2) high throughput diagnostics performed on a high throughput device, but STAT testing carried out on a low throughput device; and (3) all diagnostics performed on low throughput devices, enabling several runs a day and the grouping of assays.

However, new generations of real-time PCR platforms are becoming increasingly alike, i.e. high throughput and reduced turn-around time. Recently, the LightCycler 480 (Roche) and the ABI 7500 (Applied Biosystems) were introduced: both have a capacity of 96 samples which can be analyzed in 30 min. Higher throughputs of 384 samples are already possible, but, due to the reduced reaction volumes used, it remains to be seen whether these formats can be employed in microbiology. A reduced input volume of extracted nucleic acid is likely to generate false negative results due to the lower number of copies added. In addition, manual pipetting of 384 samples is highly error-prone and thus requires automation of pipetting. In parallel with these high throughput developments, newly integrated devices and technologies are being developed for rapid STAT real-time PCR diagnostics, handling just one or a few samples (e.g. GeneXpert®, Cepheid).

The real-time PCR platform selected will ultimately determine which probe format and fluorophores can be used for labeled oligonucleotides. The differences between the various probe technologies available are described elsewhere (Arya *et al.*, 2005; Cockerill, III, 2003). Nucleotide sequence variability/conservation will determine which probe chemistry will be most suitable. All the different probe technologies can be used for conserved sequences but variable nucleotide sequences such as those between pathogen strains or groups are better suited to either TaqMan, TaqMan-MGB or molecular beacon probes above hybridization probes, whereas TaqMan-MGB probes with reduced probe length may offer advantages over TaqMan probes (Whiley and Sloots, 2006). When additional melting temperature analyses are required to confirm the specificity of the PCR product generated, hybridization probes are needed because a single TaqMan chemistry cannot be used. For multiplex real-time PCR applications, new generation devices are able to read five to six channels, enabling the detection of up to five targets and an internal control using different fluorophores. Beside the fact that now there are many fluorescent labels for efficient multiplexing, the design of the different primer and probe sets should be standardized with special emphasis on identical melting temperatures, GC content and amplicon length. In addition to this optimized *in silico* design, the sensitivity and specificity of the multiplex real-time PCR should be extensively investigated using mixtures of dilution lines of the different pathogens to rule out mutual influence in amplification efficiency. Reduction of the sensitivity of an assay in the presence of amplification of another target should be prevented. In practice, it should be kept in mind that amplification of high levels of a target is likely to mask the presence of low levels of another target. Whether this second target detected with a high C_T value is relevant for the disease of the patients remains the question.

At present, the number of fluorescent labels that can be detected is limited, so multiplexing can currently be efficiently done using up to four different targets (Templeton *et al.*, 2004). It would be extremely valuable for routine clinical microbiology if new devices and technologies were developed that could handle more labels, enabling a higher level of multiplexing. Recently, the Luminex Corporation (Austin, TX, USA) introduced a technology enabling the parallel detection of 100 targets by using 5 μm polystyrene beads which each have a different level of fluorescence and different probes (for review: see Dunbar, 2006). However, sensitivities obtained directly in clinical specimens are not sufficient for clinical microbiology, but by using it as a detection system for multiplex PCR-amplified targets (Schmitt *et al.*, 2006) this technology seems to be promising. Applications using 30 to 40 simultaneous reactions are possible, but it remains to be seen whether the sensitivity can be sufficiently improved for microbiology applications.

Performance of real-time PCR

A newly developed real-time PCR assay has to be analyzed for its analytical sensitivity and specificity, so that the detection limit and other possible limitations, if any, are known. Before implementation in routine diagnostics, the assay should be further validated by determining its reproducibility and testing clinical specimens (Niesters, 2004; Mackay, 2004; Apfalter *et al.*, 2005). Practical guidelines for reliable performance of molecular microbiological methods for diagnosis of infectious diseases have recently been proposed by the Dutch Working Group for Molecular Diagnostics of Infectious Diseases (G. Noordhoek, M. Ieven, R. Roosendaal, personal communication). With respect to validation, it was proposed that the assay be performed on a relevant number of known samples, preferably from external quality assessment programs. The number of samples to be tested will differ for different targets. It is mandatory when using PCR to generate a diagnostic result that affects patient management to include sufficient controls to monitor extraction, amplification, and inhibition. This is of particular importance to obtain reliable negative results. For commercial assays the procedures described by the manufacturer should be followed strictly and a run validation, consisting of the determination of the analytical sensitivity and a comparison of results with data from the literature and/or description by the manufacturer, should be performed.

In diagnostic real-time PCR assays, the use of dUTP and uracil-N-glycosylase (UNG; to break down dUTP-containing amplification products) is no longer required, as no post PCR processing will be performed. To ensure reliable testing, each real-time PCR experiment should include a low positive run control, one or more negative controls (no-template controls) and an extraction/inhibition control (internal control) (Stranska *et al.*, 2004; Niesters, 2004). Laboratories should participate in regularly organized external quality control programs using proficiency panels such as the Quality Control for Molecular Diagnostics (QCMD, Glasgow, U.K.), to ensure the reliability of the laboratory procedures and result interpretation. Although internal and external quality control of real-time PCR is of great importance, the quality of specimen collection, transport, and storage of the sample are not taken into account and are still key factors in the whole process (Apfalter *et al.*, 2005).

Clinical value of real-time PCR

Qualitative real-time PCR has become standard practice in many routine microbiology laboratories. Its clinical value is generally accepted by the clinicians in cases of strong positive samples. Nevertheless it is still unclear what to do with weak positive samples (high C_T values of 38). There is much debate on whether positivity thresholds or grey areas for retesting should be defined. Interpretation of these values will be dependent on the kind of pathogen tested and the clinical symptoms of the patient. An enterovirus-positive CSF with a C_T of over 35 will be definitively accepted as positive and clinically relevant for a final diagnosis, whereas a *Streptococcus pneumoniae*-positive sample with a C_T of over 35 will likely be interpreted as clinically unimportant and regarded as a commensal presence. However, it remains very difficult and highly arbitrary to interpret the clinical value of real-time PCR positives. Clearly, these data should be considered in relation to the clinical symptoms of the patient. All the data currently generated in different patient cohorts will result in a better interpretation of C_T values in relation to the disease. It will be interesting and clinically important to investigate whether a broad range of C_T values has been obtained for a particular pathogen in a given sample type or whether specific infections are always associated with certain C_T values. In addition, new views on the epidemiology of a number of infectious agents will emerge. Implementation of respiratory virus PCR has shown that in some cases of severe pneumonia, the only pathogen detected was rhinovirus or coronavirus, but with a high viral load (= low C_T value). Although this does not prove there is an etiologic relationship, especially since these viruses can also frequently be detected in asymptomatic individuals (Gageldonk-Lafeber *et al.*, 2005), it does prompt further study. Similarly, the effect of multiple infections, which may have an important role in the severity of infection, should be studied in more detail (Templeton *et al.*, 2005).

Future clinical validation studies will greatly determine the diagnostic value of qualitative real-time PCR. In addition, cost benefit studies (Rand *et al.*, 2005) will ultimately determine the implementation of such assays in routine clinical microbiology.

Qualitative applications of real-time PCR

For the primary diagnosis of pathogens PCR assays have been shown to be superior in speed and sensitivity to conventional diagnostic methods, such as culture and immunoassays. Compared to conventional PCR, real-time PCR is faster and less prone to false positivity due to the combination of amplification and probe detection and due to the closed nature of the reaction (Gunson *et al.*, 2006; Mackay, 2004) In general; conventional PCR tests already used in routine clinical microbiology can easily be converted into real-time PCR formats. However, because of the simultaneous annealing of primers and probes in real-time PCR, optimization of real-time PCR is more complex and primers/probes need to be designed according to specific rules as described in Chapter 2. A major advantage of designing an assay for real-time PCR in contrast to conventional end-point PCR is that, by analyzing dilution series of the target, the efficiency of the PCR assay can be established (Figure 7.1).

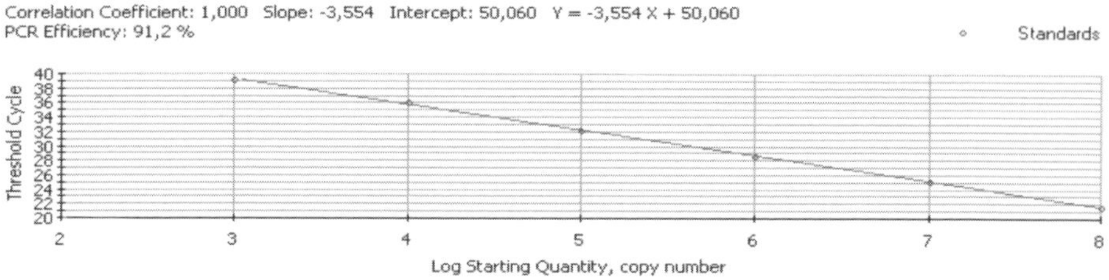

Figure 7.1 CT values obtained with a10-fold dilution series of a standard are plotted against the log value of the copy number. In a 100% efficient PCR, the slope of the standard line is –3.32, indicating that the next ten-fold dilution will become positive with a CT value of + 3.32 cycles over the 10 times more concentrated dilution. Efficiency of real-time PCR can be calculated by the equation: Y =X(1+E)n, in which Y = product generated, X = start copy number, E = efficiency of amplification, and n = number of cycles. If the efficiency of the PCR is 100%, E = 1 and the equations reads Y = X(2)n, indicating that in every PCR cycle the amount of DNA is doubled.

Optimization can be performed by aiming at an efficiency close to 100%, which means that in every PCR cycle the DNA is doubled. In practice, an efficiency of exactly 100% is difficult to achieve, especially using clinical samples. We aim at an efficiency of 95–105%. Especially when the ultimate goal is multiplex PCR, starting with efficient assays is crucial (Table 7.1). Combining efficient assays for which it has been determined that the primers and probes do not interfere with each other is the key to successful multiplex PCR (Scheltinga *et al.*, 2005; Templeton *et al.*, 2004; Templeton *et al.*, 2003). Because of the better optimization possibilities, real-time PCR assays are generally more sensitive and robust than conventional PCR.

Conventional PCR is still employed for some applications in routine clinical microbiology. Identification of bacteria in either culture isolates or directly in clinical specimens, for example, is still based on conventional 16S ribosomal DNA-PCR followed by nucleotide sequence analysis (Deutch *et al.*, 2006).

Identification of mycobacterial species and viruses such as hepatitis C virus (HCV) and HPV is efficiently performed using broad spectrum PCRs followed by reverse line-blot hybridization (van den Brule *et al.*, 2002). In this area of identification and genotyping,

Table 7.1 Amplification efficiencies and the design of real-time PCR assays: potential reasons for deviation

Efficiency	Slope	Standard line	Potential cause
100% (E = 1)	–3.32	OK	Design well done!
< 100%	> 3.32	Steeper	Primer/probe mismatches Non-optimal primer/probe concentration PCR product too long, i.e. > 200bp
> 100%	< 3.32	Flatter	outliers in CT values of standards inefficient amplification of highly diluted samples inhibitory effect on highly concentrated standard samples

DNA arrays may find an application in clinical microbiology. For the primary diagnosis of infections, DNA arrays will most likely be of limited use because the targets to be analyzed have to be labeled first. If this labeling is performed by PCR, using the array is "one step back" to end-point detection of amplicons in a fancy DNA array platform. In order to label all the potential targets in a clinical sample, large numbers of primers will be required, reducing the sensitivity of the assay. However, new developments for general labeling may eventually be successful. Nonetheless, for the sensitive diagnosis of microbial infections (multiplex) real-time PCR has already been shown to be a powerful method, resulting in a decreased turn-around-time for a diagnosis.

General applications for routine clinical microbiology

A limitation of (real-time) PCR is that the only pathogens detected are those for which specific primers and probes have been added to the reaction, whereas in culture unexpected pathogens may grow. Nevertheless, real-time PCR applications are abundantly implemented in routine clinical microbiology, i.e. clinical virology, clinical bacteriology and clinical mycology and parasitology. In Table 7.2 an overview of representative examples of pathogens is shown for which qualitative real-time PCR assays have been used to replace diagnosis by culture, direct immunoassay or microscopy for detection. In addition, real-

Table 7.2 Overview of representative qualitative real-time PCR applications for routine clinical microbiology

	Bacteria	Virus*	Fungi/parasites
Culture non/poorly slow growing fast growing	*Chlamydia trachomatis* *Bordetella pertussis* *Legionella pneumophila* *Mycobacterium tuberculosis* *Mycoplasma pneumoniae* *Clostridium difficile*	HPV VZV Metapneumovirus (HMPV) Coronaviruses (hCoV) Enterovirus HRSV CMV	*Pneumocystis jiroveci* Dermatophytes Aspergillus?
Replacement for direct immunoassay or microscopy	Group B streptococcen *Clostridium difficile* *Treponema pallidum*	HRSV Influenza A virus Rotavirus	Malaria *Toxoplasma gondii* *Enteromoeba histolytica* Fungi
Virulence/resistance genes	*MRSA* *M. tuberculosis katG/ RPO β-gene*		
Typing/identification	*Mycobacteria*	Enterovirus** HAdV**	*Dermatophytes*

*HPV: human papillomavirus; VZV: varicella zoster virus; HRSV: human respiratory syncytial virus; CMV: cytomegalovirus; HAdV: human adenovirus; MRSA: methicillin-resistant *Staphylococcus aureus*.

**Often typing is performed, but clinical relevance needs to be determined.

time PCR has been introduced for typing and identification of species and strains and the detection of virulence and resistance genes.

The decision to replace conventional diagnostics by real-time PCR (or other PCR-based techniques) is dependent on the clinical need for fast and sensitive test results and cost benefit analysis. In the case of STAT testing for detection of sepsis-related pathogens, the speed and reliability of laboratory testing will outweigh the costs. When a life-threatening infection is the result, the sooner a reliable diagnosis is made, the sooner efficient treatment can be started. In these cases, diagnostics will mostly be performed using one or a few clinical specimens from a single patient, making this diagnosis quite expensive, due to relatively increased labor requirements and the need for more controls per sample (particularly when highly priced commercial assays are being used). This also holds true for the fast testing of methicillin-resistant *Staphylococcus aureus* (MRSA) infections in newly hospitalized patients arriving from suspected MRSA high-risk countries (Warren *et al.*, 2004). However, for the screening of MRSA outbreaks in hospitals, high throughput testing is required and the costs per sample should be greatly reduced. This could be achieved by the automation of sample processing and real-time PCR, not just for high throughput testing, but also for STAT testing. Completely integrated systems (GeneXpert, Cepheid) are already available for the latter, which enable nucleic acid extraction and real-time PCR in a fully automated manner within 1 hour.

Applications for clinical virology

So far, molecular diagnostics has had the greatest impact on clinical virology, due to the fact that numerous viruses are not, or only poorly culturable, including HPV, rotavirus and the coronaviruses. There is an obvious role for real-time PCR in the specific and sensitive detection of these viruses (Gunson *et al.*, 2006; Niesters, 2002; Niesters, 2004). Another group of viruses successfully detected by real-time PCR are culturable, but sometimes slow growing like cytomegalovirus (CMV), human metapneumovirus (hMPV), and parainfluenzavirus 4 (HPIV4) (Sumino *et al.*, 2005; Templeton *et al.*, 2004). Here, a major advantage of PCR is that negative results are also known within a day, whereas this may take up to 14 days using cell culture. As an alternative, rapid agglutination and antigen detection assays have been widely used to decrease result turn-around time in diagnosis. However, despite the fact that these methods are fast and simple, sensitivity is lacking compared to real-time PCR assays. Rapid antigen-negative samples may become positive in real-time PCR with C_T values of 25–30. Nevertheless, those rapid tests with reliable positive results will still have a limited place in routine clinical microbiology. However, because of the lack of sensitivity it is only a matter of time before fast real-time PCRs will replace these rapid assays in clinical virology.

Viruses which can be propagated in cell culture and which grow relatively fast might also be detected by real-time PCR when a more rapid diagnosis is preferred in combination with the detection of other agents in the differential diagnosis (Schmutzhard *et al.*, 2004; Gunson *et al.*, 2006). PCR detection is indicated for these viruses when clinical specimens are not suitable for culture or volumes are limited, e.g. Herpes simplex virus (HSV) type 1 and 2, Varicella zoster virus (VZV) and human enterovirus detection in cerebrospinal fluid (CSF) (Stranska *et al.*, 2004; Rabenau *et al.*, 2002; Rand *et al.*, 2005; Petitjean *et al.*, 2006).

In many laboratories real-time PCR is performed in adjunct to culture. Though viral culture is more specific for certain viruses, its lack of speed and sensitivity justifies replacement by real-time PCR in routine clinical virology (Schmutzhard *et al.*, 2004; Scheltinga *et al.*, 2005; Templeton *et al.*, 2004; Boivin *et al.*, 2004). The cell culture facilities could focus on those samples where no agent has been detected using a wider range of cell lines as in 24% of the cases with a clinical diagnosis of pneumonia and 10% of severe pneumonia cases in which no etiologic agent can be detected by PCR (Figure 7.2).

In the near future, a package PCR approach to detect all the relevant pathogens for a specific illness might rapidly provide proof of a viral infection so that no antibiotic therapy needs to be given or can be stopped. Package multiplex tests for respiratory infections, gastroenteritis, sexually transmitted diseases, and infectious neurological disorders could be assembled for efficient diagnosis. However, multiplexing 10–15 different pathogens using current real-time PCR formats is not feasible in routine diagnostics but splitting them up into multiplex reactions of three to four targets is (Templeton *et al.*, 2004). Nevertheless, to improve the cost effectiveness, there is a strong need for a new generation of technology which is rapid, sensitive and can handle high-level multiplexing.

Applications for clinical bacteriology

In most clinical microbiology laboratories molecular diagnostics of bacteria was mainly restricted to non-culturable and poorly culturable bacteria, such as *Chlamydia trachomatis, Bordetella pertussis, Bordetella parapertussis, Bartonella henselae* and *Tropheryma whipplei*

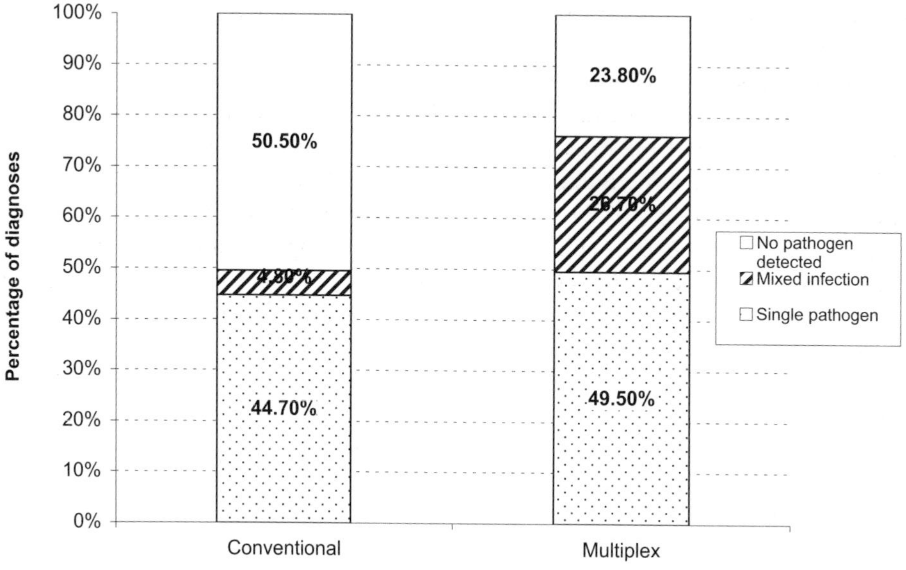

Figure 7.2 Comparison of the laboratory diagnoses in a group of 105 patients with radiograph confirmed pneumonia. With conventional methods such as culture, serology and antigen detection (left bar) a pathogen could be detected in 49.5% of the cases, whereas using multiplex real-time PCR (right bar) this number increased to 76.2%. In 93% of severe cases (grade IV and V pneumonia) a diagnosis could be made. Mixed infections were detected in 26.8% of the cases by real-time PCR and were associated with more severe disease (Templeton *et al.*, 2005).

(Sloan *et al.*, 2005). However, it has become increasingly accepted to use real-time PCR assays for the detection of culturable, but slow-growing bacteria, such as *Mycobacteria tuberculosis complex* (Ortu *et al.*, 2006; Takakura *et al.*, 2005; Espy *et al.*, 2006). Implementation of *M. tuberculosis* PCR has been hampered by the fear of false negative results due to possible PCR inhibitory factors in sputum samples. False negativity in a patient with a clinical diagnosis of tuberculosis can be greatly reduced by testing more samples from a single time point and more frequent sampling in time, resulting in improved clinical sensitivity.

For some applications, real-time PCR detection has replaced conventional diagnostic methods, e.g. the detection of the respiratory pathogens *Chlamydophila pneumoniae*, *Legionella pneumoniae* and *Mycoplasma pneumoniae* (Raggam *et al.*, 2002; Morozumi *et al.*, 2006; Pitcher *et al.*, 2006; Welti *et al.*, 2003). For other targets real-time PCR is still performed in adjunct to culture (*M. tuberculosis* complex), despite the fact that real-time PCR is more sensitive and faster. Culture remains important to determine the antibiotic susceptibility of bacteria and to detect a broader spectrum of pathogens. In the case of selective culturing for particular bacteria such as Mycobacteria or MRSA, real-time PCR can be used for faster screening. Due to the high negative predictive value of PCR, only those samples found positive in PCR will subsequently be monitored by culture for confirmation and subsequent antibiotic susceptibility testing. In other instances, culture plates have to be analyzed for weeks according to laboratory guidelines to ensure that clinical samples are really negative for all the other pathogens not detected by PCR.

Most bacteria are easily cultured and grow overnight, but in some instances more rapid diagnosis by PCR is preferred. Detection of MRSA directly in clinical specimens by real-time PCR in adjunct to culture has proven to be faster and more cost-effective in comparison to culture alone (Warren *et al.*, 2004); PCR negative patients can be released from isolation and surgery rooms can be reopened due to the high negative predictive value (NPV) of MRSA real-time PCR, ensuring that nosocomial spread of bacteria has not occurred (Nilsson *et al.*, 2005; Fang and Hedin, 2003). In the case of positive MRSA-real-time PCR, repeated testing on follow-up samples is indicated as well as confirmation by culture to prove that the MRSA is viable. A drawback of this screening assay is the false positive results due to the detection of coagulase-negative staphylococci which can also carry mecA or methicillin-resistant *S. epidermidis*. Therefore, new assays with better positive predictive value in addition to high NPVs are required. Other avenues being explored include the recently developed MRSA PCR based on the staphylococcal cassette chromosome harboring the mecA gene (Francois *et al.*, 2004; Huletsky *et al.*, 2004) and detection of additional virulence genes in MRSA (McDonald *et al.*, 2005). Unfortunately, using the scc-mec elements as target elements for MRSA detection also generates some false positive results.

As previously mentioned, real-time PCR will have a prominent role in future STAT testing. A clinically valuable application is the rapid real-time PCR detection of group B streptococci (Uhl *et al.*, 2005; Straka *et al.*, 2004) in pregnant women just before delivery, so that only infected women will receive antibiotic treatment and transmission to the infant is avoided. Cost benefit studies will ultimately determine implementation of this diagnostic assay.

Future developments in multiplexed real-time PCR assays enabling the simultaneous detection of more pathogens will most likely have an impact on the use of cultures in routine clinical microbiology. Ideal future multiplexed assays will be able to detect all clinically relevant pathogens for certain diseases such as meningitis (Uzuka *et al.*, 2004), respiratory illness (*C. pneumoniae, L. pneumoniae, M. pneumoniae*, and also including viral pathogens like Influenza virus A and B, hRSV, parainfluenzaviruses type 1–4) (Scheltinga *et al.*, 2005; Hindiyeh *et al.*, 2005; Khanna *et al.*, 2005; Templeton *et al.*, 2004), or sexually transmitted diseases (*C. trachomatis*, gonococci, *Trichomonas vaginalis*, *Treponema pallidum*, HSV and HPV). In the latter package, the targets will be divided into a "discharge" or "ulcerous" package. Multiplex real-time PCR assays that are able to predict antibiotic susceptibility or resistance as an alternative to culture are not to be expected in the near future. Therefore, culture will retain its importance for investigating positive real-time PCR results. Alternatively, automation of routine culturing as already achieved for blood cultures could lead to cost reduction, have additional value with regard to the detection of viable pathogens and to determine the antibiotic-susceptibility pattern of bacteria.

Applications for clinical mycology and parasitology

Fungi are still rarely detected by real-time PCR in routine clinical microbiology with the exception of *Aspergillus fumigatus* infections in immunocompromised patients (Halliday *et al.*, 2005; White *et al.*, 2006; Bretagne and Costa, 2005; Jordanides *et al.*, 2005). Culture of yeast is often problematic as well. Pathogens like *Pneumocystis jiroveci*, the causative agent of PCP (Pneumonia caused by *P. carinii*, its previous name) are diagnosed by real-time PCR in many laboratories and these assays have replaced direct staining and microscopy. A major problem is the high level of asymptomatic *Pneumocystis* carriers resulting in true positive results usually with high C_T values. Therefore, this assay is often used as a screening tool, where positive results are confirmed by staining as well. Again, the high NPV is what justifies the use of PCR.

In general, fungi can be cultured, but they grow slowly. Detection by real-time PCR would greatly reduce the turn-around time in diagnosing these infections. For example, real-time PCR for dermatophytes used directly on clinical specimens of skin, hair or nails, which targeted the conserved 18S ribosomal DNA and interspacer region would enable a fast selection between dermatophyte positive and negative samples and could potentially replace conventional culture and fluorescence microscopy. However, implementation of this real-time PCR is awaiting comparison studies with conventional methods. The genetic heterogeneity of the clinically important dermatophytes complicates a general approach, as many different assays may have to be designed.

Parasites like *Plasmodium falciparum*, *Toxoplasma gondii* and *Entamoeba histolytica* are to date mainly detected by rapid, direct immunoassays, but real-time PCR has also been used. Routine diagnostic applications of real-time PCR have been described in differentiating infection with *Entamoeba histolytica* from the benign *Entamoeba dispar*, which is quite difficult by microscopy (Kebede *et al.*, 2004). Recently, multiplex real-time PCR for *Giardia lamblia*, *Entamoeba histolytica* and *Cryptosporidium parvum* has been applied for simultaneous detection of these parasites (Verweij *et al.*, 2004).

Genotype and mutation analysis

Real-time PCR has been successfully used for the identification of microbial genotypes. For HSV detection, different PCR assays have been described (Corey *et al.*, 2005; Stevenson *et al.*, 2005). In assays targeting the thymidine kinase gene it is difficult to differentiate between HSV type 1 and type 2 due to the high homology between the nucleotide sequences for this gene. Other assays use type-specific primers and probes which can either be used in two single PCRs or a duplex PCR assay (Weidmann *et al.*, 2003; van Doornum *et al.*, 2003; Ryncarz *et al.*, 1999). In addition, real-time PCR assays have been developed which target the DNA polymerase gene using one set of primers and hybridization probes, after which HSV-1 and 2 differentiation can be made by melting curve analysis (Whiley *et al.*, 2004). SYBR® Green I detection can also be used instead of probes, but specificity will be reduced, as there is no probe confirmation. Although melt-curve analysis provides some specificity, it lacks the stringency of a probe. However, when a suitably heterogeneous target for a specific group of pathogens exists, real-time PCR using SYBR Green in combination with melting curve analyses will be able to differentiate genotypes, avoiding the need for different genotype-specific probes. The use of multiple probes is hampered by the limited number of fluorescent labels that can be monitored simultaneously to date. However, real-time PCR applications have been able to differentiate between seven different Candida species (White *et al.*, 2003). Extensive laboratory and clinical evaluation studies are required to prove the specificity of the melting temperatures and their diagnostic value.

Besides genotyping, real-time PCR technology is suitable for the fast and specific detection of mutations in microorganisms. This might be of particular clinical interest in the area of antimicrobial resistance (McDonald *et al.*, 2005; Lawson *et al.*, 2005) and virulence (Fitzmaurice *et al.*, 2004). A specific mutation (S315T) in the katG gene of *M. tuberculosis* has been associated with high level isoniazid resistance and mutations in rpoB gene for rifampin resistance (Kocagoz *et al.*, 2005; van Doorn *et al.*, 2003). A simple real-time PCR assay has been described using melting curve analysis of hybridization probes which can easily differentiate between the wild-type and mutated katG gene. Alternatively, assays employing TaqMan or TaqMan-MGB probes have been successfully developed for the same purpose (Lawson *et al.*, 2005; van Doorn *et al.*, 2003) (Lawson *et al.*, 2005; van Doorn *et al.*, 2003). With the increasing use of drugs for either the treatment of or prophylaxis against infection, the prevalence of resistance, or our understanding of it, may increase in the future; specific PCR assays might contribute to a better prediction of the response to therapy, allowing the timely switching to alternative therapies. For simple low number nucleotide polymorphisms, real-time PCR can be an option. However, usually more mutations are associated with resistance in both viral and bacterial targets (Gilbert *et al.*, 2002; Fluit *et al.*, 2001), which will be hard to analyze using real-time PCR probes. In those cases, DNA arrays of large multiplex systems like Luminex beads (Luminex Corporation) are better suited as multiple mutations can be detected simultaneously.

Quantitative applications of real-time PCR in routine diagnostics

One of the major advantages of the real-time PCR is the quantitative nature of the assay. Conventional PCR could be used quantitatively through a complicated competitive PCR set-up as applied in the commercial Roche COBAS AMPLICOR™ MONITOR assays for HIV-1, HCV and hepatitis B virus (HBV). The specific target and a calibrator were simultaneously amplified using the same primers, but detected using different probes. The ratio of the PCR products was used to quantify the unknown target. This was implemented in routine diagnosis and significantly improved patient management. The analytical range, however, was limited. The HBV MONITOR assay had an analytical range of 2×10^2 to 2×10^5 copies per ml. In HBV infections, viral loads of over 1×10^7 copies are not uncommon and can only be reliably quantified by diluting the samples and retesting. In real-time PCR, the C_T value is related to the original amount of target present in the sample and an analytical range of 10^8 to 10^9 can be routinely achieved. As a result, most companies are implementing real-time PCR protocols in their commercial kits for quantitation of these pathogens. For example, Roche is replacing their successful COBAS AMPLICOR lineage with the COBAS TaqMan technology that can be connected to the automated AmpliPrep extraction platform via a docking station. This provides a fully automated line from sample to result. A similar approach is employed by other companies such as Abbott, with a combination of the m1000 isolation and m2000 amplification units, and Corbett Research with the X-Tractor Gene™/Rotor-Gene™ combination.

Quantitative PCR (qPCR) is implemented in the clinical microbiology laboratory for several applications in which information is provided on the clinical consequences, the risk of contagion, and the status of antimicrobial therapy. Although most applications can still be found in virology, developments in other areas of microbiology are occurring rapidly. There are roughly three fields in which qPCR results can be applied for routine diagnosis and contribute to patient treatment. First, in patients with chronic infections; second, in immunocompromised patients with reactivation of endogenous viruses; and third, in differentiating colonization from infection in commensally present pathogens, provided the microbial load has been linked to disease severity.

Chronic infections

QPCR is widely used in monitoring chronic infections with HIV, HCV and HBV. Especially for HIV, the first target of commercially available quantitative microbial PCR assays, the benefits of monitoring the HIV RNA load in plasma have become an important part of patient management. The well-known kinetics of HIV and the $CD4^+$ cells in an HIV infected patient and the effect of highly active antiretroviral therapy (HAART) are shown in Figure 3. Patients are monitored frequently to ensure that the viral load is still undetectable or to rapidly identify increasing viral loads that may be the result of replication of a resistant virus. With proper medication and the proper attitude of the patient with respect to adhering to the treatment regimen, HIV infection can now be transformed to a chronic disease without progression to AIDS. As a result of this improved patient management the mortality rate, unfortunately in the Western world only, has decreased significantly. In the Netherlands this reduction was from 3.8 to 0.7 per 100 person-years between 1996 and 2000 (van Sighem *et al.*, 2003) (Figure 7.3).

Another virus causing chronic infection is HCV. In the acute phase of illness, the symptoms resemble those of hepatitis A and B infection but the majority of cases will subsequently convert to chronic hepatitis. In the decades that follow, about 20% of the chronic infections progress to cirrhosis and 5% to hepatocellular carcinoma (Heathcote, 2004). Recent developments, including diagnosing the load of hepatitis C virus in blood, have enabled improved therapeutic strategies in patients with chronic HCV infections and liver cirrhosis (Saadeh and Davis, 2004). Determining the initial HCV viral load in combination with the genotype, permits an optimized treatment protocol with PEG-Interferon and ribavirin. The kinetics of the viral load reduction in combination with the HCV genotype

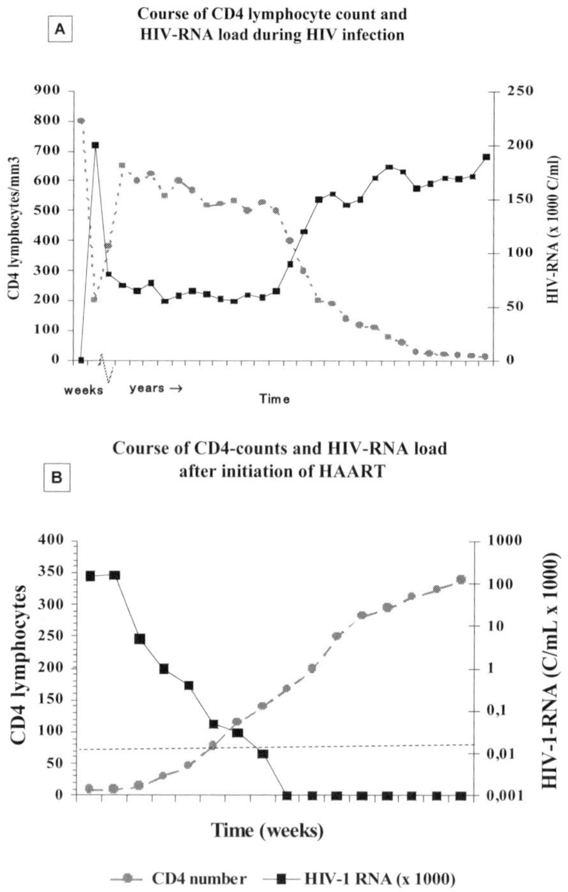

Figure 7.3 Natural course of HIV RNA level in plasma and the number of CD4 cells following infection with HIV (panel A) and effects on these parameters after initiation of HAART (panel B). NB. HIV viral loads value of 0.001x 1000 copies/ml should actually read "less than 50 C/ml," which is the limit of detection of the assay.

provides information on the chance of a sustained response in the patient, predicting a long-term suppression of the virus and thus a better prognosis. Genotype 1 viruses require 48 weeks of therapy with a 50% chance of sustained response, whereas non-genotype 1 (particularly 2 and 3) viruses have a 90% chance of a sustained viral response after only 24 weeks of therapy (Hayashi and Takehara, 2006; Thomas, 2006). The reduction of the viral load over time predicts the chance of a sustained response or a relapse of the disease. If the reduction is too slow, the costly treatment can be terminated early as success is unlikely (Lee *et al.*, 2002; Heathcote *et al.*, 2000; Wong *et al.*, 2003; Manns *et al.*, 2001).

Quantitative viral load measurements are also used in the treatment of hepatitis B virus infections (Gish and Locarnini, 2006; Suzuki *et al.*, 2006). Persistent infections can result in chronic persistent hepatitis and eventually liver cirrhosis and hepatocellular carcinoma. Treatment with lamivudine and adefovir has been used in clinical practice and its effectiveness is being monitored using viral load data. Recently, resistance to these antivirals has developed and entecavir has been added as a new treatment option in patients with lamivudine resistant viruses (Matthews, 2006; Zoulim, 2006).

Commercial interest has contributed to the availability of reliable diagnostic tests for the blood-borne viruses described above. Viral load monitoring has been available since the mid 1990s when the HIV assay was launched. Viral load assays for monitoring blood-borne viruses have since been implemented in most routine virology diagnostic laboratories.

Reactivating viruses

Introduction

The role of real-time PCR procedures in improving microbiological diagnosis is more clearly shown in a second group of viruses that reactivate from latency and cause clinical complications in immunocompromised patients. The main representatives are the members of the family *Herpesviridae*, but human adenoviruses (HAdV) and polyomaviruses have also been described as causing post-transplant complications and have therefore become important diagnostic targets. Some of these infections can be treated by effective antiviral therapy or other treatment options. Patients can be treated in different ways. One approach is to start treatment based on clinical symptoms, i.e. to start therapy once complications have been observed. Although this can be effective in some cases, sometimes it is too late as will be discussed.

CMV and EBV

A second approach to administering treatment is the prophylactic use of antiviral therapy. CMV infection in immunocompetent individuals can be asymptomatic or result in mononucleosis-like illness. The virus remains latently present in monocytes and can reactivate in immunocompromised patients, resulting in fatal disseminated disease in solid organ transplant patients with a special tropism for the transplanted organ: hepatitis after liver transplantation, gastroenteritis after transplantation of the intestines etc (Manez *et al.*, 1995; Patel and Paya, 1997). These more severe diseases are mainly observed in primary infection of seronegative transplant recipients receiving a graft from a seropositive donor. Prophylactic use of acyclovir or ganciclovir has resulted in a decrease of symptomatic CMV infections in these patients. Additional advantage of this approach is that the treatment is also effective for some other herpesviruses such as HSV (Slifkin *et al.*, 2004). Although antiviral prophylaxis is common practice in many countries, it obviously results in unnecessary, costly treatment of many patients and increases the possibility of developing resistance. Therefore, in our opinion, pre-emptive therapy is the best and most cost-effective alternative.

Until recently CMV pre-emptive therapy was guided by the pp65 antigenemia assay which detected pp65 expression on peripheral blood mononuclear cells (PBMCs). This assay has since been replaced by monitoring CMV DNA in blood or plasma using qPCR (Cortez *et al.*, 2003; Leruez-Ville *et al.*, 2003). Thresholds have been defined and included in treatment protocols and subsequently the effect of treatment can be efficiently monitored (Ikewaki *et al.*, 2005; Kalpoe *et al.*, 2004; Mori *et al.*, 2002).

For Epstein-Barr virus (EBV) pre-emptive therapy has been successfully applied as well. This virus has a seroprevalence of 90% in the normal population and latent virus remains present in B-cells. The virus can cause malignant transformation of cells and EBV DNA has been detected in Hodgkin's lymphoma (Gandhi *et al.*, 2006; Gallagher *et al.*, 1999) and nasopharyngeal carcinomas (Lin *et al.*, 2004; Hong *et al.*, 2004; Lo *et al.*, 1999). The result of chemotherapy treatment can be analyzed using qPCR as the EBV DNA load predicts survival or relapse after chemotherapy treatment. EBV reactivation does not result in infectious disease complications but has been associated with post-transplantation lymphoproliferative disorders (PTLD) in solid organ transplant patients (Hadou *et al.*, 2005; Orii *et al.*, 2000) and SCT (Aalto *et al.*, 2003; van Esser *et al.*, 2001). Pre-emptive therapy is guided by EBV DNA load determined by qPCR. Therapy consists of reduction of immunosuppression, treatment with anti-B-cell monoclonal antibodies (rituximab) or donor lymphocyte infusion and the kinetics of EBV DNA load is used to determine the efficacy (Lankester *et al.*, 2002; Wagner *et al.*, 2004; van Esser *et al.*, 2002). Frequent monitoring and reactive treatment of at-risk transplant patients has significantly reduced EBV-associated mortality.

Application of quantitative real-time PCR assays for diagnosing these two herpesviruses has been shown to be effective and has been widely implemented in routine diagnostics. What material should be tested is a matter of continuous debate. Although arguments can be provided to test whole blood, as these viruses are PBMC-associated, using plasma is common practice in many laboratories (Wagner *et al.*, 2001). Detection in plasma is more standardized, especially in SCT patients, as the number of cells in whole blood may vary. Secondly, although DNA in plasma is not a direct indication of active replication, it does reflect replication of CMV and EBV somewhere in the patient.

Other herpesviruses

After HIV, herpesviruses have the most treatment options available including the antiviral drugs, acyclovir, ganciclovir, famciclovir, cidofovir, foscarnet and other derivatives. As a result, most herpes infections will be treated after diagnosis of the virus, these days mainly by qualitative, real-time PCR. The group of alphaherpesviruses, consisting of HSV-1, HSV-2 and VZV, has a tropism for muco-epithelial cells resulting in clear clinical lesions. The viruses may also cause severe and even fatal disseminated infections and encephalitis in neonates and immunocompromised patients. So far, there has not been a role for quantitation of viral loads in the diagnosis of infections by members of the subfamily *Alphaherpesvirinae*, although some attempts have been made for HSV in CSF (Whitley, 2006).

For some other members of the herpesvirus family, the clinical significance is still unclear. Quantitative real-time PCR studies are undertaken to establish their involvement in clinical complications in the immunocompromised patient where they are frequently detected. Most data have been generated on Human herpesvirus 6 (HHV-6), discovered in 1986 in blood cells and assigned to the subfamily *Betaherpesvirinae*. HHV-6 can be found in T-lymphocytes and NK cells. HHV-6 is the etiological agent of exanthem subitum (roseola infantum) which occurs in the first two years of life. The virus can also be transmitted congenitally, crossing the placenta and can even be inherited after integration in the parental chromosomes (Daibata *et al.*, 1999). This integration may play a role in long-term clinical consequences (Clark *et al.*, 2006). Severe HHV-6 infections such as hepatitis and encephalitis have occasionally been described in immunocompetent patients

(Birnbaum *et al.*, 2005; Cacheux *et al.*, 2005; Harma *et al.*, 2003; Portolani *et al.*, 2005; Soto-Hernandez, 2004). In these cases, real-time PCR has been used in the follow-up of the patients and has shown a decrease in viral load after antiviral treatment or repeated transplantation of the liver. Obviously, with a seroprevalence of approximately 90%, HHV-6 reactivation is common and has been reported in 30–50% of SCT patients within two to four weeks after transplantation. Reactivation of HHV-6 has been suggested to be associated with pneumonitis, delayed neutrophil engraftment and graft versus host disease (GVHD). Attributing significance to viral load data has been attempted but more data are required to prove the association (Boutolleau *et al.*, 2003; Ljungman *et al.*, 2000). Other complications, such as encephalitis, have been reported in isolated cases, and the virus has been found in biopsies of renal transplant patients (Gupta *et al.*, 2003; Rayes *et al.*, 2005). However, in other studies no significant clinical symptoms have been found in association with HHV-6 reactivation (Kidd *et al.*, 2000; Ward, 2005). Quantitative real-time PCR can be used to monitor HHV-6 viral DNA load and the effects of treatment (Gautheret-Dejean *et al.*, 2002). However, further follow-up studies are required to define the clinical implications of this infection.

Human Herpesvirus 7 (HHV-7) which is closely related to HHV-6, was also initially isolated from blood cells and from saliva. Although rarely associated with clinical illness, the virus has been implicated as a co-factor for CMV disease in kidney transplant patients (Kidd *et al.*, 2000; Osman *et al.*, 1996). Clinically important reactivation has not been described in immunocompromised patients (Yamada, 2001; Griffiths *et al.*, 2000; Pascher *et al.*, 2004).

Human herpesvirus 8 (HHV-8) is a member of the subfamily *Gammaherpesvirinae* and was discovered in the biopsy material of AIDS patients with Kaposi's sarcoma (KS). In addition, it has been found in association with a type of B-cell lymphoma and with multicentric Castleman's disease. Prolonged immunodeficiency results in KS and therefore HAART prevents KS by preventing progression to AIDS in HIV infected patients (Cattelan *et al.*, 2005). A prolonged state of immunodeficiency is also induced by chemical immunosuppression after transplantation and as a result KS has been found in both solid organ transplant patients and SCT patients (Huang *et al.*, 2003; Mitxelena *et al.*, 2003; Luppi *et al.*, 2000; Szende *et al.*, 1997). HHV-8 can be detected in PBMCs from 64% of normal children and up to 80% of adults (Kikuta *et al.*, 1997). Disease development is the result of reactivation of HHV-8, and using real-time PCR, HHV-8 copy numbers and the number of positive cells have been correlated with more severe disease and disease progression (Pellet *et al.*, 2002; Song *et al.*, 2004). Quantitative real-time PCR has been used to monitor HAART treatment of KS, but also rituximab treatment of HHV-8 related lymphomas (Corbellino *et al.*, 2001; Pozo *et al.*, 2000; Song *et al.*, 2004).

Adenoviruses

Adenoviruses are a common cause of infections during childhood and in people in close communities where it causes gastrointestinal, respiratory or conjunctival disease which is usually self-limiting. HAdV result in asymptomatic, latent infections in lympho-epithelial tissues with a potential to reactivate in immunocompromised patients.

Reactivation of HAdVs has been reported to cause dissemination and subsequently severe complications in pediatric stem cell transplant recipients as recently reviewed by Ison *(Ison, 2006)*. In a retrospective study in a cohort of 328 consecutively transplanted children, 37 cases of HAdV infection (11%) were diagnosed which resulted in 17 cases of HAdV-related disease and seven (2.1%) fatal cases (van Tol *et al.*, 2005a; van Tol *et al.*, 2005b). Once HAdV dissemination results in clinical complications such as enteritis or hepatitis, therapeutic options are limited. Recently, quantitative real-time PCR has been added as a tool to improve patient management by monitoring HAdV DNA loads after transplantation (Lion *et al.*, 2003; Heim *et al.*, 2003; Leruez-Ville *et al.*, 2004). As the genus *Mastadenovirus* contains 51 serologically distinct types affecting humans, designing a general real-time PCR assay is complicated by the heterogeneity of the viral genome. Most primers used today have been selected from the hexon gene and for optimal, catch-all analyses multiplex approaches are required (Claas *et al.*, 2005; Gu *et al.*, 2003).

Another problem in applying qPCR is the lack of standardization. Due to primer and probe heterogeneity, viral loads may be subtype-dependent: the presence of mismatches in the primer or probe region from a patient with, for example HAdV serotype 31, will result in higher C_T values and thus lower viral loads if the standard is for a virus with identical sequence to the oligos such as HAdV5 (Claas *et al.*, 2005). Despite these drawbacks, real-time PCR has been proven to generate valuable information for SCT patient management.

Quantitative real-time PCR showed that once clinical disease became overt, HAdV viral DNA loads of up to 10^{11} copies per ml of plasma could be detected (Lankester *et al.*, 2002a; Claas *et al.*, 2005). Patients at risk of a disseminated fatal HAdV reactivation could be quickly identified. A very powerful antiviral agent is required for treatment of patients with a viral load greater than 10^{11} copies DNA/ml, a 99% reduction of virus load still leaves 10^9 copies per ml. The antiviral agents ribavirin and cidofovir inhibit HAdV replication *in vitro* (Morfin *et al.*, 2005; Naesens *et al.*, 2005) and in animal models (Lenaerts *et al.*, 2005). Although antiviral therapy by ribavirin or cidofovir has been advocated for treatment in SCT patients, the effectiveness of treatment has not been convincingly shown (Lankester *et al.*, 2004; Naesens *et al.*, 2005; Ison, 2006). The main omission in the studies claiming successful treatment is that reduction in viral load itself is no proof of the treatment's effectivity as shown in Figure 7.4. High viral DNA loads can be cleared with or without antiviral agent as a result of effective immune reconstitution of NK and T cells (Heemskerk *et al.*, 2005). Thus, data on immune reconstitution together with HAdV load should be used to guide optimal treatment of the patient. Detection of HAdV DNA as a tool for predicting and preventing complications by HAdV infection should be performed quantitatively. DNA viremia without symptoms is not uncommon and therefore a threshold of 10^3 copies/ml plasma in the absence of immune reconstitution has been defined as a starting point for pre-emptive therapy. In clinical practice, effective treatment consists of tapering immunosuppression or donor lymphocyte infusion (van Tol *et al.*, 2005a; Ison, 2006). In monitoring its effects, qPCR provides the objective data on the effectiveness of this treatment by showing reduction of HAdV DNA load with time.

Polyomavirus, BKV

BK virus (BKV) is another DNA virus that has been included in routine screening procedures mainly in follow-up of renal transplant patients. This polyomavirus was first isolated in 1971 from the urine of a renal transplant patient and although the mode of transmission is unknown, seroprevalence of up to 100% has been reported (Reploeg *et al.*, 2001).

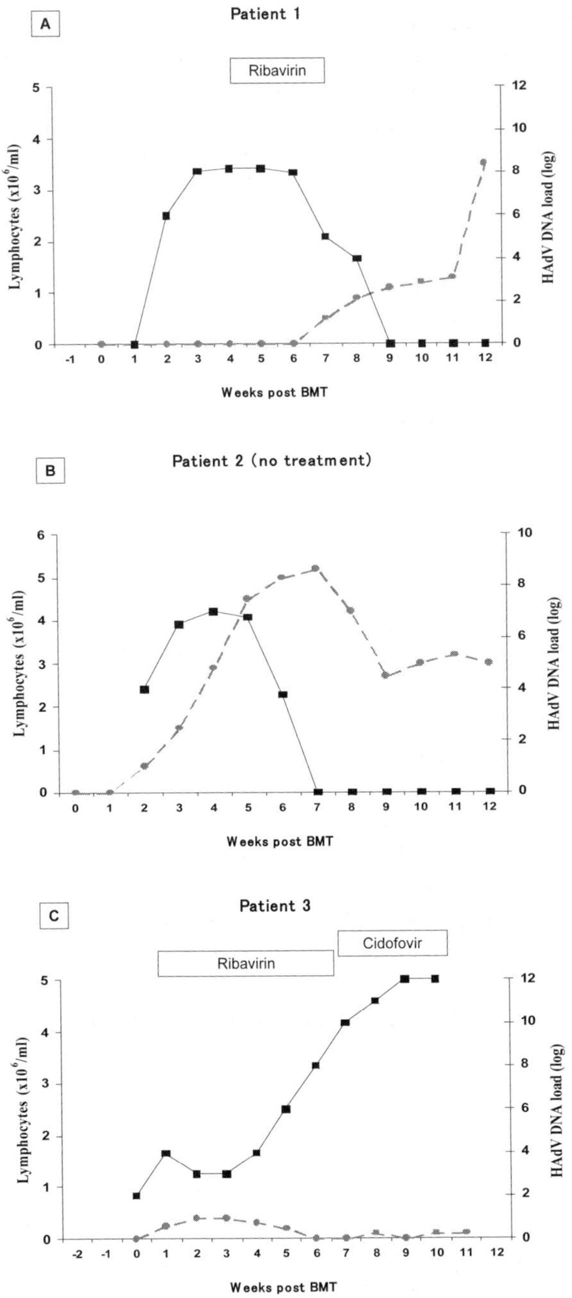

Figure 7.4 Use of qPCR to monitor HAdV viral load (solid lines) in the course of dissemination of HAdV and the effects of antiviral treatment in pediatric SCT patients. Panel A shows patient 1 who received ribavirin from week 4 to 7 which was apparently effective in decreasing the HAdV DNA load and permitted subsequent engraftment as shown by the increased number of lymphocytes (dashed line). However, patient 2 (panel B) shows clearance of HAdV without antiviral treatment but in combination with immune reconstitution and patient 3 (panel C) is a fatal case of HAdV dissemination despite treatment with ribavirin and cidofovir for 10 weeks, in the absence of immune reconstitution (Lankester *et al.*, 2002; Heemskerk *et al.*, 2005)

BKV has a clear tropism for the genitourinary tract and in immunocompetent individuals some cases of cystitis have been reported. Hemorrhagic cystitis in bone marrow transplant patients has been associated with increased BK viral loads in urine (Priftakis *et al.*, 2003; Azzi *et al.*, 1999) and plasma (Erard *et al.*, 2005) as determined by real-time PCR. A variety of predominantly renal complications have been found in immunocompromised patients (Wu *et al.*, 2006). BKV nephritis usually develops around 9 to 14 months after transplantation of the kidney and the majority of cases of BKV associated nephropathy (BKVN) develop within a year of transplantation with a prevalence of 1 to 10% (Hirsch *et al.*, 2001). BKV reactivation has been described as the cause of graft loss in kidney transplant patients, but with an unknown prevalence (Wu *et al.*, 2006).

Traditional diagnosis of BKV infection or reactivation is made by cytological detection of so-called "decoy cells" in urine. However, these cells with an enlarged nucleus and basophilic inclusions can also be present without clinical symptoms. A definitive diagnosis of BKVN relies on the detection of BKV associated pathology of the kidney (Drachenberg *et al.*, 2004; Drachenberg *et al.*, 2001).

QPCR assays have been developed for monitoring BKV DNA in urine, blood and biopsies from renal transplant patients. Viremia as well as viruria is common (Bressollette-Bodin *et al.*, 2005). BKV DNA can be detected in urine prior to detection in plasma but the presence of BKV in urine or blood does not always result in BKVN (Randhawa *et al.*, 2005; Si-Mohamed *et al.*, 2006). Again, thresholds may have to be defined to determine an increased risk of developing BKVN (Priftakis *et al.*, 2003; Azzi *et al.*, 1999; Herman *et al.*, 2004; Randhawa *et al.*, 2004). In biopsies, all BKVN cases had more than 59 copies of BKV DNA per cell but clearly biopsies are too difficult to obtain to use them for screening procedures. Although serum or plasma have been used as screening material to identify patients at risk for BKVN (Randhawa *et al.*, 2004; Limaye *et al.*, 2001), recently combined urine and plasma screening were recommended for this diagnosis. Detection of viral loads of 1×10^4 copies/ml of BKV DNA in plasma and 6.5×10^5 copies of VP1 mRNA per ng of total RNA in urine for more than three weeks is considered highly suggestive of BKV infection therefore screening every two to four weeks is recommended (Hirsch *et al.*, 2005).

QPCR has been shown to be effective in monitoring the effect of treatment. BKV viral loads decline after the reduction of immune suppression or following nephrectomy. Antiviral treatment of nephritis and nephropathy by cidofovir were accompanied by the reduction of BKV DNA loads (Hirsch *et al.*, 2001; Randhawa *et al.*, 2002). Cidofovir has *in vitro* activity against polyomaviruses (Andrei *et al.*, 1997) but its nephrotoxicity makes its use controversial in patients with renal complications. However, it has been used successfully and low dose treatment is recommended (Hirsch *et al.*, 2005). Another agent, leflunomide, also has *in vitro* activity against BKV and may be another option for antiviral treatment (Josephson *et al.*, 2006).

Standardization

Quantitative real-time PCR has proven to be a valuable addition in managing reactivating viruses in the immunocompromised patient, first by assessing the risk of infection and secondly by monitoring the effect of intervention. In addition, the assays will play a role in further elucidating a potential etiology for some of the viruses in post-transplant complications. However, in the broad implementation of these assays, better standardization of the reference materials used for quantitation is required. The World Health Organisation (WHO)-defined standards for HBV, HCV and parvovirus B19 have been made available, resulting in standardized reporting of the viral loads for these agents (Saldanha *et al.*, 2001; Saldanha *et al.*, 2005). In reporting viral loads of all other viruses, the load is calculated on the basis of many different standards including cultured virus titers, electron microscopy determined virus particle numbers and plasmid DNA copies. As a result, defining clinically important thresholds is nearly impossible and is laboratory dependent. Although some standardization can result from participation in well-defined proficiency panels, even the standards from these panels are no universal reference. This is an important issue to keep in mind when defining thresholds for initiating pre-emptive therapy.

Infection versus colonization

Real-time PCR was shown to improve the diagnosis of community acquired pneumonia, mainly by improving the detection of atypical bacterial agents and viruses (Templeton *et al.*, 2005). *Streptococcus pneumoniae* and *Haemophilus influenzae* are a main cause of severe pneumonia, as was shown in this study as well. These bacteria were diagnosed by conventional bacterial culture which is fast and simple for these pathogens. However, for a uniform approach, molecular detection of these agents has been considered. This application is complicated by the fact that there is a high level of colonization of these pathogens in the population, resulting in positive PCR results in healthy controls. Quantitation has been advocated as the key to successful diagnosis of these infections (Kais *et al.*, 2006). Based on the quantity of pathogen detected, infections may be differentiated from colonization. Because the median load in infected patients is higher than the median load in colonized controls, the diagnosis in the individual patient is complicated by the overlapping C_T values (due to a high degree of variation in C_T values in both groups) resulting in inconclusive C_T values and subsequent microbial load (K.E. Templeton and E.C.J. Claas, unpublished).

Similar problems are observed when detecting group B streptococci in pregnant women (Honest *et al.*, 2006) or *Pneumocystis jiroveci* as a cause of pneumonia (Larsen *et al.*, 2002). Quantitative detection gives a clear impression of the status of the presence of the microorganism, but it cannot be considered conclusive. In the latter applications, it is mainly the high negative predictive values of the assays that are used in the routine microbiology laboratory.

Conclusions

The famous journal *Science* named the heat-stable Taq DNA polymerase that PCR employs, as its first "Molecule of the Year" in 1989. It was not until 1992 that the first commercial tests, the COBAS AMPLICOR *Chlamydia trachomatis* Test (CT) and later the COBAS AMPLICOR HIV-1 MONITOR Test were introduced outside of the United States. Dr. Kary Mullis received the Nobel Price for Chemistry for PCR technology in 1993. Since molecular diagnostics seriously entered routine clinical microbiology only a decade ago, the developments have been rapid and revolutionary (Gilbert, 2002). In fact, in the majority of routine clinical microbiological settings, "conventional" PCR is considered out-dated and has been replaced by real-time PCR technology. This technological shift is promoted by the many industries providing test systems for molecular diagnosis and all new developments are directed toward detecting infectious diseases in real-time. Detection of any target for which the sequence is known can be achieved by real-time PCR (Espy *et al.*, 2006). The many different applications and advantages of using real-time PCR have been outlined in this and the other chapters. In this part, we have focused on molecular diagnostics in routine microbiology laboratories. Clearly, real-time PCR has become an important tool in the diagnostic laboratory and has rapidly conquered a central role in the laboratory, with increased sensitivity and results in a relevant timeframe as the major advantages.

Development of real-time PCR, like many other new developments, has been accelerated by war and terrorism, resulting in, for example, portable machines for real-time PCR analysis (Holland and Kiechle, 2005). The Cepheid SmartCycler™ is a commercially available combination of 16 individual thermocyclers designed by the US Army. The handheld advanced nucleic acid analyzer (HANAA) is a portable real time thermal cycler unit conducting rapid heating and cooling of plastic reaction tubes, sufficient for four simultaneous reactions. Although the system is specifically employed to detect bioterrorism-related outbreaks of anthrax, it is also used for the detection of other bacterial pathogens (Higgins *et al.*, 2003). However, one can easily imagine that portable machines will also be used at the bedside in clinical settings, increasing the speed of laboratory diagnosis. In fact, the developments in miniaturized portable machines are progressing rapidly and there are several portable machines (e.g. RAPID (ruggedized advanced pathogen identification device; Idaho Technology Inc., U.S.A.) on the market (Spitzack and Ugaz, 2006; Song *et al.*, 2006; Liao *et al.*, 2005). One feature of the RAPID system that makes it particularly attractive is its web-enabled software and potential for wireless broadcast that allows transmission of data in real-time from the analysis site back to a command centre (specialized laboratory).

For application in routine diagnostics, it is important that all new developments in technologies run in parallel, and include: (i) advances in DNA and RNA automated extraction and purification, (ii) improved target detection, including instrument technology and primer/probe design and (iii) high-throughput as well as small-scale analysis. An increasing number of large industries are becoming players in this market. Apart from the "old" diagnostics firms such as Roche, Abbott, bioMerieux and Bayer, less conventional industries are moving into this area, such as Philips, General Electric and Siemens, who recently acquired Bayer. Their experience and know-how in electronics and medical systems may provide new impacts and strategies for molecular diagnosis of infectious diseases. For Philips a further area of research in the field of molecular medicine is the "electronic blood test" using a chip (biosensor). Until very recently, patients who are suspected of having meningitis had to wait for up to three days for the test results. Molecular diagnostics now allow the genetic detection of the pathogen to be determined within half an hour. This means that, instead of prescribing broad-spectrum antibiotics, targeted treatment can be started immediately, resulting in a shorter treatment time.

In a large number of clinical microbiology laboratories almost 80% of clinical virology is already based around molecular diagnostics. As progress in technological development is extremely fast, the importance of molecular diagnosis for the whole field of clinical microbiology is increasing. Improved knowledge of host genetic factors that influence the severity of infection and or disease will make microbiological diagnosis a multidisciplinary field. The speed, sensitivity, specificity and information from other areas will have an enormous impact on hospital logistics, and laboratories should be prepared for these new advances in the diagnosis of infectious diseases. In fact, diagnosis of infectious diseases will become more and more complex. New technologies will provide further information on the pathogen and the patient. Diagnosis and treatment will not only be based on multiplex detection of the pathogen in relation to the clinical picture. Additional information will be collected on the virulence of the pathogen and its susceptibility to treatment options, but also the susceptibility of the patient to the pathogen and the patient's pharmacogenomic make-up and even pathogen and patient-related environmental factors. All this information will be combined together and result in optimal, individualized treatment protocols for accurate treatment of the infection and the disease. Until that can be achieved, the number of applications of real-time PCR in routine microbial diagnostics will further increase. Molecular diagnostics and especially implementation of real-time PCR has significantly contributed to the development of the medical microbiology laboratory by providing sensitive diagnostics for the patient in a clinically relevant time-frame.

Future trends

The field of molecular diagnostics is evolving rapidly. An important step in its further maturation is ongoing automation by coupling automated extraction procedures to real-time PCR amplification. Especially further standardization of extraction procedures is an important issue. Although so-called "total nucleic acid" isolations are available, different samples may still require different pre-treatment steps. For example, in a multiplex real-time PCR gastroenteritis panel, uniform procedures can simultaneously amplify viral, bacterial, and parasitic targets, once the nucleic acid is isolated. However, to recover DNA from the cysts of parasites, obviously more stringent methods are required than in releasing RNA from noroviruses. A universal protocol may be very hard to achieve. For this reason, many companies use separate kits for the isolation of RNA, DNA and for difficult samples. Standardization in this part of the procedure would mean a giant leap forward in automating and therefore streamlining molecular diagnostics.

Apart from real-time PCR technology, multitarget detection devices are ready to be used in microbiology as well. Theoretically, multiplex PCR of many more targets than the current maximum of six in real-time PCR can be achieved. Thousands of targets can be detected in microarrays (Bryant *et al.*, 2004), and in a system employing the xMAP technology (Luminex corporation) theoretically 100 different targets can be differentiated. Although microbiological applications are possible, both approaches lack sensitivity for detection of pathogens directly in clinical material. Therefore, as they both represent post-PCR, end-point detection systems after labeling/amplification, they are actually a step back from real-time PCR.

Designing a four-target multiplex real-time PCR without reducing the sensitivity for each of the individual targets requires extensive analysis of interactions between all primers and probes. Retaining sensitivity in a multiplex for 40 targets will be a huge challenge. When all the primers are present alongside DNA from clinical material including human chromosomal DNA, non-specific PCR products are likely to be generated and will reduce the efficiency of the PCR.

Nanotechnology is another promising application in molecular diagnosis. Smaller, faster platforms can be applied for real-time PCR. Reaction volumes of a few microliters or even nanoliters (BioTrove, BioTrove Inc., Woburn, USA) can be subjected to real-time PCR amplification. Most likely, these are of limited value for application in detection of pathogens due to the simple fact that, because of the small input volume of nucleic acid into the amplification, the sensitivity will be too low. Indeed, if using smaller input volumes, one needs more copies/ml to avoid false negative results. In the isolation procedure for clinical samples, a certain amount of buffer is required for the elution of nucleic acids. An input of 1 nanoliter of nucleic acid into the nano-application would require at least 1 copy to be present in every nl of the eluted nucleic acid from the clinical sample. If the elution volume is for example 50 µl (= 5×10^4 nl), 5×10^4 copies would have to be present in this 50 µl of eluted nucleic acids. Depending on the input of clinical sample, the sensitivity would be 5×10^4 C/ml if 1 ml of plasma was used, or with an input of 200 µl of plasma $5 \times 5 \times 10^4 = 2.5 \times 10^5$/ml. With the current qPCR approaches, sensitivities of 50 copies per ml can be achieved. Other, even more advanced, technological developments are in progress in which pathogens can be detected without any need for amplification using nanotechnology.

A major new direction is single molecule detection, using single molecule spectroscopy by, for example, scanning probe microscopy (SPM). In fact, new technology in which automated extraction is coupled to portable detection systems will advance us into a new diagnostic dimension.

Eventually, diagnosis of microbial infections may become a bedside activity with results available for immediate measures to be taken. Microbiological expertise remains required for a proper interpretation, i.e. the clinical impact, of these results.

Acknowledgment

We thank Yvonne Montfoort for her help in improving the manuscript.

Web resources

SPM
http://www.mobot.org/jwcross/spm/
Information on Scanning probe microscopy

RAPID
http://www.idahotech.com/rapid/index.html
The Ruggedized Advanced Pathogen Identification Device (RAPID), an instrument using LightCycler® Instrument technology. A portable, impact resistant instrument based on LightCycler technology, allowing field identification of pathogens.

Luminex corporation
http://www.luminexcorp.com/technology/
xMAP technology: high throughout detection of multiple targets.

HANAA
http://www.arrowtechinc.com/bio%20seeq.htm
Example of Handheld Advanced Nucleic Acid Analyzer

In vitro *diagnostics (IVD)*
http://www.fda.gov/cdrh/oivd/regulatory-overview.html
Overview of how the US Food and Drug Administration regulates IVD.

QCMD
http://www.qcmd.org/Index2.htm
Quality Control in Molecular Diagnostics: proficiency panels for laboratories performing routine molecular microbiological testing.

References

Aalto, S.M., Juvonen, E., Tarkkanen, J., Volin, L., Ruutu, T., Mattila, P.S., Piiparinen, H., Knuutila, S., and Hedman, K. (2003). Lymphoproliferative disease after allogeneic stem cell transplantation—pre-emptive diagnosis by quantification of Epstein-Barr virus DNA in serum. J. Clin. Virol. *28*, 275–283.

Abramson, R.D. and Myers, T.W. (1993). Nucleic acid amplification technologies. Curr. Opin. Biotechnol. *4*, 41–47.

Andrei, G., Snoeck, R., Vandeputte, M., and De Clercq, E. (1997). Activities of various compounds against murine and primate polyomaviruses. Antimicrob. Agents Chemother. *41*, 587–593.

Apfalter, P., Reischl, U., and Hammerschlag, M.R. (2005). In-house nucleic acid amplification assays in research: how much quality control is needed before one can rely upon the results? J. Clin. Microbiol. *43*, 5835–5841.

Arya, M., Shergill, I.S., Williamson, M., Gommersall, L., Arya, N., and Patel, H.R. (2005). Basic principles of real-time quantitative PCR. Expert. Rev. Mol. Diagn. *5*, 209–219.

Aslanzadeh, J. (2004). Preventing PCR amplification carryover contamination in a clinical laboratory. Ann. Clin. Lab Sci. *34*, 389–396.

Azzi, A., Cesaro, S., Laszlo, D., Zakrzewska, K., Ciappi, S., De Santis, R., Fanci, R., Pesavento, G., Calore, E., and Bosi, A. (1999). Human polyomavirus BK (BKV) load and haemorrhagic cystitis in bone marrow transplantation patients. J. Clin. Virol. *14*, 79–86.

Bekkers, R.L., Massuger, L.F., Bulten, J., and Melchers, W.J. (2004). Epidemiological and clinical aspects of human papillomavirus detection in the prevention of cervical cancer. Rev. Med. Virol. *14*, 95–105.

Birnbaum, T., Padovan, C.S., Sporer, B., Rupprecht, T.A., Ausserer, H., Jaeger, G., and Pfister, H.W. (2005). Severe meningoencephalitis caused by human herpesvirus 6 type B in an immunocompetent woman treated with ganciclovir. Clin. Infect. Dis. *40*, 887–889.

Boivin, G., Cote, S., Dery, P., De Serres, G., and Bergeron, M.G. (2004). Multiplex real-time PCR assay for detection of influenza and human respiratory syncytial viruses. J. Clin. Microbiol. *42*, 45–51.

Borst, A., Box, A.T., and Fluit, A.C. (2004). False-positive results and contamination in nucleic acid amplification assays: suggestions for a prevent and destroy strategy. Eur. J. Clin. Microbiol. Infect. Dis. *23*, 289–299.

Boutolleau, D., Fernandez, C., Andre, E., Imbert-Marcille, B.M., Milpied, N., Agut, H., and Gautheret-Dejean, A. (2003). Human herpesvirus (HHV)-6 and HHV-7: two closely related viruses with different infection profiles in stem cell transplantation recipients. J. Infect. Dis. *187*, 179–186.

Bressollette-Bodin, C., Coste-Burel, M., Hourmant, M., Sebille, V., Andre-Garnier, E., and Imbert-Marcille, B.M. (2005). A prospective longitudinal study of BK virus infection in 104 renal transplant recipients. Am. J. Transplant. *5*, 1926–1933.

Bretagne, S. and Costa, J.M. (2005). Towards a molecular diagnosis of invasive aspergillosis and disseminated candidosis. FEMS Immunol. Med. Microbiol. *45*, 361–368.

Bryant, P.A., Venter, D., Robins-Browne, R., and Curtis, N. (2004). Chips with everything: DNA microarrays in infectious diseases. Lancet Infect. Dis. *4*, 100–111.

Bustin, S.A., Benes, V., Nolan, T., and Pfaffl, M.W. (2005). Quantitative real-time RT-PCR—a perspective. J. Mol. Endocrinol. *34,* 597–601.

Cacheux, W., Carbonell, N., Rosmorduc, O., Wendum, D., Paye, F., Poupon, R., Chazouilleres, O., and Gozlan, J. (2005). HHV-6-related acute liver failure in two immunocompetent adults: favourable outcome after liver transplantation and/or ganciclovir therapy. J. Intern. Med. *258,* 573–578.

Cattelan, A.M., Calabro, M.L., De Rossi, A., Aversa, S.M., Barbierato, M., Trevenzoli, M., Gasperini, P., Zanchetta, M., Cadrobbi, P., Monfardini, S., and Chieco-Bianchi, L. (2005). Long-term clinical outcome of AIDS-related Kaposi's sarcoma during highly active antiretroviral therapy. Int. J. Oncol. *27,* 779–785.

Cheng, V.C., Yew, W.W., and Yuen, K.Y. (2005). Molecular diagnostics in tuberculosis. Eur. J. Clin. Microbiol. Infect. Dis. *24,* 711–720.

Chui, L., Drebot, M., Andonov, A., Petrich, A., Glushek, M., and Mahony, J. (2005). Comparison of 9 different PCR primers for the rapid detection of severe acute respiratory syndrome coronavirus using 2 RNA extraction methods. Diagn. Microbiol. Infect. Dis. *53,* 47–55.

Claas, E.C., Schilham, M.W., de Brouwer, C.S., Hubacek, P., Echavarria, M., Lankester, A.C., van Tol, M.J., and Kroes, A.C. (2005). Internally controlled real-time PCR monitoring of adenovirus DNA load in serum or plasma of transplant recipients. J. Clin. Microbiol. *43,* 1738–1744.

Clark, D.A., Nacheva, E.P., Leong, H.N., Brazma, D., L.i., Y.T., Tsao, E.H., Buyck, H.C., Atkinson, C.E., Lawson, H.M., Potter, M.N., and Griffiths, P.D. (2006). Transmission of integrated human herpesvirus 6 through stem cell transplantation: implications for laboratory diagnosis. J. Infect. Dis. *193,* 912–916.

Cockerill, F.R., III (2003). Application of rapid-cycle real-time polymerase chain reaction for diagnostic testing in the clinical microbiology laboratory. Arch. Pathol. Lab Med. *127,* 1112–1120.

Corbellino, M., Bestetti, G., Scalamogna, C., Calattini, S., Galazzi, M., Meroni, L., Manganaro, D., Fasan, M., Moroni, M., Galli, M., and Parravicini, C. (2001). Long-term remission of Kaposi sarcoma-associated herpesvirus-related multicentric Castleman disease with anti-CD20 monoclonal antibody therapy. Blood *98,* 3473–3475.

Corey, L., Huang, M.L., Selke, S., and Wald, A. (2005). Differentiation of herpes simplex virus types 1 and 2 in clinical samples by a real-time taqman PCR assay. J. Med. Virol. *76,* 350–355.

Cortez, K.J., Fischer, S.H., Fahle, G.A., Calhoun, L.B., Childs, R.W., Barrett, A.J., and Bennett, J.E. (2003). Clinical trial of quantitative real-time polymerase chain reaction for detection of cytomegalovirus in peripheral blood of allogeneic hematopoietic stem-cell transplant recipients. J. Infect. Dis. *188,* 967–972.

Cuzick, J., Szarewski, A., Terry, G., H.o., L., Hanby, A., Maddox, P., Anderson, M., Kocjan, G., Steele, S.T., and Guillebaud, J. (1995). Human papillomavirus testing in primary cervical screening. Lancet *345,* 1533–1536.

Daibata, M., Taguchi, T., Nemoto, Y., Taguchi, H., and Miyoshi, I. (1999). Inheritance of chromosomally integrated human herpesvirus 6 DNA. Blood *94,* 1545–1549.

Deiman, B., van Aarle, P., and Sillekens, P. (2002). Characteristics and applications of nucleic acid sequence-based amplification (NASBA). Mol. Biotechnol. *20,* 163–179.

Deutch, S., Pedersen, L.N., Podenphant, L., Olesen, R., Schmidt, M.B., Moller, J.K., and Ostergaard, L. (2006). Broad-range real time PCR and DNA sequencing for the diagnosis of bacterial meningitis. Scand. J. Infect. Dis. *38,* 27–35.

Drachenberg, C.B., Papadimitriou, J.C., Hirsch, H.H., Wali, R., Crowder, C., Nogueira, J., Cangro, C.B., Mendley, S., Mian, A., and Ramos, E. (2004). Histological patterns of polyomavirus nephropathy: correlation with graft outcome and viral load. Am. J. Transplant. *4,* 2082–2092.

Drachenberg, R.C., Drachenberg, C.B., Papadimitriou, J.C., Ramos, E., Fink, J.C., Wali, R., Weir, M.R., Cangro, C.B., Klassen, D.K., Khaled, A., Cunningham, R., and Bartlett, S.T. (2001). Morphological spectrum of polyoma virus disease in renal allografts: diagnostic accuracy of urine cytology. Am. J. Transplant. *1,* 373–381.

Dunbar, S.A. (2006). Applications of Luminex xMAP technology for rapid, high-throughput multiplexed nucleic acid detection. Clin. Chim. Acta *363,* 71–82.

Erard, V., Kim, H.W., Corey, L., Limaye, A., Huang, M.L., Myerson, D., Davis, C., and Boeckh, M. (2005). BK DNA viral load in plasma: evidence for an association with hemorrhagic cystitis in allogeneic hematopoietic cell transplant recipients. Blood *106,* 1130–1132.

Espy, M.J., Uhl, J.R., Sloan, L.M., Buckwalter, S.P., Jones, M.F., Vetter, E.A., Yao, J.D., Wengenack, N.L., Rosenblatt, J.E., Cockerill, F.R., III, and Smith, T.F. (2006). Real-time PCR in clinical microbiology: applications for routine laboratory testing. Clin. Microbiol. Rev. *19*, 165–256.

Fang, H. and Hedin, G. (2003). Rapid screening and identification of methicillin-resistant Staphylococcus aureus from clinical samples by selective-broth and real-time PCR assay. J. Clin. Microbiol. *41*, 2894–2899.

Fitzmaurice, J., Glennon, M., Duffy, G., Sheridan, J.J., Carroll, C., and Maher, M. (2004). Application of real-time PCR and RT-PCR assays for the detection and quantitation of VT 1 and VT 2 toxin genes in, *E. coli* O157:H7. Mol. Cell Probes *18*, 123–132.

Florent, M., Katsahian, S., Vekhoff, A., Levy, V., Rio, B., Marie, J.P., Bouvet, A., and Cornet, M. (2006). Prospective evaluation of a polymerase chain reaction-ELISA targeted to Aspergillus fumigatus and Aspergillus flavus for the early diagnosis of invasive aspergillosis in patients with hematological malignancies. J. Infect. Dis. *193*, 741–747.

Fluit, A.C., Visser, M.R., and Schmitz, F.J. (2001). Molecular detection of antimicrobial resistance. Clin. Microbiol. Rev. *14*, 836–71, table.

Francois, P., Renzi, G., Pittet, D., Bento, M., Lew, D., Harbarth, S., Vaudaux, P., and Schrenzel, J. (2004). A novel multiplex real-time PCR assay for rapid typing of major staphylococcal cassette chromosome mec elements. J. Clin. Microbiol. *42*, 3309–3312.

Frieden, T.R., Sterling, T.R., Munsiff, S.S., Watt, C.J., and Dye, C. (2003). Tuberculosis. Lancet *362*, 887–899.

Gageldonk-Lafeber, A.B., Heijnen, M.L., Bartelds, A.I., Peters, M.F., van der Plas, S.M., and Wilbrink, B. (2005). A case-control study of acute respiratory tract infection in general practice patients in The Netherlands. Clin. Infect. Dis. *41*, 490–497.

Gallagher, A., Armstrong, A.A., MacKenzie, J., Shield, L., Khan, G., Lake, A., Proctor, S., Taylor, P., Clements, G.B., and Jarrett, R.F. (1999). Detection of Epstein-Barr virus (EBV) genomes in the serum of patients with EBV-associated Hodgkin's disease. Int. J. Cancer *84*, 442–448.

Gandhi, M.K., Lambley, E., Burrows, J., Dua, U., Elliott, S., Shaw, P.J., Prince, H.M., Wolf, M., Clarke, K., Underhill, C., Mills, T., Mollee, P., Gill, D., Marlton, P., Seymour, J.F., and Khanna, R. (2006). Plasma Epstein-Barr virus (EBV) DNA is a biomarker for EBV-positive Hodgkin's lymphoma. Clin. Cancer Res. *12*, 460–464.

Gautheret-Dejean, A., Manichanh, C., Thien-Ah-Koon, F., Fillet, A.M., Mangeney, N., Vidaud, M., Dhedin, N., Vernant, J.P., and Agut, H. (2002). Development of a real-time polymerase chain reaction assay for the diagnosis of human herpesvirus-6 infection and application to bone marrow transplant patients. J. Virol. Methods *100*, 27–35.

Gilbert, C., Bestman-Smith, J., and Boivin, G. (2002). Resistance of herpesviruses to antiviral drugs: clinical impacts and molecular mechanisms. Drug Resist. Updat. *5*, 88–114.

Gilbert, G.L. (2002). Molecular diagnostics in infectious diseases and public health microbiology: cottage industry to postgenomics. Trends Mol. Med. *8*, 280–287.

Ginsburg, A.S., Grosset, J.H., and Bishai, W.R. (2003). Fluoroquinolones, tuberculosis, and resistance. Lancet Infect. Dis. *3*, 432–442.

Gish, R.G. and Locarnini, S.A. (2006). Chronic hepatitis B: current testing strategies. Clin. Gastroenterol. Hepatol. *4*, 666–676.

Griffiths, P.D., Clark, D.A., and Emery, V.C. (2000). Betaherpesviruses in transplant recipients. J. Antimicrob. Chemother. *45 Suppl T3*, 29–34.

Gu, Z., Belzer, S.W., Gibson, C.S., Bankowski, M.J., and Hayden, R.T. (2003). Multiplexed, real-time PCR for quantitative detection of human adenovirus. J. Clin. Microbiol. *41*, 4636–4641.

Gunson, R.N., Collins, T.C., and Carman, W.F. (2006). Practical experience of high throughput real time PCR in the routine diagnostic virology setting. J. Clin. Virol. *35*, 355–367.

Gupta, M., Diaz-Mitoma, F., Feber, J., Shaw, L., Forget, C., and Filler, G. (2003). Tissue HHV6 and 7 determination in pediatric solid organ recipients—a pilot study. Pediatr. Transplant. *7*, 458–463.

Hadou, T., Andre, J.L., Bourquard, R., Krier-Coudert, M.J., Venard, V., and Le Faou, A. (2005). Long-term follow-up of Epstein-Barr virus viremia in pediatric recipients of renal transplants. Pediatr. Nephrol. *20*, 76–80.

Halliday, C., W.u., Q.X., James, G., and Sorrell, T. (2005). Development of a nested qualitative real-time PCR assay to detect Aspergillus species DNA in clinical specimens. J. Clin. Microbiol. *43*, 5366–5368.

Harma, M., Hockerstedt, K., and Lautenschlager, I. (2003). Human herpesvirus-6 and acute liver failure. Transplantation *76*, 536–539.

Hayashi, N. and Takehara, T. (2006). Antiviral therapy for chronic hepatitis C: past, present, and future. J. Gastroenterol. *41*, 17–27.

Heathcote, E.J., Shiffman, M.L., Cooksley, W.G., Dusheiko, G.M., Lee, S.S., Balart, L., Reindollar, R., Reddy, R.K., Wright, T.L., Lin, A., Hoffman, J., and De Pamphilis, J. (2000). Peginterferon alfa-2a in patients with chronic hepatitis C and cirrhosis. N. Engl. J. Med. *343*, 1673–1680.

Heathcote, E.J. (2004). Prevention of hepatitis C virus-related hepatocellular carcinoma. Gastroenterology *127*, S294-S302.

Heemskerk, B., Lankester, A.C., van Vreeswijk, T., Beersma, M.F., Claas, E.C., Veltrop-Duits, L.A., Kroes, A.C., Vossen, J.M., Schilham, M.W., and van Tol, M.J. (2005). Immune reconstitution and clearance of human adenovirus viremia in pediatric stem-cell recipients. J. Infect. Dis. *191*, 520–530.

Heim, A., Ebnet, C., Harste, G., and Pring-Akerblom, P. (2003). Rapid and quantitative detection of human adenovirus DNA by real-time PCR. J. Med. Virol. *70*, 228–239.

Herman, J., Van Ranst, M., Snoeck, R., Beuselinck, K., Lerut, E., and Damme-Lombaerts, R. (2004). Polyomavirus infection in pediatric renal transplant recipients: evaluation using a quantitative real-time PCR technique. Pediatr. Transplant. *8*, 485–492.

Higgins, J.A., Nasarabadi, S., Karns, J.S., Shelton, D.R., Cooper, M., Gbakima, A., and Koopman, R.P. (2003). A handheld real time thermal cycler for bacterial pathogen detection. Biosens. Bioelectron. *18*, 1115–1123.

Hindiyeh, M., Levy, V., Azar, R., Varsano, N., Regev, L., Shalev, Y., Grossman, Z., and Mendelson, E. (2005). Evaluation of a multiplex real-time reverse transcriptase PCR assay for detection and differentiation of influenza viruses A and B during the 2001–2002 influenza season in Israel. J. Clin. Microbiol. *43*, 589–595.

Hirsch, H.H., Brennan, D.C., Drachenberg, C.B., Ginevri, F., Gordon, J., Limaye, A.P., Mihatsch, M.J., Nickeleit, V., Ramos, E., Randhawa, P., Shapiro, R., Steiger, J., Suthanthiran, M., and Trofe, J. (2005). Polyomavirus-associated nephropathy in renal transplantation: interdisciplinary analyses and recommendations. Transplantation *79*, 1277–1286.

Hirsch, H.H., Mohaupt, M., and Klimkait, T. (2001). Prospective monitoring of BK virus load after discontinuing sirolimus treatment in a renal transplant patient with BK virus nephropathy. J. Infect. Dis. *184*, 1494–1495.

Holland, C.A. and Kiechle, F.L. (2005). Point-of-care molecular diagnostic systems—past, present and future. Curr. Opin. Microbiol. *8*, 504–509.

Honest, H., Sharma, S., and Khan, K.S. (2006). Rapid tests for group B Streptococcus colonization in laboring women: a systematic review. Pediatrics *117*, 1055–1066.

Hong, R.L., Lin, C.Y., Ting, L.L., K.o., J.Y., and Hsu, M.M. (2004). Comparison of clinical and molecular surveillance in patients with advanced nasopharyngeal carcinoma after primary therapy: the potential role of quantitative analysis of circulating Epstein-Barr virus DNA. Cancer *100*, 1429–1437.

Huang, P.M., Chang, Y.L., Chen, J.S., Hsu, H.H., K.o., W.J., Kuo, S.H., and Lee, Y.C. (2003). Human herpesvirus-8 associated Kaposi's sarcoma after lung transplantation: a case report. Transplant. Proc. *35*, 447–449.

Huletsky, A., Giroux, R., Rossbach, V., Gagnon, M., Vaillancourt, M., Bernier, M., Gagnon, F., Truchon, K., Bastien, M., Picard, F.J., van Belkum, A., Ouellette, M., Roy, P.H., and Bergeron, M.G. (2004). New real-time PCR assay for rapid detection of methicillin-resistant Staphylococcus aureus directly from specimens containing a mixture of staphylococci. J. Clin. Microbiol. *42*, 1875–1884.

Ikewaki, J., Ohtsuka, E., Satou, T., Kawano, R., Ogata, M., Kikuchi, H., and Nasu, M. (2005). Real-time PCR assays based on distinct genomic regions for cytomegalovirus reactivation following hematopoietic stem cell transplantation. Bone Marrow Transplant. *35*, 403–410.

Ison, M.G. (2006). Adenovirus infections in transplant recipients. Clin. Infect. Dis. *43*, 331–339.

Jordanides, N.E., Allan, E.K., McLintock, L.A., Copland, M., Devaney, M., Stewart, K., Parker, A.N., Johnson, P.R., Holyoake, T.L., and Jones, B.L. (2005). A prospective study of real-time panfungal PCR for the early diagnosis of invasive fungal infection in haemato-oncology patients. Bone Marrow Transplant. *35*, 389–395.

Josephson, M.A., Gillen, D., Javaid, B., Kadambi, P., Meehan, S., Foster, P., Harland, R., Thistlethwaite, R.J., Garfinkel, M., Atwood, W., Jordan, J., Sadhu, M., Millis, M.J., and Williams, J. (2006). Treatment of renal allograft polyoma BK virus infection with leflunomide. Transplantation *81*, 704–710.

Kais, M., Spindler, C., Kalin, M., Ortqvist, A., and Giske, C.G. (2006). Quantitative detection of Streptococcus pneumoniae, Haemophilus influenzae, and Moraxella catarrhalis in lower respiratory tract samples by real-time PCR. Diagn. Microbiol. Infect. Dis. 55, 169–178.

Kalpoe, J.S., Kroes, A.C., de Jong, M.D., Schinkel, J., de Brouwer, C.S., Beersma, M.F., and Claas, E.C. (2004). Validation of clinical application of cytomegalovirus plasma DNA load measurement and definition of treatment criteria by analysis of correlation to antigen detection. J. Clin. Microbiol. 42, 1498–1504.

Kebede, A., Verweij, J.J., Endeshaw, T., Messele, T., Tasew, G., Petros, B., and Polderman, A.M. (2004). The use of real-time PCR to identify Entamoeba histolytica and, E. dispar infections in prisoners and primary-school children in Ethiopia. Ann. Trop. Med. Parasitol. 98, 43–48.

Khanna, M., Fan, J., Pehler-Harrington, K., Waters, C., Douglass, P., Stallock, J., Kehl, S., and Henrickson, K.J. (2005). The pneumoplex assays, a multiplex PCR-enzyme hybridization assay that allows simultaneous detection of five organisms, Mycoplasma pneumoniae, Chlamydia (Chlamydophila) pneumoniae, Legionella pneumophila, Legionella micdadei, and Bordetella pertussis, and its real-time counterpart. J. Clin. Microbiol. 43, 565–571.

Kidd, I.M., Clark, D.A., Sabin, C.A., Andrew, D., Hassan-Walker, A.F., Sweny, P., Griffiths, P.D., and Emery, V.C. (2000). Prospective study of human betaherpesviruses after renal transplantation: association of human herpesvirus 7 and cytomegalovirus co-infection with cytomegalovirus disease and increased rejection. Transplantation 69, 2400–2404.

Kikuta, H., Itakura, O., Ariga, T., and Kobayashi, K. (1997). Detection of human herpesvirus 8 DNA sequences in peripheral blood mononuclear cells of children. J. Med. Virol. 53, 81–84.

Kocagoz, T., Saribas, Z., and Alp, A. (2005). Rapid determination of rifampin resistance in clinical isolates of Mycobacterium tuberculosis by real-time PCR. J. Clin. Microbiol. 43, 6015–6019.

Kwok, S. and Higuchi, R. (1989). Avoiding false positives with PCR. Nature 339, 237–238.

Lankester, A.C., Heemskerk, B., Claas, E.C., Schilham, M.W., Beersma, M.F., Bredius, R.G., van Tol, M.J., and Kroes, A.C. (2004). Effect of ribavirin on the plasma viral DNA load in patients with disseminating adenovirus infection. Clin. Infect. Dis. 38, 1521–1525.

Lankester, A.C., van Tol, M.J., Claas, E.C., Vossen, J.M., and Kroes, A.C. (2002a). Quantification of adenovirus DNA in plasma for management of infection in stem cell graft recipients. Clin. Infect. Dis. 34, 864–867.

Lankester, A.C., van Tol, M.J., Vossen, J.M., Kroes, A.C., and Claas, E. (2002b). Epstein-Barr virus (EBV)-DNA quantification in pediatric allogenic stem cell recipients: prediction of EBV-associated lymphoproliferative disease. Blood 99, 2630–2631.

Larsen, H.H., Masur, H., Kovacs, J.A., Gill, V.J., Silcott, V.A., Kogulan, P., Maenza, J., Smith, M., Lucey, D.R., and Fischer, S.H. (2002). Development and evaluation of a quantitative, touch-down, real-time PCR assay for diagnosing Pneumocystis carinii pneumonia. J. Clin. Microbiol. 40, 490–494.

Lass-Florl, C., Gunsilius, E., Gastl, G., Bonatti, H., Freund, M.C., Gschwendtner, A., Kropshofer, G., Dierich, M.P., and Petzer, A. (2004). Diagnosing invasive aspergillosis during antifungal therapy by PCR analysis of blood samples. J. Clin. Microbiol. 42, 4154–4157.

Lawson, A.J., Elviss, N.C., and Owen, R.J. (2005). Real-time PCR detection and frequency of 16S rDNA mutations associated with resistance and reduced susceptibility to tetracycline in Helicobacter pylori from England and Wales. J. Antimicrob. Chemother. 56, 282–286.

Lee, S.S., Heathcote, E.J., Reddy, K.R., Zeuzem, S., Fried, M.W., Wright, T.L., Pockros, P.J., Haussinger, D., Smith, C.I., Lin, A., and Pappas, S.C. (2002). Prognostic factors and early predictability of sustained viral response with peginterferon alfa-2a (40KD). J. Hepatol. 37, 500–506.

Lenaerts, L., Verbeken, E., De Clercq, E., and Naesens, L. (2005). Mouse adenovirus type 1 infection in SCID mice: an experimental model for antiviral therapy of systemic adenovirus infections. Antimicrob. Agents Chemother. 49, 4689–4699.

Leruez-Ville, M., Ouachee, M., Delarue, R., Sauget, A.S., Blanche, S., Buzyn, A., and Rouzioux, C. (2003). Monitoring cytomegalovirus infection in adult and pediatric bone marrow transplant recipients by a real-time PCR assay performed with blood plasma. J. Clin. Microbiol. 41, 2040–2046.

Leruez-Ville, M., Minard, V., Lacaille, F., Buzyn, A., Abachin, E., Blanche, S., Freymuth, F., and Rouzioux, C. (2004). Real-time blood plasma polymerase chain reaction for management of disseminated adenovirus infection. Clin. Infect. Dis. 38, 45–52.

Liao, C.S., Lee, G.B., Liu, H.S., Hsieh, T.M., and Luo, C.H. (2005). Miniature RT-PCR system for diagnosis of RNA-based viruses. Nucleic Acids Res. 33, e156.

Limaye, A.P., Jerome, K.R., Kuhr, C.S., Ferrenberg, J., Huang, M.L., Davis, C.L., Corey, L., and Marsh, C.L. (2001). Quantitation of BK virus load in serum for the diagnosis of BK virus-associated nephropathy in renal transplant recipients. J. Infect. Dis. *183*, 1669–1672.

Lin, J.C., Wang, W.Y., Chen, K.Y., Wei, Y.H., Liang, W.M., Jan, J.S., and Jiang, R.S. (2004). Quantification of plasma Epstein-Barr virus DNA in patients with advanced nasopharyngeal carcinoma. N. Engl. J. Med. *350*, 2461–2470.

Lion, T., Baumgartinger, R., Watzinger, F., Matthes-Martin, S., Suda, M., Preuner, S., Futterknecht, B., Lawitschka, A., Peters, C., Potschger, U., and Gadner, H. (2003). Molecular monitoring of adenovirus in peripheral blood after allogeneic bone marrow transplantation permits early diagnosis of disseminated disease. Blood *102*, 1114–1120.

Lisby, G. (1999). Application of nucleic acid amplification in clinical microbiology. Mol. Biotechnol. *12*, 75–99.

Ljungman, P., Wang, F.Z., Clark, D.A., Emery, V.C., Remberger, M., Ringden, O., and Linde, A. (2000). High levels of human herpesvirus 6 DNA in peripheral blood leucocytes are correlated to platelet engraftment and disease in allogeneic stem cell transplant patients. Br. J. Haematol. *111*, 774–781.

Lo, Y.M., Chan, L.Y., L.o., K.W., Leung, S.F., Zhang, J., Chan, A.T., Lee, J.C., Hjelm, N.M., Johnson, P.J., and Huang, D.P. (1999). Quantitative analysis of cell-free Epstein-Barr virus DNA in plasma of patients with nasopharyngeal carcinoma. Cancer Res. *59*, 1188–1191.

Luppi, M., Barozzi, P., Schulz, T.F., Trovato, R., Donelli, A., Narni, F., Sheldon, J., Marasca, R., and Torelli, G. (2000). Nonmalignant disease associated with human herpesvirus 8 reactivation in patients who have undergone autologous peripheral blood stem cell transplantation. Blood *96*, 2355–2357.

Mackay, I.M. (2004). Real-time PCR in the microbiology laboratory. Clin. Microbiol. Infect. *10*, 190–212.

Manez, R., Kusne, S., Green, M., Abu-Elmagd, K., Irish, W., Reyes, J., Furukawa, H., Tzakis, A., Fung, J.J., Todo, S., and. (1995). Incidence and risk factors associated with the development of cytomegalovirus disease after intestinal transplantation. Transplantation *59*, 1010–1014.

Manns, M.P., McHutchison, J.G., Gordon, S.C., Rustgi, V.K., Shiffman, M., Reindollar, R., Goodman, Z.D., Koury, K., Ling, M., and Albrecht, J.K. (2001). Peginterferon alfa-2b plus ribavirin compared with interferon alfa-2b plus ribavirin for initial treatment of chronic hepatitis C: a randomised trial. Lancet *358*, 958–965.

Matthews, S.J. (2006). Entecavir for the treatment of chronic hepatitis B virus infection. Clin. Ther. *28*, 184–203.

McDonald, R.R., Antonishyn, N.A., Hansen, T., Snook, L.A., Nagle, E., Mulvey, M.R., Levett, P.N., and Horsman, G.B. (2005). Development of a triplex real-time PCR assay for detection of Panton-Valentine leukocidin toxin genes in clinical isolates of methicillin-resistant Staphylococcus aureus. J. Clin. Microbiol. *43*, 6147–6149.

Melchers, W.J., Claas, H.C., and Quint, W.G. (1991). Use of the polymerase chain reaction to study the relationship between human papillomavirus infections and cervical cancer. Eur. J. Clin. Microbiol. Infect. Dis. *10*, 714–727.

Mitxelena, J., Gomez-Ullate, P., Aguirre, A., Rubio, G., Lampreabe, I., and Diaz-Perez, J.L. (2003). Kaposi's sarcoma in renal transplant patients: experience at the Cruces Hospital in Bilbao. Int. J. Dermatol. *42*, 18–22.

Morfin, F., Dupuis-Girod, S., Mundweiler, S., Falcon, D., Carrington, D., Sedlacek, P., Bierings, M., Cetkovsky, P., Kroes, A.C., van Tol, M.J., and Thouvenot, D. (2005). In vitro susceptibility of adenovirus to antiviral drugs is species-dependent. Antivir. Ther. *10*, 225–229.

Mori, T., Okamoto, S., Watanabe, R., Yajima, T., Iwao, Y., Yamazaki, R., Nakazato, T., Sato, N., Iguchi, T., Nagayama, H., Takayama, N., Hibi, T., and Ikeda, Y. (2002). Dose-adjusted preemptive therapy for cytomegalovirus disease based on real-time polymerase chain reaction after allogeneic hematopoietic stem cell transplantation. Bone Marrow Transplant. *29*, 777–782.

Morozumi, M., Ito, A., Murayama, S.Y., Hasegawa, K., Kobayashi, R., Iwata, S., Kawamura, N., Kuroki, H., Nakayama, E., Tajima, T., and Ubukata, K. (2006). Assessment of real-time PCR for diagnosis of Mycoplasma pneumoniae pneumonia in pediatric patients. Can. J. Microbiol. *52*, 125–129.

Munoz, N., Bosch, F.X., de Sanjose, S., Herrero, R., Castellsague, X., Shah, K.V., Snijders, P.J., and Meijer, C.J. (2003). Epidemiologic classification of human papillomavirus types associated with cervical cancer. N. Engl. J. Med. *348*, 518–527.

Naesens, L., Lenaerts, L., Andrei, G., Snoeck, R., Van Beers, D., Holy, A., Balzarini, J., and De Clercq, E. (2005). Antiadenovirus activities of several classes of nucleoside and nucleotide analogues. Antimicrob. Agents Chemother. *49*, 1010–1016.

Niesters, H.G. (2002). Clinical virology in real time. J. Clin. Virol. *25 Suppl 3*, S3–12.

Niesters, H.G. (2004). Molecular and diagnostic clinical virology in real time. Clin. Microbiol. Infect. *10*, 5–11.

Nilsson, P., Alexandersson, H., and Ripa, T. (2005). Use of broth enrichment and real-time PCR to exclude the presence of methicillin-resistant Staphylococcus aureus in clinical samples: a sensitive screening approach. Clin. Microbiol. Infect. *11*, 1027–1034.

Orii, T., Ohkohchi, N., Kikuchi, H., Koyamada, N., Chubachi, S., Satomi, S., Kimura, H., Hoshino, Y., and Morita, M. (2000). Usefulness of quantitative real-time polymerase chain reaction in following up patients with Epstein-Barr virus infection after liver transplantation. Clin. Transplant. *14*, 308–317.

Ortu, S., Molicotti, P., Sechi, L.A., Pirina, P., Saba, F., Vertuccio, C., Deriu, A., Maida, I., Mura, M.S., and Zanetti, S. (2006). Rapid detection and identification of Mycobacterium tuberculosis by Real Time PCR and Bactec 960 MIGT. New Microbiol. *29*, 75–80.

Osman, H.K., Peiris, J.S., Taylor, C.E., Warwicker, P., Jarrett, R.F., and Madeley, C.R. (1996). "Cytomegalovirus disease" in renal allograft recipients: is human herpesvirus 7 a co-factor for disease progression? J. Med. Virol. *48*, 295–301.

Parkin, D.M., Pisani, P., and Ferlay, J. (1993). Estimates of the worldwide incidence of eighteen major cancers in 1985. Int. J. Cancer *54*, 594–606.

Pascher, A., Klupp, J., Schulz, R.J., Dignass, A., and Neuhaus, P. (2004). CMV, EBV, HHV6, and HHV7 infections after intestinal transplantation without specific antiviral prophylaxis. Transplant. Proc. *36*, 381–382.

Patel, R. and Paya, C.V. (1997). Infections in solid-organ transplant recipients. Clin. Microbiol. Rev. *10*, 86–124.

Pellet, C., Chevret, S., Frances, C., Euvrard, S., Hurault, M., Legendre, C., Dalac, S., Farge, D., Antoine, C., Hiesse, C., Peraldi, M.N., Lang, P., Samuel, D., Calmus, Y., Agbalika, F., Morel, P., Calvo, F., and Lebbe, C. (2002). Prognostic value of quantitative Kaposi sarcoma—associated herpesvirus load in posttransplantation Kaposi sarcoma. J. Infect. Dis. *186*, 110–113.

Peruski, L.F., Jr. and Peruski, A.H. (2003). Rapid diagnostic assays in the genomic biology era: detection and identification of infectious disease and biological weapon agents. Biotechniques 35, 840–846.

Petitjean, J., Vabret, A., Dina, J., Gouarin, S., and Freymuth, F. (2006). Development and evaluation of a real-time RT-PCR assay on the LightCycler for the rapid detection of enterovirus in cerebrospinal fluid specimens. J. Clin. Virol. 35, 278–284.

Pitcher, D., Chalker, V.J., Sheppard, C., George, R.C., and Harrison, T.G. (2006). Real-time detection of Mycoplasma pneumoniae in respiratory samples with an internal processing control. J. Med. Microbiol. 55, 149–155.

Portolani, M., Tamassia, M.G., Gennari, W., Pecorari, M., Beretti, F., Alu, M., Maiorana, A., and Migaldi, M. (2005). Post-mortem diagnosis of encephalitis in a 75-year-old man associated with human herpesvirus-6 variant, A.J. Med. Virol. 77, 244–248.

Pozo, F., Tenorio, A., de, l.a., M.M., de Ory, F., and Torre-Cisneros, J. (2000). Persistent human herpesvirus 8 viremia before Kaposi's sarcoma development in a liver transplant recipient. Transplantation *70*, 395–397.

Priftakis, P., Bogdanovic, G., Kokhaei, P., Mellstedt, H., and Dalianis, T. (2003). BK virus (BKV) quantification in urine samples of bone marrow transplanted patients is helpful for diagnosis of hemorrhagic cystitis, although wide individual variations exist. J. Clin. Virol. *26*, 71–77.

Rabenau, H.F., Clarici, A.M., Muhlbauer, G., Berger, A., Vince, A., Muller, S., Daghofer, E., Santner, B.I., Marth, E., and Kessler, H.H. (2002). Rapid detection of enterovirus infection by automated RNA extraction and real-time fluorescence PCR. J. Clin. Virol. *25*, 155–164.

Raggam, R.B., Leitner, E., Muhlbauer, G., Berg, J., Stocher, M., Grisold, A.J., Marth, E., and Kessler, H.H. (2002). Qualitative detection of Legionella species in bronchoalveolar lavages and induced sputa by automated DNA extraction and real-time polymerase chain reaction. Med. Microbiol. Immunol. (Berl) *191*, 119–125.

Rand, K., Houck, H., and Lawrence, R. (2005). Real-time polymerase chain reaction detection of herpes simplex virus in cerebrospinal fluid and cost savings from earlier hospital discharge. J. Mol. Diagn. *7*, 511–516.

Randhawa, P., H.o., A., Shapiro, R., Vats, A., Swalsky, P., Finkelstein, S., Uhrmacher, J., and Weck, K. (2004). Correlates of quantitative measurement of BK polyomavirus (BKV) DNA with clinical course of BKV infection in renal transplant patients. J. Clin. Microbiol. *42*, 1176–1180.

Randhawa, P., Uhrmacher, J., Pasculle, W., Vats, A., Shapiro, R., Eghtsead, B., and Weck, K. (2005). A comparative study of BK and JC virus infections in organ transplant recipients. J. Med. Virol. *77*, 238–243.

Randhawa, P.S., Vats, A., Zygmunt, D., Swalsky, P., Scantlebury, V., Shapiro, R., and Finkelstein, S. (2002). Quantitation of viral DNA in renal allograft tissue from patients with BK virus nephropathy. Transplantation *74*, 485–488.

Rayes, N., Seehofer, D., Lullius, S.G., Stein, A., May, G., Kahl, A., Frei, U., Neuhaus, P., and Meisel, H. (2005). Monitoring of human cytomegalovirus, HHV-6 and HHV-7 infection in kidney transplant recipients by molecular methods to predict HCMV disease after transplantation: a prospective study. Ann. Transplant. *10*, 23–28.

Reploeg, M.D., Storch, G.A., and Clifford, D.B. (2001). Bk virus: a clinical review. Clin. Infect. Dis. *33*, 191–202.

Ryncarz, A.J., Goddard, J., Wald, A., Huang, M.L., Roizman, B., and Corey, L. (1999). Development of a high-throughput quantitative assay for detecting herpes simplex virus DNA in clinical samples. J. Clin. Microbiol. *37*, 1941–1947.

Saadeh, S. and Davis, G.L. (2004). The evolving treatment of chronic hepatitis C: where we stand a decade out. Cleve. Clin. J. Med. *71 Suppl 3*, S3-S7.

Saiki, R.K., Bugawan, T.L., Horn, G.T., Mullis, K.B., and Erlich, H.A. (1986). Analysis of enzymatically amplified beta-globin and HLA-DQ alpha DNA with allele-specific oligonucleotide probes. Nature *324*, 163–166.

Saldanha, J., Gerlich, W., Lelie, N., Dawson, P., Heermann, K., and Heath, A. (2001). An international collaborative study to establish a World Health Organization international standard for hepatitis B virus DNA nucleic acid amplification techniques. Vox Sang. *80*, 63–71.

Saldanha, J., Heath, A., Aberham, C., Albrecht, J., Gentili, G., Gessner, M., and Pisani, G. (2005). World Health Organization collaborative study to establish a replacement WHO international standard for hepatitis C virus RNA nucleic acid amplification technology assays. Vox Sang. *88*, 202–204.

Scheltinga, S.A., Templeton, K.E., Beersma, M.F., and Claas, E.C. (2005). Diagnosis of human metapneumovirus and rhinovirus in patients with respiratory tract infections by an internally controlled multiplex real-time RNA PCR. J. Clin. Virol. 33, 306–311.

Schmitt, M., Bravo, I.G., Snijders, P.J., Gissmann, L., Pawlita, M., and Waterboer, T. (2006). Bead-based multiplex genotyping of human papillomaviruses. J. Clin. Microbiol. *44*, 504–512.

Schmutzhard, J., Merete, R.H., Zweygberg, W.B., and Grillner, L. (2004). Detection of herpes simplex virus type 1, herpes simplex virus type 2 and varicella-zoster virus in skin lesions. Comparison of real-time PCR, nested PCR and virus isolation. J. Clin. Virol. *29*, 120–126.

Segal, B.H. and Walsh, T.J. (2006). Current approaches to diagnosis and treatment of invasive aspergillosis. Am. J. Respir. Crit Care Med. *173*, 707–717.

Si-Mohamed, A., Goff, J.L., Desire, N., Maylin, S., Glotz, D., and Belec, L. (2006). Detection and quantitation of BK virus DNA by real-time polymerase chain reaction in the LT-ag gene in adult renal transplant recipients. J. Virol. Methods *131*, 21–27.

Slifkin, M., Doron, S., and Snydman, D.R. (2004). Viral prophylaxis in organ transplant patients. Drugs *64*, 2763–2792.

Sloan, L.M., Rosenblatt, J.E., and Cockerill, F.R., III (2005). Detection of Tropheryma whipplei DNA in clinical specimens by LightCycler real-time PCR. J. Clin. Microbiol. *43*, 3516–3518.

Song, J., Yoshida, A., Yamamoto, Y., Katano, H., Hagihara, K., Oka, S., Kimura, S., and Yoshizaki, K. (2004). Viral load of human herpesvirus 8 (HHV-8) in the circulatory blood cells correlates with clinical progression in a patient with HHV-8-associated solid lymphoma with aids-associated Kaposi's sarcoma. Leuk. Lymphoma *45*, 2343–2347.

Song, L., Ahn, S., and Walt, D.R. (2006). Fiber-optic microsphere-based arrays for multiplexed biological warfare agent detection. Anal. Chem. *78*, 1023–1033.

Soto-Hernandez, J.L. (2004). Human herpesvirus 6 encephalomyelitis. Emerg. Infect. Dis. *10*, 1700–1702.

Spitzack, K.D. and Ugaz, V.M. (2006). Polymerase chain reaction in miniaturized systems: big progress in little devices. Methods Mol. Biol. *321*, 97–129.

Stevenson, J., Hymas, W., and Hillyard, D. (2005). Effect of sequence polymorphisms on performance of two real-time PCR assays for detection of herpes simplex virus. J. Clin. Microbiol. *43*, 2391–2398.

Straka, M., Dela, C.W., Blackmon, C., Johnson, O., Stassen, S., Streitman, D., Golden, S., and Stamilio, D. (2004). Rapid detection of group B streptococcus and Escherichia coli in amniotic fluid using real-time fluorescent PCR. Infect. Dis. Obstet. Gynecol. *12*, 109–114.

Stranska, R., Schuurman, R., de Vos, M., and van Loon, A.M. (2004). Routine use of a highly automated and internally controlled real-time PCR assay for the diagnosis of herpes simplex and varicella-zoster virus infections. J. Clin. Virol. *30*, 39–44.

Sumino, K.C., Agapov, E., Pierce, R.A., Trulock, E.P., Pfeifer, J.D., Ritter, J.H., Gaudreault-Keener, M., Storch, G.A., and Holtzman, M.J. (2005). Detection of severe human metapneumovirus infection by real-time polymerase chain reaction and histopathological assessment. J. Infect. Dis. *192*, 1052–1060.

Suzuki, F., Kumada, H., and Nakamura, H. (2006). Changes in viral loads of lamivudine-resistant mutants and evolution of HBV sequences during adefovir dipivoxil therapy. J. Med. Virol. *78*, 1025–1034.

Szende, B., Toth, A., Perner, F., Nagy, K., and Takacs, K. (1997). Clinicopathologic aspects of 8 Kaposi's sarcomas among 1009 renal transplant patients. Gen. Diagn. Pathol. *143*, 209–213.

Takakura, S., Tsuchiya, S., Isawa, Y., Yasukawa, K., Hayashi, T., Tomita, M., Suzuki, K., Hasegawa, T., Tagami, T., Kurashima, A., and Ichiyama, S. (2005). Rapid detection of Mycobacterium tuberculosis in respiratory samples by transcription-reverse transcription concerted reaction with an automated system. J. Clin. Microbiol. *43*, 5435–5439.

Templeton, K.E., Scheltinga, S.A., Beersma, M.F., Kroes, A.C., and Claas, E.C. (2004). Rapid and sensitive method using multiplex real-time PCR for diagnosis of infections by influenza a and influenza B viruses, respiratory syncytial virus, and parainfluenza viruses 1, 2, 3, and 4. J. Clin. Microbiol. *42*, 1564–1569.

Templeton, K.E., Scheltinga, S.A., van den Eeden, W.C., Graffelman, A.W., van den Broek, P.J., and Claas, E.C. (2005). Improved diagnosis of the etiology of community-acquired pneumonia with real-time polymerase chain reaction. Clin. Infect. Dis. *41*, 345–351.

Templeton, K.E., Scheltinga, S.A., van der, Z.A., Diederen, B.M., van Kruijssen, A.M., Goossens, H., Kuijper, E., and Claas, E.C. (2003). Evaluation of real-time PCR for detection of and discrimination between Bordetella pertussis, Bordetella parapertussis, and Bordetella holmesii for clinical diagnosis. J. Clin. Microbiol. *41*, 4121–4126.

Thomas, D.L. (2006). Options for treatment of hepatitis C in HIV-infected persons. J. Hepatol. *44*, S40-S43.

Uhl, J.R., Vetter, E.A., Boldt, K.L., Johnston, B.W., Ramin, K.D., Adams, M.J., Ferrieri, P., Reischl, U., and Cockerill, F.R., III (2005). Use of the Roche LightCycler Strep B assay for detection of group B Streptococcus from vaginal and rectal swabs. J. Clin. Microbiol. *43*, 4046–4051.

Uzuka, R., Kawashima, H., Hasegawa, D., Ioi, H., Amaha, M., Kashiwagi, Y., Takekuma, K., Hoshika, A., and Chiba, K. (2004). Rapid diagnosis of bacterial meningitis by using multiplex PCR and real time PCR. Pediatr. Int. *46*, 551–554.

van den Brule, A.J., Pol, R., Fransen-Daalmeijer, N., Schouls, L.M., Meijer, C.J., and Snijders, P.J. (2002). GP5+/6+ PCR followed by reverse line blot analysis enables rapid and high-throughput identification of human papillomavirus genotypes. J. Clin. Microbiol. *40*, 779–787.

van Doorn, H.R., Claas, E.C., Templeton, K.E., van der Zanden, A.G., Te, K.V., de Jong, M.D., Dankert, J., and Kuijper, E.J. (2003). Detection of a point mutation associated with high-level isoniazid resistance in Mycobacterium tuberculosis by using real-time PCR technology with 3′-minor groove binder-DNA probes. J. Clin. Microbiol. *41*, 4630–4635.

van Doornum, G.J., Guldemeester, J., Osterhaus, A.D., and Niesters, H.G. (2003). Diagnosing herpesvirus infections by real-time amplification and rapid culture. J. Clin. Microbiol. *41*, 576–580.

van Esser, J.W., Niesters, H.G., Thijsen, S.F., Meijer, E., Osterhaus, A.D., Wolthers, K.C., Boucher, C.A., Gratama, J.W., Budel, L.M., van der, H.B., van Loon, A.M., Lowenberg, B., Verdonck, L.F., and Cornelissen, J.J. (2001). Molecular quantification of viral load in plasma allows for fast and accurate prediction of response to therapy of Epstein-Barr virus-associated lymphoproliferative disease after allogeneic stem cell transplantation. Br. J. Haematol. *113*, 814–821.

van Esser, J.W., Niesters, H.G., van der, H.B., Meijer, E., Osterhaus, A.D., Gratama, J.W., Verdonck, L.F., Lowenberg, B., and Cornelissen, J.J. (2002). Prevention of Epstein-Barr virus-lymphoproliferative disease by molecular monitoring and preemptive rituximab in high-risk patients after allogeneic stem cell transplantation. Blood 99, 4364–4369.

van Sighem, A.I., van de Wiel, M.A., Ghani, A.C., Jambroes, M., Reiss, P., Gyssens, I.C., Brinkman, K., Lange, J.M., and de Wolf, F. (2003). Mortality and progression to AIDS after starting highly active antiretroviral therapy. AIDS *17*, 2227–2236.

van Tol, M.J., Claas, E.C., Heemskerk, B., Veltrop-Duits, L.A., de Brouwer, C.S., van Vreeswijk, T., Sombroek, C.C., Kroes, A.C., Beersma, M.F., de Klerk, E.P., Egeler, R.M., Lankester, A.C., and Schilham, M.W. (2005a). Adenovirus infection in children after allogeneic stem cell transplantation: diagnosis, treatment and immunity. Bone Marrow Transplant. *35 Suppl 1*, S73-S76.

van Tol, M.J., Kroes, A.C., Schinkel, J., Dinkelaar, W., Claas, E.C., Jol-van der Zijde, C.M., and Vossen, J.M. (2005b). Adenovirus infection in paediatric stem cell transplant recipients: increased risk in young children with a delayed immune recovery. Bone Marrow Transplant. *36*, 39–50.

Verweij, J.J., Blange, R.A., Templeton, K., Schinkel, J., Brienen, E.A., van Rooyen, M.A., van Lieshout, L., and Polderman, A.M. (2004). Simultaneous detection of Entamoeba histolytica, Giardia lamblia, and Cryptosporidium parvum in fecal samples by using multiplex real-time PCR. J. Clin. Microbiol. *42*, 1220–1223.

Wagner, H.J., Wessel, M., Jabs, W., Smets, F., Fischer, L., Offner, G., and Bucsky, P. (2001). Patients at risk for development of posttransplant lymphoproliferative disorder: plasma versus peripheral blood mononuclear cells as material for quantification of Epstein-Barr viral load by using real-time quantitative polymerase chain reaction. Transplantation *72*, 1012–1019.

Wagner, H.J., Cheng, Y.C., Huls, M.H., Gee, A.P., Kuehnle, I., Krance, R.A., Brenner, M.K., Rooney, C.M., and Heslop, H.E. (2004). Prompt versus preemptive intervention for EBV lymphoproliferative disease. Blood *103*, 3979–3981.

Walboomers, J.M., Jacobs, M.V., Manos, M.M., Bosch, F.X., Kummer, J.A., Shah, K.V., Snijders, P.J., Peto, J., Meijer, C.J., and Munoz, N. (1999). Human papillomavirus is a necessary cause of invasive cervical cancer worldwide. J. Pathol. *189*, 12–19.

Ward, K.N. (2005). Human herpesviruses –6 and –7 infections. Curr. Opin. Infect. Dis. *18*, 247–252.

Warren, D.K., Liao, R.S., Merz, L.R., Eveland, M., and Dunne, W.M., Jr. (2004). Detection of methicillin-resistant Staphylococcus aureus directly from nasal swab specimens by a real-time PCR assay. J. Clin. Microbiol. *42*, 5578–5581.

Weidmann, M., Meyer-Konig, U., and Hufert, F.T. (2003). Rapid detection of herpes simplex virus and varicella-zoster virus infections by real-time PCR. J. Clin. Microbiol. *41*, 1565–1568.

Welti, M., Jaton, K., Altwegg, M., Sahli, R., Wenger, A., and Bille, J. (2003). Development of a multiplex real-time quantitative PCR assay to detect Chlamydia pneumoniae, Legionella pneumophila and Mycoplasma pneumoniae in respiratory tract secretions. Diagn. Microbiol. Infect. Dis. *45*, 85–95.

Whiley, D.M., Mackay, I.M., Syrmis, M.W., Witt, M.J., and Sloots, T.P. (2004). Detection and differentiation of herpes simplex virus types 1 and 2 by a duplex LightCycler PCR that incorporates an internal control PCR reaction. J. Clin. Virol. *30*, 32–38.

Whiley, D.M. and Sloots, T.P. (2006). Sequence variation can affect the performance of minor groove binder TaqMan probes in viral diagnostic assays. J. Clin. Virol. *35*, 81–83.

White, P.L., Linton, C.J., Perry, M.D., Johnson, E.M., and Barnes, R.A. (2006). The evolution and evaluation of a whole blood polymerase chain reaction assay for the detection of invasive aspergillosis in hematology patients in a routine clinical setting. Clin. Infect. Dis. *42*, 479–486.

White, P.L., Shetty, A., and Barnes, R.A. (2003). Detection of seven Candida species using the Light-Cycler system. J. Med. Microbiol. *52*, 229–238.

Whitley, R.J. (2006). Herpes simplex encephalitis: Adolescents and adults. Antiviral Res.

Wong, J.B., Davis, G.L., McHutchison, J.G., Manns, M.P., and Albrecht, J.K. (2003). Economic and clinical effects of evaluating rapid viral response to peginterferon alfa-2b plus ribavirin for the initial treatment of chronic hepatitis, C. Am. J. Gastroenterol. *98*, 2354–2362.

Wu, C., Randhawa, P., and McCauley, J. (2006). Transplantation: polyomavirus nephropathy and the risk of specific immunosuppression regimens. ScientificWorldJournal. 512–528.

Yamada, M. (2001). Human herpesviruses 6 and 7: effects on hematopoiesis and mode of transmission. Jpn. J. Infect. Dis. *54*, 47–54.

Yang, S. and Rothman, R.E. (2004). PCR-based diagnostics for infectious diseases: uses, limitations, and future applications in acute-care settings. Lancet Infect. Dis. *4*, 337–348.
Zhang, T. and Fang, H.H. (2006). Applications of real-time polymerase chain reaction for quantification of microorganisms in environmental samples. Appl. Microbiol. Biotechnol. *70*, 281–289.
Zoulim, F. (2006). Entecavir: a new treatment option for chronic hepatitis B. J. Clin. Virol. *36*, 8–12.

Challenges Facing Real-time PCR Characterization of Acute Respiratory Tract Infections

8

Ian M. Mackay, Katherine E. Arden, Michael D. Nissen, and Theo P. Sloots

Abstract

The age of reliance upon *in vitro* cell culture for routine laboratory diagnosis of respiratory virus infections has well and truly passed. We are much more comfortable with the application of molecular methods to detect and characterize the most common and frequent infectious agents of humans. The increased acceptance of molecular tools is not at the complete expense of other biological or serological methods; they will always have a place in microbiology, however the era of the high-throughput laboratory has driven the use of faster, more sensitive and more specific methods to diagnose viral pathogens. Unfortunately these methods have some inherent limitations and the use of molecular techniques poses a number of serious problems to overcome, especially apparent in the area of respiratory virus detection and characterization. The scope and diversity of respiratory viruses mean that scientists in this field have to design and evaluate their own assays "in-house" because commercial options are extremely limited.

The question of our ability to reliably detect so many viruses and so much subtle nucleotide variation is at the forefront of assay design and implementation problems, and we must address this aggressively. We are also faced with the extreme difficulty of quantifying respiratory viruses from essentially acellular fluids, secreted from within the host onto its surface (albeit a humid, highly convoluted invagination housed within our body) which we sample using a variety of collection methods. Quantification is further complicated by issues of specimen quality, handling and storage. Recently, the appearance of newly identified viruses, or NIVs, have both challenged and stymied respiratory virus real-time PCR assay designers and will undoubtedly continue to do so into the foreseeable future. We also see that real-time PCR cannot be used in isolation; for maximum success it must be accompanied by nucleotide sequencing, phylogenetic analyses and constant vigilance over the relevant literature. There is also a driving need to delve into the increased co-detection of multiple viral sequences and the ill-defined impact of quasispecies variation on oligonucleotide design. Real-time PCR is a mature technology and an extremely useful one for the study of acute respiratory tract infections yet it is not the perfect tool. In the meantime, we have much work to do in order to make best use of what we have available today.

Introduction

Acute respiratory tract infections (ARTIs) are the most common infection of humans and have been recorded throughout history. ARTIs most frequently occur among children where they cause an average of two to twelve episodes of illness per annum with a 20 to 30% chance of lower respiratory tract infection (LRTI) developing in any particular year (Glezen *et al.*, 1971; Wright *et al.*, 1989; Kusel *et al.*, 2006). Viruses are the most frequently detected microorganism associated with ARTIs and yet the most common, and also some of the first discovered, remain poorly characterized. Rapid diagnosis of infection by "respiratory viruses" can mean the difference between limiting the severity of an illness by rapidly deploying supportive measures in a hospital environment, or having to undertake invasive clinical and laboratory investigations and toxic "shotgun" therapies to manage advanced illness. It is clear that ARTIs are responsible for a sizable drain on health care budgets, both at the level of the individual and the state. The financial burden is recognized not simply in terms of patient and family costs (e.g. prescription and non-prescription medications and general practitioner's visits) but also health care sector costs (e.g. drug and consultation subsidies, specialized hospital equipment purchases and emergency department resources) and other, no less important costs (e.g. time spent away from the carer's usual activities or a reduction in the efficiency of their performance). As any parent of a sick child will know, it is not long before the household's little index case has company. However the parent often does not know the identity of the virus(es) associated with the illness, the likely method of transmission, how long ago the infection was acquired, how often this or a similar virus may infect their children and when a drug to treat the infection or a vaccine to prevent its acquisition will be available. Addressing these questions hinges upon the comprehensive characterization of every virus involved in ARTI and molecular methods play an essential role in this task.

Laboratory diagnostics and retrospective and prospective epidemiology studies follow characterization, aided by a mix of other molecular, serological and *in vitro* cell culture techniques. If a thoroughly characterized virus attracts enough attention due to its high prevalence, the severity of its associated illness or the rapidity of its emergence and spread, it may become the focus for antiviral or vaccine development projects. But everything relies upon the accurate initial detection of the virus(es).

Most of the viruses present in patients with ARTI have an RNA genome, e.g. the human coronaviruses (HCoVs; 229E, OC43, NL63 and HKU1), picornaviruses (rhinoviruses and enteroviruses) and paramyxoviruses (human respiratory syncytial virus, human metapneumovirus, measles virus, and parainfluenza viruses). Most prominent are more than 100 antigenically distinct human rhinoviruses (HRVs). The respiratory viruses have been assigned pathologic characteristics in ARTI based on the diagnostic techniques of the day and these roles have since become dogma; "HRVs and HCoVs cause upper respiratory tract infection (URTI; e.g. the common cold)," "human respiratory syncytial virus (HRSV) is the most common cause of more severe and costly LRTIs resulting in illness," particularly bronchiolitis, but also including wheezing, croup, bronchitis and pneumonia; influenzavirus type A (IFAV), influenzavirus type B (IFBV) and many other viruses induce a clinical syndrome—"influenza-like illness"—basically described as fever, cough and fatigue. Culture and immunofluorescence methods are seemingly still used by some as the front line diagnostic approach for studies of acute respiratory illness, supplemented by PCR assays in an apparently *ad hoc* fashion (Garbino *et al.*, 2004b; Chiu *et al.*, 2005; Scheltinga *et al.*, 2005; Sumino *et al.*, 2005). In practice such a spectrum of testing may simply indicate that additional investigations for a new pathogen employing the most modern techniques available at the time made use of a previously tested, well-defined frozen specimen population. But there are also those studies which have not looked for one or more important respiratory viruses, weakening associations made between the targeted virus and any illness in the study population (Heim *et al.*, 2003; Oosterheert *et al.*, 2005; van der Hoek *et al.*, 2005; Bouscambert-Duchamp *et al.*, 2005; Agapov *et al.*, 2006). Occasionally studies focus on the function or novelty of the particular real-time PCR assay, which is laudable, unless they also extrapolate from the data an association between viral load and signs and symptoms of illness for a specimen population that has been incompletely screened for all likely microorganisms (Heim *et al.*, 2003; Hamelin *et al.*, 2005).

It would be worthwhile re-examining partially tested specimen populations with modern molecular methods to update these data for all other possible respiratory viruses, no matter how innocuous we may currently perceive them to be. If such testing is not performed either in retrospect or prospectively, it will forever remain difficult to apply an overall clinical severity ranking to respiratory viruses since their prevalence of infection, seasonal distribution, spectrum of associated symptoms and thus the precise role of the virus in disease will not have been accurately defined.

The literature is clear in defining classical diagnostic assays as less sensitive than the current molecular tests and this can be further seen in Table 8.1 where real-time PCR assays can be up to 100-fold more sensitive than *in vitro* culture methods (Garbino *et al.*, 2004a; Templeton *et al.*, 2005; Gunson *et al.*, 2005; van Kraaij *et al.*, 2005). As the frequency of these assertions increases, it is now time to address the question of what impact additional detections have on patient outcome, i.e. to re-visit our definition of clinical relevance for all respiratory viruses in light of the increasing rate of laboratory diagnoses using PCR technologies.

In this chapter we aim to list and discuss those troublesome issues specific to the development and application of real-time PCR methods to detect viruses associated with ARTI, whether their role as pathogens is clearly demonstrated or not. Most of these assays are developed "in-house" since the vast number of respiratory viruses, the assignment of many viruses to a category of mild disease-causing potential, and the rate at which new viruses are being found has limited the development of comprehensive commercial diagnostic solutions. The mission of the current generation of researchers and assay developers should be to create and employ new tools to enhance our ability to diagnose all respiratory viral infections, without preconception, and to better understand the impact of each virus on clinical illness. If we do not update our knowledge and our diagnostic capabilities, we may be forced to play catch-up after the discovery or emergence of more life-threatening respiratory viruses.

Pedantry and PCR

In every speciality there are those who like to ensure the correct use of language in order to convey the proper message. Some reject such focused, academic correctness as unnecessary. We believe that real-time PCR for respiratory virus study engenders a number

Table 8.1 Real-time PCR assays used to investigate suspected cases of ARTI

Labels	Fluorogenic chemistry	cDNA priming	Reference	Internal Control	Sensitivity	Targets
FAM/TAMRA—Virus VIC/TAMRA—IC	TM	RH	(van Kraaij *et al.*, 2005)	EMCV (RNA) and PHV (DNA)	3x more sensitive than culture	PV (Nijhuis *et al.*, 2002), HRSV A and B (van Elden *et al.*, 2003), HCoV-229E and OC43 (van Elden *et al.*, 2004), HAdV, IFAV, IFBV, HPIV 1–4
FAM/TAMRA	TM	RH	(van Elden *et al.*, 2004)	None		229E and OC43
	TM	RH	(Garbino *et al.*, 2004a)	None	0.3–10^{-3} $TCID_{50}$ 2copies/μL—HMPV 1copie/μL (NL63). 3x more sensitive than culture	HRV, HEV, HRSV A and B, HCoV-229E, HCoV-NL63 and OC43, HAdV, IFAV, IFBV, HPIV 1 and 3, HMPV
FAM/NFQ	TM-MGB	OS	(Vijgen *et al.*, 2005)	None	10^1–10^2 copies/rxn	229E, OC43
	SYBR®	RH	(Lau *et al.*, 2005b)	None, FMCA	10^2 copies/mL	Bat-CoV
	SYBR	RH	(Poon *et al.*, 2003)	None, FMCA	10^1 copies/rxn	SARS-CoV
	SYBR/TM	Yes	(Shimizu *et al.*, 2005)	Cellular target	10^0–10^2 copies/rxn	NL63
FAM/TAMRA	TM	OS	(Bouscambert-Duchamp *et al.*, 2005; Jacques *et al.*, 2006) Derived from my assay.	GAPDH-Applied Biosystems kit	0.01 $TCID_{50}$	HMPV
	SYBR	Yes, oligo-$(dT)_{12-18}$	(Dagher *et al.*, 2004)	None, FMCA	10^3 infectious particles	HRV

FAM/BHQ CY5/BHQ VIC/TAMRA	4 × 3, TM	OS	(Gunson *et al*., 2005)	None	More sensitive than conventional or nPCR, also assessed by quality control molecular diagnostics (QCMD) panels	IFAV, IFBV, HMPV, HRSVA, HRSVB, HRV, HPIV-1, HPIV-2, HPIV-3, 229E, OC43, NL63
FL/LC RED 640	SYBR PRB + PRIMER	Yes, GSP	(Kares *et al*., 2004)	None, FMCA	10^{-5}–10^{-6}ng/rxn	HEV/HRV
FAM/VIC/BHQ/TAMRA	TM	RH	(Lee *et al*., 2006)	None	ND	NL63, 229E, OC43, HMPV
	TM	OS	(Chang *et al*., 2006)	NL63-positive spike into NL63-NEG specimens	$\leq 10^1$ copies IVT RNA/rxn	NL63
FAM/MGB/NFQ FAM/TAMRA	TM-MGB	RH	(Chiu *et al*., 2005) from (Fouchier *et al*., 2004)	None	ND	NL63, 229E, OC43
	SYBR	OS	(van der Hoek *et al*., 2005) from (van der Hoek *et al*., 2004)	None, FMCA	10^0/rxn	NL63
FAM/BHQ1,HEX/BHQ1 YAK/BHQ1 and FAM/BHQ1	TM,MB TM	OS	(Templeton *et al*., 2005) from (Scheltinga *et al*., 2005)	PHV, EAV	0.01 $TCID_{50}$ HMPV/HRV	HRV/HMPV OC43/229E
UNK	UNK	RH	(van de Pol *et al*., 2006) using (van Elden *et al*., 2004)	EMCV (RNA), PHV (DNA)	Increased diagnostic yield by 25%	HRV/229E/NL63/OC43
FAM/NFQ	TM-MGB	OS	(Keyaerts *et al*., 2006)	None	10^2 copies/rxn	SARS-CoV

Table 8.1 continued

Labels	Fluorogenic chemistry	cDNA priming	Reference	Internal Control	Sensitivity	Targets
	SYBR	RH	(Ma *et al*., 2005)	None	ND	HMPV
FAM/TAMRA	TM		(Yamamoto *et al*., 2005), (Tanaka *et al*., 2000)	β-actin	10^0/rxn	HHV-7/HHV-6
FAM/TAMRA	TM	PCR	(Heim *et al*., 2003)	None	10^{0-1}/rxn	HAdV
HEX/BHQ1, FAM/BHQ1 CY5/BHQ2	MB/TM MB	OS	(Scheltinga *et al*., 2005)	EAV	0.01 $TCID_{50}$ each	HMPV/HRV EAV
	SYBR	Yes, GSP	(Côté *et al*., 2003)	None	10^2 copies/rxn	HMPV
FAM/TAMRA	TM	RH	(Deffrasnes *et al*., 2005) L almost identical to 2403	None	10^2 copies/rxn	HMPV
FAM/TAMRA	TM	OS	(Mackay *et al*., 2003b)	None	-	HMPV
FAM/TAMRA	TM	Yes, GSP	(Maertzdorf *et al*., 2004)	None	5–10 copies/rxn or $\leq$ 0.01 $TCID_{50}$	HMPV
	TM	OS	(Sumino *et al*., 2005)	None	10^1 copies/rxn	HMPV A-specific

LC FL/RED640	HybProbe and SYBR	Yes, GSP	(Kares *et al*., 2004)	None	10^{-6} to 10^{-5} ng/rxn	HEV PV
FAM/TAMRA	TM	RH	(Oosterheert *et al*., 2005)	Virus	-	IFAV, IFBV, HRSV, OC43, 229E, HPIV1–4, HAdV, MP, CP, LP
FAM/TAMRA	TM	OS, GSP	(Agapov *et al*., 2006)	None	5 copies/rxn	HMPV
FAM/TAMRA	TM	RH	(Carrat *et al*., 2006) using (Maertzdorf *et al*., 2004)	None	-	HMPV, IFAV and other bugs
	SYBR	RH	(Hamelin *et al*., 2005) similar to (Deffrasnes *et al*., 2005)	None	250copies/rxn	HMPV (L, confirmed by N)
FAM/TAMRA VIC/TAMRA	TM	RH	(Corless *et al*., 2002)	GAPDH	100x more sensitive than culture	HEV and HPeV
	SYBR	RH	(Boivin *et al*., 2004)	HSV-2 DNA pol	10–50copies/rxn	IFAV, IFBV, HRSV
FAM/TAMRA and FAM/ MGB+NFQ	TM and TM-MGB	OS, GSP	(Kuypers *et al*., 2006)	EXO	10 copies/rxn	IFAV, HRSV, HMPV, HPIV-1, HPIV-2, HPIV-3
FAM/BHQ2	TM	OS, GSP	(Lu *et al*., 2006)	None	10 copies/reaction	HBoV

TM—TaqMan® oligoprobe; TM-MGB—TaqMan-minor groove binding moiety; CP—*Chlamydia pneumoniae*; EAV—Equine arteritis virus; EMCV—encephalomyocarditis virus; EXO—Jellyfish sequence (Limaye *et al*., 2001); HCoV—human coronavirus; HHV—human herpesvirus; HMPV—Human metapneumovirus; IFAV—Influenzavirus type A; IFBV—influenzavirus type B; LP—*Legionella pneumophila*; MP—*Mycoplasma pneumoniae*; ^PHV—phocine herpesvirus; PV—Picornaviruses (human enteroviruses, HEV and human rhinoviruses, HRV); GSP—gene specific primers; IC—internal control; OS—One step RT-PCR; RH—random hexamers.

of situations where clarity is very important, especially because the assays generate data that are interpreted by both the scientific and clinical communities; terminology must be identifiable for each group. Therefore to aid the process we will touch on some terms that can be a cause for dispute when preparing or reviewing manuscripts.

As a result of the PCR process we produce a single amplified species representing the target nucleic acid sequence which we call the "amplicon." We have no way of determining the replicative status of the virus from the production of an amplicon. We cannot even say with certainty that the template was detected from an intact virion; the target may have originated from intracellular viral fragments phagocytosed into a host cell, or amplified from nucleocapsid protein-associated free RNA. In short we cannot say from a positive PCR result that a patient is currently "infected" (invaded by and hosting the replication of a virus; without regard to the host's clinical status) by the respiratory virus of interest. What we may state is that the viral RNA or DNA has been "detected" in the patient's specimen. Similarly, if we have multiple PCR positives from the one patient specimen, we cannot say that the patient has a "co-infection," rather the results indicate "co-detection" of the viral RNA or DNA templates. We can also say little about the clinical impact of that detection when it occurs from a patient with illness that manifests at a different anatomical site. For example, nasopharyngeal aspirates and swabs are specimens frequently used for respiratory virus screening, but a positive in one of these upper respiratory tract (URT) specimens collected from a patient with a clinical diagnosis of bronchiolitis or some other illness associated with the lower respiratory tract (LRT; Figure 8.1), may not alone be used to define an association between bronchiolitis and the target virus. Such an association may be suggested but until the virus is found to replicate in that particular tissue, preferably causing some sort of disruption of the tissue that would attract the attention of a local immune response, no stronger link can be made. It is important to discriminate correctly since it is possible that an immune reaction triggered by viral infection in the URT could indirectly affect the LRT, eliciting signs and symptoms of infection in the absence of acute viral replication in that tissue (Vile *et al.*, 1995).

A confusing if not exasperating habit in the real-time PCR literature is to define two pairs of primers used in the same reaction vessel, or even one pair with two oligoprobes, as a "multiplex PCR assay." While strictly true, it is not in keeping with the spirit of the nomenclature established by conventional multiplex PCR assays which earned the multiplex title by detecting and discriminating six to eight microbial targets from a single reaction (Quereda *et al.*, 2000; Coiras *et al.*, 2003; Mackay *et al.*, 2003a; Syrmis *et al.*, 2004; Coiras *et al.*, 2004; Mackay *et al.*, 2006b). Further detracting from the term is the fact that some multiplex real-time PCR assays only detect one target microorganism and an unrelated internal control. We propose and support the use of "duplex" to describe such assays, with three or more microbial targets truly deserving of the multiplex real-time PCR assay label. The distinction is not a trivial one; there have been few truly multiplexed assays reported for respiratory viral diagnostics, and seriously addressing this deficit is long overdue.

The next issue concerns the large number of "quantitative" real-time PCR assays that are performed in the absence of an internal control or without a method to account for inter-tube fluorescence variation due to pipetting errors, reagent batch changes during assay setup or non-specific fluorescence quenchers present in the purified specimen. Because of these short-comings, such assays should really be called semi-quantitative because un-

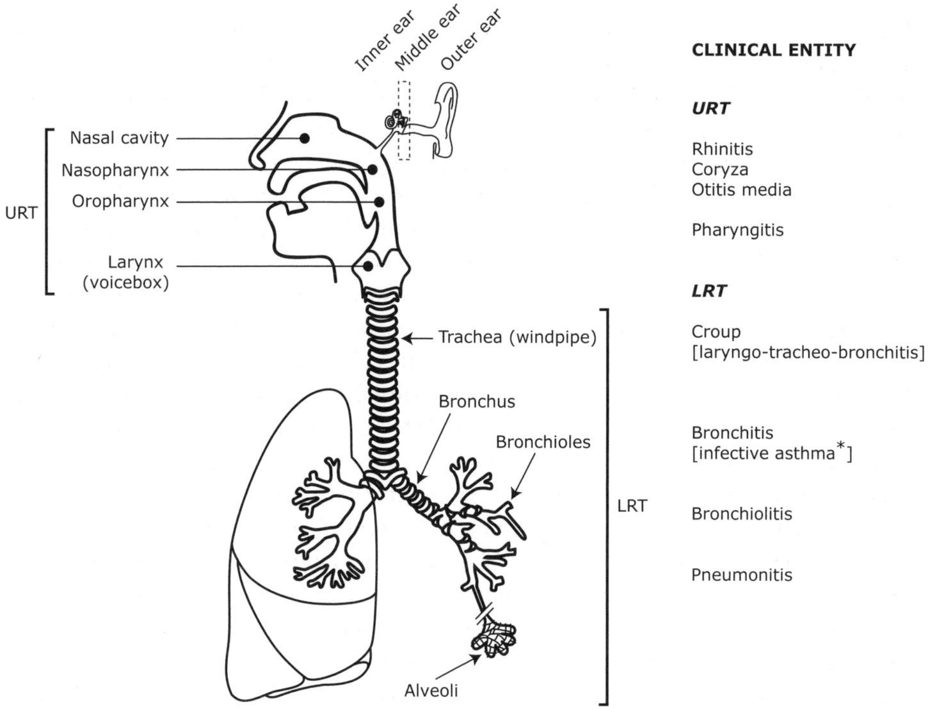

Figure 8.1 A schematic representation of the human respiratory tract. The upper and lower respiratory tract (URT/LRT) and the components of the ear are indicated as are major anatomical sites of interest. Beside the schematic are the approximate locations of upper and lower respiratory tract diseases associated with infection by respiratory viruses. *Recurrent attacks of shortness of breath and wheezing due to spasmodic contraction of the bronchi attributed to infection.

controlled real-time PCR is not an innately quantitative technique. The truly quantitative assays are well-controlled with standards chosen to reflect the target organism. As we will discuss later, quantifying respiratory virus templates may not be possible in any case so this is a contentious point, nonetheless it should be made clear from the outset.

Phylogeny is increasingly being used to subdivide respiratory viruses into groups which may circulate differently or have unique disease-causing potential. However the terminology to describe these groupings is often confusing. Over recent years we have tried to standardize the terms we use in publications as follows. A virus "type" and "subtype" are so called because there is some phenotypic evidence of difference such as serologically confirmed antigenic distinctiveness or a difference in cell tropism or *in vitro* cultivation efficiency. A nucleotide sequence-derived distinctiveness can better be characterized using the terms "lineage" and "sub-lineage" to suggest unproven groupings that are visualized by molecular epidemiology studies employing phylogenetic trees. We are also defining a viral "strain" for use in this chapter, as that instance of a virus represented by consensus sequence(s) obtained from a single patient specimen or culture isolate. As we will discuss later, a strain may itself represent a swarm of subtlety different quasispecies. Multiple strains can be grouped into sub-lineages and lineages which form the viral species. Of course, one should always remember that lineages, sub-lineages, strains, branching patterns and even viral species are only artefactual aids to satisfy our human need to impose classification schemes. The virus is quite unaware of its place in the world—and of everything else.

Respiratory virus discovery: new identification of old viruses

Since the turn of the recent century, human respiratory virus discovery has undergone a renaissance with new identifications among the families *Paramyxoviridae*, *Coronaviridae*, *Polyomaviridae* and *Parvoviridae* revitalizing an area of research at its most vital more than half a century ago. Additional studies of many of these newly described viruses have identified them in the majority of geographic locations they have been sought, often demonstrating discrete circulation patterns which have surely been recurring for a long time.

Most of the recent respiratory viral discoveries do not meet the extensive criteria defined by Morse and others to describe an "emerging" virus and therefore in a general sense they remain untitled (Murphy and Nathanson, 1994; Morse, 1995; Morse, 1997). Specifically, the viruses have not arisen from a breakdown in public health measures nor as a result of high risk, but age-old practices (such as exotic animal markets); they have not appeared due to changes in technology (apart from that used for viral diagnosis); they have not been generated by recent mutation, recombination or reassortment, nor because of ecologic or environmental reasons. Population movement and changes in density do not explain the appearance of these viruses and they have not appeared from obscurity into the human population because of recent transmission and adaptation of an animal disease from its natural host to a human host (a zoonotic event; although they may have been "remote" zoonoses (Twiselton-Wykeham-Fiennes, 1978)). The significance of a link to a zoonotic event is that most emerging viruses are zoonotic pathogens (Taylor *et al.*, 2001; Woolhouse and Gowtage-Sequeria, 2005). Since many of the circumstances which lead to zoonoses are increasing, emerging viruses will also continue to appear and play a significant role in human infection, creating more challenges for real-time PCR design (Morse, 2004). Nevertheless, unlike the emerging viruses, interest in recently identified viruses was not sparked by a sudden increase in local or global prevalence, nor was it due to the release of bioterror agents; in fact indications suggest that these viruses are endemic within the human population (Taylor *et al.*, 2001). Because of these many differences, recently identified respiratory viruses deserve a more discerning title which better describes their endemic nature and implies a method of discovery, while also discriminating them from the truly emerging respiratory viruses. So how should we define these viruses and is there sufficient need to add new terminology to a field scattered with conjoined words and three and four letter acronyms? Most certainly there is a need and we propose that such viruses should be grouped into a category called "newly identified viruses" (NIVs), a description which also hints at the dependence on modern, non-classical techniques to identify and characterize them while avoiding any suggestion of recent zoonotic involvement. Some viruses previously called "emerging" could easily be re-classified as NIVs, including human herpesvirus (HHV)-6 and 7 and parvovirus B19 (Morse, 1997). How long a virus should be labeled a NIV is probably the topic for another book but new terminology may be useful to avoid incorrect use of the emerging virus label as occurs now (Su, 2004; Williams, 2005). The distinction would also lend greater weight to those viruses that truly are emerging and which often cause acute, clinically distinctive and worrying disease outbreaks. A new definition would also better describe the renewed interest in searching for undiscovered endemic respiratory viruses.

Many NIVs associated with the respiratory tract (respiratory NIVs) have attracted the attention of scientists from all around the world and none more so than the human metapneumovirus (HMPV), which rocketed from an unknown pathogen until 2001, to an acknowledged major cause of ARTIs, in particular of LRTI, among children (van den Hoogen *et al.*, 2001; Fouchier *et al.*, 2005). PCR has played the pre-eminent role in studies of HMPV and other NIVs including HCoV-NL63, HCoV-HKU1 and the human bocavirus (HBoV) mostly because NIVs cannot be simply cultivated but also because it is a more objective, specific and less specialized technique than the isolation of respiratory viruses using *in vitro* cell culture, making it more accessible to a greater number of diagnostic laboratories. The explosion in reports following the first description of each NIV in recent times can be in part attributed to wider access to, and an increased understanding of, real-time PCR technologies. PCR is frequently employed to determine the prevalence of NIVs in study populations (where the homogeneous nature of real-time PCR is especially appreciated to avoid contamination that may falsely raise the apparent prevalence), for nucleotide sequencing and to amplify genes and gene fragments for cloning, reverse engineering and for individual gene expression studies. Sequencing has proven particularly useful to describe genetic relationships between respiratory NIV strains and other viruses through the use of phylogeny and is now an essential co-factor for successful design and implementation of real-time PCR assays.

Clinically, the illnesses associated with respiratory NIVs do not usually produce reliably novel or distinctive signs and symptoms and they cannot yet be treated with specific antiviral therapies. However preliminary characterization would suggest the respiratory NIVs are often associated with LRTI as well as URTI, indicating that they are pathogens to be taken seriously rather than unimportant oversights, thus suitably rapid, sensitive and specific detection methods should be sought.

Identifying past mistakes and future challenges for respiratory virus real-time PCR

Until relatively recently the diagnostic laboratory's method of choice for detection of respiratory viruses was incubation of patient respiratory secretions (commonly obtained from nasopharyngeal aspirates, nose and/or throat swabs) with permissible primary cells or cell lines. This *in vitro* process of virus isolation by cell culture was instrumental to respiratory virology in the halcyon days of the 1950s and 1960s. While it appears from the literature that this may still be the favored method of some laboratories publishing diagnostic and research studies, molecular techniques have in fact become the method of choice for the laboratories able to upgrade their facilities and expertise, and those in the business of rapid result turnaround. One early and obvious consequence of the move away from cell culture is the dwindling of specialized classical diagnostic virology sites and the accompanying loss of the expertise that comes with such lengthy and labor-intensive techniques.

To date, serological methods have proven the most diversely commercialized and robust diagnostic format, either for the indirect detection of a host response to a respiratory virus, or the detection of viral antigen in culture or from infected cells in specimens such as nasopharyngeal aspirates or bronchoalveolar lavage. Serological results augment both general diagnostic molecular data and those data provided by research studies aiming to better characterize NIVs or other respiratory viruses. Apart from speed, cost–benefit and familiarity, an obvious advantage derived from use of a protein-based system is the existence of conserved antigenic regions among related viruses i.e. regions that do not vary significantly among strains of the same species or taxonomic grouping. Such antigenic conservation is infrequently reflected at the nucleotide level making these regions troublesome targets for nucleic acid-based systems. Unfortunately antigenic conservation can also be represented as cross-reaction; difficulty discriminating between infections caused by closely related but distinct viruses. Such discrimination is a desirable outcome when searching for the role of each individual respiratory virus in illness (Relman, 2003).

Antigen detection methods can be performed with or without a biological amplification step such as *in vitro* cell culture. If culture is not being employed then it is necessary to collect cellular specimens since the cells confine virions to a small, easily identified space that aids immunofluorescent detection; but such cellular specimens are not common to all clinical settings. Unfortunately rapid respiratory virus antigen detection is relatively insensitive and one may need to confirm negatives using another assay, which abrogates any benefit of speed for those patients still suspected of harboring a respiratory viral infection (Falsey and Walsh, 2006; Kuypers *et al.*, 2006). In addition to detection, serological methods have traditionally been used to classify respiratory viruses into groups or "serotypes" either by targeting antigenic markers using characterized antibodies or by investigating whether viral infection can be neutralized. The latter examines whether binding of a characterized antibody can preclude cellular entry and replication of a virus. PCR methods can also confirm serotypes at the genetic level but with increased objectivity and speed compared to the complex and lengthy neutralization methods (Oberste *et al.*, 1999; Ledford *et al.*, 2004).

Where have all the multiplexes gone?

Multiplex PCR would seem to be an ideal tool for the detection of respiratory viruses and yet relatively few applications have been reported for real-time PCR. When employed, PCR has permitted the rapid and specific discrimination of multiple amplicons representing different viral genes or strains, resulting in an increasing number of reports of microbial co-detections among 25% or more of study specimens (Brouard *et al.*, 2000; Juvén *et al.*, 2000; Papadopoulos *et al.*, 2002; Maggi *et al.*, 2003; Peltola *et al.*, 2006). This is not surprising considering it is often extremely difficult to achieve even single virus isolations from clinical specimens using *in vitro* cell culture methods, let alone discriminate co-infection of a cell-line. Mixed infections have been reported to play important roles in some disease states, as has been shown for community-acquired pneumonia where the sensitivity of real-time PCR could more readily detect multiple infections among 105 adults than attempted isolation of the same viruses from throat swab and wash specimens using *in vitro* cell culture (Templeton *et al.*, 2005). However real-time PCR has not always found multiple detections to be associated with discernibly different clinical symptoms as was reported from a study of 23 children under five years of age with suspected LRTI who provided NPA and sputa specimens (van de Pol *et al.*, 2006).

When expertly rolled out in multiplex format, real-time PCR has the potential to be a formidable screening tool for viruses in ARTI (Gunson *et al.*, 2005). Unfortunately very few comprehensive and truly multiplexed real-time PCR assays have been published. In one report HMPV, HRV and an internal control could be co-amplified without impacting on assay sensitivity (Scheltinga *et al.*, 2005). Inhibition of amplification was detected in 8 of 358 (2%) mostly nasal swab and throat wash specimen extracts provided by patients with ARTI. Technical issues seem mostly to blame for so few respiratory multiplexes and include the time investment required to optimize a multiplex system, the interference caused by overlapping fluorescence emission spectra and, most importantly, the impact on assay sensitivity caused by competition between multiple templates. Anecdotally, multiplex capability is often a deciding factor when purchasing a real-time instrument; but the literature would suggest this pre-purchase requirement is rarely utilized.

Relevant issues for clinical relevance

Characterization of the causative agent of an infectious disease is the foundation upon which treatment and control strategies are built. Possibly the biggest obstacle to widespread implementation of real-time PCR for respiratory virus screening is the unknown clinical relevance of some positive respiratory virus PCR results. Koch's postulates, developed during his studies of anthrax and tuberculosis, were established to guide the identification of a specific microbial cause for infectious disease, and were additionally employed to convince sceptics that microorganisms were capable of causing such disease (Falkow, 2004). Not surprisingly, after more than a century the postulates are difficult to apply when seeking to define relationships between molecularly detected respiratory viruses and clinical illness in the host. However, as we're reminded in an expert review by Fredricks and Relman, strict adherence was never Koch's intention; rather the postulates were meant to be a prompt for discussion of the best way to implement the technology of the day to identify and define an association between microorganism and illness (Fredricks and Relman, 1996). Koch's original postulates required that the microorganism could be found in every case of the disease but not in non-diseased tissues, thus excluding a fortuitous or non-pathogenic role. The postulates also required that a microorganism isolated and purified from a diseased host be capable of inducing that disease upon infection of a healthy body. However a number of viruses are not described by these circumstances, including many respiratory viruses. Infectious clones may address this issue in the future. In the mean-time, if the virus cannot be isolated, it cannot be purified and used to infect another subject. Furthermore, if the virus in question causes severe respiratory illness in certain populations, e.g. the very young or the immunocompromised, it would be ethically unconscionable to infect subjects and replicate such disorders.

The HRVs are a particularly confounding group for molecular diagnostics and when applying Koch's postulates, since they are reportedly detected by PCR in the absence of symptoms (Pitkäranta *et al.*, 2005). However, Nokso-Koivisto *et al.* (2002) presented a patient follow-up study that demonstrated symptoms within three to six weeks of detection (Turner, 1998; Jartti *et al.*, 2004a). For the HRVs and perhaps other infectious agents, PCR data may prove to be a more direct indicator of infection in the absence of illness, than symptoms. In this context, symptoms are the likely result of an immune response to infection and so only indirectly identify infection; their absence does not preclude the possibility of viral replication while their presence does not assure replication has continued after an initial stimulation of the immune system. Real-time PCR will play an important role in identifying all types of illness but to determine clinical relevance, PCR should be performed alongside assays that quantify the subject's immune responses to determine the role of host immunity in moderating viral replication post-infection and in producing symptoms.

Well-controlled studies are clearly required to address the issue of relevance, particularly since other reports suggest that the presence of viral nucleic acid is uncommon among normal populations. HCoV RNA has been detected for up to 14 days after symptomatic infection but no detections could be made from an asymptomatic control group (van Elden *et al.*, 2004). Similarly, respiratory viruses were only detected from 4% of the bronchoalveolar lavage specimens of asymptomatic controls using real-time PCR but were detected in a significantly higher 55% of symptomatic lung transplant patients (Garbino *et al.*, 2004a). Furthermore, in the latter study detection of a virus among this hospitalized and immunocompromised population was often associated with a worsening of pulmonary function, lasting for months afterward (Nokso-Koivisto *et al.*, 2002). Without knowing the modifying impact on clinical illness or the degree of cross-protection provided by pre-existing immunity to HRV species among older children and adults, it is difficult to concede that sub-clinical carriage is a proven feature of the HRVs or for the HCoVs. Largely because of molecular data on shedding, it has been proposed that picornavirus PCR is too sensitive to be a clinically relevant tool; by comparison, successful isolation of HRV by *in vitro* culture is commonly associated with symptomatic illness (Gwaltney Jr, 1982). Unfortunately no studies have quantified individual viral loads in HRV-shedding patients using real-time PCR and such simple studies would go a long way to address the issue. High viral load in asymptomatic patients would be extremely surprising whereas low viral load in immunocompetent patients with subclinical illness would be less remarkable.

The divided opinion on the role of some respiratory viruses in ARTIs is also affected by the types of populations that have been sampled, the different time-points in an illness at which sampling has been performed and, importantly, the nature of the screening assays employed. Again the HRVs serve as a good example since their isolation came at a time when the hunt for common cold agents was progressing rapidly. Today, HRVs are still mostly considered to be agents of relatively inconsequential human illness, largely because previous study results have framed our understanding of the etiology of their infections. It should be acknowledged that older children are sometimes included in the control groups for molecular studies of viral shedding and again the presence of pre-existing immunity plays an unquantified role in moderating the signs and symptoms of infection (Jartti *et al.*, 2004a; Pitkäranta *et al.*, 2005). Interestingly, our group and others using PCR have described HRVs as the sole microorganism detected in specimens from patients with suspected LRT involvement (Papadopoulos *et al.*, 2002; Hayden, 2004; Arden *et al.*, 2006; Kusel *et al.*, 2006). Far from comprehensive, the use of populations that most likely had established immunity to at least some HRVs might have led to a significant but long-held under-estimation of the role of HRVs in serious illnesses located in the URT and LRT. It is noteworthy that even HRSV, a long-known cause of LRTI and associated illnesses, has also been reported in patients with upper respiratory tract infection, yet we do not doubt the importance of HRSV as a significant pathogen (van Benten *et al.*, 2003). It would be most interesting to re-visit the role of HRVs among very young immunocompetent children who experience infection for the first time. Hopefully in time the same importance will be attributed to the HRVs, classical HCoVs and NIVs. The role of real-time PCR will be an important one for completing the necessary studies to address this issue and applications should be considered a priority for the most frequently identified microorganisms in ARTIs.

Another virus that escapes definition by Koch's postulates is the most recently described coronavirus, HCoV-HKU1, which has not yet been isolated by *in vitro* cell culture (Woo *et al.*, 2005a; Woo *et al.*, 2005b; Vabret *et al.*, 2006; Lau *et al.*, 2006; Sloots *et al.*, 2006; Arden *et al.*, 2006). Koch's postulates are also difficult to apply when faced with PCR detection of two, three or four microbial targets in the one patient specimen. If these detections are in fact all from active infections then it seems likely that there will be some interplay between the organisms, if not a combined effect that impacts on the host. Perhaps in the near future we will find that the extent of co-infection dictates the severity of respiratory illness rather than the role of individual viruses as is the commonly held belief now.

Fredricks and Relman proposed additional thoughts and guidelines for the establishment of causal relationships between microorganisms and disease (Fredricks and Relman, 1996). Using these as a core, we have drafted our own guidelines that may better apply to respiratory NIV infections investigated using real-time PCR technology. These, as with other variants of Koch's postulates, are meant to aid our struggle to understand what role, in particular, respiratory viruses play in clinical illness based on our current, potentially limited understanding and how we employ the diagnostic tools we have available.

1. In the majority of cases, viral sequences of interest should be found in clinically ill subjects and not in clinically well subjects. The definition of illness should be broad enough to encompass low-level symptoms including headaches and general malaise.
2. A positive respiratory virus PCR result should be repeatable using at least one additional assay targeting a different region of the viral genome.
3. Viral sequences should be sought in tissues other than those appropriate to their currently understood biological characteristics because the complete biology of a virus may not be known.
4. At the least, amplicon sequences should be sought to permit comparison with similar viruses as an aid for taxonomic placement. A complete genome sequence should be attempted.
5. Viral load in a particular patient at a particular time should usually be proportional to the degree of clinical illness. Exceptions include immune over-reaction to infection which manifests as continuing signs and symptoms, even after viral load has begun to decline.
6. A single positive PCR result in an asymptomatic patient should not by itself be taken as an indication of asymptomatic infection. Complete patient history should be sought and follow-up testing undertaken.
7. A causal relationship may be more likely with the detection of microbial sequences in an asymptomatic patient who goes on to develop respiratory illness.

8 If multiple viruses are detected in a single specimen, additional specimens should be collected for testing. Quantitative PCR should also be employed to determine the most actively replicating virus(es). Additional cases of both single and multiple infections should be sought for comparative study of clinical outcomes.

9 Nucleotide sequences should be sought at the cellular level by demonstrating specific *in situ* hybridization of microbial sequence to areas of tissue pathology and to visible microorganisms or to areas where microorganisms are presumed to be located (adapted from Fredricks and Relman, 1996).

10 All attempts to isolate the molecularly detected virus in culture should be made. Successful isolation of the virus can be identified using quantitative PCR to demonstrate at least a two $\log_{10}$ increase (≥6-fold decrease in C_T value) in template concentration compared to the value at day 0, post-infection. Viral load should increase with increased cytopathic effects.

11 All other known respiratory viruses should be excluded, or their role in ARTI accounted for, prior to making an association between a particular respiratory NIV and clinical illness

Quantification of respiratory viruses: what's the hold-up?

The application of real-time PCR technology has been relatively slow to quantify the concentration of respiratory viruses in patient specimens, at least by comparison with conventional PCR methods or similar techniques employed for quantification in HIV or in human gene research. The paucity of new developments and of commercial support is most likely due to the variability of clinical specimen collection methods, nucleic acid extraction systems, the improper care and storage of nucleic acid extracts and the infrequency with which the quality of the extracted template is assessed prior to testing. This last was performed by only a third of questioned quantifiers who wished to avoid false negatives or unusual differences in RNA load caused by poor quality template (Bustin, 2005). Quality testing can be performed using a spectrophotometer, bioanalyser or by gel electrophoresis (Fleige and Pfaffl, 2006). Internal controls, as described in detail in Chapter 4, hold an essential position in real-time PCR. Or at least they should; unfortunately they are only now becoming more widely described and are still poorly implemented. As we saw from Table 8.1, approximately one third of real-time applications employ such a control. The use of an inhibition control would not only identify false negatives (complete amplification failure due to inhibition rather than an absence of template), but also partial inhibition which could render viral load estimates equally invalid. In ARTIs, an additional control is essential to address the variability in specimen collection; a sampling control. Unfortunately this control does not yet exist. A sampling control would consist of a marker present in the wash or buffer fluid that could account for the dilution effect of patient fluids on the material used to buffer the collected specimen or that used to wash the sampled area and it would permit determination of the recovery efficiency of wash material. The marker would be quantified as a normalizing agent before or after extraction. Such a marker should not interfere with the screening assay and would clearly need to be innocuous and easily detectable.

We make the argument that microbiology, which includes the study of viruses, bacteria, fungi and parasites, and in our case respiratory viruses in particular, uniquely challenges molecular diagnostics due to the sources of template. For example, to sample blood or tissue you do so by breaching the body's front line defensive tegument to collect from an essentially sterile environment. Blood consists of cells, and these are easily sampled, concentrated and quantified permitting the use of cell number as a denominator for gene quantification studies. Respiratory specimen collection is performed both figuratively and literally, in the dark. The respiratory tract is exposed to the outside environment; essentially a deep invagination of the tegument, but one with an enhanced secretory function that creates an humidified and, in different regions, mucous, cilia or surfactant-coated environment whose function, among other things, is to trap all foreign objects. There is no question that we currently obtain enough specimen to detect infection and even to isolate the organism when possible, but we lack precise collection methodology. The impact is that comparison of viral load between samplings of the same patient, different patients or between separate laboratories is not currently possible. This could be addressed using a sample control which would permit collection volume to be used as a denominator for virus quantification. Unfortunately another hurdle exists; infectious virus can exist both inside and outside the cell as intact virions and as nucleic acid associated with capsid protein in a non-particulate form. While we could perform a separate quantification to determine cell number in each specimen, we are next left asking ourselves whether we should express viral copy number per cell or per unit volume, or both? Since no studies have been done to assign meaning to any of these approaches and since we cannot determine the percentage of virus associated with either the intracellular or extracellular fraction, it seems impossible to accurately determine respiratory viral load in a reproducible manner.

Several more addressable technical problems can be found among the examples presented in Table 8.2. These examples illustrate that most quantitative real-time PCR assays for the study of respiratory viruses seem to perform absolute quantification, meaning that they report the specific number of genome copies or equivalents per unit of measure. The assays achieve this by employing previously quantified and characterized synthetic templates amplified at the same time (i.e. in the same instrument run) but as an external standard or calibration curve. However, claims of absolute quantification should be examined carefully because most fall short of meeting the stringent requirements for true quantification. For example, on some occasions DNA templates are used as the standard for RNA virus quantitative assays, in other instances a standard curve is amplified externally without employing replicates of each concentration or without a calibrator or normalizing control that can monitor and correct for inter-tube, inhibition-unrelated differences in amplification. Other studies do not employ an internal control to monitor for the presence of amplification inhibitors at all. Perhaps it is time to re-visit the co-amplification format which was the hallmark of competitive quantitative PCR but adapt it for real-time PCR as was recently reported for a piscine nodavirus (Grove *et al.*, 2006). This was by far the most accurate way to monitor the amplification of a target template in its day.

Quantification by PCR requires very careful handling of the RNA to ensure no loss or degradation of template. In contrast, experience has taught us that this does not hold

Table 8.2 Applications of quantitative real-time PCR to study viruses in ARTIs

Level of quantification	Viral target(s)	Quantification standard	Reference
Not stated	HCoV-229E and HCoV-OC43	None	(van Elden *et al.*, 2004)
Absolute	HCoV-229E and HCoV-OC43	External*, T7 cRNA	(Vijgen *et al.*, 2005)
Absolute	Bat-CoV	External, plasmid DNA	(Lau *et al.*, 2005b)
Absolute	SARS-CoV	External, plasmid DNA	(Poon *et al.*, 2003)
Absolute	HCoV-NL63	External, plasmid DNA	(Chiu *et al.*, 2005)
Absolute	HCoV-NL63	External, IVT RNA	(van der Hoek *et al.*, 2005)
Absolute	HMPV	External, plasmid cDNA	(Ma *et al.*, 2005)
Absolute	HHV-7/HHV-6	External, plasmid DNA	(Yamamoto *et al.*, 2005) / (Tanaka *et al.*, 2000)
Absolute	HAdV	External, plasmid DNA	(Heim *et al.*, 2003)
Absolute	HMPV	External, plasmid RNA	(Deffrasnes *et al.*, 2005)
Absolute	HMPV	External, plasmid RNA	(Maertzdorf *et al.*, 2004)
Absolute	HMPV	External, plasmid RNA	(Sumino *et al.*, 2005)
Absolute	HMPV	External, plasmid RNA	(Agapov *et al.*, 2006)
Absolute	HRSV, HPIV-1, HPIV-2, HPIV-2, IFAV, HMPV, HAdV	External, plasmid RNA	(Kuypers *et al.*, 2006)

*External standards are amplified in separate tubes from the viral templates; IVT- *in vitro* transcribed; CoV-coronavirus; SARS-Severe acute respiratory syndrome; HMPV-Human metapneumovirus; HAdV-Human adenovirus; HHV-human herpesvirus; HPIV-human parainfluenzavirus; IFAV-influenzavirus

true for qualitative respiratory virus real-time PCR. While most studies highlight the need to carefully collect, extract and store nucleic acid extracts; in practice, even with RNA viruses, PCR lets one get away with a multitude of short cuts, while still retaining the required qualitative result. One reason for this flexibility is provided by the tendency for real-time PCR assays to employ very small target regions (down to essentially the length of two primers and an oligoprobe) such that genome or even gene-length respiratory viral RNA need not be present for successful amplification. Of course one should never take these short cuts when patient diagnoses are the outcome or for quantitative applications. However research laboratories are sometimes faced with stretching resources, be they reagents or specimens; we have on several occasions resuspended an "empty," previously frozen specimen tube in the desperate hope that we can amplify just one more time—and been rewarded with success. In our experience, with care and luck, an entire (and hitherto unknown) viral genome can be amplified from less than 100 μL of nucleic acids, representing less than 300 μL of a patient's diluted nasopharyngeal aspirate.

Succesful monitoring of viral replication during attempts to isolate or propagate respiratory viruses using *in vitro* cell culture can result from the application of real-time PCR.

For viruses that are fastidious, grow very slowly or to low titers, qualitative or quantitative detection of the viral genome is a much faster and more sensitive method to follow the progress of viral growth than monitoring cultures for cytopathic effects (Freymuth *et al.*, 2005). The technique is especially valuable when attempting to culture a NIV, for which the precise nature of the cytopathic effect is uncharacterized, making an objective measure of virus growth essential. The benefit of a quantitative or semi-quantitative approach is that the initial viral inoculum (day 0 post-infection) can be used to define a baseline threshold cycle (C_T) value (Schildgen *et al.*, 2006). In this way, true viral replication can be identified using purified nucleic acids collected on subsequent days post infection by identifying reduced C_T values compared to the inoculum. This is an important observation since in our experience the inoculating viral RNA can linger in culture even after multiple culture medium changes and even following the passage of infected cultures to new flasks.

Of relevance to quantification is the need for careful consideration when using *in vitro* cell culture-derived material for PCR standardization. Culture can be a very efficient amplification system and the high rate of virus production under near ideal conditions can surpass the levels of virus produced within an immune competent host. This creates the possibility that an increased spectrum of quasispecies (discussed below), adaptive mutants or incomplete viral particles will accumulate which, because of sequence variation, may or may not be detected by PCR. The excessive amount of RNA present in culture is exemplified by reports of real-time PCR detection limits described in fractional terms e.g. 10^{-2} to 10^{-4} cell culture infective doses (TCID) or plaque forming units (Scheltinga *et al.*, 2005; Chui *et al.*, 2005). Clearly there is more amplifiable template present than is suggested by indicators of infectivity. This has negative implications when claims of real-time PCR assay sensitivity and quantification are based on the use of a titrated virus stock. Accordingly we prefer and recommend employing a quantified synthetic template to determine the analytical sensitivity of a real-time PCR assay. Diluting this template in a matrix similar to that of the original patient specimen should permit identical amplification efficiency.

Respiratory virus template variation: can assay design keep pace with viral variation?

A theme of inadequacy runs throughout this chapter due to the many holes in our understanding of respiratory viruses and ARTIs. In particular we do not know the impact of multiple active infections on the host's immune system. We're not even certain that multiple active infections occur; highly sensitive PCR assays may be detecting the remnants of a previous infection. We also have no accurate grasp of how many more distinct respiratory NIVs and emerging respiratory viruses we will find in the future. Neither can we be sure how our understanding of the classical respiratory viruses and their contribution to illness will change as a result of our increased implementation of real-time PCR. The future may show us that current estimates of 20 to 30% of specimens from patients with suspected ARTI containing more than one microorganism were the thin edge of the wedge. Perhaps we will one day be able to say with certainty that infection by one respiratory pathogen permits subsequent infection by another, or the development of more severe illness caused by another respiratory pathogen, as has been suggested from real-time PCR investigations of pneumococcal disease (Peltola *et al.*, 2006). Whether a virus is first or last to infect, or present in both roles will also be interesting to learn. Real-time PCR could play an essential role in contributing answers to these and other issues in respiratory virology if we can surmount some important design and implementation obstacles, which we will discuss below.

The task of designing real-time PCR assays capable of detecting the extensive viral variety inherent to respiratory viruses and then ensuring that the assays remain capable of detecting all those variants for more than a single respiratory season is a daunting one, especially for the poorly characterized historical viruses and the NIVs (Gardner *et al.*, 2003). There are benefits in using "conserved regions," referring to those regions evolutionarily maintained within a viral group, because they encode products or provide structures that are essential for viral infection, replication or propagation. Unfortunately, regions conserved at the amino acid level and therefore useful for serological methods, rarely translate (pardon the pun) to equivalent conservation of the encoding nucleotide sequences. Because amino acid coding is a redundant process, synonymous changes often arise; variation occurs in the terminal position of the nucleotide triplet that does not result in a change to the encoded amino acid. These nucleotide changes will obviously cause hybridization problems for primers and, unless part of a typing assay, for oligoprobes, which is the main reason conserved protein regions are only worth our attention if they are proven to reflect conservation of the underlying sequence. A poignant example is provided by our four year study of HMPV in Brisbane, Australia (Mackay *et al.*, 2006a). Using our first HMPV real-time PCR assay (Mackay *et al.*, 2003b) we would most likely have failed to detect 19% to 76% of HMPV positives in a particular year. In practice this would have resulted in the average prevalence dropping from 7% during each of the four years studied to 4.0%, 4.1%, 3.0% and 1.6% for 2001, 2002, 2003 and 2004 respectively. These changes reflected the drift in predominant viral lineages and sub-lineages away from the lineage A strains that predominated initially. Because we learned early on in the testing process of the existence of lineage B strains and additional sub-lineages, we were able to employ more inclusive real-time PCR assays to detect all the known HMPV sub-lineages and more comprehensively determine the role of HMPV in ARTI (discussed in more detail in the applications section).

An area slow to take up real-time PCR applications is the genotyping of respiratory viruses. While single nucleotide polymorphism (SNP) detection has been widely accepted and has proven very useful for human molecular studies, it has not proven as useful to respiratory virology. This is certainly not for lack of characterized targets; HRSV is well known for its division into two serologically distinct types with multiple sub-lineages and the situation is similar for the human adenoviruses (HAdVs), HRVs, human enteroviruses (HEVs) and the human parainfluenzaviruses. The jury is still out on many of the NIVs, but our studies and those of others suggest that dichotomous lineages of HCoV-NL63 and HCoV-HKU1, HMPV, HBoV and even novel lineages among the HRVs do exist. The virological and clinical relevance of most of these genetic groupings remains to be determined using serological techniques and epidemiological studies. Nonetheless abstract phylogenetic groupings help us to improve our molecular tools permitting detection of the complete spectrum of variants that exist for each viral species. Unfortunately it seems that much of the viral variability we know of is not reproducible enough to be useful for the foundation of molecular genotyping assays for respiratory viruses.

Nucleotide sequencing: the essential accompaniment for improved assay performance

Grouping related viruses into genetic lineages is the most common outcome of sequencing of respiratory virus amplicon. While very few of these data have proven to be of clinical relevance, a phylogenetic tree has become an obligatory figure in publications presenting epidemiological data using molecular methods. In this context such figures serve to alert the assay designer to the scope of sequence variation in the region under study. Identifying viral variability provides us with the option of avoiding it when designing real-time PCR assays. The benefits derived from this approach depend entirely on the target region chosen and its role in the viral replication process. The point should be made that not all genes or gene fragments provide acceptable targets for phylogeny, although in our experience, many do. Highly conserved nucleotide sequences are well suited for situating real-time PCR primers and oligoprobes; however they make poor targets for phylogeny studies. To be of use, molecular epidemiology targets should encompass all variants of the virus under study and this can be shown on a phylogenetic tree as diverse branching patterns. However if the region chosen does not represent all variants, the designer may be left with the false impression that little or no variation exists and design their assays accordingly. A practical example is provided in Figure 8.2 whereby two nucleotide sequences of the HMPV polymerase protein gene, of very different lengths, are used to define lineages and sub-lineages using prototype viral strains. The use of a gene fragment of insufficient length

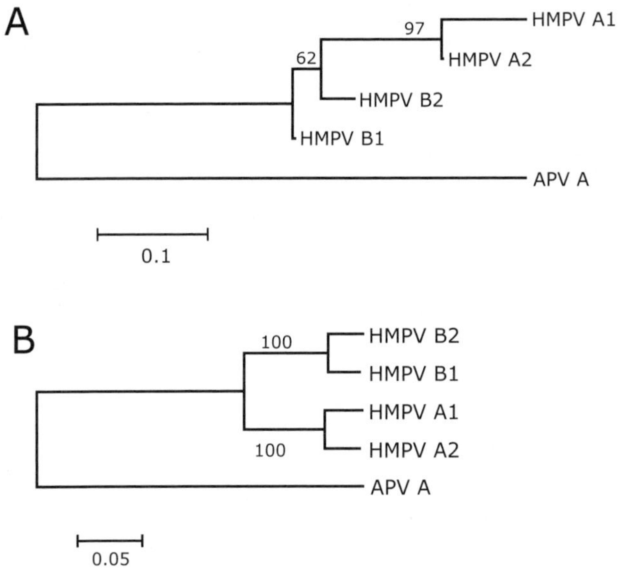

Figure 8.2 Comparison of amplicon lengths for use in molecular epidemiology studies. A less defined phylogenetic pattern is apparent when a 92 nt amplicon derived from the HMPV large/polymerase gene (A) is aligned and used in Neighbour joining analysis to determine patterns of viral strain similarity, compared to using a longer 6018 nt amplicon (B). This is indicated by improved nodal values, indicating better subdivisions in the latter tree. Only the prototype virus for each sub-lineage is included; HMPV A1 (GenBank accession number AF371337), HMPV B2 (AY297748), HMPV A2 (NC_004148), HMPV B1 (AY525843) and Avian pneumovirus type A (APV A; AY640317).

leads to non-representative, or at least unclear, branching patterns for some or all of the four sub-lineages of HMPV, whereas the larger fragment demonstrates each sub-lineage clearly. This is not to say that small amplicons are of no value, however phylogenetic trees resulting from their use should, whenever possible, include additional characterized reference sequences to ensure the branching patterns are valid. We prefer to use phylogeny targets of at least 150 nt in length—preferably 200 nt to 500 nt. Otherwise, and perhaps additionally, these patterns should be compared to the results of similar studies using different gene targets from the same respiratory virus strains to ensure similar patterns occur. The fragment length we recommend has the additional benefit of permitting easy sequencing of both complete amplicon strands with only a single primer pair.

Nucleotide sequencing is also being more frequently employed to support assertions that real-time PCR positives obtained during a respiratory virus study are not due to amplicon carry-over contamination. However, for such reasoning to be believable a caveat needs to be introduced and addressed. The region chosen for sequencing must exhibit enough variability between strains of the target virus to permit convincing discrimination from a clonal contaminating amplicon. If we take as an example our own sequencing studies of the HBoV NS1 gene (Figure 8.3), it is clear that across the 245 nt gene fragment, very little nucleotide variation occurred, in fact only a single nucleotide difference differentiated some Australian strains from one of the prototype HBoV strains. It is therefore conceivable that we detected a single instance of HBoV and then went on to contaminate our PCR mix-making laboratory with traces of its amplicon manifesting as subsequent

(A)

```
                     10        20        30        40        50        60        70        80
St1         TATGGGTGTGTTAATCATTTGAACAAAGGATTTGTATTTAATGACTGCAGACAACGCTTAGTTGTTTGGTGGGAGGAGTG
St2         .........................................................C......................
QPID04-0007 .........................................................C......................
QPID04-0009 .........................................................C......................
QPID04-0011 .........................................................C................A.....

                     90       100       110       120       130       140       150       160
St1         CTTAATGCACCAGGATTGGGTGGAACCTGCAAAGTGTATCTTGGGCGGGACAGAATGCAGAATTGACGTCAAGCATAGAG
St2         ................................................................................
QPID04-0007 ................................................................................
QPID04-0009 ................................................................................
QPID04-0011 ................................................................................

                    170       180       190       200       210       220       230       240
St1         ACAGTGTACTTTTAACTCAAACACCTGTAATTATATCCACTAACCACGATATCTACGCGGTTGTTGGTGGCAATTCTGTT
St2         ................................................................................
QPID04-0007 ................................................................................
QPID04-0009 ................................................................................
QPID04-0011 ................................................................................

St1         TCTCA
St2         .....
QPID04-0007 .....
QPID04-0009 .....
QPID04-0011 .....
```

(B)

	St1	St2	QPID04-0007	QPID04-0009	QPID04-0011
St1		0.995	0.995	0.995	0.991
St2	0.995		1.000	1.000	0.995
QPID04-0007	0.995	1		1	0.995
QPID04-0009	0.995	1	1		0.995
QPID04-0011	0.991	0.995	0.995	0.995	

Figure 8.3 A nucleotide alignment (A) of two prototype HBoV strains (St1, St2) and three Brisbane, Queensland strains (QPID) and (B), the corresponding nucleotide sequence identity matrix for the HBoV NS1 amplicon (values represent fractional identity). GenBank accession numbers for HBoV prototype strains St1—DQ000495 and St2—DQ000496. Prepared using BioEdit version 7.

positives, resulting in the production of identical sequences. However this was not the case as we proved by detecting and sequencing additional, more variable gene fragments to confirm that these detections were in fact true HBoV positives. Without those additional data however, the sequenced region did not vary enough to exclude the possibility that the positives were caused by contamination.

A sequencing approach to confirm true positivity is not possible when targets are derived from highly conserved viral genes. An example of this is provided by the coronaviruses which have regions of extremely conserved nucleotide sequence within the *1ab* open reading frame (ORF); so conserved, in fact, that one could again be forgiven for thinking that amplicons from multiple strains had arisen from an amplicon contamination event. Another downside to targeting these islands of conservation becomes apparent if one tries to reduce expenses and target a region that is ideal both for detection and for molecular epidemiology. Extensive conservation will stymie that approach, as will the propensity for real-time PCR assay designers to develop very small, sub-100 nt amplicons. Ideally a dual-purpose target region would contain conserved sites for primer binding which span a region of variable sequence. Alternatively, the target should be long enough to encompass at least some region of variability, permitting accurate phylogenetic results. We utilized such a region for our HMPV studies encompassing the 3′ terminus of the phosphoprotein coding sequence through to the 5′ region of the matrix coding sequence (Mackay *et al.*, 2004; Mackay *et al.*, 2006a).

Highly conserved regions of sequence may also lure the designer into a false sense of security. The previous coronavirus ORF *1ab* example appears ideal for the location of a comprehensive pan-HCoV assay (i.e. an assay capable of detecting all known HCoVs) however the region also exhibits extensive similarity to human genomic DNA sequences, a common feature among viral genes, which can result in many unwanted amplicons of various sizes when employed as a diagnostic screening tool, especially for cellular specimens.

Secondary structure: tying knots in assay design

Once a suitable target region of the respiratory virus has been identified, a search for putative secondary structures can easily be performed to examine a single strand of the RNA or DNA sequence using mfold (http://www.bioinfo.rpi.edu/applications/mfold/). If the presence of secondary structures is predicted, and if those predictions represent reality (remember the program, just gives probable and possible outcomes, with no guarantee these occur in practice) then such structures may confound the progress of reverse transcriptase or PCR polymerase and/or the hybridization of oligonucleotides, resulting in lowered amplification efficiencies, decreased oligoprobe-generated fluorescence or even complete assay failure (Gunson *et al.*, 2006; Hoebeeck *et al.*, 2006). Among the picornaviruses, multiple stem-loop secondary structures are predicted in the native RNA form of the 5′ untranslated region (Figure 8.4d). The virus makes use of these structures for ribosome-independent transcription initiation but this region is also frequently employed as a diagnostic PCR target (Halonen *et al.*, 1995; Lönnrot *et al.*, 1999; Jartti *et al.*, 2004b). It is evident from the following example that even after cDNA and amplicon are produced, secondary structure will remain an obstacle for oligonucleotide hybridization in some regions thus care should be taken when locating primers and choosing cycling temperatures (Figure 8.4a-c). The use of this region may be contributing to underestimation of the prevalence, and therefore

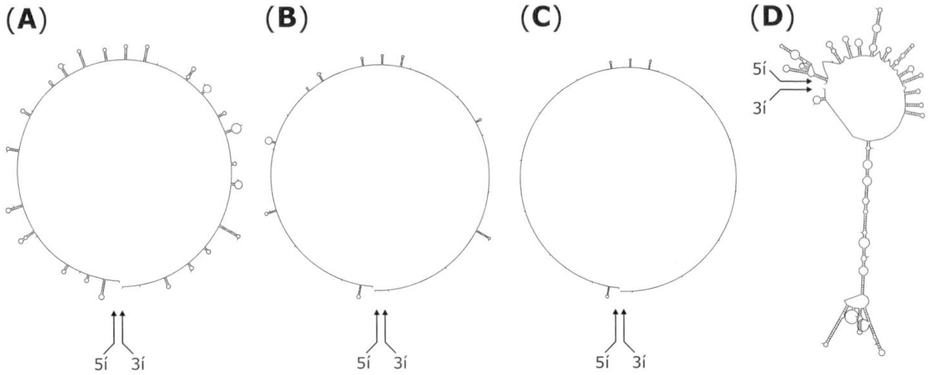

Figure 8.4 A single-stranded DNA or RNA representation of the Coxsackievirus A21 5′ untranslated region demonstrating the secondary structures which form at (A) 42°C (possible reverse transcription temperature), (B) 55°C (possible primer annealing temperature) and (C) 72°C (possible polymerase extension temperature) using a Na^+ concentration of 50 mM and Mg^{2+} concentration of 2.5 mM, compared to (D), the predicted RNA structure at 42°C in 1 M Na^+. These structures could confound oligonucleotide hybridization and the reannealing of double-stranded amplicon. The termini of the sequence are indicated. Derived from GenBank accession number D00538. Predicted DNA structures generated by the mfold web server (http://www.bioinfo.rpi.edu/applications/mfold/).

importance, of these viruses in ARTI. Unfortunately little sequence is available from additional genetic regions to investigate the possibility of better PCR targets and for the picornaviruses at least, designing a genus-specific primer pair is next to impossible using the sequences available.

At the time of writing, respiratory NIVs were genetically better characterized than many classical respiratory viruses. There were four complete HMPV genomes, three HBoV, two HCoV-NL63 and more than 20 derived from HCoV-HKU1 strains, housed on the GenBank database. The emergent SARS coronavirus was represented by more than 100 genomes. Astonishingly, only five complete HRV genomes existed on GenBank, despite the discovery of these viruses nearly half a century ago. Since these viruses constitute the largest single genus of viruses causing human respiratory illnesses, as far as we know to date, this lack of sequencing seems to indicate that we have a great deal of work to do to better detect and properly characterize the viruses we now know exist, let alone what lurks beyond our current knowledge.

Primer selection: as simple as another pair of primers, another assay?

When selecting primers from the literature, there may be significant concerns for their reliability, especially if it is clear that they have been incompletely evaluated, i.e. the oligonucleotide and magnesium concentration-optimized assay has been tested on some known target virus positive and negative specimens but not against all possible strains. Alternatively, an optimized and evaluated real-time PCR assay may not have been validated or "put into the field" in comparison to a gold standard assay where it will face amplifying template from many different tissues, possibly extracted, stored and handled in different ways, that may even have been combined to make screening easier (see Pooling resources below). Nowhere is this more obvious than when reviewing respiratory picornavirus RT-PCR assays. Because of the difficult, expensive and tiresome issue of obtaining virus or viral templates for almost 200 antigenically distinct viruses, a smattering of the members of the genus *Rhinovirus* or *Enterovirus* is usually all that is examined during evaluation and to date most of these studies have employed conventional RT-PCR (Gama *et al.*, 1988; Gama *et al.*, 1989; Aruda and Hayden, 1993; Santti *et al.*, 1997; Lönnrot *et al.*, 1999; Papadopoulos *et al.*, 1999a; Loens *et al.*, 2003; Kares *et al.*, 2004; Grutke *et al.*, 2004; Scheltinga *et al.*, 2005). A cursory glance at sequence alignments for these viruses highlights the extreme nucleotide diversity between members of these genera but also the paucity of sequence data available for each virus, making accurate assay design in many genomic regions next to impossible. It is these issues which make it so essential to interrogate a new molecular assay thoroughly to ensure efficient amplification of all the expected targets. If, for example, an assay detected only 80% of strains, how could we say what role the remaining 20% of strains play in clinical illness, their impact on effective therapeutic design or their involvement in co-infection? We must address all viruses using the same basic diagnostic technologies or else we perpetuate potentially false beliefs rather than continually question them and adjust them wherever necessary. Molecular assays will be useful only if dynamically maintained through a cyclical process of design, optimization, evaluation, validation, performance monitoring and then re-design as required.

We and others often choose to employ a conventional PCR approach when commencing the study of a new respiratory virus because this is a less expensive method for trialing a range of primers in a variety of combinations (Vabret *et al.*, 2006). Once the first target has been shown to be appropriate, we move on to develop more assays targeting other regions of the genome in order to confirm our preliminary findings, and then develop the best of them into real-time PCR assays. This approach works better if the virus in question has been precisely characterized. For respiratory NIVs or poorly characterized classical viruses the process can be long and involved because discrepant results between assays may require additional, confirmatory assays to be developed. Nucleotide sequencing is also required throughout the procedure. Nonetheless this approach begins to address the challenges presented by viral variation, particularly the complicated role played by viral quasispecies.

When designing a new real-time assay it is important to analyze one's oligonucleotides using the Basic Local Alignment Search Tool (BLAST) via the internet (http://www.ncbi.nlm.nih.gov/blast/). This *in silico* process compares the oligonucleotide sequences against the GenBank database maintained by the National Center for Biotechnology Information (NCBI, USA), as if performing a gigantic multiple sequence alignment. The GenBank database contains more than 130 gigabases of volunteered sequence and permits a peek into the specificity of a primer. The first problem often identified by this process is the striking similarities between the complete or partial respiratory virus oligonucleotide sequence and human genomic DNA sequences which may manifest practically as non-specific amplification. We have frequently encountered problems with assay design among the paramyxoviruses, picornaviruses and particularly when designing pan-tropic assays for the coronaviruses. The problem may be reduced when acellular or poorly cellular material is used but in respiratory specimens, many cells may be sloughed, scraped, washed or otherwise abraded out of this anatomically complex system. In Figure 8.5 we use the entire HCoV-HKU1 nucleocapsid gene as an example of how difficult it can be to find

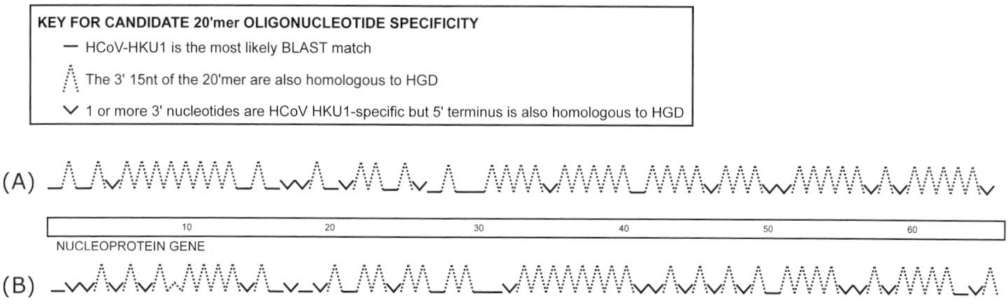

Figure 8.5 A schematic representation of the HCoV-HKU1 nucleoprotein (N) gene. Short oligonucleotide sequences (each 20mer is indicated by one of three symbols) were subjected to BLAST analysis starting from nucleotide position 2 (A) or shifted downstream and started again from nucleotide position 5 (B) in order to test the effect of shifting each 20mer. The analysis found few sites that were solely specific for HCoV-HKU1 oligonucleotide hybridization but many that were likely to cross-react with human genomic DNA (HGD) to some extent.

coronavirus-specific oligonucleotide sequences for a real-time PCR assay design. The easiest approach to avoid trouble would be to move the oligonucleotide sequence up or down the target sequence and continue to do so until only suitably specific BLAST matches result, but as shown in the figure using two different starting points to test the impact of different sequence compositions, this outcome may not always eventuate. Of course, the suitable oligonucleotide must then satisfy all the necessary design constraints (see Chapter 2). Briefly, the occasions when one is locked into a particular viral target sequence or gene region include:

1. When adding an oligoprobe sequence between an existing, validated primer pair
2. When amplicon size constraints are important, e.g. if BLAST results situate the most specific primer pair 1000 nt rather than the intended 100 nt apart, then the relative importance of cross-reactivity must be re-considered
3. When it is the least non-specific region of viral sequence
4. When secondary structures in adjacent template regions and in the resultant oligonucleotide sequences preclude good assay design elsewhere
5. When overly GC-, AT- or sequence repeat-rich regions in adjacent sequences preclude assay design

It must be noted that no amount of *in silico* preparation can guarantee the desired performance of a primer pair and/or oligoprobe, nor should it completely deter one from applying a primer pair despite bad reports from software. Our laboratory has many examples of assays designed to detect respiratory viruses in which the oligonucleotide primers match only the intended target by BLAST but non-specific amplification is apparent when viewing the amplicon. Nucleotide sequencing reveals it to be of human origin. Ultimately, new oligonucleotides must be tested against all likely viral targets during the assay's evaluation phase to determine suitability.

One may be forgiven for expecting that non-specific amplification, especially that associated with human genomic DNA can be dismissed by simply examining the amplicon size; this is not always the case. Not only can a non-specific product of the size expected to represent the viral target result from human genomic DNA amplification but single oligonucleotides can also function as both forward and reverse primers (by binding to different strands), increasing the chance of producing non-specific amplicon. Such a process may also be intentionally used to detect unknown targets in a technique called arbitrarily primed (AP) PCR (Williams *et al.*, 1990; Welsh and McClelland, 1990; Welsh *et al.*, 1992). While AP-PCR studies usually employ primers that are not designed with a specific target in mind, in our experience a virus-specific primer can also produce the same result as using AP-PCR. Unfortunately the major benefit of a homogenous PCR system has encouraged a less welcome outcome; a suitable fluorescence curve is often taken as sufficient confirmation of specific amplification (Chui *et al.*, 2005). If one does not examine an agarose gel during the optimization and evaluation of a new real-time PCR assay, it is impossible to know whether unintentional products have been produced. While the influence of unrelated product formation may have little or no impact on the qualitative capacity of an oligoprobe-based real-time PCR assay, there is no question that serious consequences result when the assay is deployed for quantitative purposes. The unwanted and unmeasured drain on PCR reagents and the production of additional amplification-inhibiting by-products will affect how the PCR begins and progresses and these phases need to be performing optimally for real-time assays to produce valid data for quantitative respiratory virus real-time PCR.

These examples serve to illustrate that PCR is a tool that occasionally stumbles with the reliable detection of even those sequence which we have previously characterized. For certain, we can use sequence degeneracy, reduced stringency and a number of other molecular approaches (Gao and Moore, 1996) to tease our oligonucleotides into amplifying sequences that we do not precisely know, but these approaches will not serve to reliably detect virus in a high-throughput diagnostic laboratory setting, where patient treatments are reliant upon the results.

Pooling resources to save money

One approach to reduce expenses for respiratory testing is to batch specimens; taking aliquots from a number of extracts and combining them in a single tube for testing by PCR. As shown by the example in Figure 8.6, the reduction in testing may be significant, especially during "off peak" respiratory virus seasons. However during peak seasons for the virus of interest, the benefit may be lost as there are so many specimens to test that it may simply be easier and of more clinical benefit to test them as they arrive. An unwanted side-effect of pooling is the possibility of an increase in the total amount of inhibitors or human genomic DNA in each tested pool when compared to testing each specimen individually. For this to occur some of the specimen extracts would need to be free of inhibitors and human genomic DNA in order for their amplification to be affected by an average increase in these substances as a result of pooling. However, if all specimens have some amount of one or both substances, the dilution effect (1:9 in the example below, and our preferred ratio) should negate any cumulative effect. More likely is that pooling specimens dilutes out either factor. The exact number of specimens which can be combined depends on the specimen matrix and requires careful investigation to ensure that diluting an extract by approximately one $\log_{10}$ will not leave weak positives below the assay's limit of detection. While reagent expenses can be reduced using a specimen pooling approach there is an

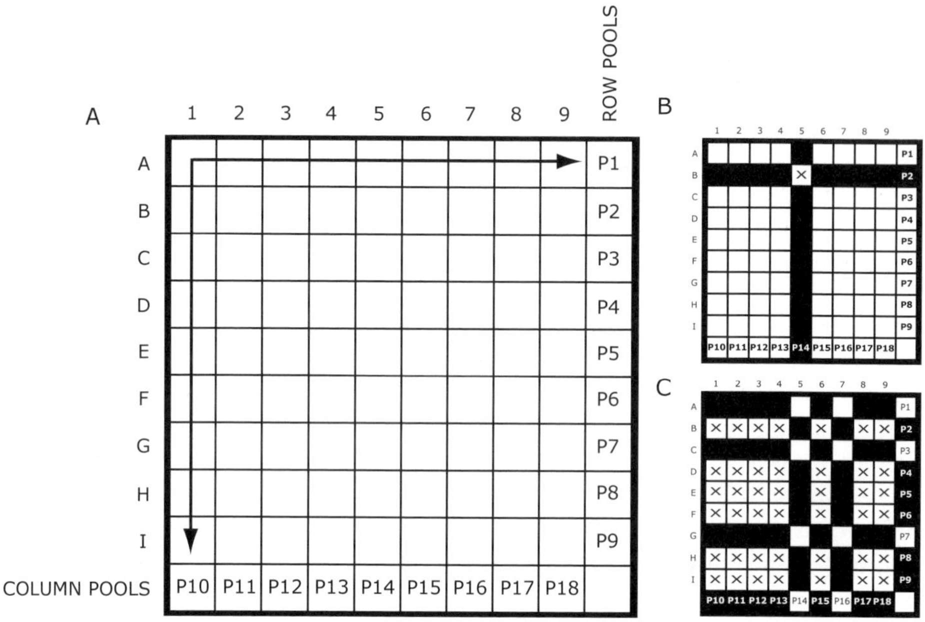

Figure 8.6 (A) Schematic representation of a hypothetical pooling system employed to minimize the specimen testing load. Specimens are placed into an X by Y grid, in this example, 9 × 9. An aliquot from each specimen in row A to I is added into a tube at the end of each respective row (P1 to P9). Similarly, columns 1 to 9 are pooled (P10 to P18). Only the pools go on to be tested. In the "off" season for a particular respiratory viral target, or for viruses with low prevalence in the study population or the general community, a result pattern such as that in (B) may occur. Pools P2 and P14 were positive, so it is likely that the specimen present at the junction (marked with an "x") of the two coordinates will be the cause of the positive PCRs. This result required testing of 18 pools plus a confirmatory test of the one suspect positive specimen versus 81 individually screened specimens. In the peak season for a particular respiratory virus or for ubiquitous viruses, a much higher prevalence may occur as depicted in (C). In this example, 18 pools were tested revealing 13 positives (P2, P4–6, P8–13, P15, and P17 and 18) that correlated with 42 possible positive specimens requiring confirmatory testing. Of these only 7 are required to be positive to satisfy the pattern; however there may be more than a single positive per pool so all 42 must be screened. Without pooling, only 21 more specimens would have required tested, making the time required to construct, document and test the pools and then perform the confirmatory pool testing of questionable benefit to the routine laboratory.

associated increase in personnel time for pool construction unless a liquid handling robot is available (Johnson *et al.*, 2002). Also, the laboratory needs to have a sufficiently high specimen throughput to ensure delays due to pooling do not detrimentally impact on result turnaround times. Since high throughput is often the reason cited for sample pooling in the first place, this issue should not be a major concern.

Points to consider when taking a respiratory virus real-time PCR assay to the publisher

Anecdotal evidence would suggest it is increasingly difficult to publish one's respiratory virus study data. In practice, limited space in the handful of relevant journals with respectable impact factors means more than three quarters of manuscripts are rejected after internal review. To be acceptable one must therefore write as tightly as possible about one's novel assay or study—excluding the discussion of anything beyond strictly relevant detail. When a real-time PCR that detects respiratory NIVs or poorly characterized respiratory viruses is published, the precise function of that assay must be made clear because that role may influence how others employ the assay. Its role should also dictate the necessary level of detail to be included. The main point to consider is whether the assay is to be used as a tool for preliminary investigations of a respiratory virus or whether the assay will perform as a frontline prospective diagnostic tool. If publishing a novel assay for preliminary investigations, authors need to include data on the usefulness of the assay for retrospective prevalence and molecular epidemiology studies. If presenting a diagnostic tool, the results of which may have direct clinical relevance, possibly even affecting patient management, additional data are required from comprehensive sensitivity testing in addition to that from optimization and evaluation testing and *in silico* and *in vitro* investigations, to demonstrate the minimal likelihood of cross-reaction with human genomic DNA or other microorganisms. Specificity testing is common to both applications of a real-time PCR assay and occurs during the evaluation phase when the primers are tested against all genetically related viruses as well as those microorganisms likely to occur during the same season, or in the same host tissues.

Clearly defining the assay's role also raises another issue unique to the respiratory virus hunter: How does one validate a real-time PCR assay for a respiratory NIV, especially if yours is the only laboratory to have made the discovery? With luck, similar viruses exist and can be used as a scientific and clinical model to aid assay and study design. If it is impossible to obtain positive material then one should develop several assays targeting and sequencing distinct genetic regions of the virus to provide the best possible backup to ensure positives are true, and to help develop second generation assays capable of detecting more strains. A useful next step is to make synthetic templates appropriate to the viral target region and nucleic acid type. These can take the form of one or more large oligonucleotides if viral template is unavailable (Horton *et al.*, 1990; Dillon and Rosen, 1990), otherwise purified amplicon or amplicon inserted into plasmid and biologically expanded are equally useful. The latter have the added benefit of being renewable and the PCR template is less likely to suffer from storage-related degradation of the termini than is amplicon alone. Additional measures must be employed to ensure false-positives do not occur via contamination from such highly concentrated template sources.

For studies addressing questions of viral prevalence, such as "what is the detection rate of new respiratory virus X in specimens from the LRT" (Lee *et al.*, 2006), it seems sensible to us to apply fewer, but no less stringent, criteria to the assay development component and focus more on the results generated and to further characterize the virus and patient population since there may be little or no precedent in the literature. Recent experiences has shown us and others that early assays will detect only a portion of the total respiratory NIV strains circulating at the time and site of a particular study and they are likely to provide only a partial snapshot of endemic prevalence. It should not be necessary to over-develop an assay to address the example question, since in all likelihood the assay will be discarded once further studies of virus X are reported, which applies equally to poorly characterized respiratory viruses and NIVs. For example, inter-seasonal variation occurring due to the cycling of respiratory viral variants within a community is most likely caused by the hosts developing strong immune memory and thus some degree of protection to recently acquired viral strains which may manifest as sub-clinical infection, or entirely protect from infection. Such herd immunity may preclude virus X variant "y" from circulating one year because it predominated in the previous year. However since all variants will contribute to the prevalence of virus X, the ultimate real-time PCR assay should be capable of detecting each of them with equal efficiency.

Permitting manuscripts that describe prevalence or virus characterization studies some assay evaluation leeway might aid the increased reporting of more descriptive investigations, rather than dismissing such studies because they do not present the ultimate diagnostic assay or the most well-optimized, evaluated and validated real-time PCR assay. However, this is a topic of debate, and one's choices are eventually influenced by one's peers, should a manuscript eventuate. The conclusions from studies using first generation assays permit validation or further development of subsequent assays with the caveat that none of the assays may be capable of detecting all strains of the respiratory NIV under investigation, and that a suitable assay may not be available to do so for some time. In this way, an incremental, and hopefully global process of comprehensive assay development can occur for each respiratory virus be it NIV or historical.

Study populations: better to look forward or backward?

Because there have been so few truly comprehensive studies to detect respiratory viruses by real-time PCR we have little to no data to confirm how many concurrent infections a host may harbor nor the clinical impact to a patient harboring an active population of multiple respiratory viruses. Complicating factors for the design of studies to address these issues include the choice of retrospective or prospective specimen populations and the uneven number of hospital-based versus community-wide studies. There is an urgent need for more community-based studies to re-visit the role of respiratory viruses in ARTI using real-time PCR in order to get the most comprehensive picture of patient outcomes and health-care impacts. However, studies of both populations are equally essential. To perform more of one study than the other is to poorly define the scope of illness associated with infection by a particular respiratory virus.

Prospective studies can be considered to provide more reliable clinical and demographic data, since the data are not influenced by recall bias i.e. the assembly of reviewed data with a view to a desired outcome. These studies may also have the benefit of ongoing access to the patient which helps if further sampling is required. Retrospective studies mean trying to piece together laboratory and clinical data from previously collected samples, different clinicians who may ask different questions, take down different notes, recommend different therapies and have a greater or lesser tendency to admit a patient to hospital rather than send them home without identifying a cause for disease. Study bias is also introduced by holes in retrospective data caused by a lack of access to some specimen records or because specimens were misplaced, stored incorrectly or completely consumed during previous testing, resulting in variable specimen quality and thus inconclusive or absent results.

However, retrospective testing is invaluable to identify whether emerging respiratory viruses such as the SARS-CoV have occurred previously in humans and for respiratory NIV studies to characterize their prevalence and to define parameters for the design of future prospective epidemiology studies. The term "retrospective" may itself be a confusing term if used in relation to testing of a specimen population. For example, one may look for a respiratory NIV among a frozen specimen population that was previously collected in a prospective manner thus the clinical, demographic and socioeconomic data should still be sound so long as the original study was for a related illness. Is this a retrospective or prospective population when used for respiratory NIV screening? To clarify this terminology it might be useful to talk about the chronology of the study discriminating between the clinical component and the testing component.

Additional prejudice can arise from snapshot or poorly planned respiratory virus PCR-based studies exemplified by reports that draw conclusions from tightly defined time-periods, usually the peak respiratory season, when prevalence values are well above the annual mean. While these data need reporting, they should always be clearly put into context. Alternatively, a population with a particular clinical illness may be the study target. Once again viral prevalence may be disproportionately elevated or lowered. Such data are not always helpful when determining the overall role of a respiratory virus in ARTI and can lead to over or under-estimation of the virus's prevalence and therefore importance.

How many primers does it take to detect one virus: the quasispecies conundrum

When designing assays for respiratory viruses (or any human pathogen) it is easy to forget that our intended target for PCR is not the only genome in town. In particular an RNA virus infection is the result of receiving and incubating an evolving population of quasispecies (quasi-; resembling or having a likeness to) of variable genome sequence also known as "mutant swarms" (Jenkins *et al.*, 2001; Mullan *et al.*, 2004). The practical extent of this variation is currently unclear for respiratory viruses but RNA virus quasispecies populations have been described as the most rapidly mutating, highly variable and genetically versatile life forms on earth (Holland, 2006). For simplicity's sake, drawing upon our own experiences and without data to the contrary, we will not consider respiratory virus quasispecies variation to affect the lineage or sub-lineage classifications, but they may hinder attempts to further sub-classify viral sequences e.g. the sub-, sub-lineage. When considering quasispecies, one should first adjust one's thinking of a "wild-type" virus or a newly identified virus from something that has a single, often-detected nucleotide sequence to that of a distribution of genomes which can be statistically defined but not individually determined (Domingo, 1998). The potential for a detrimental impact on real-time PCR of predominantly RNA respiratory viruses is concerning. A worst case scenario could be that sequence variation affects, to some extent, the way oligonucleotides hybridize to any viral gene target, with the scope of quasispecies variety depending on host immune factors, the severity of illness and the spectrum of inoculating quasispecies introduced into the new host (human or cell). At best there will only be a few genes with sequence that varies from the consensus, and such variation could be overcome by designing assays for additional distinctive targets as we discussed earlier. It is certainly clear that since molecular biology was first applied to virus characterization, we have been able to detect the same viral species, lineages or sub-lineages repeatedly over time and between locations, suggesting a certain amount of conservation of these artefactual groupings. We should also acknowledge that this pattern could simply be perpetuated by our choice of target and manifested in the primers we employ.

There can be no doubt that the constant evolution of RNA viruses produces completely new genome permutations, aside from quasispecies variation, that will necessitate the constant development and upgrading of molecular diagnostic tools (Steinhauer and Holland, 1987). These new variants may also exist as populations with greater or lesser virulence. Mullan *et al.* described the following strategies to increase the likelihood that the amplicon produced by PCR originated from a single virus strain; they may also prove useful to investigate the role of quasispecies in ARTI (Mullan *et al.*, 2004):

1 Use RNA from a plaque-purified viral isolate.
2 Sequence numerous RT-PCR products obtained by limiting dilution of cDNA or RNA.
3 Sequence 20 clones obtained from a plasmid library constructed by shotgun cloning of the PCR amplicon.

For several reasons these options are unsuited to the clinical microbiology laboratory. Firstly, time and resources are at a premium and it is not usually the mandate of the diagnostic laboratory to embark upon research studies. Secondly it is not always possible, even for a research laboratory, to isolate virus using *in vitro* cell culture since little original material may remain and what does remain may have been over-handled, improperly stored or repeatedly freeze-thawed. It should also be remembered that the main reason NIVs have not been characterized until recently is because they are not easy to grow using *in vitro* culture techniques, if they grow at all. Thus the third approach may be the best for researchers.

If the virus does grow it is possible that the strain which adapts best to the *in vitro* cell culture environment is not the same molecular strain predominating and causing illness within the host. Additionally, a plaque-purified "clone" may still develop quasispecies variation. One should also remember that a predominant viral strain suited to growth in one cell type, may not be suited to growth in another. That example could presumably be extended to different human hosts with differently primed immune responses and to cases of multiple infection by different virus species each having a specific tropism (Novella *et al.*, 1995). The impact of these points suggests:

1 An isolate propagated from a patient's specimen may not have an identical sequence to that determined from the original PCR-positive specimen.
2 Primers designed to a culture-adapted virus may amplify patient strains of the virus with reduced efficiency and could even amplify a group of quasispecies with enhanced efficiency resulting in a limited view of the overall variability of the respiratory virus species under study.
3 A single sequence derived from RNA respiratory virus strains will most likely represent an average or consensus of the quasispecies present in the extract (Domingo *et al.*, 2006).

The result could be considerable for PCR-based diagnostics, with the possibility of confusing our understanding of the target virus's contribution to respiratory illness by underestimating its prevalence. These points serve to further indicate why accurate respi-

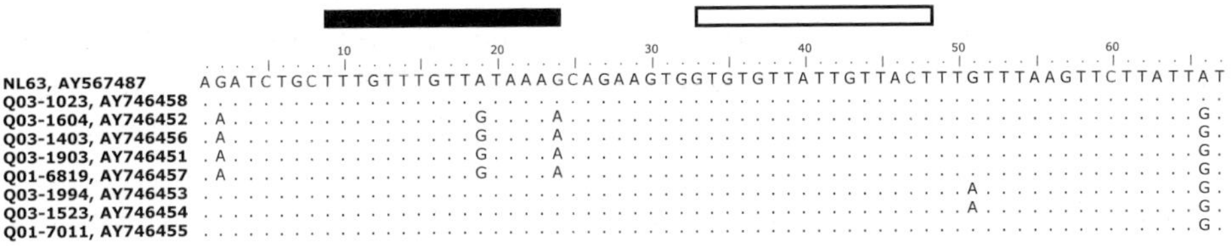

Figure 8.7 Sequence variation and the impact on PCR efficiency. A multiple alignment of fragments of the HCoV-NL63 *1a* coding sequence. The NL63 prototype sequence (top) is compared to strains detected in Queensland, Australia (Q). If the prototypical sequence were used to design an oligonucleotide primer (filled box), it is apparent that variation in some Australian strains (Q03–1604, Q03–1403, Q03–1903 and Q03–6819) would cause inefficient primer hybridization and possibly under represent these strains. However, a primer designed further downstream (open box) would overcome these effects. Another example could be a sense primer (filled box) used in combination with an hydrolysis oligoprobe (open box) for the purpose of quantitative PCR. The reduced efficiency of amplification due to the nucleotide variation at the 3′ end of the sense primer would make accurate quantification extremely difficult. Even worse would be the case whereby an antisense primer (open box) and hydrolysis oligoprobe (filled box) were located as shown leading to poor oligoprobe hybridization and subsequent hydrolysis of the probe if targeting the same strand. GenBank accession numbers are indicated.

ratory virus quantification by real-time PCR is virtually impossible. In Figure 8.7 we provide some hypothetical examples of the impact of sequence-related bias whereby primers designed to the first published sequence of a newly identified coronavirus, HCoV-NL63, could have been foiled by sequence variants we detected in Brisbane, Australia.

A major implication of viral quasispecies and the unknown extent to which respiratory NIVs may be affected by them is that we may not be able to make concrete assertions linking clinical severity to infection by a particular genetic sub-variant; at least without defining the scope of quasispecies sequence variation and finding that it is relatively insignificant compared to that in a sub-lineage. Furthermore, today's molecular approaches are likely to be under-representing the scope of viral variation and we may be laying the framework for ill-informed rational design of antiviral drugs. Only time will tell, but addressing these issues might also help to explain why some respiratory viruses such as HPIV-4, IFCV and HCoV-229E reportedly circulate with such low prevalence (Bellau-Pujol *et al.*, 2005; Vijgen *et al.*, 2005; Lau *et al.*, 2005a). Perhaps these prototype viruses represent the tip of a pyramid of representative strains that have not yet been isolated by *in vitro* cell culture or escaped molecular detection because of overly specific methods.

Applications of real-time PCR for characterization of viral ARTI

Respiratory NIVs have been both a testing and a proving ground for real-time PCR technology during investigations of respiratory virus infection, to the extent that many "better characterized" or historical respiratory viruses have yet to be given the attention they deserve from modern molecular tools. Real-time PCR is every bit as useful as conventional PCR for detecting viruses present in any tissue, which the literature tells us means that virus detection may be achieved from all types of infectious respiratory illnesses. Confusingly, some studies leave the impression that real-time techniques are innately more sensitive than conventional PCR, however these studies are rarely performed using the same primer pairs, reagents or thermal cycling conditions, leaving such claims in serious doubt. Nonetheless, there is no argument that detection of a particular respiratory virus by PCR is more sensitive than isolation of that virus using *in vitro* culture. In the following sections we will review how real-time PCR has been used together with or instead of *in vitro* cell culture to re-examine the roles of some historical respiratory viruses and to begin the definition of roles for the NIVs in ARTI.

Where real-time PCR has made its first significant impact on diagnostic respiratory virology is result turnaround time. This increased speed is essential for quick onset, rapidly disseminating illnesses such as ARTIs which cause outbreaks in hospitals, schools, aged care facilities and among the military. Speed will continue to be important in the future when rapid cohorting of infected patients will be required to quickly terminate the transmission cycle of zoonotic or bioterror-related viruses in community and hospital settings, or at least to exclude the involvement of known viruses in such disease outbreaks. Little benefit is gained from using *in vitro* culture to test for respiratory viruses and respiratory NIVs under outbreak conditions since culture results are not available in a timeframe clinically relevant for patient treatment (Adcock *et al.*, 1997). Real-time PCR is often touted as providing the ability to quickly generate results thus avoiding the unnecessary use, cost and side-effects of shotgun antibiotic therapy. However it may be that it does not in practice achieve this outcome because the results do not yet commonly impact on clinical decision-making (Oosterheert *et al.*, 2005). Evidence that diagnostic speed does matter has been available even before real-time PCR's expanded role e.g. comparison of rapid antigen testing to *in vitro* cell culture found the faster method reduced antibiotic usage by 52% and saved 26% of the cost of hospital care (Woo *et al.*, 1997). However, this study also showed that even rapid HRSV testing did not prevent the continuation of antibiotic treatment for infants less than two months of age.

As presented in Table 8.1, published real-time PCR assays to detect RNA respiratory viruses are divided in the way they prime reverse transcription. Assays either employ a two-step process using a separate reverse transcriptase and polymerase, or a single-tube (one-step) approach combining reverse transcription with PCR amplification in an optimized and proprietary buffer. The former approach is approximately three $\log_{10}$ more sensitive and less likely to produce variable results using known positive specimens, however very few investigators have described the best way to proceed for respiratory virus detection (Corless *et al.*, 2002; Kares *et al.*, 2004). Because of the heavy reliance upon RT-PCR in respiratory virus real-time PCR, it is important to ensure the reverse transcriptase system one chooses works well since reported efficiencies can range from 90% down to less than 25% depending on target, enzyme and reaction conditions (Ståhlberg *et al.*, 2004a; Ståhlberg *et al.*, 2004b). Many applications of real-time PCR in detecting a respiratory virus have used either SYBR® green I or dual-labeled hydrolysis probes (Gunson *et al.*, 2006; Table 8.1). Increasingly, the applications apply a third or fourth oligonucleotide in the form of a fluorogenic oligoprobe, most often the TaqMan® or TaqMan®-MGB oligoprobes (Table 8.1). As we have seen, subtle sequence variation or more dramatic genetic variants will stymie, to varying degrees, the hybridization of one or more of the oligonucleotides and in particular, those fluorogenic chemistries which specialize in detecting polymorphisms such as the TaqMan-MGB® and molecular beacons. However our experience with respiratory viruses indicates that even a well optimized assay may occasionally amplify, by varying degrees, spurious fragments and primer dimers which make the use of non-specific fluorogenic chemistries undesirable. The same table also shows that overall, the use of non-fluorescent quenchers is replacing that of the florescent quenchers, bringing with them the added benefit of improved multiplexing, but that the range of fluorophore and quencher combinations is diverse.

Since real-time PCR is frequently the frontline diagnostic tool for fastidious and newly identified viruses, it is easy to see how prevalence data could be significantly underestimated by developing or adopting a molecular assay too early in the characterization process of such viruses. We will next review the literature as it applies to the real-time PCR detection and characterization of some historically under-characterized viruses and the respiratory NIVs.

Family *Picornaviridae*

Whilst relatively few NIVs currently populate this virus family, it does contain the most frequent infectious agents of humans, the HRVs. Among children, at least six infections and five illnesses per year can be attributed to these viruses (Winther *et al.*, 2006) and because they are so ubiquitous they contribute significantly to data describing co-detections (Tsolia *et al.*, 2004; Arden *et al.*, 2006). Nevertheless, the epidemiology of individual members of this family is woefully under characterized, except to place many of them into a category of mild pathogens generally associated with simple head colds or other mild forms of ARTI. This is despite their abilities to replicate in tissues obtained from the lower respiratory tract, under similar conditions, *in vitro* (Papadopoulos *et al.*, 1999b; Papadopoulos *et al.*, 2000). Real-time PCR has quickly confirmed the ubiquity of the picornavirus positive-sense, single-stranded RNA genome, particularly the HRVs, in both URTIs and LRTIs, and has demonstrated that, using suitably designed primers, the historical requirement for a range of permissive cell lines and cell culture techniques can be overcome (van Kraaij *et al.*, 2005). Real-time RT-PCR has been reported to be 10-fold more sensitive than conventional RT-PCR utilizing the same primer set for the detection of HRV plasmid DNA or viral RNA (Dagher *et al.*, 2004). Unfortunately, interpretation of this comparison was complicated by the use of different magnesium concentrations and a proprietary buffer for the SYBR Green I real-time PCR.

Because of the relative paucity of protein coding sequences on GenBank and the small number of sequences from variants of each prototype virus, the few real-time assays detecting picornaviruses to date have targeted the 5′ untranslated region (Kares *et al.*, 2004; Scheltinga *et al.*, 2005). Unfortunately this region suffers from extensive similarity among all members of the picornaviruses, so in the past, conventional PCR assays targeting this region have been used for all-round detection of picornaviruses, but not for discriminating between HEV and HRV (Kares *et al.*, 2004). Some respiratory virus studies employing real-time PCR do not include testing for HRV at all (Chiu *et al.*, 2005; van der Hoek *et al.*, 2005) while others may demonstrate some "weak," generally undesirable signals contributed by the sequences of members of the genus *Enterovirus* (Kares *et al.*, 2004; Scheltinga *et al.*, 2005). Real-time PCR assays that reportedly detect all HEVs sometimes discriminate between polio- and non-polio enteroviruses without providing greater speciation (Kares *et al.*, 2004). Such approaches may even require a second assay to detect HRV where a positive HRV result excludes the presence of HEV. This approach does not allow for the possibility of dual picornavirus infections from each genus, or even from the same genus.

Currently, clinical management for respiratory infections by members of either genus is similar, so there is little to be gained clinically from a discriminating assay. However there is much to be said from an epidemiological standpoint for clarifying the role of these viruses in ARTI, although the implied need for real-time PCR testing of each serotype individually represents significant time, technical effort and expense. We have found that for selected individual HRVs a prevalence of 2% may be common (unpublished data).

Findings generated by studies of bronchoalveolar specimens from patients hospitalized with ARTIs show HRVs to be the predominant infectious agent and that infection may be associated with high morbidity, particularly in lung transplant patients and in patients with community acquired pneumonia (Garbino *et al.*, 2004a; Templeton *et al.*, 2005). Pneumonia is the most common infectious complication among patients undergoing stem cell transplants, and HRVs also contribute the greatest number of these infections (van Kraaij *et al.*, 2005).

Three serotypes of the human parechoviruses (HPeV) have been described to date and assigned to the genus *Parechovirus* in the family *Picornaviridae* (Ito *et al.*, 2004). Two of the serotypes were formerly called human echoviruses 22 and 23, and the other is a NIV. A range of respiratory symptoms have been reported among pediatric patients infected by these viruses (Boivin *et al.*, 2005; Abed and Boivin, 2006). Real-time PCR has successfully detected HPeV from the throat swabs of young culture-negative children (Corless *et al.*, 2002). These and several other studies (Heikkinen *et al.*, 1999; Ruohola *et al.*, 2000; Faustini *et al.*, 2006) suggest that HEVs play a greater role in ARTI than is currently thought.

Because real-time PCR applications are so infrequently employed to study picornaviruses, research is needed to focus in earnest on the role of this family in all the clinical manifestations of ARTI.

Family *Coronaviridae*

The genus *Coronavirus* has doubled the number of endemic respiratory viruses it includes thanks to recent discoveries. Real-time PCR serves an obvious role for these extremely fastidious viruses which cannot practically be isolated from primary clinical material (van Kraaij *et al.*, 2005). It took the emergence of the severe acute respiratory syndrome coronavirus (SARS-CoV) to spark interest in the historical respiratory HCoVs, a group largely forgotten by modern testing. Real-time PCR studies permitted the first detection of HCoV positive-sense, single-stranded RNA from LRT specimens and have proved ideal for rapid detection of HCoV-229E and HCoV-OC43, which is especially important when dealing with and treating patients suffering from pneumonia (Garbino *et al.*, 2004a; Templeton *et al.*, 2005). Viral loads in hospitalized patients of all ages reportedly ranged from 10^4 to 10^7 copies/mL in HCoV-OC43 infections and 10^4 copies/mL in the single HCoV-229E infection (Vijgen *et al.*, 2005). Real-time PCR detected several more positives than conventional PCR, although different assays were used for each format making comparison difficult.

Mean viral loads of 2.1×10^6 detected during HCoV-NL63 sole infections were associated with croup (laryngotracheobronchitis) in one study (van der Hoek *et al.*, 2005) but no association was found with severity of disease by another, despite viral loads peaking at the beginning of clinical illness (Chiu *et al.*, 2005). Definitive associations with specific disease entities could help the clinician request appropriate laboratory testing menus and generally aids clinical diagnosis in an area where many infections result in similar symptoms. The latter study also demonstrated that a specific real-time PCR assay was more useful to diagnose HCoV-NL63 than was a conventional pan-CoV assay which failed to detect half of all instances of HCoV-NL63. Furthermore, the study also demonstrated significantly lower HCoV loads from patients with co-incident HRSV or HPIV-3 detections, possibly suggesting competition between multiple viruses for the same target cells, or an enhancement of innate immunity resulting from previous viral infection which moderated subsequent viral replication. It might also suggest that HCoV-NL63 is simply an opportunistic pathogen making use of a depleted or recovering immune system following primary infection by HRSV or HPIV-3. During one early study of HCoV-NL63, a molecular variant from New Haven in Connecticut, USA entitled "NH" was associated with Kaposi's sarcoma. However a number of subsequent publications have cast the association in doubt due to lack of corroborating evidence, also highlighting that there was no evidence to suggest that HCoV-NH differed in any significant way from the prototype virus which might have been an explanation for its proposed unique clinical association (Shimizu *et al.*, 2005; Chang *et al.*, 2006).

HCoV-NL63 has proven to be extremely difficult to isolate using *in vitro* culture so the successful application of real-time PCR to identify two suitably permissive cell lines (LLC-MK2 and Vero-B4) was a welcome result (Schildgen *et al.*, 2006).

The SARS-CoV is an emerging virus rather than a NIV and one that has been the subject of more publications than any other emerging virus in such a short time period. This chapter will not seek to list these studies as the SARS-CoV literature has itself been extensively reviewed elsewhere (Lau and Peiris, 2005; Kahn, 2006; Sørensen *et al.*, 2006). However, real-time PCR methods have proven useful to quantify SARS-CoV load in patient specimens and these tools continue to improve upon the sensitivity of earlier real-time PCR assays (Keyaerts *et al.*, 2006; Petrich *et al.*, 2006). The initial and continuing characterization of SARS-CoV benefited from automated nucleic acid extraction and liquid handling systems which enhanced the already rapid real-time PCR result turnaround and further reduced the chance of false positive results. The pathogenesis of SARS-CoV was clarified with the determination that most viral RNA was present in the lower respiratory tract (Poon *et al.*, 2004). Therefore it followed that correct sampling was required to maximize the sensitivity of PCR assays for this target. Also, real-time PCR permitted detailed evaluation of some of the pre-PCR steps, showing that kit-based automated systems could be imprecise and yield less RNA than manual extraction procedures and highlighed the need for further validation and development in this area (Chiu *et al.*, 2006). Similarly, quantitative real-time PCR identified that the very high viral loads in feces made it the specimen of choice during the second week of illness. The association was then made between early positive, high viral load specimens, and more severe clinical outcomes. Real-time assays permitted the detection of SARS-CoV prior to seroconversion but also highlighted that very low loads of this virus could be present (Poon *et al.*, 2003). The missing host for the SARS-CoV has been hypothesized to be bats and closely related bat-CoVs have since been found in a broad range of bat species, including the bat-CoV, HKU3 where replication attains levels of 10^7 copies/mL (Lau *et al.*, 2005b).

Family *Paramyxoviridae*

The human metapneumovirus fits the criteria for a respiratory NIV perfectly; a recently detected virus associated with a range of unremarkable respiratory illnesses ranging from mild to severe. The virus was largely characterized by a range of molecular means and is endemic within the community (van den Hoogen *et al.*, 2001). To date, HMPV has had more studies published about it than any other respiratory NIV and since its first report there have been many detection, molecular epidemiology and pure virology papers published, using a raft of in-house real-time PCR assays (Table 8.1). Our own foray into the study of NIVs began here and we learned some hard lessons about developing real-time PCR assays early on in the characterization process for this negative sense, single-stranded RNA virus. While our first real-time PCR assay was designed to detect all known sequences present on GenBank (Mackay *et al.*, 2003b), we soon learned that these sequences represented only a fraction of the HMPV genetic sub-lineages that we now know exist. Thereafter, Maertzdorf *et al.* presented a new real-time PCR assay and compared its performance to our own and to a modification of the original RT-PCR assay described in the first report identifying HMPV (Maertzdorf *et al.*, 2004). The comparison demonstrated for the first time that both early-generation assays were indeed less efficient at detecting the strains for which no sequences were available when design was undertaken, and we were later able to confirm this for our own assay (Mackay *et al.*, 2004). Nonetheless, the early implementation of a molecular assay for the detection of a NIV should still be a goal we strive for; so long as we understand the perils faced by the early-adopter. It is important to identify a NIV in the local population so we may begin to examine its clinical impact among patients with ARTI and to encourage clinical microbiology laboratories to consider adding this, or a second-generation assay to their diagnostic menus. Waiting for the "perfect" assay is analogous to postponing the purchase of a computer until a future-proof system is available; it will be a long and non-productive wait. However, once an assay's failings are identified it should not be employed as a frontline screening tool. Unfortunately the use of first-generation assays, or a variant of them, is still being published, despite their reported shortcomings (Scheltinga *et al.*, 2005; Bouscambert-Duchamp *et al.*, 2005; Jacques *et al.*, 2006; Choi *et al.*, 2006). The result of using an assay incapable of detecting all four sub-lineages with equal efficiency may be a world-wide underestimation of HMPV prevalence, particularly for the type B subtypes, thus providing the perfect example of the problems faced when attempting to diagnose NIVs using PCR. One approach proposed to side-step this problem might be to study specimens using a lineage-specific assay from a time-period when only that lineage circulated (Sumino *et al.*, 2005)—although that presumably requires a foreknowledge of which lineage is circulating at what time. Since we now know that all four sub-lineages can circulate each season, that approach will undoubtedly contribute to further underestimation of the total HMPV prevalence. The Maertzdorf assay described above was recently validated by our own extensive screening and sequencing project investigating over 10 000 specimens across a four year period (Maertzdorf *et al.*, 2004; Mackay *et al.*, 2004; Mackay *et al.*, 2006a). Examination of the 724 HMPV positives we identified indicated that the lineages and sub-lineages of this NIV, like HRSV before it (Peret *et al.*, 1998; Choi and Lee, 2000) and we suspect most respiratory viruses, varied in their prevalence usually with a different sub-lineage predominating each year. Such predominance patterns suggest that antigenicity plays a likely role in the "survival" of the sub-lineages but no data yet exist to support this. Implementing a well-optimized and evaluated real-time PCR assay can also yield unexpected new results with the potential to make us question dogma; in our case, HMPV was unexpectedly detected more frequently than HRSV during some years among our large, predominantly pediatric, hospital and clinic-based population.

Since we now know that the prevalence of each HMPV sub-lineage does cycle over time, we can better understand how these viral sequence sub-clusters, which are detected every year, and in different locations might maintain their existence. Our data would argue against brand new HMPV sub-lineages routinely arising each season either as a result of quasispecies variation, immune selection or due to import from another country. The evidence indicates that all strains circulate to some degree each year, even at a single study site like ours. Very long-term international collaborative studies could be used to identify whether completely new lineages do arise over time and if the old lineages fade away.

An unusual application of real-time PCR was reported for monitoring the neutralizing effects of newly developed monoclonal antibodies directed toward HMPV (Ma *et al.*, 2005). Results indicated suppression of full blown replication despite 10^3 viral copies/μg of RNA remaining in an *in vitro* cultured monkey kidney cell line that was viral-protein negative. Real-time PCR assays have also proven useful for following the replication of HMPV in culture to identify which cell line supports the notoriously slow growth and mild cytopathicity of this NIV (Deffrasnes *et al.*, 2005). The results indicated viral loads reaching up to 10^{10} copies/mL. By comparison, children less than 12 months of age are more likely to have respiratory viral loads greater than 10^6 copies/mL than specimens from older children (Kuypers *et al.*, 2006).

When different real-time PCR assays employing SYBR Green I were compared, the HMPV N and L genes appeared to be the best diagnostic targets. However the compared assays targeted five different genes, used two or three primers per target, and the relatively large amplicons varied in size from 928 nt (nucleoprotein), 876 nt (phosphoprotein), 778 nt (matrix), 758 nt (fusion) to 549 nt (large/polymerase) which may have had an impact on the comparative amplification efficiency of the assays (Côté *et al.*, 2003). The nucleoprotein-gene assay was found to be the best and was subjected to further optimization. It could detect 10^2 copies of synthetic RNA per reaction with a dynamic range of 4 to 5 $\log_{10}$ of input template. The Maertzdorf assay has also contributed data describing that HMPV infections occur three-fold less often than HRSV infections among patients with chronic cardiac or pulmonary disorders (Carrat *et al.*, 2006). Another study found that HMPV occurred half as often as HRSV in patients presenting to the emergency department for pneumonia or acute exacerbation of chronic obstructive pulmonary disease, but with similar prevalence to IFAV (Hamelin *et al.*, 2005).

Future trends

The immediate future for real-time respiratory virus PCR is relatively clear. The wide variety of viral targets will require a variety of assays. And those will have to be kept updated. We hope that multiplexed real-time PCR assays will become more commonplace as this will reduce the number of individual reactions required, the subsequent expenses involved and the wear on instruments due to constant use. It would also seem sensible that a dedicated research and development capacity become the norm for clinical microbiology laboratories, ensuring that all molecular assays perform optimally and that new in-house assays are efficiently and reliably developed, evaluated, validated and deployed as required.

The use of microarrays will undoubtedly have a significant impact in the future, but in the meantime efforts must be made to increase their sensitivity and reduce their cost, perhaps by mass marketing or by reducing their range to tailored arrays such as a "respiratory array." Also in the near future is the increased use of xMap® technology employing a spectrally diverse population of fluorogenically labeled polystyrene beads coupled to different protein- or nucleic acid-based probes (Luminex® corporation) to solve the complexities of detecting multiple respiratory viruses. Template amplification of some sort is still required but if sensitivity could be increased this technology would permit the direct screening of a single patient specimen for a multitude of respiratory viruses and these technologies will quickly come into their own. Perhaps "sneeze-on-a-stick" diagnostics are just around the corner.

The role of the antigen could also rise to dominance again if a concerted effort was to be made to develop reliable, recombinant antibodies to detect multiple targets from all the respiratory viruses. More reliable applications for antigen detection would help to temper the volatility of PCR resulting from its vulnerability to nucleotide sequence variation.

Conclusion

Respiratory viruses are associated with the most common (number of afflicted people) and frequent (afflictions per person) infections of humans world wide. Severe illness predominantly occurs in young children, the elderly and certain "at risk" populations. Despite this, the majority of serologically distinct viral agents, most notably the picornaviruses, have not been investigated using molecular testing by real-time PCR. Considering the vast benefits of increased diagnostic speed and specificity, this is somewhat surprising although problems abound for this molecular technology when it is applied to respiratory viruses. The issues for concern can be largely defined as those related to the sequence, quality, origin and amount of the template, those relating to slow rate of development of multiplex real-time PCRs and those issues that could be addressed by the routine incorporation of internal controls in these assays.

Respiratory viruses are the most diverse target population yet encountered by real-time PCR assay designers and it seems that the more we characterize them, the more difficult it becomes to be confident in our choice of oligonucleotides and fluorogenic chemistries. Many respiratory NIVs cannot be classified as emerging viruses, but rather are endemic viruses due a separate classification. But whatever we call them, these NIVs are directly and significantly contributing to an already overwhelming diagnostic menu and new diagnostic approaches would be welcomed. Indirectly, the NIVs have taught us that the application of real-time PCR assays should follow an incremental process, and that no assay for an RNA virus will ever be entirely complete. One thing made very apparent however, is that real-time PCR technologies have matured along with our understanding of their use. Real-time PCR is an ideal tool for better structured screening studies and now it only remains for us to better implement our tools to glean the answers we need. Hopefully five years from now we could re-visit this subject and find these issues addressed.

Acknowledgments

Thanks must go to several sources of inspiration. Firstly, those who recognized the benefits of delivering the fruits of the *Coffee arabica* plant in its many and varied liquid forms. Also, SG, ST, BG and Firefly. Peter McErlean's efforts were very much appreciated in critically reviewing parts of the chapter and for working hard to provide proof for some of the topics discussed. Ron Fouchier, Lia van der Hoek, Domingo Esteban and John F. Mackay are thanked for very useful eMail exchanges, some spanning years. Thanks also to Stephen Lambert for helpful discussions. Special thanks go to Corin Mackay for being small and sleeping long.

Web references

Luminex corporation
http://www.luminexcorp.com/
Creators of xMAP technology. A high throughout system employing color-coded beads (microspheres) which can be manipulated for detection of multiple targets. Detection occurs through a flow cytometry-based approach using compact analysers.

GenBank
http://www.ncbi.nlm.nih.gov/Genbank/index.html
A massive genetic sequence database housed at the National Centre for Biotechnology Information run by the National Institutes of Health, USA.

mfold
http://www.bioinfo.rpi.edu/applications/mfold/rna/form1.cgi
A free-access internet web-server housing software that predicts possible DNA and RNA secondary structures and represents those predictions in graphical form. Emphasis on "predictions."

Picornaviridae online
http://www.picornaviridae.com/
Housing background information about members of the family and linking to study groups and news.

Eurogentec
http://www.eurogentec.be/code/en/what.asp?pk_id_what = 103
A useful description of black hole quenchers (BHQ); beneficial for design of multiplex real-time PCR assays.

Validated real-time PCR primer database
http://www.realtimeprimers.org/

An online database of oligonucleotides, including some for microorganisms, that may be a useful starting point for developing a new real-time PCR assay.

References

Abed, Y. and Boivin, G. (2006). Human parechovirus types 1, 2 and 3 infections in Canada. Emerg. Infect. Dis. *12*, 969–975.

Adcock, P.M., Stout, G.G., Hauck, M.A., and Marshall, G.S. (1997). Effect of rapid viral diagnosis on the management of children hospitalized with lower respiratory tract infection. Pediatr. Infect. Dis. J. *16*, 842–846.

Agapov, E., Sumino, K.C., Gaudreault-Keener, M., Storch, G.A., and Holtzman, M.J. (2006). Genetic variability of human metapneumovirus infection: Evidence of a shift in viral genotype without a change in illness. J. Infect. Dis. *193*, 396–403.

Arden, K.E., McErlean, P., Nissen, M.D., Sloots, T.P., and Mackay, I.M. (2006). Frequent detection of human rhinoviruses, paramyxoviruses, coronaviruses, and bocavirus during acute respirtaory tract infections. J. Med. Virol. *78*, 1232–1240.

Aruda, E. and Hayden, F.G. (1993). Detection of human rhinovirus RNA in nasal washings by PCR. Mol. Cell. Probes *7*, 373–379.

Bellau-Pujol, S., Vabret, A., Legrand, L., Dina, J., Gouarin, S., Petitjean-Lecherbonnier, J., Pozzetto, B., Ginevra, C., and Freymuth, F. (2005). Development of three multiplex RT-PCR assays for the detection of 12 respiratory RNA viruses. J. Virol. Methods *126*, 53–63.

Boivin, G., Abed, Y., and Boucher, F.D. (2005). Human parechovirus 3 and neonatal infections. Emerg. Infect. Dis. *11*, 103–107.

Boivin, G., Côté, S., Déry, P., De Serres, G., and Bergeron, M.G. (2004). Multiplex real-time PCR assay for detection of influenza and human respiratory syncytial viruses. J. Clin. Microbiol. *42*, 45–51.

Bouscambert-Duchamp, M., Lina, B., Trompette, A., Moret, H., Motte, J., and Andréoletti, L. (2005). Detection of human metapneumovirus RNA sequences in nasopharyngeal aspirates of young French children with acute bronchiolitis by real-time reverse transcriptase PCR and phylogenetic analysis. J. Clin. Microbiol. *43*, 1411–1414.

Brouard, J., Freymuth, F., Vabret, A., Jokic, M., Guillois, B., and Duhamel, J.F. (2000). Viral co-infections in immunocompetent infants with bronchiolitis: prospective epidemiologic study. Arch Peditr *7*, 531s-535s.

Bustin, S.A. (2005). Real-time, fluorescence-based quantitative PCR: a snapshot of current procedures and preferences. Expert Rev Mol Diagn 5, 493–498.

Carrat, F., Leruez-Ville, M., Tonnellier, M., Baudel, J.-L., Deshayes, J., Meyer, P., Maury, E., Galimand, J., Rouzioux, C., and Offenstadt, G. (2006). A virologic survey of patients admitted to a critical care unit for acute cardiorespiratory failure. Intensive Care Med 32, 156–159.

Chang, L.-Y., Chiang, B.-L., Kao, C.-L., W.u., M.-H., Chen, P.-J., Berkhout, B., Yang, H.-C., and Huang, L.-M. (2006). Lack of association between infection with a novel human coronavirus (HCoV), HCoV-NH, and Kawasaki disease in Taiwan. J. Infect. Dis. *193*, 283–286.

Chiu, R.W.C., Jin, Y., Chung, G.T.Y., Lui, W.-B., Chan, A.T.C., Lim, W., and, L.o., Y.M.D. (2006). Automated extraction protocol for quantification of SARS-Coronavirus RNA in serum: an evaluation study. BMC Infect. Dis. 6.

Chiu, S.S., Chan, K.H., Chu, K.W., Kwan, S.W., Guan, Y., Poon, L.L.M., and Peiris, J.S.M. (2005). Human coronavirus NL63 infection and other coronavirus infections in children hospitalized with acute respiratory disease in Hong Kong, China. Clin. Inf. Dis. *40*, 1721–1729.

Choi, E.H. and Lee, H.J. (2000). Genetic diversity and molecular epidemiology of the G protein of subgroups A and B of respiratory syncytial viruses isolated over 9 consecutive epidemics in Korea. J. Infect. Dis. *181*, 1547–1556.

Choi, E.H., Lee, H.J., Kim, S.J., Eun, B.W., Kim, N.H., Lee, J.A., Lee, J.H., Song, E.K., Kim, S.H., Park, J.Y., and Sung, J.Y. (2006). The association of newly identified respiratory viruses with lower respiratory tract infections in Korean children, 2000–2005. 43.

Chui, L., Drebot, M., Andonov, A., Petrich, A., Glushek, M., and Mahony, J. (2005). Comparison of 9 different PCR primers for the rapid detection of severe acute respiratory syndrome coronavirus using 2 RNA extraction methods. Diagn Microbiol. Infect Dis.

Coiras, M.T., Aguilar, J.C., García, M.L., Casas, I., and Pérez-Breña, M.P. (2004). Simultaneous detection of fourteen respiratory viruses in clinical specimens by two multiplex reverse transcription nested-PCR assays. J. Med. Virol. *72*, 484–495.

Coiras, M.T., Pérez-Breña, P., García, M.L., and Casas, I. (2003). Simultaneous detection of influenza, A., B and C viruses, respiratory syncytial virus, and adenoviruses in clinical samples by multiplex reverse transcriptase nested-PCR assay. J. Med. Virol. *69*, 132–144.

Corless, C.E., Guiver, M., Borrow, R., Edwards-Jones, V., Fox, A.J., Kaczmarski, E.B., and Mutton, K.J. (2002). Development and evaluation of a "real-time" RT-PCR for the detection of enterovirus and parechovirus RNA in CSF and throat swab samples. J. Med. Virol. *67*, 555–562.

Côté, S., Abed, Y., and Boivin, G. (2003). Comparative evaluation of real-time PCR assays for detection of the human metapneumovirus. J. Clin. Microbiol. *41*, 3631–3635.

Dagher, H., Donninger, H., Hutchinson, P., Ghildyal, R., and Bardin, P. (2004). Rhinovirus detection: comparison of real-time and conventional PCR. J. Virol. Methods *117*, 113–121.

Deffrasnes, C., Côté, S., and Boivin, G. (2005). Analysis of replication kinetics of the human metapneumopvirus in different cell lines by real-time PCR. J. Clin. Microbiol. *43*, 488–490.

Dillon, P.J. and Rosen, C.A. (1990). A rapid method of the construction of synthetic genes using the polymerase chain reaction. Biotechniques *9*, 298–300.

Domingo, E. (1998). Quasispecies and the implications for virus persistence and escape. Clin. Diag. Virol. *10*, 97–101.

Domingo, E., Martín, V., Perales, C., Grande-Pérez, A., García-Arriaza, J., and Arias, A. (2006). Viruses as quasispecies: Biological implications. Curr Top Microbiol. Immunol *299*, 51–58.

Falkow, S. (2004). Molecular Koch's postulates applied to bacterial pathogenicity—a personal recollection 15 years later. Nat Rev Microbiol. *2*, 67–72.

Falsey, A.R. and Walsh, E.E. (2006). Viral pneumonia in older adults. Clin. Inf. Dis. *42*, 518–524.

Faustini, A., Fano, V., Muscillo, M., Zaniratti, S., La Rosa, G., Tribuzi, L., and Perucci, C.A. (2006). An outbreak of aseptic meningitis due to echovirus 30 associated with attending school and swimming in pools. Int. J. Infect. Dis. *10*, 291–297.

Fleige, S. and Pfaffl, M.W. (2006). RNA integrity and the effect on the real-time qRT-PCR performance. Mol. Aspects Med. *27*, 126–139.

Fouchier, R.A., Hartwig, N.G., Bestebroer, T.M., Niemeyer, B., de Jong, J.C., Simon, J.H., and Osterhaus, A.D. (2004). A previously undescribed coronavirus associated with respiratory disease in humans. Proc. Natl. Acad. Sci. USA *101*, 6212–6216.

Fouchier, R.A.M., Rimmelzwaan, G.F., Kuiken, T., and Osterhaus, A.D.M.E. (2005). Newer respiratory virus infections: human metapneumovirus, avian influenza virus, and human coronaviruses. Curr. Opin. Infect. Dis. *18*, 141–146.

Fredricks, D.N. and Relman, D.A. (1996). Sequence-based identification of microbial pathogens: a reconsideration of Koch's postulates. Clin. Microbiol. Rev. *9*, 18–33.

Freymuth, F., Vabret, A., Rozenberg, F., Dina, J., Petitjean, J., Gouarin, S., Legrand, L., Corbet, S., Brouard, J., and Lebon, P. (2005). Replication of respiratory viruses, particularly influenza virus, rhinovirus, and coronavirus in HuH7 hepatocarcinoma cell line. J. Med. Virol. *77*, 295–301.

Gama, R.E., Horsnell, P.R., Hughes, P.J., North, C., Bruce, C.B., Al-Nakib, W., and Stanway, G. (1989). Amplification of rhinovirus specific nucleic acids from clinical samples using the polymerase chain reaction. J. Med. Virol. *28*, 73–77.

Gama, R.E., Hughes, P.J., Bruce, C.B., and Stanway, G. (1988). Polymerase chain reaction amplification of rhinovirus nucleic acids from clinical material. Nucleic Acids Res. *16*, 9346.

Gao, S.-J. and Moore, P.S. (1996). Molecular approaches to the identification of unculturable infectious agents. Emerg. Infect. Dis. *2*, 159–167.

Garbino, J., Gerbasse, M.W., Wunderli, W., Deffernez, C., Thomas, Y., Rochat, T., Ninet, B., Schrenzel, J., Yerly, S., Perrin, L., Soccal, P.M., Nicod, L., and Kaiser, L. (2004a). Lower respiratory viral illnesses: Improved diagnosis by molecular methods and clinical impact. Am. J. Respir. Crit. Care Med. *170*, 1197–1203.

Garbino, J., Gerbasse, M.W., Wunderli, W., Kolarova, L., Nicod, L.P., Rochat, T., and Kaiser, L. (2004b). Respiratory viruses and severe lower respiratory tract complications in hospitalized patients. Chest *125*, 1033–1039.

Gardner, S.N., Kuczmarski, T.A., Vitalis, E.A., and Slezak, T.R. (2003). Limitations of TaqMan PCR for detecting divergent viral pathogens illustrated by hepatitis, A., B., C and E viruses and human immunodeficiency virus. J. Clin. Microbiol. *41*, 2417–2427.

Glezen, W.P., Loda, F.A., Clyde, W.A., Senior, R.J., Sheaffer, C.I., Conley, W.G., and Denny, F.W. (1971). Epidemiologic patterns of acute lower respiratory disease of children in a pediatric group practice. J. Pediatr. *78*, 397–406.

Grove, S., Faller, R., Soleim, K.B., and Dannevig, B.H. (2006). Absolute quantitation of RNA by a competitive real-time RT-PCR method using piscine nodavirus as a model. J. Virol. Methods *132*, 104–112.

Grutke, P., Glas, A.S., Dierdorp, M., Vreede, W.B., Pilon, J.-W., and Bruisten, S.M. (2004). Practical implementation of a multiplex PCR for acute respiratory tract infections in children. J. Clin. Microbiol. *42*, 5596–5603.

Gunson, R.N., Collins, T.C., and Carman, W.F. (2005). Real-time RT-PCR detection of 12 respiratory viral infections in four triplex reactions. J. Clin. Virol.

Gunson, R.N., Collins, T.C., and Carman, W.F. (2006). Practical experience of high throughput real time PCR in the routine diagnostic virology setting. J. Clin. Virol. *35*, 355–367.

Gwaltney, J.r., J.M. (1982). Rhinoviruses. In Viral Infection of Humans: Epidemiology and Control, A.S. Evans and, R.A. Kaslow, eds. (New York: Plenum Press), pp. 491–517.

Halonen, P., Rocha, E., Hierholzer, J., Holloway, B., Hyypiä, T., Hurskainen, P., and Pallansch, M. (1995). Detection of enteroviruses and rhinoviruses in clinical specimens by PCR and liquid-phase hybridization. J. Clin. Microbiol. *33*, 648–653.

Hamelin, M.-E., Côté, S., Laforge, J., Lampron, N., Bourbeau, J., Weiss, K., Gilca, R., DeSerres, G., and Boivin, G. (2005). Human metapneumovirus infection in adults with community-acquired pneumonia and exacerbation of chornic obstructive pulmonary disease. Clin. Inf. Dis. *41*, 498–502.

Hayden, F.G. (2004). Rhinovirus and the lower respiratory tract. Rev. Med. Virol. *14*, 17–31.

Heikkinen, T., Thint, M., and Chonmaitree, T. (1999). Prevelance of various respiratory viruses in the middle ear during acute otitis media. N. Engl. J Med *340*, 260–264.

Heim, A., Ebnet, C., Harste, G., and Pring-Åkerblom, P. (2003). Rapid and quantitative detection of human adenovirus by real-time PCR. J. Med. Virol. *70*, 228–239.

Hoebeeck, J., Speleman, F., and Vandesompele, J. (2006). Real-time quantitative PCR as an alternative to Southern blot or fluorescence *in situ* hybridization for detection of gene copy number changes. In Protocols for Nucleic Acid Analysis by Non-Radioactive Probes, E. Hilario and, J. Mackay, eds. (Totowa: Humana Press), pp. 205–226.

Holland, J.J. (2006). Transitions in understanding of RNA viruses: A historical perspective. Curr. Top. Microbiol. Immunol. *299*, 371–401.

Horton, R.M., Cai, Z., H.o., S.N., and Pease, L.R. (1990). Gene splicing by overlap extension: Tailor-made genes using the polymerase chain reaction. Biotechniques *8*, 528–535.

Ito, M., Yamashita, T., Tsuzuki, H., Takeda, N., and Sakae, K. (2004). Isolation and identification of a novel human parechovirus. J Gen Virol *85*, 391–398.

Jacques, J., Bouscambert-Duchamp, M., Moret, H., Carquin, J., Brodard, V., Lina, B., Motte, J., and Andréoletti, L. (2006). Association of respiratory picornaviruses with acute bronchiolitis in French infants. J. Clin. Virol.

Jartti, T., Lehtinen, P., Vuorinen, P., Koskenvuo, M., and Ruuskanen, O. (2004a). Persistence of rhinovirus and enterovirus RNA after acute respiratory illness in children. J. Med. Virol. *72*, 695–699.

Jartti, T., Lehtinen, P., Vuorinen, T., Österback, R., van den Hoogen, B., Osterhaus, A.D.M.E., and Ruuskanen, O. (2004b). Respiratory picornaviruses and respiratory syncytial virus as causative agents of acute expiratory wheezing in children. Emerg. Infect. Dis. *10*, 1095–1101.

Jenkins, G.M., Worobey, M., Woelk, C.H., and Holmes, E.C. (2001). Evidence for the non-quasispecies evolution of RNA viruses. Mol. Biol. Evol. *18*, 987–994.

Johnson, R.E., Newhall, W.J., Papp, J.R., Knapp, J.S., Black, C.M., Gift, T.L., Steece, R., Markowitz, L.E., Devine, O.J., Walsh, C.M., Wang, S., Gunter, D.C., Irwin, K.L., DeLisle, S., and Berman, S.M. (2002). Screening tests to detect *Chlamydia trachomatis* and *Neisseria gonorrhoeae* infections –2002. MMWR 38.

Juvén, T., Mertsola, J., Waris, M., Leinonen, M., Meurman, O., Roivainen, M., Eskola, J., Saikku, P., and Ruuskanen, O. (2000). Etiology of community-acquired pneumonia in 254 hospitalized children. Pediatr. Infect. Dis. J. *19*, 293–298.

Kahn, J.S. (2006). The widening scope of coronaviruses. Curr Opin Pediatr *18*, 42–47.

Kares, S., Lönnrot, M., Vuorinen, P., Oikarinen, S., Taurianen, S., and Hyöty, H. (2004). Real-time PCR for rapid diagnosis of entero- and rhinovirus infections using LightCycler. J. Clin. Virol. *29*, 99–104.

Keyaerts, E., Vijgen, L., Maes, P., Duson, G., Neyts, J., and Van Ranst, M. (2006). Viral load quantitation of SARS-coronavirus RNA using a one-step real-time RT-PCR. Int. J. Infect. Dis. *10*, 32–37.

Kusel, M.M.H., de Klerk, N.H., Holt, P.G., Kebadze, T., Johnston, S.L., and Sly, P.D. (2006). Role of respiratory viruses in acute upper and lower respiratory tract illness in the first year of life. Pediatr. Infect. Dis. J. *25*, 680–686.

Kuypers, J., Wright, N., Ferrenberg, J., Huang, M.-L., Cent, A., Corey, L., and Morrow, R. (2006). Comparison of real-time PCR assays with fluorescent-antibody assays for diagnosis of respiratory virus infections in children. J. Clin. Microbiol. *44*, 2382–2388.

Lau, S.K.P., T.o., W.-K., Tse, P.W.T., Chan, A.K.H., Woo, P.C.Y., Tsoi, H.-W., Leung, A.F.Y., L.i., K.S.M., Chan, P.K.S., Lim, W.W.L., Yung, R.W.H., Chan, K.-H., and Yuen, K.-Y. (2005a). Human parainfluenza virus 4 outbreak and the role of diagnostic tests. J. Clin. Microbiol. *43*, 4515–4521.

Lau, S.K.P., Woo, P.C.Y., L.i., K.S.M., Huang, Y., Tsoi, H.-W., Wong, B.H.L., Wong, S.S.Y., Leung, S.-Y., Chan, K.-H., and Yuen, K.-Y. (2005b). Severe acute respiratory syndrome coronavirus-like virus in Chinese horsehoe bats. Proc. Natl. Acad. Sci. USA *102*, 14040–14045.

Lau, S.K.P., Woo, P.C.Y., Yip, C.C.Y., Tse, H., Tsoi, H.-W., Cheng, V.C.C., Lee, P., Tang, B.S.F., Cheung, C.H.Y., Lee, R.A., S.o., L.-Y., Lau, Y.-L., Chan, K.-H., and Yuen, K.-Y. (2006). Coronavirus HKU1 and other coronavirus infections in Hong Kong. J. Clin. Microbiol. *44*, 2063–2071.

Lau, Y.L. and Peiris, J.S.M. (2005). Pathogenesis of severe acute respiratory syndrome. Curr. Opin. Immunol. *17*, 404–410.

Ledford, R.M., Patel, N.R., Demenczuk, T.M., Watanyar, A., Herbertz, T., Collett, M.S., and Pevear, D.C. (2004). VP1 sequencing of all human rhinovirus serotypes: Insights into genus phylogeny and susceptibility to antiviral capsid-binding compounds. J. Virol. *78*, 3663–3674.

Lee, B.E., Robinson, J.L., Khurana, V., Pang, X.L., Preiksaitis, J.K., and Fox, J.D. (2006). Enhanced identification of viral and atypical bacterial pathogens in lower respiratory tract samples with nucleic acid amplification tests. J. Med. Virol. *78*, 702–710.

Limaye, A.P., Huang, M.-L., Leisenring, W., Stensland, L., Corey, L., and Boeckh, M. (2001). Cytomegalovirus (CMV) DNA load in plasma for the diagnosis of CMV disease before engraftment in hematopoietic stem-cell transplant recipients. J. Infect. Dis. *183*, 377–382.

Loens, K., Ieven, M., Ursi, D., de Laat, C., Sillekens, P., Oudshoorn, P., and Goossens, H. (2003). Improved detection of rhinoviruses by nucleic acid sequence-based amplification after nucleotide sequence determination of the 5′ noncoding regions of additional rhinovirus strains. J. Clin. Microbiol. *41*, 1971–1976.

Lönnrot, M., Sjöroos, M., Salminen, K., Maaronen, M., Hyypiä, T., and Hyöty, H. (1999). Diagnosis of enterovirus and rhinovirus infections by RT-PCR and time-resolved fluorometry with lanthanide chelate labelled probes. J. Med. Virol. *59*, 378–384.

Lu, X., Chittaganpitch, M., Olsen, S.J., Mackay, I.M., Sloots, T.P., Fry, A.M., and Erdman, D.D. (2006). Real-time PCR assays for detection of bocavirus in human specimens. J. Clin. Microbiol. *44*, 3231–3235.

Ma, X., Endo, R., Ebihara, T., Ishiguro, N., Ishiko, H., and Kikuta, H. (2005). Production and characterization of neutralizing monoclonal antibodies against human metapneumovirus F protein. Hybridoma *24*, 201–205.

Mackay, I.M., Bialasiewicz, S., Jacob, K.C., McQueen, E., Arden, K.E., Nissen, M.D., and Sloots, T.P. (2006a). Genetic diversity of human metapneumovirus over 4 consecutive years in Australia. J. Infect. Dis. *193*, 1630–1633.

Mackay, I.M., Gardam, T., Arden, K.E., McHardy, S., Whiley, D.M., Crisante, E., and Sloots, T.P. (2003a). Co-detection and discrimination of six human herpesviruses by multiplex PCR-ELAHA. J. Clin. Virol. *28*, 291–302.

Mackay, I.M., Harnett, G.B., Jeoffreys, N., Bastian, I., Sriprakash, K.S., Siebert, D., and Sloots, T. (2006b). Detection and discrimination of herpes simplex viruses, Haemophilus ducreyi, Treponema pallidum, and Calymmatobacterium (Klebsiella) granulomatis from genital ulcers. Clin. Infect. Dis. *42*, 1431–1438.

Mackay, I.M., Jacob, K.C., Woolhouse, D., Waller, K., Syrmis, M.W., Whiley, D.M., Siebert, D.J., Nissen, M.N., and Sloots, T.P. (2003b). Molecular assays for the detection of human metapneumovirus. J. Clin. Microbiol. *41*, 100–105.

Mackay, I.M., Waliuzzaman, Z., Chidlow, G.R., Fegredo, D.C., Laingam, S., Adamson, P., Harnett, G.B., Rawlinson, W., Nissen, M.D., and Sloots, T.P. (2004). Use of the P gene to genotype human metapneumovirus identifies 4 viral subtypes. J. Infect. Dis. *190*, 1913–1918.

Maertzdorf, J., Wang, C.K., Brown, J.B., Quinto, J.D., Chu, M., de Graaf, M., van den Hoogen, B.G., Spaete, R., Osterhaus, A.D.M.E., and Fouchier, R.A.M. (2004). Real-time reverse transcriptase PCR assay for detection of human metapneumoviruses from all known genetic lineages. J. Clin. Microbiol. *42*, 981–986.

Maggi, F., Pifferi, M., Vatteroni, M., Fornai, C., Tempestini, E., Anzilotti, S., Lanini, L., Andreoli, E., Ragazzo, V., Pistello, M., Specter, S., and Bendinelli, M. (2003). Human metapneumovirus associated with respiratory tract infections in a 3-year study of nasal swabs from infants in Italy. J. Clin. Microbiol. *41*, 2987–2991.

Morse, S.S. (1995). Factors in the emergence of infectious diseases. Emerg. Infect. Dis. *1*, 7–15.

Morse, S.S. (1997). The public health threat of emerging viral disease. J. Nutr. *127*, 951S-957S.

Morse, S.S. (2004). Factors and determinants of disease emergence. Rev Sci. Tech *23*, 443–451.

Mullan, B., Sheehy, P., Shanahan, F., and Fanning, L. (2004). Do *Taq*-generated RT-PCR products from RNA viruses accurately reflect viral genetic heterogeneity? J. Viral Hepatitis *11*, 108–114.

Murphy, F.A. and Nathanson, N. (1994). Introduction: New and emerging virus diseases. Sem. Virol. *5*, 85–86.

Nijhuis, M., van Maarseveen, N., Schuurman, R., Verkuijlen, S., de Vos, M., Hendricksen, K., and van Loon, A.M. (2002). Rapid and sensitive routine detection of all members of the genus *Enterovirus* in different clinical specimens by real-time PCR. J. Clin. Microbiol. *40*, 3666–3670.

Nokso-Koivisto, J., Kinnari, T.J., Lindahl, P., Hovi, T., and Pitkäranta, A. (2002). Human picornavirus and coronavirus RNA in nasopharnyx of children without concurrent respiratory symptoms. J. Med. Virol. *66*, 417–420.

Novella, I.S., Duarte, E.A., Elena, S.F., Moya, A., Domingo, E., and Holland, J.J. (1995). Exponential increases of RNA virus fitness during large population transmissions. Proc. Natl. Acad. Sci. USA *92*, 5841–5844.

Oberste, M.S., Maher, K., Kilpatrick, D.R., and Pallansch, L.A. (1999). Molecular evolution of the human enteroviruses: Correlation of serotype with VP1 sequence and application to picornavirus classification. J. Virol. *73*, 1941–1948.

Oosterheert, J.-J., van Loon, A.M., Shuurman, R., Hoepelman, A.I.M., Hak, E., Thijsen, S., Nossent, G., Schneider, M.M.E., Hustinx, W.M.N., and Bonten, M.J.M. (2005). Impact of rapid detection of viral and atypical bacterial pathogens by real-time polymerase chain reaction for patients with lower respiratory tract infection. Clin. Inf. Dis. *41*, 1438–1444.

Papadopoulos, N.G., Bates, P.J., Bardin, P.G., Papi, A., Leir, S.H., Fraenkel, D.J., Meyer, J., Lackie, P.M., Sanderson, G., Holgate, S.T., and Johnston, S.L. (2000). Rhinoviruses infect the lower airways. J. Infect. Dis. *181*, 1875–1884.

Papadopoulos, N.G., Hunter, J., Sanderson, G., Meyer, J., and Johnston, S.L. (1999a). Rhinovirus identification by *Bgl*I digestion of picornavirus RT-PCR amplicons. J. Virol. Methods *80*, 179–185.

Papadopoulos, N.G., Moustaki, M., Tsolia, M., Bossios, A., Astra, E., Prezerakou, A., Gourgiotis, D., and Kafetzis, D. (2002). Association of rhinovirus infection with increased disease severity in acute bronchiolitis. Am. J. Respir. Crit. Care Med. *165*, 1285–1289.

Papadopoulos, N.G., Sanderson, G., Hunter, J., and Johnston, S.L. (1999b). Rhinoviruses replicate effectively at lower airway temperatures. J. Med. Virol. *58*, 100–104.

Peltola, J., Waris, M., Hyypiä, T., and Ruuskanen, O. (2006). Respiratory viruses in children with invasive pneumococcal disease. Clin. Inf. Dis. *43*, 266–268.

Peret, T.C.T., Hall, C.B., Schnabel, K.C., Golub, J.A., and Anderson, L.J. (1998). Circulation patterns of genetically distinct group A and B strains of human respiratory syncytial virus in a community. J. Gen. Virol. *79*, 2221–2229.

Petrich, A., Mahony, J., Ching, S., Broukhanski, G., Gharabaghi, F., Johnson, G., Louie, L., Luinstra, K., Willey, B., Akhaven, P., Chui, L., Jamieson, F., Louie, M., Mazzulli, T., Tellier, R., Smieja, M., Cai, W., Chernesky, M., and Richardson, S.E. (2006). Multicenter comparison of nucleic acid extraction methods for detection of severe acute respiratory syndrome coronavirus RNA in stool specimens. J. Clin. Microbiol. *44*, 2681–2688.

Pitkäranta, A., Roivainen, M., Blomgren, K., Peltola, J., Kaijalainen, T., Räty, R., Ziegler, T., Rönkkö, E., Hatakka, K., Korpela, R., Poussa, T., Leinonen, M., and Hovi, T. (2005). Presence of viral and bacterial pathogens in the nasopharynx of otitis-prone children. A prospective study. Int. J. Pediatr. Otorhinolaryngol. *70*, 647–654.

Poon, L.L.M., Nicholls, J.M., Yuen, K.Y., and Peiris, J.S.M. (2004). The aetiology, origins, and diagnosis of severe acute respiratory syndrome. Lancet *4*, 663–671.

Poon, L.L.M., Wong, O.K., Luk, W., Yuen, K.Y., Peiris, J.S.M., and Guan, Y. (2003). Rapid diagnosis of a coronavirus associated with severe acute respiratory syndrome (SARS). Clin. Chem. *49*, 953–955.

Quereda, C., Corral, I., Laguna, F., Valencia, M.E., Tenorio, A., Echevarria, J.E., Navas, E., Martín-Davilla, P., Moreno, A., Moreno, V., Gonzalez-Lahoz, J.M., Arribas, J.R., and Guerrero, A. (2000). Diagnostic utility of a multiplex herpesvirus PCR assay performed with cerebrospinal fluid from human immunodeficiency virus-infected patients with neurological disorders. J. Clin. Microbiol. *38*, 3061–3067.

Relman, D.A. (2003). Shedding light on microbial detection. N. Engl. J. Med. *349*, 2162–2163.

Ruohola, A., Heikkinen, T., Waris, M., Puhakka, T., and Ruuskanen, O. (2000). Intranasal fluticasone propionate does not prevent acute otitis media during viral upper respiratory infection in children. J. Allergy Clin. Immunol. *106*, 467–471.

Santti, J., Hyypiä, T., and Halonen, P. (1997). Comparison of PCR primer pairs in thre detection of human rhinoviruses in nasopharyngeal aspirates. J. Virol. Methods *66*, 139–147.

Scheltinga, S.A., Templeton, K.E., Beersma, M.F.C., and Claas, E.C.J. (2005). Diagnosis of human metapneumovirus and rhinovirus in patients with respiratory tract infections by an internally controlled multiplex real-time RNA PCR. J. Clin. Virol. *33*, 306–311.

Schildgen, O., Jebbink, M.F., de Vries, M., Pyrc, K., Dijkman, R., Simon, A., Müller, A., Kupfer, B., and van der Hoek, L. (2006). Identification of cell lines permissive for human coronavirus NL63. J. Virol. Methods.

Shimizu, C., Shike, H., Baker, S.C., Garcia, F., van der Hoek, L., Kuijpers, T.W., Reed, S.L., Rowley, A.H., Shulman, S.T., Talbot, H.K.B., Williams, J.V., and Burns, J.C. (2005). Human coronavirus NL63 is not detected in the respiratory tracts of children with acute Kawasaki disease. J. Infect. Dis. *192*, 1767–1771.

Sloots, T.P., McErlean, P., Speicher, D.J., Arden, K.E., Nissen, M.D., and Mackay, I.M. (2006). Evidence of human coronavirus HKU1 and human bocavirus in Australian children. J. Clin. Virol. *35*, 99–102.

Sørensen, M.D., Sørensen, B., Gonzalez-Dosal, R., Melchjorsen, C.J., Weibel, J., Wang, J., Jun, C.W., Huanming, Y., and Kristensen, P. (2006). Severe acute respirtaory syndrome (SARS): Development of diagnostics and antivirals. Ann. N. Y. Acad. Sci. *1067*, 500–505.

Ståhlberg, A., Håkansson, J., Xian, X., Semb, H., and Kubista, M. (2004a). Properties of the reverse transcription reaction in mRNA quantification. Clin. Chem. *50*, 509–515.

Ståhlberg, A., Kubista, M., and Pfaffl, M. (2004b). Comparison of reverse transcriptases in gene expression analysis. Clin. Chem. *50*, 1678–1680.

Steinhauer, D.A. and Holland, J.J. (1987). Rapid evolution of RNA viruses. Annu. Rev. Microbiol. *41*, 409–433.

Su, J.R. (2004). Emerging viral infections. Clin. Lab. Med. *24*, 773–795.

Sumino, K.C., Agapov, E., Pierce, R.A., Trulock, E.P., Pfeifer, J.D., Ritter, J.H., Gaudreault-Keener, M., Storch, G.A., and Holtzman, M.J. (2005). Detection of severe human metapneumovirus infection by real-time polymerase chain reaction and histopathological assessment. J. Infect. Dis. *192*, 1052–1060.

Syrmis, M.W., Whiley, D.M., Thomas, M., Mackay, I.M., Williamson, J., Siebert, D.J., Nissen, M.D., and Sloots, T.P. (2004). A sensitive, specific and cost-effective multiplex reverse-transcriptase-PCR assay for the detection of seven common respiratory viruses in respiratory samples. J. Mol. Diagn. *6*, 125–131.

Tanaka, N., Kimura, H., Hoshino, Y., Kato, K., Yoshikawa, T., Asano, Y., Horibe, K., Kojima, S., and Morishima, T. (2000). Monitoring four herpesviruses in unrelated cord blood transplantation. Bone Marrow Transplant *26*, 1193–1197.

Taylor, L.H., Latham, S.M., and Woolhouse, M.E.J. (2001). Risk factors for human disease emergence. Philos. Trans. R. Soc. Lond. B Biol. Sci. *356*, 983–989.

Templeton, K.E., Scheltinga, S.A., van den Eeden, W.C.J.F.M., Graffelman, A.W., van den Broek, P.J., and Claas, E.C.J. (2005). Improved diagnosis of the etiology of community-acquired pneumonia with real-time polymerase chain reaction. Clin. Inf. Dis. *41*, 345–351.

Tsolia, M.N., Psarras, S., Bossios, A., Audi, H., Paldanius, M., Gourgiotis, D., Kallergi, K., and Kafetzis, D.A. (2004). Etiology of community-acquired pneumonia in hospitalized school-aged children: Evidence for high prevalance of viral infections. Clin. Inf. Dis. *39*, 681–686.

Turner, R.B. (1998). The common cold. Pediatr Annals *27*, 790–795.

Twiselton-Wykeham-Fiennes, R.N. (1978). In Zoonoses and the origins and ecology of human disease (London: Academic Press).

Vabret, A., Dina, J., Gouarin, S., Petitjean, J., Corbet, S., and Freymuth, F. (2006). Detection of the new human coronavirus HKU1: A report of 6 cases. Clin. Inf. Dis. *42*, 634–639.

van Benten, I., Koopman, L., Niesters, B., Hop, W., vsn Middelkoop, B., de Waal, L., van Drunen, K., Osterhaus, A., Neijens, H., and Fokkens, W. (2003). Predominance of rhinovirus in the nose of symptomatic and asymptomatic infants. Pediatr. Allergy Immunol. *14*, 363–370.

van de Pol, A.C., Wolfs, T.F.W., Jansen, N.J.G., van Loon, A.M., and Rossen, J.W.A. (2006). Diagnostic value of real-time polymerase chain reaction to detect viruses in young children admitted to the paediatric intensive care unit with lower respiratory tract infection. Crit. Care *10*, R61.

van den Hoogen, B.G., de Jong, J.C., Groen, J., Kuiken, T., de Groot, R., Fouchier, R.A.M., and Osterhaus, A.D.M.E. (2001). A newly discovered human pneumovirus isolated from young children with respiratory tract disease. Nat. Med. *7*, 719–724.

van der Hoek, L., Pyrc, K., Jebbink, M.F., Vermeulen-Oost, W., Berkhout, R.J.M., Wolthers, K.C., Wertheim-van DIllen, P.M.E., Kaandorp, J., Spaargaren, J., and Berkhout, B. (2004). Identification of a new human coronavirus. Nat. Med.

van der Hoek, L., Sure, K., Ihorst, G., Stang, A., Pyrc, K., Jebbink, M.F., Petersen, G., Forster, J., Berkhout, B., and Überla, K. (2005). Croup is associated with the novel coronavirus NL63. PloS Medicine *2*, e240.

van Elden, L.J.R., van Loon, A.M., van Alphen, F., Hendriksen, K.A.W., Hoepelman, A.I.M., van Kraaij, M.G.J., Oosterheert, J.-J., Schipper, P., Schuurman, R., and Nijhuis, M. (2004). Frequent detection of human coronaviruses in clinical specimens from patients with respiratory tract infection by use of a novel real-time reverse-transcription polymerase chain reaction. J. Infect. Dis. *189*, 652–657.

van Elden, L.J.R., van Loon, A.M., van der Beck, A., Hendricksen, K.A.W., Hoepelman, A.I.M., van Kraaij, M.G.J., Schipper, P., and Nijhuis, M. (2003). Applicability of a real-time quantitative PCR assay for diagnosis of respiratory syncytial virus infection in immunocompromised adults. J. Clin. Microbiol. *41*, 4378–4381.

van Kraaij, M.G.J., van Elden, L.J.R., van Loon, A.M., Hendriksen, K.A.W., Laterveer, L., Dekker, A.W., and Nijhuis, M. (2005). Frequent detection of respiratory viruses in adult recipients of stem cell transplants with the use of real-time polymerase chain reaction, compared with viral culture. Clin. Inf. Dis. *40*, 662–669.

Vijgen, L., Keyaerts, E., Moës, E., Maes, P., Duson, G., and Van Ranst, M. (2005). Devlopment of one-step, real-time, quantitative reverse transcriptase PCR assays for absolute quantitation of human coronaviruses OC43 and 229E. J. Clin. Microbiol. *43*, 5452–5456.

Vile, R.G., Diaz, R.M., Miller, N., Mitchell, S., Tuszyanski, A., and Russell, S.J. (1995). A model of viral wheeze in nonasthmatic adults: symptoms and physiology. Virology *214*, 307–313.

Welsh, J., Chada, S., Dalal, S.S., Cheng, R., Ralph, D., and McClelland, M. (1992). Arbitrarily primed PCR fingerprinting of RNA. Nucleic Acids Res. *20*, 4965–4970.

Welsh, J. and McClelland, M. (1990). Fingerprinting genomes using PCR with arbitrary primers. Nucleic Acids Res. *18*, 7213–7218.

Williams, J.G.K., Kubelik, A.R., Livak, J., Rafalski, J.A., and Tingey, S.V. (1990). DNA polymorphisms amplified by arbitrary primers are useful as genetic markers. Nucleic Acids Res. *18*, 6531–6535.

Williams, J.V. (2005). The clinical presentation and outcomes of children infected with newly identified respiratory tract viruses. Inf. Dis. Clin. Nth. Am. *19*, 569–584.

Winther, B., Hayden, F.G., and Hendley, J.O. (2006). Picornavirus infections in children diagnosed by RT-PCR during longitudinal surveillance with weekly sampling: Association with symptomatic illness and effect of season. J. Med. Virol. *78*, 644–650.

Woo, P.C.Y., Chiu, S.S., Seto, W.H., and Peiris, M. (1997). Cost-effectiveness of rapid diagnosis of viral respiratory tract infections in pediatric patients. J. Clin. Microbiol. *35*, 1579–1581.

Woo, P.C.Y., Lau, S.K.P., Chu, C.-M., Chan, K.-H., Tsoi, H.-W., Huang, Y., Wong, B.H.L., Poon, R.W.S., Cai, J.J., Luk, W.-K., Poon, L.L.M., Wong, S.S.Y., Guan, Y., Peiris, J.S.M., and Yuen, K.-Y. (2005a). Characterization and complete genome sequence of a novel coronavirus, coronavirus HKU1, from patients with pneumonia. J. Virol. *79*, 884–895.

Woo, P.C.Y., Lau, S.K.P., Tsoi, H.-W., Huang, Y., Poon, R.W.S., Chu, C.-M., Lee, R.A., Luk, W.-K., Wong, G.K.M., Wong, B.H.L., Cheng, V.C.C., Tang, B.S.F., W.u., A.K.L., Yung, R.W.H., Chen, H., Guan, Y., Chan, K.-H., and Yuen, K.-Y. (2005b). Clinical and molecular epidemiological features of coronavirus HKU1-associated community-acquired pneumonia. J. Infect. Dis. *192*, 1898–1907.

Woolhouse, M.E.J. and Gowtage-Sequeria, S. (2005). Host range and emerging and reemerging pathogens. Emerg. Infect. Dis. *11*, 1842–1847.

Wright, A.L., Taussig, L.M., Ray, C.G., Harrison, H.R., and Holberg, C.J. (1989). The Tucson children's respiratory study. II. Lower respiratory tract illness in the first year of life. Am. J. Epidemiol. *129*, 1232–1246.

Yamamoto, K., Yoshikawa, T., Okamoto, S., Yamamoto, K., Shimokata, K., and Nishiyama, Y. (2005). HHV-6 and HHV-7 DNA loads in lung tissues collected from patients with interstitial pneumonia. J. Med. Virol. *75*, 70–75.

Rapid Detection of Bioterror Agents

9

Andreas Nitsche

Abstract

The detection of infectious agents is one of the most important tasks in the modern diagnostic laboratory. Many PCR-based methods for clinically relevant infectious agents have been implemented in routine diagnostics during recent years. However, due to the discussions about biowarfare and possible bioterrorist attacks, an arsenal of additional agents now has to be identified in a rapid and reliable manner by the diagnostic laboratory. In recent years, natural infections with pathogens such as *Yersinia pestis* or *Bacillus anthracis* have occurred only sporadically in humans and their diagnosis was restricted to selected expert laboratories. Since the anthrax letter attacks in 2001 the further abuse of these pathogens for bioterror is feared and their identification has become a routine task in expert laboratories. Similarly, *variola virus,* which was eradicated by a global vaccination program at the end of the last century, has not been considered as a realistic infectious threat for the last 25 years. But today, variola virus is the most dangerous viral candidate for bioterrorism and diagnosis and the treatment and prevention of poxvirus infections are of heightened public interest.

These re-emerging pathogens have posed no global danger to the human population until recently therefore methods to verify their presence have not been developed during recent decades. Real-time PCR based methods are promising to fill this gap rapidly. This chapter gives a brief summary of the present applications of real-time PCR for the diagnosis of infectious agents of bioterrorism.

Introduction

Since the anthrax-containing letters were sent to US-Americans in autumn 2001, the problem of bioterrorism has attracted increasingly more attention in politics, public forums and science. In total five people died and it became clear that bioterrorism had become a reality (Jernigan *et al.*, 2001). Several potential agents and pathogens, including viruses, bacteria and toxins, have since been described as bioterrorism (BT)-relevant and in addition to countermeasures such as vaccination, prophylaxis and therapy, a rapid and reliable diagnosis is a prerequisite for a reasonable risk assessment of suspected specimens (Hicks, 2003).

Classical biochemical or culture-based methods have been well established in expert laboratories for the direct detection of several BT-relevant pathogens (Lim *et al.*, 2005). However, these methods have not been updated and may be laborious and time consuming. On the other hand, methods that include propagation or culturing of a pathogen are the best way to prove not only its presence but also its replication competence which cannot be addressed by other methods. In any case, for successful culture propagation to be used diagnostically there must be at least a hint of the identity of the putative pathogen for the selection of adequate culture conditions. Otherwise, the presence of a pathogen can only be excluded by an absence of growth. Propagation of a pathogen may take days which can be much too long for BT scenarios.

By comparison, electron microscopy is a rapid method that profits from its broad specificity and can identify many different pathogens so long as they are present in sufficient concentrations or "loads" (Hazelton and Gelderblom, 2003). In clinical specimens the pathogen loads are generally high, but for the analysis of low load specimens, such as environmental specimens, reduced diagnostic sensitivity may be crucial. Because electron microscopy can only draw conclusions about pathogen families, method variations such as immuno-electron microscopy are essential to permit more detailed pathogen-typing (Gelderblom and Bannert, 2005). For example, electron microscopy alone could only confirm or exclude the presence of orthopoxvirus particles in an environmental specimen suspected to contain variola virus but these could either be highly pathogenic variola virus particles or absolutely harmless mousepox virus particles (Miller, 2003). The implications of failing to detect the more lethal microbe during a BT-attack are obvious.

A third direct pathogen detection approach is the polymerase chain reaction (PCR). PCR is an important diagnostic tool to detect nucleic acids with high specificity and sensitivity, and with the introduction of rapid real-time PCR methods, PCR has increased its reputation within the diagnostic field (Mackay, 2004). PCR cannot prove the replication competence of a pathogen or the presence of proteinaceous toxins however in a BT setting, the very presence of a pathogen is often alarming and in these instances PCR is a reliable and valuable method for microbial detection. Of all the benefits of real-time PCR, the most striking advantage for the detection of BT-relevant agents is the increased specificity by the use of additional hybridizing probes that can be employed for rapid subsequent genotyping of the detected pathogen. Unlike electron microscopy in the variola virus and mousepox virus example above, PCR cannot identify morphological criteria, but special real-time PCR techniques can easily identify significant differences in the genomic sequence. Moreover, since real-time PCR works in a closed tube format, the risk of carry-over contamination with already amplified PCR product is minimized due to the lack of post PCR handling (Mackay *et al.*, 2002). This significantly improves the degree of reliability of results generated for the analysis of such important agents. Consequently, during recent years several real-time PCR assays have been published that can detect BT-relevant pathogens by using different real-time PCR formats and various reaction platforms.

Although this book focuses on real-time PCR, it must be mentioned that each of the previously summarized techniques has its advantages and a good diagnostic facility should always apply a combination of real-time PCR, electron microscopy and pathogen propagation to rapidly and accurately identify microbial agents of bioterror.

Bioterrorism

What is bioterrorism?

A bioterrorist attack is defined as the deliberate release of viruses, bacteria, or other agents that can cause illness or death in people, animals, or plants.

The ideal bioterror agent is lethal, easy and inexpensive to produce in large quantities, stable in aerosol with the ability to be dispersed over wide areas. For ideal bioterror agents there is no effective treatment or vaccine, and the agent may be communicable from person to person.

Typically the agents employed can be found in nature, but technically or genetically modified variants with increased pathogenicity, resistance to current medicines or improved ability to be spread into the environment through air, water, or in food must also be considered. Release of those agents can be extremely difficult to detect and may not cause illness for several hours to several days. While some BT-relevant agents, such as the smallpox virus, can be spread from person to person, others, such as anthrax or toxins, cannot.

Short outline of BT history

Although it could be assumed that bioterrorism is a problem of today and is certainly much more present in the public conscience since the anthrax letters sent in October 2001, the use of biological agents as weapons is not a novelty. Even several hundred years ago, biological agents were used to support warfare (Klietmann and Ruoff, 2001) although today's use of bioweapons stretches the conventional definition of war.

Existing evidence suggests that ancient civilizations such as the Greeks or the Romans attempted to pollute the drinking-water supplies of their enemies by contamination with foul-smelling dead animals, a strategy that was still employed in the 19th century during the American Civil War (Poupard and Miller, 1992). In the 18th century a British commander learned that smallpox had broken out among British troops at Fort Pitt and he suggested that the disease could be used as a biological weapon against the native Americans. The plan was to pass along blankets used by smallpox victims to the native Americans. An epidemic of smallpox did occur among these tribes, but it was not proven that the outbreak resulted from biological warfare activities.

These early attempts of biological warfare occurred long before the concept of microorganisms as agents of infectious diseases was accepted. In parallel with increasing skills in isolation and propagation of pathogenic microorganisms, knowledge about prevention and eradication of infectious diseases grew, and so did their potential use for nefarious purposes.

While the use of biological weapons was minimal during World War I, chemical weapons were developed and applied on military personnel. This led to the formulation of the Geneva Protocol for the Prohibition of the Use in War of Asphyxiating, Poisonous or Other Gases, and of Bacteriological Methods of Warfare, in 1925. This treaty banned the use of chemical as well as biological weapons in war. However, in the period between World Wars I and II the Japanese effort to develop biological weapons was notable. It is estimated that during 13 years of Japanese biological warfare research in Manchuria and China, 10 000 unwilling human "subjects" lost their lives (Harris, 1992). Other countries attempted to developing vaccines and antibiotic drugs to combat such efforts. The United States began a small biological weapons program in 1942 but in 1950 they began a more extensive program (Christopher *et al.*, 1997) establishing a research and development institute at Fort Detrick. Both the US Army and Air Force worked on biological weapons projects until 1969, the year of President Nixon's biological weapons disarmament declaration. This decision was followed in 1972 by the Biological and Toxic Weapon Convention, after which Western governments stopped biological weapons development. However, the Soviet Union started a clandestine program which reached beyond our imagination in its scientific, technological, and production capacity. Of particular interest was an organization that was created between 1973 and 1974 under the name Biopreparat (Domaradskij and Orent, 2006; Roffey *et al.*, 2002b). It comprised 52 sites of extensive research, development and production capability for bioweapons employing an estimated 50 000 people over 25 years, many of them highly trained scientists and technicians. Biopreparat created a variety of different agents for specific strategic situations and demands, such as pathogens resistant to degradation by heat, light, cold, UV, and ionizing radiation (Davis, 1999) and the production levels were impressive. The Soviet Union stored more than 30 metric tons of *Bacillus anthracis* spores and more than 20 metric tons of smallpox virus. This potential was reported by a former deputy director of Biopreparat, Ken Alibek. His revelations described a gruesome picture of this elaborate death factory (Alibek and Handelmann, 1999). According to his book, the former Soviet Union's warfare program included the aiming of SS-18 intercontinental missiles armed with anthrax and other agents at New York, Los Angeles, Seattle, and Chicago in 1988.

Several bioterrorist attacks have occurred in the more recent past. In 1984, salad bars at two restaurants in Oregon, USA were contaminated with *Salmonella* species by the Bhagwan Shree Rajneesh group attempting to sicken citizens and prevent them from voting in an upcoming election (Torok *et al.*, 1997). In the late 1990s, a number of anthrax threats were made in the United States, but all were hoaxes (Moran, 1999). One of the most frightening recent terrorist attacks involved release of the nerve gas sarin in the Tokyo subway system in 1995. Aum Shinrikyo, the cult responsible for killing 12 people and injuring approximately 3800 in the sarin attack, has also attempted to develop botulinum toxin, anthrax, cholera, and Q fever for bioterrorist use (Olson, 1999).

Use of PCR as diagnostic tool in bioterrorism

The most crucial factor in diagnosing infectious diseases is the reliability of the result obtained. This applies to routine clinical specimens and even more to the diagnosis of BT-relevant agents. But what does diagnostic "reliability" exactly mean in this context and what can PCR contribute to its achievement? There are three main criteria that determine the degree of reliability of the method of choice:

1. Specificity
2. Sensitivity/detection limit
3. Speed

Specificity

The detection of BT-relevant pathogens first of all demands a high level of specificity. It seems obvious that both possibilities of a false diagnostic result will have dramatic consequences for the community. A false negative result would give the agent time to be distributed, first in close vicinity to the place of the attack but as seen for the rapid distribution of SARS in 2003, pathogens that are transmittable from human to human could in the worst case travel around the globe in a matter of hours. Countermeasures such as post-exposure vaccination or antimicrobial treatment could not be initiated in time and as a consequence many fatalities would be expected. A false positive result would not cause harm to people directly, but would certainly lead to panic, the initiation of appropriate countermeasures and finally would cause economic damage in hardly predictable dimensions. Therefore both cases have to be avoided by application of reliable diagnostic assays. The difficulty of developing such reliable assays is that nearly each of the known highly pathogenic agents potentially used for BT purposes has a close relative that is either significantly less pathogenic or even innocuous. Unfortunately, in most cases there are only a few differences in the nucleic acid sequences of such closely related agents and a BT risk assessment depends on the accurate differentiation of these minimal changes.

For example, *variola virus*, one of the most pathogenic viruses for humans, shares high genome homology with camelpoxvirus, a virus that is not able to infect humans at all, but causes high morbidity and mortality in its host, the camel. Also monkeypox viruses are much less pathogenic to humans than *variola virus*, but display only minor genetic differences. Similarly, *B. anthracis* has a non-pathogenic close relative called *B. cereus*, which can hardly be distinguished from *B. anthracis* based on genomic sequences thus bacterial plasmid sequences must be determined to enable reliable differentiation of the two microorganisms. But how can such a high degree of specificity be achieved? Two approaches have been used to create real-time PCR assays with the required specificity for infectious agents.

Specific amplification of the target of interest

Oligonucleotide primers are designed to bind exclusively to the target of interest and amplification by PCR is monitored by an additional probe that may not necessarily be as specific as the primers. If significant sequence differences exist at the primer binding sites then the perfectly matched target is amplified while similar targets are not. However, once non-perfectly matching primers are elongated, they produce amplified DNA that is a perfectly matching target for primer hybridization in subsequent rounds of amplification, leading to false positive results. This may happen to primers targeting only single specific bases most commonly at their 3′ end.

The generic amplification of a whole group or species of pathogens

Several members of a species are amplified with the same primers indicating the presence of at least one member of this group. They are then discriminated using selective binding of a type-specific HybProbe showing amplification curves only for the perfect matching variant, or probes that bind to all species of the group but identify single nucleotide polymorphisms (SNPs) by fluorescence curve melting analysis. The latter approach has a lower risk of obtaining false positive results as long as the chosen SNPs are valid for all possible pathogens. However, it may fail in situations when the specimen contains mixtures of pathogens. Especially in melting curve analysis, the more abundant target will be clearly identifiable, while low amounts of a second target will not.

Since both approaches have their strengths and drawbacks, the combination of a specific and a generic approach should be implemented to achieve maximum reliability. Real-time PCR seems to be a promising technique to enable the required specificity, although for complex genomes more than one SNP should be considered to draw final conclusions.

Sensitivity/detection limit

Beside specificity, the detection limit (often understood as sensitivity) of an assay is an important criterion. It may be assumed that attacks would be carried out using high amounts of pathogens therefore requiring only a moderate assay sensitivity. However, the intent of a terrorist can only be presumed. It is unclear what effective concentration of the pathogen would remain for analysis after isolation and preparation of the specimen. In this context, the makeup of the specimen or its "matrix" and the volume in which the pathogen is delivered will influence detection limits dramatically; the proven concentration of the detected pathogen therefore would not at all reflect the true number of infectious particles. Care must be taken to remove inhibitory compounds prior to analysis. On the other hand, low amounts of an infectious agent may be employed by a terrorist so that they intentionally fall below the detection limit of commonly applied assays and could be distributed unnoticed and spread later from human to human. In any case, the reliable exclusion of suspected pathogens requires a high sensitivity. And moreover, besides quantitative considerations, the detection of only minute amounts of *variola virus* particles would be alarming, since *variola virus* is restricted to laboratory use only in two centers worldwide and unavailable to non-authorized persons. Whatever the reason might be, sensitivity is important and recently published real-time PCR assays can usually detect below 10 genomic equivalents per reaction (Lim *et al.*, 2005).

The detection limit can be improved by the use of suitable specimen volumes, preparation and purification methods and the implementation of enrichment steps. However, various scenarios can be assumed where this sensitivity is insufficient. On the other hand, the number of pathogen genomes is not identical to the number of infectious particles and does not at all reflect replication competence. For example, the ratio of genome equivalents to infectious particles for orthopoxviruses is at least 1:10, e.g. within a specimen that contained ten detectable genomes, there would be statistically one infectious virus particle present. This ratio varies with the virus and with the specimen matrix which may significantly damage the pathogen's integrity.

Speed

Compared to classical direct microbial detection methods, such as growth using selective media or propagation in susceptible cell lines, real-time PCR is extremely fast. While propagation may take up to several days, most of the newly developed real-time PCR platforms allow rapid amplification and on-line monitoring in less than one hour after the preparation of nucleic acid. This is impressively fast, but considering the consequences of a positive result in BT diagnostics, there is no method too fast and further improvements in amplification speed and efficiency are needed.

In summary, BT-diagnostics require rapid methods that combine high specificity with excellent detection limits. These demands are best met by real-time PCR.

Assay evaluation

The first steps in the development of a new assay are usually the selection of primers and hybridizing probes based on the known target sequences. These can be deposited in public databases or identified from other sources. Primers and probes are selected by targeting the sequence that is specific for the target of interest. Reliable discrimination can be obtained following general design rules. Despite a careful design, the specificity of an assay must always be proven experimentally to ensure the significance of the chosen sequence.

Another important question is whether further pathogen isolates exist that are not identical to the considered sequences and could therefore lead to false negative results. Unknown isolates may lack the exact sequence positions their identification should be based on. Therefore, every assay has to be evaluated practically with as many isolates as possible. Unfortunately for comprehensive assay development, access to certain BT-relevant pathogens is highly restricted and even nucleic acids from these pathogens are difficult to obtain from specialized laboratories due to safety regulations. Thus, external quality assessment studies including BT agents have evolved in recent years that are either defined by the detection method or the pathogen (Niedrig *et al.*, 2006; Niedrig *et al.*, 2004).

To optimize the assay's sensitivity, a lot of helpful rules for primer and probe design are already defined and the design of accurate, functional real-time PCR assays is no secret. We have described these steps in detail in Chapter 2. If the basic rules are followed, the common sensitivity of 10 genome equivalents per reaction can be reliably achieved with a number of formats under appropriate conditions.

Considering this, it seems obvious that the establishment of acceptably reliable assays is not possible for every individual laboratory. Home-brew assays, although they may be evaluated perfectly, need to be tested regularly to ensure their functionality. This is only possible for facilities that actually work with the agent of interest. Therefore, to shorten the evaluation procedure and to enable more laboratories to perform a reliable diagnosis for a broad spectrum of pathogens, several ready-to-use commercial kits are now on the market. These kits eliminate the need for extensive primer/probe design, optimization and evaluation and facilitate rapid detection and monitoring studies. The kits often include controls and only lack the specimen DNA to be tested and are a good solution for non-specialized laboratories to perform preliminary diagnostics. Additionally, there are innumerable sets of probes and primers in the literature that can be used with a number of pathogens or potential bioweapon agents.

Methods for specimen processing

The specimen processing and preparation of BT-relevant specimens cannot be compared to that of common human clinical specimens. For clinical specimens such as blood or serum there are numerous commercial kits available that enable the rapid purification of nucleic acids for use in any PCR reaction. Moreover, nucleic acid preparations of clinical specimens can easily be controlled for quality by amplification of reference genes present in the specimen (Radonic *et al.*, 2004).

In addition to the clinical specimens of suspected persons, BT diagnostics should cover the analysis of powders, soils, solids and air. Since the compositions of these specimen matrices are manifold and usually unknown, different methods for processing are required.

Ideally the universal method of choice concentrates the target BT agent and purifies the nucleic acid allowing analysis of all types of targets, e.g. DNA or RNA from viruses, bacteria or spores. In addition substances potentially inhibiting detection should be removed. Since a PCR positive result reflects the presence of the nucleic acid only, cell viability should be maintained in a portion of the specimen so that the target organism can be cultured for confirmation of viability, an important component of the final BT risk assessment. To date, there is definitely no single method that can address these requirements but various specimen preparation methods are currently under investigation (Cheun *et al.*, 2003; Luna *et al.*, 2003; Makino and Cheun, 2003).

However, for the diagnosis of environmental samples, which are prone to contain contaminating traces of PCR-inhibitory substances even after appropriate preparation procedures, the implementation of internal controls is vitally important. Endogenous internal controls can prove the presence of amplifiable DNA in each reaction vessel adding a higher degree of reliability to negative PCR results.

BT-relevant pathogens

Classification

Twelve agents have been categorized as the most dangerous if abused in BT attacks. These agents are colloquially termed the "dirty dozen" and the rapid diagnosis of these agents is of the highest priority. The dirty dozen includes viruses, bacteria and also toxins as listed in Table 9.1.

A classification of these and further agents was proposed by the Centers for Disease Control and Prevention (CDC, USA) and has become widely accepted. According to this classification all BT-relevant agents can be assigned to three categories; A, B and C, where category A comprises agents of highest threat and category C those of lowest concern (Table 9.2).

Table 9.1 The dirty dozen of bioterrorism

Viruses	Bacteria	Toxins
Smallpox	*Bacillus anthracis*	Botulinum toxin
Ebola-Lassa virus	*Francisella tularaemia*	Ricin
Equine encephalitis viruses	*Yersinia pestis*	*Stapylococcus enterotoxin* B (SEB)
	Brucella spp	
	Coxiella burnetii	
	Burkholderia spp	

Table 9.2 Classification of putative bioterror agents

Agent	Category A	Category B§	Category C
	Variola virus Viral hemorrhagic fevers Filoviruses Arenaviruses	Viral encephalitis viruses Alphaviruses	Nipah virus Hantavirus
Bacteria	*Bacillus anthracis* *Yersinia pestis* *Francisella tularensis*	*Brucella species* *Burkholderia mallei* *Burkholderia pseudomallei* *Chlamydia psittaci* *Coxiella burnetii* *Rickettsia prowazekii*	
Toxin	Botulinum toxin	Ricin toxin *Clostridium perfringens* Epsilon toxin *Staphylococcal enterotoxin* B	

§ Food and water safety threats such as *Salmonella* spp, *Shigella* spp, *Vibrio cholerae*

Category A Diseases/Agents

Highest-priority agents include microorganisms that pose a risk to national security because they can:

- be easily disseminated or transmitted from person to person;
- result in high mortality rates and have the potential for major public health impact;
- might cause public panic and social and financial disruption;
- require special action for public health preparedness.

Category B Diseases/Agents

Second highest priority agents include those that:

- are moderately easy to disseminate;
- result in moderate morbidity and low mortality rates;
- require specific enhancements of CDC's diagnostic and disease surveillance capacity

Category C Diseases/Agents

Third highest priority agents include emerging pathogens that could be engineered for mass dissemination in the future because of:

- availability;
- ease of production and dissemination;
- potential for high morbidity and mortality rates and major health impact.

In each of the following sections we will first give a brief overview of relevant BT pathogen or agent and then summarize the real-time PCR approaches that have been developed and/or applied to permit their diagnosis. Conventional PCR methods as well as classical microbiological methods will not be discussed here.

Bacteria

Anthrax

Anthrax is a serious disease caused by *Bacillus anthracis,* a Gram-positive bacterium that forms spores. A spore is a robust cell that is dormant but can revive under special conditions. Consequently, spores must also be detected by diagnostic approaches. *B. anthracis* is classified as a category A bioterror agent that can infect humans by handling products from infected animals or by breathing in anthrax spores from infected animal products resulting in anthrax (Inglesby *et al.*, 2002). Infection with gastrointestinal anthrax can occur by eating undercooked meat from infected animals. The spread of anthrax from person-to-person has not been reported. However, as happened in the United States in 2001 when anthrax was deliberately spread through the postal system by sending letters with weaponized spores, anthrax can also be used as a weapon. These attacks caused 22 cases of infection and five people died of anthrax (Jernigan *et al.*, 2001).

There are three types of anthrax presenting with different symptoms depending on the type of the disease. Symptoms can appear within seven days after contact with the bacterium, although inhalation anthrax may take up to six weeks to appear (Moran and Talan, 2003).

1. Cutaneous anthrax: A small sore develops initially develops into a blister and later into a skin ulcer with a black central area. The sore, blister and ulcer do not hurt.
2. Gastrointestinal anthrax: The first symptoms are nausea, loss of appetite, bloody diarrhea, and fever, followed by bad stomach pain.
3. Inhalation anthrax: The first symptoms of inhalation anthrax resemble cold or flu-like illness including a sore throat and mild fever. Later symptoms include cough, shortness of breath, tiredness and muscle aches.

There is a vaccine that is not yet available for the general public (Chabot *et al.*, 2004; Splino *et al.*, 2005). However, people probably exposed to anthrax by entering contaminated areas or those exposed to anthrax by a bioterror attack would get the vaccine (Little, 2005). Antibiotics including ciprofloxacin, levofloxacin, doxycycline, or penicillin are employed for a 60-day period to treat all three types of anthrax (Brouillard *et al.*, 2006). Success depends on the type of anthrax and how soon treatment begins. In most cases, early treatment with antibiotics can cure cutaneous anthrax since 80 percent of untreated people who become infected with cutaneous anthrax do not die. Gastrointestinal anthrax leads to death in 25% to 50% of cases while inhalation anthrax is much more severe. In 2001, about half of the cases of inhalation anthrax ended in death (Jernigan *et al.*, 2001).

The recent use of anthrax in bioterrorist attacks has demonstrated the need for rapid and specific diagnostics. Routine culture and biochemical testing methods for *B. anthracis* have been well-established for many years, but a definitive identification may take 24h to 48 h or longer. PCR is a more rapid alternative, but only one anthrax specific real-time PCR assay had been published before the bioterrorism attacks in October 2001 in Washington DC. Meanwhile, in response to that event, several real-time PCR assays have been published for different platforms. Because virulent isolates of *B. anthracis* contain two plasmids, the 182 kb plasmid pX01 and the 96 kb plasmid pX02, there is general consent that identification of genomic sequences is insufficient and should augmented by the amplification of unique plasmid sequences.

The first real-time PCR assay to detect *B. anthracis* was published in 1999 (Lee *et al.*, 1999). This assay was based on the minor-groove DNA intercalating fluorescence dye SYBR® Gold that exhibits increased fluorescence when bound to double stranded DNA. Unfortunately, SYBR Gold cannot guarantee a sequence-specific signal and sometimes produces false positive results even in negative specimens, as was reported for this assay. Nonetheless, this study confirmed *B. anthracis* infection by melting curve analysis of the amplified *capC* gene and presented an artificial internal control, which is of considerable importance to achieve good diagnostic reliability. Recently published real-time PCR assays avoid the use of intercalating dyes and prefer the use of hybridizing probes, such as 5′ nuclease® or HybProbes probes.

Qi *et al.* 2001 used a unique nucleotide sequence of the *B. anthracis rpoB* gene for differentiation of *B. anthracis* by LightCycler PCR (Qi *et al.*, 2001). Primers were selected to have one *B. anthracis*-specific nucleotide at each of their 3′ ends, whereas two nucleotides were specific to the probe region. No melting curve analysis was performed however the assay was found to be specific for 144 *B. anthracis* strains and did not cross-react with other related bacilli (175 strains), with the exception of one strain.

In 2002 a LightCycler-based assay was published that can now be purchased as a ready-to-use kit (Bell *et al.*, 2002), LightCycler Bacillus anthracis kit; Roche Applied Science, Indianapolis, Ind.) which complete the whole analysis in less than one hour. PCR primers and probes were designed to identify gene sequences specific for both the protective antigen (*pagA*) located on plasmid pX01 and the encapsulation B protein (*capB*) located on plasmid pX02. The pagA gene was detected in 29 of 29, and the capB gene in 28 of 29 virulent *B. anthracis* strains respectively. Three avirulent strains containing only pX01 or pX02, and therefore only *pagA* or *pagB*, could be detected and differentiated from virulent strains. The assays were negative for 57 bacterial strains representing a broad range of organisms, including Bacillus species other than anthracis (n = 31) and other non-Bacillus species (n = 26). The analytical sensitivity demonstrated with target DNA cloned into control plasmids was 1 copy/μL of specimen.

A 5′ nuclease probe-based real-time PCR for identification of *B. anthracis* was also based on the detection of the two pathogenicity plasmids, as well as one chromosomally encoded marker (Ellerbrok *et al.*, 2002). Both *pagA* on plasmid pX01 and *capC* on plasmid pX02 were targeted and amplified with identical reaction conditions to the chromosomally located RNA polymerase gene (*rpoB*). Importantly, there were low cross-reactions of the *rpoB* primers with *B. cereus* DNA, which could give false positive results if extremely high amounts of *B. cereus* were present. However, the plasmid targets were highly specific for *B. anthracis* allowing its reliable identification. The crucial benefit of this publication is that not only purified DNA, but also complete bacteria picked from colonies or pelleted from liquid cultures were subjected to PCR, avoiding time-consuming DNA preparation and minimizing specimen handling. To prove replication competence of the bacteria contained in the suspect specimen and to increase the target amount for PCR analysis, spores of *B. anthracis* were cultivated for a few hours in enrichment broth before PCR analysis, or used directly for real-time PCR. This approach took approximately two to three hours to confirm or exclude a potential attack following specimen arrival in the laboratory.

As described above, the identification of chromosomal markers for rapid detection of *B. anthracis* is difficult because significant chromosomal homology exists between it, *B. cereus*, and *B. thuringiensis*. Beside *rpoB*, a second chromosomally located *B. anthracis*-specific gyrA gene, was evaluated as a potential chromosomal marker for *B. anthracis* (Hurtle *et al.*, 2004). A 5′ nuclease based real-time PCR assay was developed for this region using a minor groove binding probe (MGB). The 5′ nuclease-MGB probes are modified to bind very tightly to the target DNA which allows the design of shorter probes. Such a short probe does not tolerate mismatches in the probe binding region and a pair of differently labeled probes can therefore be used for discrimination of similar target sequences. The MGB assay was tested with 171 organisms including 102 Bacillus strains and was found to be specific for all 43 strains of *B. anthracis* tested, demonstrating the usefulness of *gyrA* as a specific chromosomal marker for *B. anthracis*. Further *B. anthracis*-specific chromosomally located sequences are presently under investigation for their usefulness in diagnostics (Bode *et al.*, 2004).

Kim *et al.* published a multitarget multiplex PCR to differentiate *B. anthracis* from *B. cereus* by SYBR® Green melting curve analysis (Kim *et al.*, 2005). *B. anthracis* specific sspE gene sequences as well as a shorter version of sspE representing the *B. cereus* group were selected as targets. In addition the lef gene on plasmid pXO1 and *capC* were amplified. A multitarget PCR amplifies multiple targets of one pathogen by separate PCR reactions that can ideally be performed under identical reaction conditions in parallel in one PCR run. A combination of the PCR product melting temperatures was employed to allow a classification of the bacillus species. Purified DNAs from 38 bacterial strains including 11 strains of *B. anthracis* and 18 *B. cereus* group strains were analyzed. All virulent genotypes could be identified correctly by this assay.

With the increasing skills in the design of real-time PCR assays and the improved reagents and PCR machines available to many laboratories, sensitive analysis is more and more a question of establishing an efficient preparation method for the specimens of interest.

Soil is certainly one type of specimen matrix that often impairs PCR analysis. To detect *B. anthracis* DNA from soil various amounts of *B. anthracis* spores were added artificially to soil and DNA was prepared by several preparation and enrichment steps. Nested and real-time PCR could detect a single *B. anthracis* cell in 1 g of soil (Cheun *et al.*, 2003). Other preparation approaches revealed a sensitivity of < 10 spores for various specimen matrices. Heat treatment of specimens suspended with sucrose plus non-ionic detergent was considered an effective spore disruption method (Ryu *et al.*, 2003). Similarly, the detection of one cell of *B. anthracis* in 100 liters of air was shown by a simple and rapid procedure combining filtration and real-time PCR (Makino *et al.*, 2001).

Independent of the specimen matrix, the identification of *B. anthracis* should always be based on the multitarget amplification of genomic sequences in addition to the virulence plasmids. The chosen format may be adapted to the real-time PCR instrument available.

Yersinia pestis

The infectious agent of bubonic, pneumonic or septicemic plague is the Gram-negative bacterium *Yersinia pestis*. The organism is motile when isolated, but becomes non-motile in the mammalian host. All three forms of plague have been responsible for high mortality rates in epidemics throughout human history. Beside *Y. pestis*, the genus *Yersinia* consists of 10 species, but only *Y. pestis*, *Y. pseudotuberculosis* and *Y. enterocolitica* are considered to be pathogenic for humans (Perry and Fetherston, 1997). *Y. pestis* is classified as a category A bioterror agent.

Naturally, fleas become infected by feeding on rodents and transmit the plague bacteria to humans and other mammals. Transmission from human to human is possible by airborne infection when a person suffering from pneumonic plague coughs droplets containing the plague bacteria. Large outbreaks in humans can occur where sanitary facilities are poor. These outbreaks are usually associated with infected rats and rat fleas that live nearby.

A person usually becomes ill with bubonic plague two to six days after being infected. When bubonic plague is left untreated, plague bacteria invade the bloodstream, spread rapidly throughout the body and cause a severe and often fatal disease (Koirala, 2006).

Infection of the lungs with the plague bacterium causes the pneumonic form of plague, a severe respiratory illness. The infected person may experience high fever, chills, cough, and breathing difficulties, and expel bloody sputum. The typical sign of bubonic plaque, the most common form of human plague, is a swollen and very tender lymph gland.

A formalin-inactivated vaccine was available for adults at high risk of contracting the plague. It was subsequently taken off the market by the Food and Drug Administration of the USA due to reduced efficiency and the risk of causing severe inflammation. First experiments with genetically engineered vaccines are promising (Welkos *et al.*, 2002).

Yersinia pestis is highly susceptible to antibiotic treatment with streptomycin. Gentamicin can be used if Streptomycin is unavailable. Tetracyclines and chloramphenicol are also effective. Unfortunately strains resistant to one or two antibiotics have already been isolated. Without therapy the disease progresses rapidly to death (Hernandez *et al.*, 2003).

Possible target genes are either located on the *Y. pestis* chromosome or on one of the three virulence plasmids that most *Y. pestis* strains contain (Brubaker, 1972). The *Y. pestis*-specific 9.5-kb plasmid pPla (also pPCP1 or pPst) codes for a plasminogen activator (Pla). Pla is important for the dissemination of *Y. pestis* transmitted by flea bites, because pPla deficient strains are avirulent in mice infected via the subcutaneous route. The 64-kb pCD1 plasmid contains a number of genes coding for several virulence factors. These proteins are referred to as Yops (Yersinia outer proteins) and are required for full virulence in the mouse (Hinnebusch *et al.*, 2000). This plasmid is shared by all pathogenic members of the species *Y. pestis*, *Y. enterocolitica* and *Y. pseudotuberculosis* (Cornelis *et al.*, 1998). The caf1 locus on the *Y. pestis*-specific 100–110-kb plasmid pMT1 (pFra) encodes the highly immunogenic fraction 1 capsule antigen (F1). The *Y. pestis* murine toxin gene encodes a 61-kDa phospholipase protein (Hinnebusch *et al.*, 2000), which is highly toxic for mice when injected.

The first real-time PCR assay for the detection of *Y. pestis* was published in 1998 (Higgins *et al.*, 1998). This assay was based on a 5′ nuclease probe and was established to rapidly detect and quantitate DNA from clinical specimens by targeting Pla. To prove its applicability *Y. pestis* was quantified in experimentally infected *Xenopsylla cheopis* fleas and in experimentally infected monkeys.

In 2000 Iqbal *et al.* published a *Y. pestis* specific assay targeting a stretch of the Pesticin (pst) gene (Iqbal *et al.*, 2000). This gene is a species-specific gene located on the unique plasmid pPla and encodes a bacteriocin. The assay was shown to detect as few as three copies of pst from total *Y. pestis* genomic DNA. The pst probe used was positive only for pesticinogenic isolates and did not show complementarity with Yersiniae nor with other bacteria targeted and was postulated to be highly specific for *Y. pestis*.

A further aspect beside basic detection and quantification of *Y. pestis* was presented in 2001 (Lindler *et al.*, 2001). Based on FRET probes, an assay was presented to detect ciprofloxacin resistant (Cp(r)) mutants of *Y. pestis*. The detection of antibiotic resistance in agents of biowarfare is extremely critical in order to start appropriate therapy in a timely manner. First, spontaneous mutants of the attenuated *Y. pestis* KIM 5 strain were selected that showed resistance to ciprofloxacin concentration of > 1 μg/ml. DNA sequencing of gyrA encoded by 65 of these mutants revealed that all resistant isolates contained one of four different point mutations within the quinolone resistance-determining region of gyrA. The FRET-based assay was designed to detect all of these mutations by using a single pair of fluorescent probes with sequences complementary to the wild-type *Y. pestis* gyrA sequence. As expected, fluorescence melting curve analysis revealed decreased melting temperatures for the four mutant templates of between 4°C and 11°C. Optimized reaction conditions lead to the detection of approximately 10 pg of purified wild-type template DNA, or of approximately four CFU of wild-type *Y. pestis* KIM 5 or Cp(r) mutants, in crude lysates.

The same group published a 5′ nuclease probe based assay which was performed using the LightCycler for the identification of resistant mutants of *Y. pestis* only two years later (Lindler and Fan, 2003). Two groups of fluorogenic probes were developed. The first group included a probe homologous to the wild type *Y. pestis* gyrA sequence with two corresponding probes that were homologous with two different mis-sense mutations in codon 81 of gyrA. The second group of probes included a wild-type probe and two corresponding probes that recognized mis-sense mutations in codon 83 of gyrA. These probes reacted only with the homologous DNA sequences. The 5′ nuclease assay was sensitive to 1 pg (approximately one colony forming unit, CFU) of starting template and could be used on semi-purified DNA.

Two complex multitarget multiplex PCR assay were described in 2003 for the detection of *Y. pestis* on the LightCycler (Tomaso *et al.*, 2003). Targets for PCR amplification included the 16S rRNA gene as a chromosomal marker, Pla, the murine toxin gene and the fraction 1 antigen gene (both located on the 100 kb plasmid) and bacteriophage lambda-DNA as an internal control. Twenty-five strains of *Y. pestis*, 94 strains of other Yersinia species and 33 clinically relevant bacteria were investigated but *Y. pestis* was selectively identified by displaying all characteristic melting temperatures. Other bacteria yielded no amplification products. The lower limit of detection in a background of rat or flea DNA was approximately 0.1 genome equivalents.

A comparison of different real-time PCR detection formats for the identification of a *Y. pestis*-specific genomic sequence was published in 2005 (Chase *et al.*, 2005). According to the idea that a chromosomally located marker should always be included in the PCR diagnosis of Yersinia, the gene region yp48 was utilized to construct a real-time PCR assay specific for *Y. pestis* or *Y. pseudotuberculosis*. The recently used 16S rRNA and 23S rRNA sequences differed very little compared to the gene homologues of *Y. pseudotuberculosis* and could therefore lead to difficult data interpretation in some cases. The yp48 gene displays homology to an E. coli gene which encodes an ATP-binding component of the maltose transport system. However, this gene has a 25-bp insertion that is present in *Y. pseudotuberculosis*, but not in *Y. pestis*. Several real-time 5′ nuclease-MGB assays, SimpleProbe® and MGB Eclipse™ probe assays for the selective differentiation of *Y. pseudotuberculosis* from *Y. pestis* were evaluated. The *Y. pestis*-specific-MGB assay distinguished *Y. pestis* from other Yersinia species but showed an unacceptable level of false positive detection of *Y. pseudotuberculosis*. This was probably due to the tight binding of the MGB moiety at the 3′ end of the MGB-probe that is also present in *Y. pseudotuberculosis*. The SimpleProbe and MGB Eclipse probes specific for the 25-bp insertion detected only *Y. pseudotuberculosis* DNA. Probes that spanned the deletion site detected both *Y. pestis* and *Y. pseudotuberculosis* DNA, and the two species were clearly differentiated by a post-PCR fluorescence melting curve analysis. The SimpleProbe assay produced an almost 7°C melting temperature difference and the MGB Eclipse probe a slightly more than 4°C difference. These data for the newly introduced real-time PCR formats may be helpful for the development of assays for the diagnosis of other BT-relevant agents.

Francisella tularensis

The Gram-negative bacteria *Francisella tularensis* is the causative agent of tularemia or rabbit fever (Oyston and Isherwood, 2002). Four subspecies of *Francisella tularensis* have been classified. The most virulent, Type A or *Francisella tularensis tularensis* is found predominantly in North America and is associated with lethal pulmonary infections. Type B or *Francisella tularensis palearctica* is found predominantly in Europe and Asia but rarely leads to fatal disease. Subspecies *Francisella tularensis novicida* primarily found in North America was characterized as relatively non-virulent while *Francisella tularensis mediasiatica* is found primarily in central Asia. Little is currently known about its ability to infect humans. *Francisella tularensis* is classified as a category A bioterror agent (Cronquist, 2004).

Infection by *Francisella tularensis* occurs after being bitten by infected ticks or insects, handling infected animal carcasses, eating or drinking contaminated food or water or inhaling bacteria. Only 10 to 50 organisms are required to cause disease. People who inhale an infectious aerosol generally experience severe respiratory illness, including life-threatening pneumonia and systemic infection, if they are not treated. Human-to-human transmission has not been reported. People who have been exposed to the tularemia bacteria should be treated as soon as possible. The disease can be fatal if it is not treated with the right antibiotics.

Symptoms usually appear three to five days after exposure to the bacteria, but can take as long as 14 days. Infection with *Francisella tularensis* can occur via several routes but most commonly via skin contact, yielding an ulceroglandular form of the disease. Inhalation of bacteria, particularly biovar *Francisella tularensis*, leads to potentially lethal pneumonic tularemia. Other rare routes of inoculation have been described and include oropharyngeal infection due to consumption of contaminated food and conjunctival infection due to inoculation at the eye.

Symptoms of tularemia include sudden fever, chills, headaches, diarrhea, muscle aches, dry cough and progressive weakness. Pneumonia with chest pain, bloody sputum and breathing trouble is also possible. Other symptoms of tularemia depend on how a person was exposed to the tularemia bacteria, and may include ulcers on the skin or mouth, swollen and painful lymph glands and a sore throat. An attenuated live vaccine strain of subspecies *palearctica* has been described, though it is not yet fully licensed by the Food and Drug Administration as a vaccine (Isherwood *et al.*, 2005).

The diagnosis of human cases of tularemia often relies upon the demonstration of an antibody response to *Francisella tularensis* or the direct culturing of bacteria from the patient. While the antibody response to *Francisella tularensis* usually takes two weeks to develop and cannot be applied to the identification of *Francisella tularensis* in environmental specimens, culturing needs special media and some suspicion of tularemia. Moreover, the handling of live bacteria poses a risk to laboratory personnel due to the highly infectious nature of this pathogen. The detection of nucleic acids promised to accelerate reliable diagnosis.

Therefore, in 2000 the first real-time PCR assay for *Francisella tularensis* was published (Higgins *et al.*, 2000). The 5′ nuclease assay directed against the subspecies *tularensis* outer membrane protein gene (*FopA*) was used to determine the bacterial load in experimentally infected mice. This assay had a detection limit of 1 pg of genomic DNA corresponding to approximately 100 CFU. This publication included the use of Flinders Technology Associates (FTA) filter cards (Whatman, Germany) for inexpensive and easy DNA preparation out of tick extracts or mouse tissue.

A similar 5′ nuclease assay targeting *fopA* was applied in a study to determine the pathogen load (including subspecies *tularensis*) in ~6000 ticks in Switzerland (Wicki *et al.*, 2000). DNA was extracted from pools of ten ticks grouped by gender. It was shown that ticks harbor subspecies *tularensis* in Switzerland at a rate of 0.12%. In this study 2300 PCR reactions were carried out in only five days, indicating the potential of real-time PCR to be used as a high throughput method.

To evaluate the mobile detection of *Francisella tularensis*, a hand-held portable fluorescence thermocycler designed for use in the field was employed (BioSeeq; Smiths Detection-Edgewood BioSeeq) and compared to the laboratory-based Applied Biosystems 7900 real-time PCR instrument (Emanuel *et al.*, 2003). Two independent 5′ nuclease-based PCR assays targeting *fopA* and *tul4* of *Francisella tularensis* were developed to detect the organism in tissues of infected mice with respiratory tularemia. A comparison of culturing and PCR for detection of *Francisella tularensis* in infected tissues showed that culturing was more sensitive than PCR, but took 72 h compared to 4 hours for completion. Template preparation may be the reason for the loss of sensitivity by PCR compared to culturing techniques. The results of both the ABI 7900 and the BioSeeq detection platforms were in agreement when presented with the same specimen, making the hand-held BioSeeq thermocycler a promising tool of diagnosis of infection in the field.

There is general agreement that the amplification of multiple targets would enhance specificity by reducing the likelihood of obtaining false positives due to reliance on a single target, and this led to the publication in 2003 of a four target, 5′ nuclease real-time PCR assay capable of rapidly and accurately detecting *Francisella tularensis* in complex specimens (Versage *et al.*, 2003). With an analytical sensitivity of one organism per reaction, the likelihood of detecting *Francisella tularensis* in low microbial load environmental specimens was significantly increased. The four regions targeted were specific for *Francisella* and included the *ISFtu2* element, a newly described insertion element-like sequence present in multiple copies (Thomas *et al.*, 2003), the 23kDa gene, which encodes a protein that is expressed upon macrophage infection (Golovliov *et al.*, 1997), *tul4* and *fopA* (Nano, 1988; Sjostedt *et al.*, 1991). The *ISFtu2*, 23 kDa, and *tul4* assays were each capable of detecting one organism per reaction and therefore were combined into a single multitarget assay for further analysis. This multitarget assay was species-specific, permitting differential diagnosis of both *Francisella tularensis* and *Francisella philomiragia*. Comparison of the *Francisella tularensis* 5′ nuclease assay to culturing, using environmentally contaminated specimens, demonstrated that the combined 5′ nuclease PCR assay was significantly more sensitive than culturing.

A SYBR Green-based real-time PCR assay confirmed the first recovery of *Francisella tularensis* in Denmark in an 8-year-old boy (Bystrom *et al.*, 2005). The strain was identified as *Francisella tularensis* subsp. *holarctica* (type B) by melting curve analysis. The different melting temperatures between the strains were due to a 30-bp deletion at a genomic locus designated Ft-M19 of *Francisella tularensis* subsp. *holarctica* strains and by the nucleotide substitutions in the *Francisella tularensis* subsp. *mediasiatica* strain. This approach for typing of *Francisella* species is rapid, but due to the fact that the maximal difference in melting temperature between all tested strains was 2.5°C, reliable differentiation may be difficult. However, GeneScan analysis (Johansson *et al.*, 2004) confirmed the results from SYBR Green melting curve analysis.

A *Francisella tularensis*-specific HybProbe assay located in *fopA* was established for use on the LightCycler platform (Fujita *et al.*, 2006). While 25 *Francisella tularensis* strains were tested positive by this assay, *Francisella philomiragia* and other bacterial species did not show any specific fluorescent signal or amplification product by agarose gel analysis, indicating high primer specificity under the optimized reaction conditions. Linear detection of genomic DNA was observed from 1.2 to 1.2×10^5 bacteria. This was the first assay to enable clear discrimination of *Francisella tularensis* from *Francisella philomiragia* by careful selection of the primer and probe binding region.

Brucella *species*

Brucellosis is an infectious disease caused by bacteria of the genus *Brucella*. These bacteria are primarily passed among animals, and they cause disease in many different vertebrates (Sarinas and Chitkara, 2003). *Brucella* spp. are classified as category B bioterror agents.

Various *Brucella* spp. affect sheep, goats, cattle, deer, elk, pigs, dogs, and several other animals. In the United States 100 to 200 cases occur each year but it can be very common in countries where animal disease control programs have not reduced the amount of disease among animals. Areas currently listed as high risk are the Mediterranean basin (Portugal,

Spain, Southern France, Italy, Greece, Turkey, North Africa), South and Central America, Eastern Europe, Asia, Africa, the Caribbean, and the Middle East (Pappas *et al.*, 2005a).

Humans are generally infected in one of three ways: eating or drinking something that is contaminated with *Brucella* spp., inhaling the organism, or having the bacteria enter the body through skin wounds. The most common way to be infected is by eating or drinking contaminated milk products. Direct person-to-person spread of brucellosis is extremely rare. In humans brucellosis can cause a range of flu-like symptoms and may include fever, sweats, headaches, back pain, and physical weakness (Sauret and Vilissova, 2002). Severe infection of the central nervous system or of the heart may occur. Brucellosis can also cause long-lasting or chronic symptoms that include recurrent fevers, joint pain, and fatigue (Cutler *et al.*, 2005).

There is no vaccine available for humans. The main way of preventing Brucellosis is by pasteurization of all milk that is to be ingested by human beings, either in its pure form or as a derivate, such as cheese. Antibiotics such as tetracyclines, chloramphenicol and rifampicin, streptomycin and gentamicin are effective against *Brucella* spp. However, treatment can be difficult and the use of more than one antibiotic is needed for several weeks, due to the fact that the bacteria grow within cells. Depending on the timing of treatment and severity of illness, recovery may take a few weeks to several months. Mortality is low (< 2%), and is usually associated with endocarditis (Pappas *et al.*, 2006; Pappas *et al.*, 2005b).

A 5′ nuclease probe-based real-time PCR that is genus-specific for Brucella spp was presented in 2004 (Bogdanovich *et al.*, 2004). Amplification of a 322-bp fragment of the per gene enabled identification of Brucella spp. in less than 2h, directly from agar plates. The assay included an internal amplification control, identified 23 Brucella strains and did not detect 174 non-Brucella strains.

In 2005 Queipo-Ortuno *et al.* presented a Sybr Green-based real-time PCR assay for the detection of *Brucella* spp. DNA in serum specimens (Queipo-Ortuno *et al.*, 2005). Amplification of a fragment of the immunogenetic membrane protein gene (BCSP31) specific for all species and biovars of the genus *Brucella* was followed by melting curve analysis to verify the specificity of the PCR products. The detection limit was 1 genome equivalent of *Brucella* spp. DNA. The assay was found to be 91.9% sensitive and 95.4% specific when tested with 65 negative control specimens and 62 serum specimens from 60 consecutive patients with active brucellosis.

A quite similar assay was published the same year (Debeaumont *et al.*, 2005). Again, the assay was based on the amplification of a portion of *bcsp31* but employed hybridization probes instead of Sybr Green. Species specificity and selectivity of this real-time PCR assay were evaluated using genomic DNA from 15 *Brucella* spp. strains and 42 non-*Brucella* spp. strains, and the results were 100%. Among 17 culture-proven brucellosis patients, sera from 11 gave a positive amplification signal, corresponding to a sensitivity of 64.7%. In contrast, negative results were obtained for all sera from 60 control patients, corresponding to a specificity of 100%. This specificity was superior to the results published by Queipo-Ortuno *et al.* which may be attributed to the use of Sybr Green giving false positive results in some cases. However, the investigation of different probes as well as the use of different primers in the same gene will have an impact on these results.

Coxiella burnetii

Coxiella burnetii is an intracellular, pathogenic bacterium that is distributed globally and the causative agent of Q fever. The genus *Coxiella* is morphologically similar to the Rickettsia, but with a variety of genetic and physiological differences. C. *burnetii* are small Gram-negative bacteria which form spores. They can survive standard disinfectants, and are resistant to many other environmental changes including heat and drying (Franz *et al.*, 1997) *Coxiella burnetii* is classified as a category B bioterror agent.

Coxiella burnetii is a highly infectious agent that can become airborne and inhaled by humans. One organism is all that is required to yield disease in 50% of an infected population (ID_{50} = 1). Cattle, sheep, and goats are the primary reservoirs of C. *burnetii* but infection has been noted in a wide variety of other animals but does not usually cause clinical disease in these animals. Organisms are excreted in the milk, urine, and feces of infected animals. Most importantly, during birthing the organisms are shed in high numbers within the amniotic fluids and the placenta. Moreover, the bacteria to survive for long periods in the environment. Infection of humans usually occurs by inhalation of these organisms from air that contains airborne barnyard dust contaminated by dried placental material, birth fluids, and excreta of infected herd animals. Humans are often very susceptible to the disease, and very few organisms may be required to cause infection. Less common modes of transmission include ingestion of contaminated milk, followed by regurgitation and inspiration of the contaminated food but in humans tick bites and human to human transmission are rare (Kazar, 2005).

Most patients become ill within two to three weeks after exposure but many human infections are unapparent. Only about half of all people infected with C. *burnetii* show signs of clinical illness. Acute cases of Q fever begin with sudden onset of high fevers, severe headache, general malaise, myalgia, confusion, sore throat, chills, sweats, non-productive cough, nausea, vomiting, diarrhea, abdominal pain, or chest pain. Thirty to fifty percent of patients with a symptomatic infection will develop pneumonia, liver malfunction and some will develop hepatitis. In general, most patients will recover to good health within several months without any treatment. Only 1% to 2% of people with acute Q fever die of the disease. Chronic Q fever persists for more than 6 months and is uncommon, but is much more serious. Patients who have had acute Q fever may develop the chronic form as soon as one year or as long as 20 years after initial infection. Those who recover fully from infection may possess lifelong immunity against re-infection (Kazar, 2005).

A whole-cell vaccine for Q fever has successfully protected humans in Australia (Kermode *et al.*, 2003). However, this vaccine is not commercially available in other countries. A vaccine for use in animals has also been developed. Doxycycline is the treatment of choice for acute Q fever and most effective when initiated within the first three days of illness, and continued for 15 to 21 days. Therapy should be started again if the disease relapses.

To determine the antibiotic susceptibility of C. *Burnetii*, Brennan *et al.* established a SYBR Green based assay for bacterial quantification located in the com1 gene (Brennan and Samuel, 2003), which is highly conserved among C. *burnetii* strains (Zhang *et al.*, 1997). Due to the inability to cultivate this organism on axenic medium, calculation of infectious units is challenging and prevents the use of conventional antibiotic susceptibility assays. The real-time PCR assay was successfully used to quantify the efficacy of several antibiotics. A similar SYBR Green PCR amplified a stretch of the superoxide dismutase gene (Boulos *et al.*, 2004). The method was described as sensitive and specific enough to allow the evaluation of the doubling time of the Coxiella strains Nine Mile and Q212, respectively.

In 2006 Klee *et al.* published a multitarget PCR targeting the singular isocitrate dehydrogenase gene (icd) and the transposase of the IS1111a element present in multiple copies in the *C. burnetii* genome (Klee *et al.*, 2006). As few as 10 and 6.5 genome equivalents per reaction of the *icd* and IS markers could be detected with a 95% probability, respectively. Both assays showed no cross-reactivity to several closely related bacteria, such as *Legionella* spp. or *Francisella* spp. In total, 75 different *C. burnetii* isolates originating from all over the world were subjected to both PCR assays and the number of IS1111 elements was quantified. For the Nine Mile strain the number of IS1111 elements in the genome was determined to be 23, in other isolates, the number of IS1111 elements varied between seven and 110. These data may be useful for further phylogenetic studies.

Burkholderia mallei/pseudomallei

Glanders is an infectious disease that is caused by the bacterium *Burkholderia mallei*. It is still commonly seen among domestic animals in Africa, Asia, the Middle East, and Central and South America. Melioidosis, also called Whitmore's disease, is an infectious disease caused by the bacterium *B. pseudomallei*. Melioidosis is clinically and pathologically similar to glanders disease, but the ecology and epidemiology of melioidosis are different (Raja *et al.*, 2005). *B. mallei* and *B. pseudomallei* are classified as category B bioterror agents (Cheng *et al.*, 2005).

Glanders is primarily a disease affecting horses, but it also affects donkeys and mules and can be naturally contracted by goats, dogs, and cats. Human infection has occurred rarely among those in direct contact with infected, domestic animals through the skin and through mucosal surfaces. The symptoms of glanders depend upon the route of infection with the organism. Generalized symptoms include fever, muscle aches, chest pain, muscle tightness, and headache. If there is a cut or scratch in the skin, a localized infection with ulceration will develop within one to five days at the site where the bacteria entered the body. In pulmonary infections, pneumonia and pleural effusion can occur. Glanders bloodstream infections are usually fatal within seven to 10 days. The chronic form of glanders involves multiple abscesses within the muscles of the arms and legs or in the spleen or liver (Cheng and Currie, 2005).

The bacteria causing melioidosis are found in contaminated water and soil and are spread to humans and animals, such as sheep, goats, horses, swine, cattle, dogs and cats through direct contact with the contaminated source. Melioidosis is endemic in Southeast Asia and in many countries the prevalence of *B. pseudomallei* is so high, that it is a common contaminant found on laboratory cultures. Person-to-person transmission can occur. Illness from melioidosis can be categorized as acute or localized; acute pulmonary infection, acute bloodstream infection, and chronic suppurative infection. Unapparent infections are also possible. The incubation period is not clearly defined, but may range from two days to many years (White, 2003).

There is no vaccine available for glanders. In countries where glanders is endemic in animals, prevention of the disease in humans involves identification and elimination of the infection in the animal population. Because human cases of glanders are rare, there is limited information about antibiotic treatment of the organism in humans. *B. mallei* is usually sensitive to tetracyclines, ciprofloxacin, streptomycin, novobiocin, gentamicin, imipenem, ceftazidime, and the sulfonamides. Resistance to chloramphenicol has been reported.

There is also no vaccine for melioidosis. Prevention of the infection in endemic-disease areas can be difficult since contact with contaminated soil is so common. *B. psuedomallei* is usually sensitive to imipenem, penicillin, doxycycline, amoxicillin-clavulanic acid, azlocillin, ceftazidime, ticarcillin-vulanic acid, ceftriaxone, and aztreonam. Treatment should be initiated early in the course of the disease (Chaowagul, 2000).

Burkholderia pseudomallei and *B. mallei* are closely related and can also be mistaken for non-pathogenic species such as *B. thailandensis*. A 5′ nuclease probe-based real-time PCR was established analyzing the uneven distribution of different type III secretion system (TTS) genes among these three species, named TTS1 and TTS2 (Thibault *et al.*, 2004). The TTS is a toxin delivery mechanism that allows pathogenic bacteria to inject toxic substances into the cytoplasm of the host's cells, like a "toxin gun." *Orf11* and *orf13* from TTS1 and bp*SCU2* from TTS2 were chosen as target genes. bp*SCU2* was amplified from *B. pseudomallei*, *B. mallei*, and *B. thailandensis*. Only the two highly pathogenic species, *B. pseudomallei* and *B. mallei*, appeared positive for *orf13*. Only *B. pseudomallei* generated the *orf11* amplicon. No amplification of the three targets occurred with any of the other species tested. Among the three genes studied, *orf13* can be regarded as specific for the two highly pathogenic species and *orf11* is a specific marker for *B. pseudomallei*. bp*SCU2* is shared by the three species, but, when it is the only marker detected, it provides a means for identification of *B. thailandensis*. However, three separate PCR reactions are necessary for reliable analysis.

Three other genes were included for the development of a multitarget real-time PCR described by Tomaso *et al.* (Tomaso *et al.*, 2006). Using fluorescent HybProbes the 16S rDNA, the flagellin C (*fliC*) and the ribosomal protein subunit S21 (*rpsU*) genes were targeted. Fluorescence melting curve analysis revealed characteristic melting temperatures for each gene that allowed an unequivocal identification of *B. mallei*/*B. pseudomallei* from other *Burkholderia* spp. The test sensitivity and specificity were assessed with a representative panel of 39 *B. pseudomallei*, 9 *B. mallei*, 126 other Burkholderia strains of 29 species, and 45 clinically relevant non-Burkholderia organisms.

Another 5′ nuclease PCR approach used 5′ nuclease-MGB probes for the allelic discrimination of *B. pseudomallei* and *B. mallei* (U'Ren *et al.*, 2005). A synonymous SNP in a putative antibiotic resistance gene, termed *P27*, at base 2 351 851 of chromosome 1 fit the criterion of differentiating *B. pseudomallei* (G) from *B. mallei* (A). Primers were designed to amplify these two species exclusively due to mismatches in the primer binding region for other *Burkholderia* spp. Such an approach was important for differentiation from *B. cepacia* which shares the 5′ nuclease-MGB probe binding region sequence with *B. pseudomallei*. The resulting assay rapidly identified and discriminated between *B. pseudomallei* and *B. mallei* and did not cross-react with genetically similar neighbors, such as *B. thailandensis* and *B. cepacia*.

A discriminatory 5′ nuclease-based real-time assay for the rapid identification of *B. pseudomallei* isolates was designed to amplify a region within *orf2* of the *B. pseudomallei* type III secretion system gene cluster (Novak *et al.*, 2006). Assay performance was evaluated with 224 *B. pseudomallei* isolates and in addition with 23 *B. mallei*, 5 *B. thailandensis*, and 35 other *Burkholderia* spp. and 76 non-*Burkholderia* spp. for cross-reactivity. The limit of detection was found to be 5.2×10^3 genome equivalents/ml.

A highly sensitive assay specific for *B. mallei* was developed to overcome the low bacterial load in infected tissues that hamper diagnosis (Tomaso *et al.*, 2005). The 5′-nuclease real-time PCR assay targeted the *fliP* of *B. mallei* and included an internal amplification control. Specificity was assessed with 19 *B. mallei* strains, 27 *B. pseudomallei* strains, other *Burkholderia* spp. strains of 29 species, and clinically relevant non-*Burkholderia* spp. This assay presents a useful tool for differential diagnosis of *B. pseudomallei*.

Viruses

Variola major

Smallpox is a highly contagious viral disease unique to humans caused by the two virus variants called *Variola major* and *Variola minor*. *Variola major* is the more deadly form, with a typical mortality of 20 to 40% of those infected, while *Variola minor* "only" kills 1% of its victims. Smallpox was responsible for an estimated 300–500 million deaths in the 20th century with more than two million deaths each year (Moore *et al.*, 2006). Except for laboratory stockpiles, the *variola viruses* have been eliminated. However, there is heightened concern that the *variola viruses* might be used as an agent of bioterrorism (Halloran *et al.*, 2002). *Variola viruses* are classified as category A bioterror agents.

Humans are the only natural hosts of *variola virus* which is not known to be transmitted by insects or animals. Smallpox is contracted by infection of the skin or the lungs after direct face-to-face contact with infected body fluids or contaminated objects. Compared to chickenpox, smallpox is only moderately infectious. The incubation period to obvious disease is approximately 12 to 15 days when lysis of many infected cells occurs and the virus is found in the bloodstream in large numbers. In the prodromal period smallpox is not notably infectious and viral shedding is delayed until the appearance of rash. At this stage the infected person is usually very sick and not able to move around in the community. By days 15 to 16 the condition worsens and the disease can take two courses. Classical ordinary smallpox manifests as discrete or confluent smallpox, of which the latter form is usually fatal. The infected person is contagious until the last smallpox scab falls off. The second course, hemorrhagic smallpox, has a mortality of approximately 96% and presents with an entirely different set of symptoms. The skin does not blister, but instead, bleeding will occur under the skin (black pox). The eyes will turn deep red and bleeding simultaneously begins in the organs, leading to death. Opportunistic infections can lead to multiorgan failure (Moore *et al.*, 2006).

There is no specific treatment for smallpox disease, and the only prevention is a very effective vaccination. Since there was no animal reservoir for *variola virus*, successful vaccination campaigns by the WHO lead to certification of the eradication of smallpox in 1979 (Henderson, 1980). The last naturally occurring case in the world was in Somalia in 1977 (Henderson, 1980). Routine vaccination against smallpox among the general public was stopped because it was no longer necessary for prevention and the morbidity and mortality of vaccination exceeded the risk of infection by a disease extinct in the wild (Letai, 2003). Cultures of the virus are kept by the CDC in the United States and at the Institute of Virus Preparations in Siberia, Russia (WHO, 2002).

Since *variola virus* was regarded as eradicated, the increase of nucleotide sequence information has been limited. The first real-time PCR assays were therefore based on only a handful of sequences and no one knew at that point of time if the sequence characteristics chosen for differentiation of *variola virus* from other orthopoxviruses (OPV) were reliable and also conserved among unknown isolates. The most important problem for the development of *variola virus* specific assays is that genomic *variola virus* DNA is only available in the two institutions mentioned above. All other laboratories lack *variola virus* DNA and so began to construct synthetic genes which were identical to the known *variola virus* sequences. However, without WHO permission one is **not** allowed to synthetically construct *variola virus* DNA fragments that are longer than 500 bp in size, which is, however, of sufficient length for an efficient real-time PCR assay.

The first real-time PCR assay that was published to identify *variola virus* appeared in 2002 (Espy *et al.*, 2002). The assay was located in the *hemagglutinin* (HA) gene, a gene exhibiting moderate sequence conservation among strains of the orthopoxviruses, and used two generic primers to amplify all orthopoxviruses and *variola virus* along with specific HybProbes. These probes should identify *variola virus* by fluorescence melting curve analysis on the LightCycler as shown by comparison of four orthopoxvirus isolates with an artificial *variola virus* fragment of the HA gene. The primer and probe sequences were presented only later in an erratum and revealed that the HybProbes designed to be *variola virus* specific were also complementary to recently published sequences of individual strains of camelpoxvirus and cowpoxvirus. This assay was therefore not specific for *variola virus*.

The same gene was chosen for a similar LightCycler PCR approach by Panning *et al.* in 2004 (Panning *et al.*, 2004). Once again, *variola virus* should have been identified by fluorescence melting curve analysis of specific HybProbes after generic amplification of orthopoxviruses. Interestingly, these HybProbes were not specific for *variola virus* and the variola specific melting temperature was between the melting temperatures for camelpoxvirus, vaccinia virus and cowpoxvirus respectively. It is a matter of discussion if *variola virus*-specific HybProbes that display the highest melting temperature for *variola virus* and reduced temperatures for mismatching virus strains, permit a better interpretation of the results. Unfortunately this assay gave false positive results for *variola virus* with several cowpox virus strains.

In 2003 a 5′ nuclease-based real-time PCR assay was developed for the HA gene and performed on the SmartCycler® and the LightCycler (Sofi Ibrahim M. *et al.*, 2003). The assay used one generic primer and one primer that had two *variola virus*-specific bases in the middle of the primer. The 5′ nuclease probe also had two *variola virus*-specific bases close to the 5′ and the 3′ ends. In the middle of the probe were two additional positions that were not conserved among the available *variola virus* sequences, possibly resulting in a reduced detection efficiency of different *variola virus* strains. In total, DNA from 48 *variola virus* isolates and 27 poxvirus isolates were repeatedly subjected to this assay and the results showed a high specificity for *variola virus*. However, low-intensity false positive results were obtained for cowpox virus DNA. This assay was improved by the use of 5′ nuclease-MGB probes that allow a better discrimination of different genotypes compared to regular 5′ nuclease probes therefore reducing the risk of false positive reactions (Kulesh *et al.*, 2004). PCR assays for *B9R*, the *B10R* and *HA* were established that could be performed under identical reaction conditions in a multitarget PCR. While the *B9R* and the *B10R* gene targets were positive for individual cowpoxvirus isolates (unpublished data), the re-designed *HA* assay showed a specificity of 100% in detecting *variola virus*. It has to be mentioned, that the primers are identical to the preceding publication of Ibrahim and amplified DNA of several orthopoxvirus species. Only the 5′ nuclease probe was shortened and the MGB modification added. Beside a reliable 5′ nuclease assay this publication shows the advantages of MGB-modified probes for discriminating different genotypes.

In 2003 we published a further extremely reliable fluorescence melting curve analysis method based on LightCycler identification of *variola virus* (Nitsche *et al.*, 2004). Two assays were described based on generic amplification of orthopoxviruses and discrimination of variola by melting curve analysis of the *rpo18* and *VETF* genes. A third assay was developed to specifically amplify only DNA of *variola virus* using an *A13L* target, but confirmed *variola virus* additionally by melting curve analysis. The specific amplification may have benefits when mixed populations of orthopoxviruses have to be analyzed. An alternative to this assay was presented with a primer that was not *variola virus* specific but detected all orthopoxviruses except *variola virus*. Both the *A13L* assays were designed as internal LightCycler assays, meaning that one primer is labeled with a fluorescent dye that is incorporated in the extended primer strand and FRET occurs after binding of a fluorescein-labeled probe to the complementary DNA strand. This approach can be applied when available gene regions are too short to permit the design of two HybProbes. All assays used *variola virus* specific HybProbes showing the highest possible melting temperature for *variola virus* only. The *rpo18* assay resulted in one melting temperature for *variola virus* and a single, decreased melting temperature for all other tested orthopoxviruses. The *VETF* and the orthopoxvirus-specific *A13L* assay showed various, discernable melting temperatures for different non-variola virus orthopoxviruses. Combination of these melting temperatures allowed a rough identification of monkeypox virus, camelpoxvirus, ectromelia and vaccinia virus. Most importantly, the polymorphisms chosen for *variola virus* identification were conserved among all tested variola strains and not present in non-variola virus isolates.

To complement the published *variola virus* specific assays, we developed a LightCycler assay for the differential diagnosis of vaccinia virus that was designed to detect vaccinia virus specifically by fluorescence melting curve analysis (Nitsche *et al.*, 2005). The assay made use of a vaccinia virus-specific SNP in the IFNγ-receptor gene of orthopoxviruses. Interestingly, this assay presented amplification curves exclusively for vaccinia virus strains, although agarose gel analysis also revealed amplification products for other orthopoxviruses. In addition, various melting temperatures for non-vaccinia virus orthopoxviruses were obtained but these were below the annealing temperature of the PCR reaction. This indicates that probe binding was not occurring at this temperature, making the amplification signal vaccinia virus-specific.

Probably the best evaluated real-time PCR assay for orthopoxvirus detection was published in 2004 (Olson *et al.*, 2004). The assay targeted the gene for the 14 kD fusion protein of orthopoxviruses using degenerate generic primers to successfully amplify DNA from all tested species. A SNP was covered by a non-*variola virus*, orthopoxvirus-specific HybProbe, resulting in a characteristic melting temperature for *variola virus* (57°C) and an increased melting temperature for all other non-variola orthopoxviruses (62°C). This assay was evaluated with 46 different *variola virus* strains and 190 non-variola virus orthopoxvirus strains, resulting in 100% specificity for *variola virus*. This assay together with an internal reaction control was implemented into a kit that can be purchased from Qiagen (RealArt™ Orthopox PCR Kit). It is currently the method of choice to detect orthopoxviruses and perform primary identification of *variola virus*, even by non-expert laboratories. Of special interest is the design of that assay. One conventional typing HybProbe is labeled at its 3′ end with LC640. FRET occurs between this fluorescent dye and the downstream FAM fluorophore attached to a 5′ nuclease-MGB probe that is also labeled with a non-fluorescent quencher at its 3′ end. Therefore, the amplification signal of orthopoxvirus DNA can either be monitored by the FAM signal of the 5′ nuclease-MGB probe or by the LC640 signal of the typing probe. Amplification curves can therefore be obtained with any real-time PCR instrument while the identification of *variola virus* needs an instrument capable of discriminating the LC640 signal.

Very recently, a new assay was published that used LUX™ primers targeting the *HA* of *variola virus* (Aitichou *et al.*, 2005a). LUX primers are also called self-quenched primers and need no additional hybridizing probe. Therefore the additional specificity information that usually comes from the probes" signal is absent from this approach. A pair of primers for orthopoxviruses and for *variola virus* can be purchased from Invitrogen™ but no details for their application have been published to date.

We employed a completely new approach to prove not only the presence, but also the replication competence of orthopoxviruses using real-time PCR (Nitsche *et al.*, 2006). This approach combined virus culture and one-step real-time RT-PCR detection of poxvirus genes expressed immediately after infection. Since the mRNA levels of the selected genes increase significantly following infection of the cell, elevated target amounts indicated that the virus was replicating within the cell. Using dilution experiments we could detect as few as 3 PFU in less than five hours. This new approach could be applied to additional pathogens.

Fortunately, today there are real-time PCR assays published that promise to be *variola virus* specific (Nitsche *et al.*, 2004; Olson *et al.*, 2004; Sofi Ibrahim M. *et al.*, 2003). However, for a final BT risk assessment of suspected samples, one should always consider the use of a combination of the published assays and additionally the use of techniques that provide exact sequence information, like Sanger sequencing or rapid Pyrosequencing.

Viral hemorrhagic fever viruses

Viral hemorrhagic fevers (VHFs) refer to a group of illnesses that are caused by several distinct families of viruses, including the arenaviruses, filoviruses, bunyaviruses, and flaviviruses (Table 9.3). While some types of hemorrhagic fever viruses can cause relatively mild illnesses, many of these viruses cause severe, life-threatening disease with a high case fatality rate. In general, the term "viral hemorrhagic fever" is used to describe a severe multisystem syndrome often accompanied by hemorrhage (Cleri *et al.*, 2006).

The agents causing VHFs are all enveloped RNA viruses. The infections are zoonotic and are geographically restricted to the areas where their natural animal or insect reservoirs live. However, the hosts for some viruses such as Ebola or Marburg virus remain unknown. Human to human transmission has been reported for some of these viruses

Table 9.3 Viral families and the viral hemorrhagic fevers they cause

Arenaviridae	Bunyaviridae	Filoviridae	Flaviviridae	Paramyxoviridae
Argentine HF@	Crimean-Congo HF (CCHF)	Ebola HF	Kyasanur Forest disease	Hendra virus disease
Bolivian HF	Hantavirus pulmonary syndrome (HPS)	Marburg HF	Tick-borne encephalitis	Nipah virus encephalitis
Sabia-associated HF	HF with renal syndrome (HFRS)		Omsk HF	
Venezuelan HF	Rift Valley fever		Dengue virus	
Lassa fever	Hantavirus pulmonary syndrome (HPS)			
Lymphocytic choriomeningitis (LCM)				

*HF—hemorrhagic fever

including Ebola virus, Marburg virus, Lassa virus and Crimean-Congo hemorrhagic fever virus, but humans are usually infected when they come into contact with infected hosts. Human outbreaks occur sporadically and unpredictably. While people usually become infected only in areas where the host lives, occasionally people become infected by a host that has been exported from its native habitat. For example, the first outbreaks of Marburg hemorrhagic fever, in Marburg and Frankfurt, Germany, and in Yugoslavia, occurred when laboratory workers handled imported monkeys infected with Marburg virus (Lacy and Smego, 1996).

Specific signs and symptoms vary by the type of VHF, but initial signs and symptoms often include marked fever, fatigue, dizziness, muscle aches, loss of strength, and exhaustion. Patients with severe cases of VHF often show signs of bleeding under the skin, in internal organs, or from body orifices such as the mouth, eyes, or ears. However, although they may bleed from many sites around the body, patients rarely die because of blood loss but from organ failure. Severely ill patient cases may also show shock, nervous system malfunction, coma, delirium, and seizures (Jeffs, 2006).

With the exception of yellow fever and Argentine hemorrhagic fever, for which vaccines have been developed (Barrera Oro and McKee, Jr., 1991), no vaccines exist that can protect against VHF. Therefore, prevention efforts must concentrate on avoiding contact with host species. With a few noteworthy exceptions, there is no cure or established drug treatment for VHFs. The anti-viral drug Ribavirin has been effective in treating some individuals with Lassa fever. Treatment with convalescent-phase plasma has been used with success in some patients with Argentine hemorrhagic fever.

As this group of viruses is very diverse, the sequence of each virus is understandably variable. In contrast to DNA viruses that show high genome stability and do not vary significantly in sequence over time, VHF viruses are without exception RNA viruses that are in general prone to higher sequence variation. Hence, it seems clear that there is no PCR that can generically detect hemorrhagic fever viruses. But because VHFs are clinically difficult to diagnose and to distinguish; a rapid and reliable laboratory diagnosis is essential in suspected cases. To date there are some real-time PCR assays published for individual viruses of the VHF group which are helpful for diagnosing a viral role in cases of VHF. But due to sequence variability and the steadily increasing availability of sequence information from more strains, one should consider employing only the most recently published assays that can detect more strains. A selection of good assays for some of the VHF viruses is presented below.

In 2002 Drosten *et al.* presented a complete set of one-step real-time RT-PCR assays for the detection of VHF viruses (Drosten *et al.*, 2003). These assays included the detection of important VHF agents such as Ebola and Marburg viruses, Lassa virus, Crimean-Congo hemorrhagic fever virus, Rift Valley fever virus, dengue virus, and yellow fever virus. For Rift Valley fever virus, dengue virus and yellow fever virus 5′ nuclease probes were designed, while the assays for Marburg and Ebola virus, Lassa virus and Crimean-Congo hemorrhagic fever virus used SYBR Green dye. Reaction conditions were optimized using *in vitro* transcribed RNA and up to three assays could be performed in parallel, facilitating rapid testing for several pathogens. This publication demonstrates the convenience of SYBR Green for situations where the target sequences do not allow the design of an additional 5′ nuclease probe. However, for some of these viruses it is also difficult to find suitable regions for primers and degenerated primers must be considered.

Dengue virus

Dengue virus infections pose a huge problem to public health in South America and Asia. For example, in the Rio de Janeiro area about 290 000 dengue fever cases were reported in the summer months of 2002 alone. In 2001 Callahan *et al.* published five 5′ nuclease reverse transcriptase real-time PCR assays. Four assays were specific for each of the serotypes 1 to 4 while one assay was group-specific (Callahan *et al.*, 2001). All assays were evaluated with 67 viremic human sera received from Peru, Indonesia, and Taiwan. The serotype-specific 5′ nuclease RT-PCR assay detected 62 of 67 serologically confirmed dengue virus-positive specimens, while the group-specific assay detected 66 of 67 confirmed dengue virus-positive specimens. This assay system offers the detection of individual serotypes, which is an important factor for the differential diagnosis of Dengue virus infections. Similar assays were developed by Shu *et al.* in 2003 using SYBR Green instead of 5′ nuclease probes. Primers were located in the nucleocapsid gene and although SYBR Green was used all assays showed a sensitivity of at least 10 PFU/ml.

Recently, we developed a 5′ nuclease-MGB probe assay that detects all four serotypes in one reaction. Typing of positive specimens can be performed by rapid Pyrosequencing (unpublished data).

Hantaviruses

Hantaviruses are emerging viruses and infections have been reported from various regions in Asia, Europe and North America. Rodents or insectivores are natural hosts of hantaviruses and transmit the virus to humans chiefly by aerosolization. A promising approach detecting the various hantaviruses was recently published (Aitichou *et al.*, 2005b). Four assays for specifically identifying the Dobrava (DOB), Hantaan (HTN), Puumala (PUU), and Seoul (SEO) hantaviruses were developed as one-step real-time RT-PCRs with the glycoprotein gene encoded in the small segment used as the target sequence. The detection limits of DOB, HTN, PUU, and SEO assays were 25, 25, 25, and 12.5 PFU, respectively.

We developed a comparable set of 5′ nuclease-MGB probe assays to detect all known hantaviruses, including Sin Nombre virus and Andes virus. Identification of the respective hantavirus could be achieved by rapid Pyrosequencing (unpublished data).

Crimean Congo hemorrhagic fever virus

Recently, an assay for the detection of Crimean Congo hemorrhagic fever virus was published which employed degenerate primers to overcome problems due to virus variation between strains (Duh *et al.*, 2006). A degenerated 5′ nuclease probe was designed to catch all virus variants that are endemic in the Balkans. The detection limit of this assay was 100 copies of RNA/ml.

Toxins

Biologic toxins like ricin and botulinum toxin, staphylococcal enterotoxin B or T-2 mycotoxin possess all the necessary properties to make them a bioweapon. However, PCR amplifies nucleic acids and not toxin proteins. Hence, the presence of toxins which are usually bacterial protein products cannot be directly identified by PCR. However, the toxin genes of the toxin producing bacteria can be used as targets for PCR and can at least show the potential for expression of the toxin protein. For example the identification of the appropriate bacterial genes in food will be sufficient for a risk assessment of botulinum contamination. The crucial question will be whether the presence of the encoding gene found in environmental specimens can be regarded as an indicator for the production of the toxin. The assumption is that even pure preparations of toxin proteins contain contaminating bacterial nucleic acids of their producers in concentrations that will be found using sensitive real-time PCR. In BT-suspected specimens, the proof of the respective genes would at least give rise to detailed toxin analysis using ELISA techniques or a toxicity test in mice.

An interesting approach is the combination of highly toxin-specific antibodies and the sensitive real-time PCR, a method called immuno-PCR (Wu *et al.*, 2001). The crucial trick is the conjugation of the antibody with a small fragment of an irrelevant DNA, the reporter DNA. After binding of the antibody and extensive washing, the reporter DNA is detected by real-time PCR. This technique allows an enhancement of the detection sensitivity by a factor of 10^2 to 10^5 (Barletta, 2006).

Botulinum toxin

Botulism is a muscle-paralyzing disease caused by a toxin that is produced by a bacterium called Clostridium botulinum. *This spore-forming* group of bacteria is commonly found in soil growing best in low oxygen conditions. There are seven types of Botulinum toxins (BoNT) designated by the letters A through G; only types A, B, E and F cause illness in humans.

Botulinum toxin specifically targets the peripheral nervous system, where it is internalized into the presynaptic nerve terminal of motor neurons and blocks release of the neurotransmitter acetylcholine. This prevents nerve impulses reaching the muscle resulting in the characteristic bilateral flaccid paralysis of botulism (Patocka *et al.*, 2005).

There are three main types of botulism that can be fatal and are considered medical emergencies:

1 Food-borne botulism occurs when a person ingests pre-formed toxin that leads to illness within a few hours to days. Food-borne botulism is a public health emergency because the contaminated food may still be available to other persons besides the already identified patient.
2 Infant botulism occurs in a small number of susceptible infants each year who harbor *C. botulinum* in their intestinal tract.
3 Wound botulism occurs when wounds are infected with *C. botulinum* that secretes the toxin.

Food-borne botulism can occur in all age groups, but is not spread from one person to another. Symptoms of food-borne botulism begin most commonly within 12 to 36 hours after ingestion of toxin-containing food. Symptoms include double vision, blurred vision, slurred speech, difficulty swallowing, dry mouth, and most importantly muscle weakness that always descends through the body. Paralysis of breathing muscles can cause death, unless mechanical ventilation is provided for several weeks until the paralysis slowly improves. If diagnosed early, food-borne and wound botulism can be treated with an antitoxin which blocks the action of toxin circulating in the blood. The antitoxin is effective in reducing the severity of symptoms if administered early in the course of the disease. Over the past 50 years the proportion of patients who die from botulism has fallen from about 50% to 8% (Horowitz, 2005).

The most direct way to confirm the diagnosis is to demonstrate the botulinum toxin in the patient's serum or stool by injecting serum or stool into mice and looking for signs of botulism. The bacteria can also be isolated from the stool of persons with food-borne and infant botulism.

Prevention of food-borne botulism can be achieved by boiling the food for 10 minutes before eating it, because the botulism toxin is destroyed by high temperatures. Wound botulism can be prevented by rapidly seeking medical care for infected wounds.

The first published real-time PCR assay for botulinum toxins was a 5′ nuclease assay (Akbulut *et al.*, 2005). PCR assays for detection of botulinum types A, B, and E were developed as alternatives to the mouse bioassay. The expected specificities of the PCR assays were demonstrated by *in silico* analysis as well as empirical testing of target DNA extracted from 83 pure cultures of *C. botulinum*, and 44 bacteria from other species. The sensitivities of the assays were found to be equivalent to 16, 10, and 141 genomes for BoNT A, B, and E, respectively. The assays were shown to be applicable to both purified DNA, as well as crude DNA extracted from cultures and enrichment broths. Depending on the purity of the toxin preparation, this may be sufficient to detect toxins in environmental specimens.

In 2005 a 5′ nuclease real-time PCR method for the quantitative detection of *C. botulinum* type A was developed (Yoon *et al.*, 2005). The validity of this assay was verified by using 10 genera of 20 strains, including reference strains of *C. botulinum* types A, B, C, D, E and F. The detection limit of this assay was evaluated on *C. botulinum* type A, using a 10-

fold dilution series of DNA and spores. The DNA and spores were detected up to level of 0.1 ng/ml and 10^2 spores/ml, respectively. Two methods were applied for preparation: heat treatment and guanidinium isothiocyanate (GuSCN) treatment. GuSCN precipitation proved to be the method of choice for the quantification of *C. botulinum* type A because it showed identical C_T values in both pure spore solutions and food slurries.

Ricin

Ricin is a poison that can be made from processing castor beans. Castor beans are processed throughout the world to make castor oil. Ricin is part of the waste "mash." It is a stable, heat sensitive powdery substance that can be employed in either powder-form or dissolved in water or weak acid. Ricin consists of two protein chains of approximately 30 kDa each, which are linked by disulphide bonds. One chain facilitates entrance into the cell and the other binds to ribosomes and inhibits protein synthesis (Spivak and Hendrickson, 2005).

Accidental exposure to ricin is highly unlikely and it would need a deliberate act to poison people by injection or inhalation. Depending on the route of exposure as little as 500 µg of ricin could kill an adult, which would occupy a space about the size of the head of a pin. Ricin poisoning is not contagious and cannot spread from human to human (Bigalke and Rummel, 2005).

In 1978, Georgi Markov, a Bulgarian writer and journalist who was living in London, died after he was attacked by a man with an umbrella. The umbrella had been rigged to inject a poison ricin pellet under Markov's skin (Crompton and Gall, 1980).

Initial symptoms of ricin poisoning by inhalation occur within eight hours of exposure and include respiratory distress, fever, cough, nausea, and tightness in the chest. Pulmonary edema make breathing difficult and the skin might turn blue. Finally, low blood pressure and respiratory failure may occur, leading to death. Following ingestion of ricin, initial symptoms typically occur in less than six hours and include vomiting and diarrhea that may become bloody. Severe dehydration may then result, followed by low blood pressure. Other signs or symptoms may include hallucinations, seizures, and blood in the urine. Within several days, the person's liver, spleen, and kidneys might stop working, and the person could die (Bigalke and Rummel, 2005).

If death has not occurred in three to five days, the victim usually recovers. Because no antidote exists for ricin poisoning, the most important factor is avoiding ricin exposure. After contact, the most important factor is getting the ricin out of the body as quickly as possible (Audi *et al.*, 2005).

Although an unpublished real-time PCR assay exists at the Robert Koch-Institute, interestingly, there are no real-time PCR assays published for ricin to date. Recently, an immuno-PCR detection assay for ricin was published (Lubelli *et al.*, 2006). This indicates that PCR is not the method of choice for detecting toxins; however, it may be used as a supportive technique.

Future trends

Today, real-time PCR represents the best diagnostic tool for the identification of a BT-relevant pathogen's nucleic acid. In terms of speed and sensitivity there is no method that can compete against real-time PCR.

Although unrivaled, the adequate preparation of nucleic acids may still take between 30 and 90 minutes depending on the kind of specimen but producing DNA of high purity that can be amplified with high sensitivity in 60 to 90 minutes (Espy *et al.*, 2006). Improved preparation protocols and mobile cycling instrumentation are required to further shorten the generation time for reliable results. AlphaHelix's rapid QuanTyper™ (www.alphahelix.com) uses centrifugational forces to reduce the real-time PCR reaction for 96 samples to 15 minutes and the mobile R.A.P.I.D.® (Idaho Technologies) and SmartCycler systems (Cepheid) have been described as promising innovations (Christensen *et al.*, 2006). Unfortunately, there is a lack of reliable ready-to-use real-time PCR reagents that need no freezing or cooling and have long shelf-lives, making a real field-diagnostic kit a difficult concept. Recently, Cepeid introduced the GenExpert with ready-to-use dry reagents that are promising for mobile real-time PCR detection. However, future developments are necessary to stabilize PCR reagents for room temperature storage.

Broader detection systems that allow the identification of a whole group or family of pathogens are necessary to reduce the number of PCR reactions applied to a specimen lacking any hint of a pathogen. Such systems are also essential to catch genetically modified pathogen variants possibly escaping established diagnostic tools. To this end, additional chemistries that reduce the need for long conserved DNA stretches are under development. These include LUX primers (Nazarenko, 2006), MGB modifications (Kutyavin *et al.*, 2000), and the incorporation of modifications such as locked nucleic acid (LNA) nucleotides (Simeonov and Nikiforov, 2002) or proprietary superbases (Cloud *et al.*, 2005) that promise a certain degree of specificity in BT diagnostic assays.

An important requirement for robust assay design is the steady increase of sequence information for BT-relevant pathogens. To maintain maximum reliability for detection assays of BT-relevant agents, any appearance of new sequences for a certain target gene must be monitored carefully. To ensure that sequence information essential to BT diagnostic real-time PCR assay design is current, future developments should focus on automation of sequence updating and automatic alignments of the respective sequences.

Conclusions

The problem of bioterrorism is not new and several examples from history show the application of biological agents for obtaining diverse strategic purposes (Roffey *et al.*, 2002a). However, bioterrorism was always regarded as very unlikely to happen and arrangements to counteract BT attacks were marginal in most countries. The anthrax containing letters sent in 2001 within the USA changed the perception of the possibilities of misuse of biological agents. Since then, the risk of future abuses of biological agents is considered to be significantly increased (Gottschalk and Preiser, 2005; Lane *et al.*, 2001). Hence, programs developing catalogs of countermeasures to allow a dependable risk assessment of suspected BT scenarios have sprung up like mushrooms. These programs include the establishment of diagnostic tools for any BT-relevant agents as well as the development of new strategies for treatment and prophylaxis for the respective diseases. First, the most dangerous agents including *variola virus* or *Bacillus anthracis* were the focus of specific diagnostic assay developments. However, the unambiguous detection of modified pathogens or those agents that are important for differential diagnosis has become more and more important. Soon it became clear that for most BT-suspected situations, the first priority was the exclusion of the most dangerous agents followed by the accurate identification of the agent present. This requires a broad spectrum of analytical tools and its successful implementation in the future by expert laboratories. However, the primary diagnosis, excluding a suspected pathogen or proving its possible presence, can nowadays be performed by real-time PCR in several well-equipped diagnostic laboratories.

In cases of human infections the clinical presentation usually offers sufficient information for an experienced physician to find the accurate diagnosis without laboratory support, in others this laboratory support is essential for the identification or typing of the pathogen. This support can either be offered from individual, specialized, expert laboratories or even better, from any laboratory close to the patient.

When comparing the classical three methods of cell culture, electron microscopy and PCR for the direct detection of pathogens, the latter seems to be the one that finds most application in the modern diagnostic laboratory and particularly in specialized laboratories dealing with the identification of BT agents. There are many reasons for the popularity of PCR but especially for the popularity of real-time PCR. Real-time PCR is a very sensitive method that can detect minute amounts of a pathogen's nucleic acid in a short time, for several specimens in parallel, without the demand for highly experienced personnel.

In BT scenarios involving environmental specimens that are suspected of containing infectious agents, there are no clinical symptoms that can be interpreted and therefore laboratory diagnosis is the only way to obtain a reliable risk assessment, and ultimately a "diagnosis" of a possible bioterrorist attack. The more dangerous the suspected specimen is, the faster it must be diagnosed. Several techniques for the rapid detection and identification of infectious agents have been developed in recent years. Each method has its power but most methods also suffer from certain drawbacks. As a consequence, no method should be regarded as a stand-alone technique or as the optimal solution for every diagnostic problem. This also applies to PCR.

The nucleic acids of BT-suspected specimens that are subjected to the PCR reaction are usually non-infectious. After an inactivation step that is included in most nucleic acid preparation protocols, the specimen can be treated like regular non-infectious material and no special equipment or safety cabinets are required for its further analysis. Especially for the diagnosis of BT-relevant specimens this is very important to establish some degree of diagnostic capacity even in non-specialized laboratories. In addition, inactivation enables transport of the specimens to expert laboratories for confirmation of preliminary results, however, it must be clear that these inactivated specimens can no longer be used for pathogen propagation.

None of the other listed techniques can offer such enormous benefits. Electron microscopic analysis requires a lot of experience and very specialized equipment for the accurate identification of pathogen particles while cell culture techniques are usually laborious, also require considerable experience but still permits pathogen discrimination using PCR or antibody staining techniques. With the increasing number of commercial kits available for real-time PCR detection of rarely occurring infectious diseases, the preliminary diagnosis of these agents is possible in nearly every diagnostic lab.

Whatever the method used to detect a BT agent, results should be confirmed by a reference laboratory and should be based on additional techniques. However, for an initial screening, PCR seems to be the ideal tool.

Web resources

1 Robert Koch-Institute: www.rki.de
2 Centers for Disease Control and Prevention: www.cdc.gov
3 European Center for Disease Prevention and Control: www.ecdc.eu.int
4 Health Protection Agency: www.hpa.org.uk
5 World Health Organisation: www.who.int
6 National Center for Biotechnology Information: http://www.ncbi.nlm.nih.gov
7 Poxvirus Bioinformatics resource Center: www.poxvirus.org

References

Aitichou, M., Javorschi, S., and Ibrahim, M.S. (2005a). Two-color multiplex assay for the identification of orthopox viruses with real-time LUX- PCR. Mol. Cell Probes. *19*, 323–328.

Aitichou, M., Saleh, S.S., McElroy, A.K., Schmaljohn, C., and Ibrahim, M.S. (2005b). Identification of Dobrava, Hantaan, Seoul, and Puumala viruses by one-step real-time RT-PCR. J. Virol. Methods. *124*, 21–26.

Akbulut, D., Grant, K.A., and McLauchlin, J. (2005). Improvement in laboratory diagnosis of wound botulism and tetanus among injecting illicit-drug users by use of real-time PCR assays for neurotoxin gene fragments. J. Clin. Microbiol. *43*, 4342–4348.

Alibek, K. and handelmann, S. (1999). Biohazard: the chilling true story of the largest covert biological weapons program in the world. To from inside by the man who ran it. (New York: Dell Publishing), p. 281.

Audi, J., Belson, M., Patel, M., Schier, J., and Osterloh, J. (2005). Ricin poisoning: a comprehensive review. JAMA *294*, 2342–2351.

Barletta, J. (2006). Applications of real-time immuno-polymerase chain reaction (rt-IPCR) for the rapid diagnoses of viral antigens and pathologic proteins. Mol. Aspects Med. *27*, 224–253.

Barrera Oro, J.G. and McKee, K.T., Jr. (1991). Toward a vaccine against Argentine hemorrhagic fever. Bull. Pan Am. Health Organ 25, 118–126.

Bell, C.A., Uhl, J.R., Hadfield, T.L., David, J.C., Meyer, R.F., Smith, T.F., and Cockerill, F.R., III (2002). Detection of Bacillus anthracis DNA by LightCycler PCR. J. Clin. Microbiol. *40*, 2897–2902.

Bigalke, H. and Rummel, A. (2005). Medical aspects of toxin weapons. Toxicology *214*, 210–220.

Bode, E., Hurtle, W., and Norwood, D. (2004). Real-time PCR assay for a unique chromosomal sequence of Bacillus anthracis. J. Clin. Microbiol. *42*, 5825–5831.

Bogdanovich, T., Skurnik, M., Lubeck, P.S., Ahrens, P., and Hoorfar, J. (2004). Validated 5′ nuclease PCR assay for rapid identification of the genus Brucella. J. Clin. Microbiol. *42*, 2261–2263.

Boulos, A., Rolain, J.M., Maurin, M., and Raoult, D. (2004). Measurement of the antibiotic susceptibility of Coxiella burnetii using real time PCR. Int. J. Antimicrob. Agents. *23*, 169–174.

Brennan, R.E. and Samuel, J.E. (2003). Evaluation of Coxiella burnetii antibiotic susceptibilities by real-time PCR assay. J. Clin. Microbiol. *41*, 1869–1874.

Brouillard, J.E., Terriff, C.M., Tofan, A., and Garrison, M.W. (2006). Antibiotic selection and resistance issues with fluoroquinolones and doxycycline against bioterrorism agents. Pharmacotherapy *26*, 3–14.

Brubaker, R.R. (1972). The genus Yersinia: biochemistry and genetics of virulence. Curr. Top. Microbiol. Immunol. 57, 111–158.

Bystrom, M., Bocher, S., Magnusson, A., Prag, J., and Johansson, A. (2005). Tularemia in Denmark: identification of a Francisella tularensis subsp. holarctica strain by real-time PCR and high-resolution typing by multiple-locus variable-number tandem repeat analysis. J. Clin. Microbiol. *43*, 5355–5358.

Callahan, J.D., W.u., S.J., on-Schultz, A., Mangold, B.E., Peruski, L.F., Watts, D.M., Porter, K.R., Murphy, G.R., Suharyono, W., King, C.C., Hayes, C.G., and Temenak, J.J. (2001). Development and evaluation of serotype- and group-specific fluorogenic reverse transcriptase PCR (TaqMan) assays for dengue virus. J. Clin. Microbiol. 39, 4119–4124.

Chabot, D.J., Scorpio, A., Tobery, S.A., Little, S.F., Norris, S.L., and Friedlander, A.M. (2004). Anthrax capsule vaccine protects against experimental infection. Vaccine 23, 43–47.

Chaowagul, W. (2000). Recent advances in the treatment of severe melioidosis. Acta Trop. *74*, 133–137.

Chase, C.J., Ulrich, M.P., Wasieloski, L.P., Jr., Kondig, J.P., Garrison, J., Lindler, L.E., and Kulesh, D.A. (2005). Real-time PCR assays targeting a unique chromosomal sequence of Yersinia pestis. Clin. Chem. *51*, 1778–1785.

Cheng, A.C. and Currie, B.J. (2005). Melioidosis: epidemiology, pathophysiology, and management. Clin. Microbiol. Rev. *18*, 383–416.

Cheng, A.C., Dance, D.A., and Currie, B.J. (2005). Bioterrorism, Glanders and melioidosis. Euro. Surveill *10*, E1-E2.

Cheun, H.I., Makino, S.I., Watarai, M., Erdenebaatar, J., Kawamoto, K., and Uchida, I. (2003). Rapid and effective detection of anthrax spores in soil by PCR. J. Appl. Microbiol. *95*, 728–733.

Christensen, D.R., Hartman, L.J., Loveless, B.M., Frye, M.S., Shipley, M.A., Bridge, D.L., Richards, M.J., Kaplan, R.S., Garrison, J., Baldwin, C.D., Kulesh, D.A., and Norwood, D.A. (2006). Detection of biological threat agents by real-time PCR: comparison of assay performance on the, R.A.P.I.D., the LightCycler, and the Smart Cycler platforms. Clin. Chem. *52*, 141–145.

Christopher, G.W., Cieslak, T.J., Pavlin, J.A., and Eitzen, E.M., Jr. (1997). Biological warfare. A historical perspective. JAMA *278*, 412–417.

Cleri, D.J., Ricketti, A.J., Porwancher, R.B., Ramos-Bonner, L.S., and Vernaleo, J.R. (2006). Viral hemorrhagic fevers: current status of endemic disease and strategies for control. Infect. Dis. Clin. North Am. *20*, 359–93, x.

Cloud, J.L., Hoggan, K., Belousov, E., Cohen, S., Brown-Elliott, B.A., Mann, L., Wilson, R., Aldous, W., Wallace, R.J., Jr., and Woods, G.L. (2005). Use of the MGB Eclipse system and SmartCycler PCR for differentiation of Mycobacterium chelonae and, M. abscessus. J. Clin. Microbiol. *43*, 4205–4207.

Cornelis, G.R., Boland, A., Boyd, A.P., Geuijen, C., Iriarte, M., Neyt, C., Sory, M.P., and Stainier, I. (1998). The virulence plasmid of Yersinia, an antihost genome. Microbiol. Mol. Biol. Rev. *62*, 1315–1352.

Crompton, R. and Gall, D. (1980). Georgi Markov—death in a pellet. Med. Leg. J. *48*, 51–62.

Cronquist, S.D. (2004). Tularemia: the disease and the weapon. Dermatol. Clin. *22*, 313-vii.

Cutler, S.J., Whatmore, A.M., and Commander, N.J. (2005). Brucellosis—new aspects of an old disease. J. Appl. Microbiol. *98*, 1270–1281.

Davis, C.J. (1999). Nuclear blindness: An overview of the biological weapons programs of the former Soviet Union and Iraq. Emerg. Infect. Dis. *5*, 509–512.

Debeaumont, C., Falconnet, P.A., and Maurin, M. (2005). Real-time PCR for detection of Brucella spp. DNA in human serum samples. Eur. J. Clin. Microbiol. Infect. Dis. *24*, 842–845.

Domaradskij, I.V. and Orent, L.W. (2006). Achievements of the soviet biological weapons programme and implications for the future. Rev. Sci. Tech. *25*, 153–161.

Drosten, C., Kummerer, B.M., Schmitz, H., and Gunther, S. (2003). Molecular diagnostics of viral hemorrhagic fevers. Antiviral Res. *57*, 61–87.

Duh, D., Saksida, A., Petrovec, M., Dedushaj, I., and vsic-Zupanc, T. (2006). Novel one-step real-time RT-PCR assay for rapid and specific diagnosis of Crimean-Congo hemorrhagic fever encountered in the Balkans. J. Virol. Methods *133*, 175–179.

Ellerbrok, H., Nattermann, H., Ozel, M., Beutin, L., Appel, B., and Pauli, G. (2002). Rapid and sensitive identification of pathogenic and apathogenic Bacillus anthracis by real-time PCR. FEMS Microbiol. Lett. *214*, 51–59.

Emanuel, P.A., Bell, R., Dang, J.L., McClanahan, R., David, J.C., Burgess, R.J., Thompson, J., Collins, L., and Hadfield, T. (2003). Detection of Francisella tularensis within infected mouse tissues by using a hand-held PCR thermocycler. J. Clin. Microbiol. *41*, 689–693.

Espy, M.J., Cockerill III, F.R., Meyer, R.F., Bowen, M.D., Poland, G.A., Hadfield, T.L., and Smith, T.F. (2002). Detection of smallpox virus DNA by LightCycler PCR. J. Clin. Microbiol. *40*, 1985–1988.

Espy, M.J., Uhl, J.R., Sloan, L.M., Buckwalter, S.P., Jones, M.F., Vetter, E.A., Yao, J.D., Wengenack, N.L., Rosenblatt, J.E., Cockerill, F.R., III, and Smith, T.F. (2006). Real-time PCR in clinical microbiology: applications for routine laboratory testing. Clin. Microbiol. Rev. *19*, 165–256.

Franz, D.R., Jahrling, P.B., Friedlander, A.M., McClain, D.J., Hoover, D.L., Bryne, W.R., Pavlin, J.A., Christopher, G.W., and Eitzen, E.M., Jr. (1997). Clinical recognition and management of patients exposed to biological warfare agents. JAMA *278*, 399–411.

Fujita, O., Tatsumi, M., Tanabayashi, K., and Yamada, A. (2006). Development of a real-time PCR assay for detection and quantificaition of Francisella tularensis. Jpn. J. Infect. Dis. *59*, 46–51.

Gelderblom, H.R. and Bannert, N. (2005). Diagnostic electron microscopy in infectious diseases emergencies and bio-terrorism. In Yearbook 2005, European Microscopy Society, Peter Hawkes, ed., pp. 47–54.

Golovliov, I., Ericsson, M., Sandstrom, G., Tarnvik, A., and Sjostedt, A. (1997). Identification of proteins of Francisella tularensis induced during growth in macrophages and cloning of the gene encoding a prominently induced 23-kilodalton protein. Infect. Immun. *65*, 2183–2189.

Gottschalk, R. and Preiser, W. (2005). Bioterrorism: is it a real threat? Med. Microbiol. Immunol. (Berl). *194*, 109–114.

Halloran, M.E., Longini, I.M., Jr., Nizam, A., and Yang, Y. (2002). Containing bioterrorist smallpox. Science. *298*, 1428–1432.

Harris, S. (1992). Japanese biological warfare research on humans: a case study of microbiology and ethics. Ann. N. Y. Acad. Sci. *666*, 21–52.

Hazelton, P.R. and Gelderblom, H.R. (2003). Electron microscopy for rapid diagnosis of infectious agents in emergent situations. Emerg. Infect. Dis. *9*, 294–303.

Henderson, D.A. (1980). Smallpox eradication. Public Health Rep. *95*, 422–426.

Hernandez, E., Girardet, M., Ramisse, F., Vidal, D., and Cavallo, J.D. (2003). Antibiotic susceptibilities of 94 isolates of Yersinia pestis to 24 antimicrobial agents. J. Antimicrob. Chemother. *52*, 1029–1031.

Hicks, J.M. (2003). An overview of terrorism and its impact on biomedical research facilities. Lab Anim (NY) *32*, 41–48.

Higgins, J.A., Ezzell, J., Hinnebusch, B.J., Shipley, M., Henchal, E.A., and Ibrahim, M.S. (1998). 5′ nuclease PCR assay to detect Yersinia pestis. J. Clin. Microbiol. *36*, 2284–2288.

Higgins, J.A., Hubalek, Z., Halouzka, J., Elkins, K.L., Sjostedt, A., Shipley, M., and Ibrahim, M.S. (2000). Detection of Francisella tularensis in infected mammals and vectors using a probe-based polymerase chain reaction. Am. J. Trop. Med. Hyg. *62*, 310–318.

Hinnebusch, J., Cherepanov, P., D.u., Y., Rudolph, A., Dixon, J.D., Schwan, T., and Forsberg, A. (2000). Murine toxin of Yersinia pestis shows phospholipase D activity but is not required for virulence in mice. Int. J. Med. Microbiol. *290*, 483–487.

Horowitz, B.Z. (2005). Botulinum toxin. Crit Care Clin. *21*, 825–39, viii.

Hurtle, W., Bode, E., Kulesh, D.A., Kaplan, R.S., Garrison, J., Bridge, D., House, M., Frye, M.S., Loveless, B., and Norwood, D. (2004). Detection of the Bacillus anthracis gyrA gene by using a minor groove binder probe. J. Clin. Microbiol. *42*, 179–185.

Inglesby, T.V., O'Toole, T., Henderson, D.A., Bartlett, J.G., Ascher, M.S., Eitzen, E., Friedlander, A.M., Gerberding, J., Hauer, J., Hughes, J., McDade, J., Osterholm, M.T., Parker, G., Perl, T.M., Russell, P.K., and Tonat, K. (2002). Anthrax as a biological weapon, 2002: updated recommendations for management. JAMA. *287*, 2236–2252.

Iqbal, S.S., Chambers, J.P., Goode, M.T., Valdes, J.J., and Brubaker, R.R. (2000). Detection of Yersinia pestis by pesticin fluorogenic probe-coupled PCR. Mol. Cell Probes. *14*, 109–114.

Isherwood, K.E., Titball, R.W., Davies, D.H., Felgner, P.L., and Morrow, W.J. (2005). Vaccination strategies for Francisella tularensis. Adv. Drug Deliv. Rev. *57*, 1403–1414.

Jeffs, B. (2006). A clinical guide to viral haemorrhagic fevers: Ebola, Marburg and Lassa. Trop. Doct. *36*, 1–4.

Jernigan, J.A., Stephens, D.S., Ashford, D.A., Omenaca, C., Topiel, M.S., Galbraith, M., Tapper, M., Fisk, T.L., Zaki, S., Popovic, T., Meyer, R.F., Quinn, C.P., Harper, S.A., Fridkin, S.K., Sejvar, J.J., Shepard, C.W., McConnell, M., Guarner, J., Shieh, W.J., Malecki, J.M., Gerberding, J.L., Hughes, J.M., and Perkins, B.A. (2001). Bioterrorism-related inhalational anthrax: the first 10 cases reported in the United States. Emerg. Infect. Dis. *7*, 933–944.

Johansson, A., Farlow, J., Larsson, P., Dukerich, M., Chambers, E., Bystrom, M., Fox, J., Chu, M., Forsman, M., Sjostedt, A., and Keim, P. (2004). Worldwide genetic relationships among Francisella tularensis isolates determined by multiple-locus variable-number tandem repeat analysis. J. Bacteriol. *186*, 5808–5818.

Kazar, J. (2005). Coxiella burnetii infection. Ann. N. Y. Acad. Sci. *1063*, 105–114.

Kermode, M., Yong, K., Hurley, S., and Marmion, B. (2003). An economic evaluation of increased uptake in Q fever vaccination among meat and agricultural industry workers following implementation of the National Q Fever Management Program. Aust. N. Z. J. Public Health *27*, 390–398.

Kim, K., Seo, J., Wheeler, K., Park, C., Kim, D., Park, S., Kim, W., Chung, S.I., and Leighton, T. (2005). Rapid genotypic detection of Bacillus anthracis and the Bacillus cereus group by multiplex real-time PCR melting curve analysis. FEMS Immunol. Med. Microbiol. *43*, 301–310.

Klee, S.R., Tyczka, J., Ellerbrok, H., Franz, T., Linke, S., Baljer, G., and Appel, B. (2006). Highly sensitive real-time PCR for specific detection and quantification of Coxiella burnetii. BMC. Microbiol.19;6:2., 2.

Klietmann, W.F. and Ruoff, K.L. (2001). Bioterrorism: implications for the clinical microbiologist. Clin. Microbiol. Rev. *14*, 364–381.

Koirala, J. (2006). Plague: disease, management, and recognition of act of terrorism. Infect. Dis. Clin. North Am. *20*, 273–87, viii.

Kulesh, D.A., Baker, R.O., Loveless, B.M., Norwood, D., Zwiers, S.H., Mucker, E., Hartmann, C., Herrera, R., Miller, D., Christensen, D., Wasieloski, L.P., Jr., Huggins, J., and Jahrling, P.B. (2004). Smallpox and pan-orthopox virus detection by real-time 3′-minor groove binder TaqMan assays on the roche LightCycler and the Cepheid smart Cycler platforms. J. Clin. Microbiol. *42*, 601–609.

Kutyavin, I.V., Afonina, I.A., Mills, A., Gorn, V.V., Lukhtanov, E.A., Belousov, E.S., Singer, M.J., Walburger, D.K., Lokhov, S.G., Gall, A.A., Dempcy, R., Reed, M.W., Meyer, R.B., and Hedgpeth, J. (2000). 3′-minor groove binder-DNA probes increase sequence specificity at PCR extension temperatures. Nucleic Acids Res. *28*, 655–661.

Lacy, M.D. and Smego, R.A. (1996). Viral hemorrhagic fevers. Adv. Pediatr. Infect. Dis. *12*, 21–53.

Lane, H.C., Montagne, J.L., and Fauci, A.S. (2001). Bioterrorism: a clear and present danger. Nat. Med. *7*, 1271–1273.

Lee, M.A., Brightwell, G., Leslie, D., Bird, H., and Hamilton, A. (1999). Fluorescent detection techniques for real-time multiplex strand specific detection of Bacillus anthracis using rapid PCR. J. Appl. Microbiol. *87*, 218–223.

Letai, A.G. (2003). Smallpox and smallpox vaccination. N. Engl. J. Med. *348*, 1920–1925.

Lim, D.V., Simpson, J.M., Kearns, E.A., and Kramer, M.F. (2005). Current and developing technologies for monitoring agents of bioterrorism and biowarfare. Clin. Microbiol. Rev. *18*, 583–607.

Lindler, L.E. and Fan, W. (2003). Development of a 5′ nuclease assay to detect ciprofloxacin resistant isolates of the biowarfare agent Yersinia pestis. Mol. Cell Probes. *17*, 41–47.

Lindler, L.E., Fan, W., and Jahan, N. (2001). Detection of ciprofloxacin-resistant Yersinia pestis by fluorogenic PCR using the LightCycler. J. Clin. Microbiol. *39*, 3649–3655.

Little, S.F. (2005). Anthrax vaccines: a development update. BioDrugs. *19*, 233–245.

Lubelli, C., Chatgilialoglu, A., Bolognesi, A., Strocchi, P., Colombatti, M., and Stirpe, F. (2006). Detection of ricin and other ribosome-inactivating proteins by an immuno-polymerase chain reaction assay. Anal. Biochem. *355*, 102–109.

Luna, V.A., King, D., Davis, C., Rycerz, T., Ewert, M., Cannons, A., Amuso, P., and Cattani, J. (2003). Novel sample preparation method for safe and rapid detection of Bacillus anthracis spores in environmental powders and nasal swabs. J. Clin. Microbiol. *41*, 1252–1255.

Mackay, I.M. (2004). Real-time PCR in the microbiology laboratory. Clin. Microbiol. Infect. *10*, 190–212.

Mackay, I.M., Arden, K.E., and Nitsche, A. (2002). Real-time PCR in virology. Nucleic Acids Res. *30*, 1292–1305.

Makino, S. and Cheun, H.I. (2003). Application of the real-time PCR for the detection of airborne microbial pathogens in reference to the anthrax spores. J. Microbiol. Methods. 53, 141–147.

Makino, S.I., Cheun, H.I., Watarai, M., Uchida, I., and Takeshi, K. (2001). Detection of anthrax spores from the air by real-time PCR. Lett. Appl. Microbiol. 33, 237–240.

Miller, S.E. (2003). Bioterrorism and electron microscopic differentiation of poxviruses from herpesviruses: dos and don'ts. Ultrastruct. Pathol. *27*, 133–140.

Moore, Z.S., Seward, J.F., and Lane, J.M. (2006). Smallpox. Lancet *367*, 425–435.

Moran, G.J. (1999). Update on emerging infections from the Centers for Disease Control and Prevention. Bioterrorism alleging use of anthrax and interim guidelines for management—United States, 1998. Ann. Emerg. Med. *34*, 229–232.

Moran, G.J. and Talan, D.A. (2003). Update on emerging infections: news from the Centers for Disease Control and Prevention. Syndromic surveillance for bioterrorism following the attacks on the World Trade Center—New York City, 2001. Ann. Emerg. Med. *41*, 414–418.

Nano, F.E. (1988). Identification of a heat-modifiable protein of Francisella tularensis and molecular cloning of the encoding gene. Microb. Pathog. *5*, 109–119.

Nazarenko, I. (2006). Homogeneous detection of nucleic acids using self-quenched polymerase chain reaction primers labeled with a single fluorophore (LUX primers). Methods Mol. Biol. 335, 95–114.

Niedrig, M., Meyer, H., Panning, M., and Drosten, C. (2006). Follow-up on diagnostic proficiency of laboratories equipped to perform orthopoxvirus detection and quantification by PCR: the second international external quality assurance study. J. Clin. Microbiol. *44*, 1283–1287.

Niedrig, M., Schmitz, H., Becker, S., Gunther, S., ter, M.J., Meyer, H., Ellerbrok, H., Nitsche, A., Gelderblom, H.R., and Drosten, C. (2004). First international quality assurance study on the rapid detection of viral agents of bioterrorism. J. Clin. Microbiol. *42*, 1753–1755.

Nitsche, A., Ellerbrok, H., and Pauli, G. (2004). Detection of orthopoxvirus DNA by real-time PCR and identification of variola virus DNA by melting analysis. J. Clin. Microbiol. *42*, 1207–1213.

Nitsche, A., Steger, B., Ellerbrok, H., and Pauli, G. (2005). Detection of vaccinia virus DNA on the LightCycler by fluorescence melting curve analysis. J. Virol. Methods. *126*, 187–195.

Nitsche, A., Stern, D., Ellerbrok, H., and Pauli, G. (2006). Detection of Infectious Poxvirus Particles. Emerg. Infect. Dis. *12*, 1139–1141.

Novak, R.T., Glass, M.B., Gee, J.E., Gal, D., Mayo, M.J., Currie, B.J., and Wilkins, P.P. (2006). Development and evaluation of a real-time PCR assay targeting the type III secretion system of Burkholderia pseudomallei. J. Clin. Microbiol. *44*, 85–90.

Olson, K.B. (1999). Aum Shinrikyo: once and future threat? Emerg. Infect. Dis. *5*, 513–516.

Olson, V.A., Laue, T., Laker, M.T., Babkin, I.V., Drosten, C., Shchelkunov, S.N., Niedrig, M., Damon, I.K., and Meyer, H. (2004). Real-time PCR system for detection of orthopoxviruses and simultaneous identification of smallpox virus. J. Clin. Microbiol. *42*, 1940–1946.

Oyston, P.C. and Isherwood, K.E. (2002). Yersinia: an update. Trends Microbiol. *10*, 550–551.

Panning, M., Asper, M., Kramme, S., Schmitz, H., and Drosten, C. (2004). Rapid detection and differentiation of human pathogenic orthopox viruses by a fluorescence resonance energy transfer real-time PCR assay. Clin. Chem. *50*, 702–708.

Pappas, G., Akritidis, N., Bosilkovski, M., and Tsianos, E. (2005a). Brucellosis. N. Engl. J. Med. *352*, 2325–2336.

Pappas, G., Akritidis, N., and Tsianos, E. (2005b). Effective treatments in the management of brucellosis. Expert. Opin. Pharmacother. *6*, 201–209.

Pappas, G., Christou, L., Akritidis, N., and Tsianos, E.V. (2006). Quinolones for brucellosis: treating old diseases with new drugs. Clin. Microbiol. Infect. *12*, 823–825.

Patocka, J., Splino, M., and Merka, V. (2005). Botulism and bioterrorism: how serious is this problem? Acta Medica. (Hradec. Kralove) *48*, 23–28.

Perry, R.D. and Fetherston, J.D. (1997). Yersinia pestis—etiologic agent of plague. Clin. Microbiol. Rev. *10*, 35–66.

Poupard, J.A. and Miller, L.A. (1992). History of biological warfare: catapults to capsomeres. Ann. N. Y. Acad. Sci. *666*, 9–20.

Qi, Y., Patra, G., Liang, X., Williams, L.E., Rose, S., Redkar, R.J., and Delvecchio, V.G. (2001). Utilization of the rpoB gene as a specific chromosomal marker for real-time PCR detection of Bacillus anthracis. Appl. Environ. Microbiol. *67*, 3720–3727.

Queipo-Ortuno, M.I., Colmenero, J.D., Reguera, J.M., Garcia-Ordonez, M.A., Pachon, M.E., Gonzalez, M., and Morata, P. (2005). Rapid diagnosis of human brucellosis by SYBR Green I-based real-time PCR assay and melting curve analysis in serum samples. Clin. Microbiol. Infect. *11*, 713–718.

Radonic, A., Thulke, S., Mackay, I.M., Landt, O., Siegert, W., and Nitsche, A. (2004). Guideline to reference gene selection for quantitative real-time PCR. Biochem. Biophys. Res. Commun. *313*, 856–862.

Raja, N.S., Ahmed, M.Z., and Singh, N.N. (2005). Melioidosis: an emerging infectious disease. J. Postgrad. Med. *51*, 140–145.

Roffey, R., Lantorp, K., Tegnell, A., and Elgh, F. (2002a). Biological weapons and bioterrorism preparedness: importance of public-health awareness and international cooperation. Clin. Microbiol. Infect. *8*, 522–528.

Roffey, R., Tegnell, A., and Elgh, F. (2002b). Biological warfare in a historical perspective. Clin. Microbiol. Infect. *8*, 450–454.

Ryu, C., Lee, K., Yoo, C., Seong, W.K., and, Oh, H.B. (2003). Sensitive and rapid quantitative detection of anthrax spores isolated from soil samples by real-time PCR. Microbiol. Immunol. *47*, 693–699.

Sarinas, P.S. and Chitkara, R.K. (2003). Brucellosis. Semin. Respir. Infect. *18*, 168–182.

Sauret, J.M. and Vilissova, N. (2002). Human brucellosis. J. Am. Board Fam. Pract. *15*, 401–406.

Simeonov, A. and Nikiforov, T.T. (2002). Single nucleotide polymorphism genotyping using short, fluorescently labeled locked nucleic acid (LNA) probes and fluorescence polarization detection. Nucleic Acids Res. *30*, e91.

Sjostedt, A., Tarnvik, A., and Sandstrom, G. (1991). The T-cell-stimulating 17-kilodalton protein of Francisella tularensis LVS is a lipoprotein. Infect. Immun. *59*, 3163–3168.

Sofi Ibrahim, M., Kulesh, D.A., Saleh, S.S., Damon, I.K., Esposito, J.J., Schmaljohn, A.L., and Jahrling, P.B. (2003). Real-time PCR assay to detect smallpox virus. J. Clin. Microbiol. *41*, 3835–3839.

Spivak, L. and Hendrickson, R.G. (2005). Ricin. Crit Care Clin. *21*, 815–24, viii.

Splino, M., Patocka, J., Prymula, R., and Chlibek, R. (2005). Anthrax vaccines. Ann. Saudi. Med. *25*, 143–149.

Thibault, F.M., Valade, E., and Vidal, D.R. (2004). Identification and discrimination of Burkholderia pseudomallei, B. mallei, and, B. thailandensis by real-time PCR targeting type III secretion system genes. J. Clin. Microbiol. *42*, 5871–5874.

Thomas, R., Johansson, A., Neeson, B., Isherwood, K., Sjostedt, A., Ellis, J., and Titball, R.W. (2003). Discrimination of human pathogenic subspecies of Francisella tularensis by using restriction fragment length polymorphism. J. Clin. Microbiol. *41*, 50–57.

Tomaso, H., Pitt, T.L., Landt, O., A.l., D.S., Scholz, H.C., Reisinger, E.C., Sprague, L.D., Rathmann, I., and Neubauer, H. (2005). Rapid presumptive identification of Burkholderia pseudomallei with real-time PCR assays using fluorescent hybridization probes. Mol. Cell Probes. *19*, 9–20.

Tomaso, H., Reisinger, E.C., A.l., D.S., Frangoulidis, D., Rakin, A., Landt, O., and Neubauer, H. (2003). Rapid detection of Yersinia pestis with multiplex real-time PCR assays using fluorescent hybridization probes. FEMS Immunol. Med. Microbiol. *38*, 117–126.

Tomaso, H., Scholz, H.C., A.l., D.S., Eickhoff, M., Treu, T.M., Wernery, R., Wernery, U., and Neubauer, H. (2006). Development of a 5′-nuclease real-time PCR assay targeting fliP for the rapid identification of Burkholderia mallei in clinical samples. Clin. Chem. *52*, 307–310.

Torok, T.J., Tauxe, R.V., Wise, R.P., Livengood, J.R., Sokolow, R., Mauvais, S., Birkness, K.A., Skeels, M.R., Horan, J.M., and Foster, L.R. (1997). A large community outbreak of salmonellosis caused by intentional contamination of restaurant salad bars. JAMA *278*, 389–395.

U'Ren, J.M., Van Ert, M.N., Schupp, J.M., Easterday, W.R., Simonson, T.S., Okinaka, R.T., Pearson, T., and Keim, P. (2005). Use of a real-time PCR TaqMan assay for rapid identification and differentiation of Burkholderia pseudomallei and Burkholderia mallei. J. Clin. Microbiol. *43*, 5771–5774.

Versage, J.L., Severin, D.D., Chu, M.C., and Petersen, J.M. (2003). Development of a multitarget real-time TaqMan PCR assay for enhanced detection of Francisella tularensis in complex specimens. J. Clin. Microbiol. *41*, 5492–5499.

Welkos, S., Pitt, M.L., Martinez, M., Friedlander, A., Vogel, P., and Tammariello, R. (2002). Determination of the virulence of the pigmentation-deficient and pigmentation-/plasminogen activator-deficient strains of Yersinia pestis in non-human primate and mouse models of pneumonic plague. Vaccine *20*, 2206–2214.

White, N.J. (2003). Melioidosis. Lancet *361*, 1715–1722.

WHO (2002). Smallpox eradication: destruction of variola virus stocks. Can. Commun. Dis. Rep. *28*, 40–44.

Wicki, R., Sauter, P., Mettler, C., Natsch, A., Enzler, T., Pusterla, N., Kuhnert, P., Egli, G., Bernasconi, M., Lienhard, R., Lutz, H., and Leutenegger, C.M. (2000). Swiss Army Survey in Switzerland to determine the prevalence of Francisella tularensis, members of the Ehrlichia phagocytophila genogroup, Borrelia burgdorferi sensu lato, and tick-borne encephalitis virus in ticks. Eur. J. Clin. Microbiol. Infect. Dis. *19*, 427–432.

Wu, H.C., Huang, Y.L., Lai, S.C., Huang, Y.Y., and Shaio, M.F. (2001). Detection of Clostridium botulinum neurotoxin type A using immuno-PCR. Lett. Appl. Microbiol. *32*, 321–325.

Yoon, S.Y., Chung, G.T., Kang, D.H., Ryu, C., Yoo, C.K., and Seong, W.K. (2005). Application of real-time PCR for quantitative detection of Clostridium botulinum type A toxin gene in food. Microbiol. Immunol. *49*, 505–511.

Zhang, G.Q., T.o., H., Yamaguchi, T., Fukushi, H., and Hirai, K. (1997). Differentiation of Coxiella burnetii by sequence analysis of the gene (com1) encoding a 27-kDa outer membrane protein. Microbiol. Immunol. *41*, 871–877.

Experts' Roundtable: Real-Time PCR and Microbiology 10

M.G.H.M. Beld, C. Birch, P.A. Cane, W. Carman, E.C.J. Claas, J.P. Clewley, E. Domingo, J. Druce, C. Escarmis, R.A.M. Fouchier, V. Foulongne, M.G. Ison, L.C. Jennings, B. Kaltenboeck, I.D. Kay, M. Kubista, O. Landt, I.M. Mackay, J. Mackay, H.G.M. Niesters, M.D. Nissen, S. Palladino, N.G. Papadopoulos, A. Petrich, M.W. Pfaffl, W. Rawlinson, U. Reischl, N.A. Saunders, C. Savolainen-Kopra, O. Schildgen, G.M. Scott, M. Segondy, R. Seibl, T.P. Sloots, Y.-W. Tang, R. Tellier, and P.C.Y. Woo

Introduction

This chapter attempts to document informal discussions about topics of interest to those of us who develop, employ or make decisions based upon the use of real-time PCR in microbiology. It is also intended to complement the preceding chapters. The layout for this chapter was loosely based on the format of published conference discussions and popular magazine interviews.

The authors were most generous with their time and very responsive to badgering. The result is a wander through the minds of those whose papers we read, whose work we admire and whose methods we often employ.

Please enjoy reading the opinions provided by a diverse selection of the foremost scientists and clinicians actively working in and around the field of microbiology.

Ian M. Mackay, Editor

What impact does the existence of quasispecies have on modern day microbial diagnostics? What should we be doing to address this impact?

Genotypes, subtypes, or quasispecies have a huge impact on diagnostics in general. Good screening assays target the most conserved regions but even within these regions changes in sequence can and do occur due to lack of proof-reading by RNA viruses, the result of immune pressure, or emerging resistance profiles during treatment.

M.G.H.M. Beld

If discussing qualitative diagnostics, probably not too much of an issue if a conserved region is targeted. But if it's a treatment issue, such as the presence of HIV drug resistant mutants, sensitivity for individual "quasispecies" is important. To address the impact we should submit sequences with novel mutations to GenBank so that subsequent primer design is optimized. Diagnostic laboratories need to pay particular attention to updating/improving their assays based on new sequence information.

Chris Birch and Julian Druce

Quasispecies detection and quantitation using real-time PCR are just starting to be translated from the research laboratories to some specialist reference units. One example of this application is the detection of drug resistance associated mutations present as minority populations. Routine drug resistance genotyping uses population sequencing of PCR products with a maximum sensitivity for detection of minority species of about 25%. In contrast, allele-specific PCR can be combined with real-time assays to allow highly sensitive tests for such mutations, for example increasing the detection of mutations in transmitted drug resistance in HIV and allowing the early detection of emerging drug resistance in hepatitis B virus[1,2]. However, the quasispecies nature of those viruses which use an error-prone reverse transcriptase (or RNA-dependent RNA polymerase) step in their lifecycle means that there will be a background of 0.01–0.1% mutation prevalence existing even in the absence of any drug pressure. Thus the sensitivity of assays will always be limited by background "noise," and it will be challenging to determine the clinical utility of such assays.

1 Johnson JA, Li J-F, Bennett D *et al.* Real-time PCR assays identify transmitted drug resistant HIV-1 previously undetected by conventional nucleotide sequencing. Antivir. Ther 2004; 9:S87

2 Punia P, Cane PA, Teo CG, Saunders N. Quantitation of hepatitis B lamivudine resistant mutants by real-time amplification refractory mutation system PCR. J. Hepatology 2004; 40:986–992

Patricia A. Cane

I am unsure whether this has any real impact on our viral practice. We certainly see mixed populations on resistance profiles in particular for HIV (or, we do not see them, as our primers tend to amplify one population in preference to another). We do see variants that do not get detected by a specific primer or probe, but either by culturing or setting up a scheme where one does a longer PCR and then sequences, variants can be discovered. Whether quasispecies have any real effect on antiviral outcome is a moot point. I have probably missed something really important here, but we never think about it in our lab.

William Carman

The presence of quasi-species complicates the specificity and reproducibility of molecular diagnostics. Only the predominant variant(s) will be detected.

E.C.J. Claas

It is important to recognize the existence of minority species sequence variants when doing antiviral drug resistant testing by sequencing, e.g. for HIV-1 and HBV. For example, when sequence chromatograms are examined, peaks representing bases responsible for drug resistance may be visible below the main peak, and these should be taken into account when assigning resistance profiles and deciding on drug treatment. A good way to monitor this is to use quality control panels of mixed sequence DNA targets (e.g. a mixture of plasmids containing inserts differing by the bases at the sites of resistance). Some such panels are available, but more could be made and distributed by specialist quality control labs.

Jon Clewley

Components of the mutant spectrum of a quasispecies may not hybridize with one of the primers chosen for the amplification. Therefore, those "hybridization-defective" sequences will not be represented in the population of amplified sequences. Since the composition of the mutant spectra of the quasispecies of pathogenic microbes are largely unknown, the extent of this problem is also largely indeterminate. When suspecting a possible bias that may affect the problem under study, ways to address the possible impact are to use two or more sets of independent primers targeting different sequences on the genome under analysis, or to expand the primer repertoire for any given sequence with a mixture of degenerate primers designed on the basis of the corresponding consensus sequence. In the latter case, the problem of finding an adequate annealing temperature to permit hybridization of as many primer sequences as possible should be a concern.

In addition to these technical aspects, microbiologists should be aware that two populations displaying identical consensus sequences may nevertheless show a different phenotypic behavior due to a different composition of the mutant spectrum. Therefore, two indistinguishable results of diagnosis may correspond to different biological manifestations (particularly in disease symptoms or severity). These situations may be encountered mainly with RNA viruses due to their extensive exploitation of quasispecies dynamics.

Esteban Domingo and Cristina Escarmís

We must ensure that our tests detect all variants of the pathogen. This can be challenging in particular in molecular diagnostics as several pathogens are notoriously variable genetically. This also requires that sequence information is rapidly made available to the community, which is sometimes a problem (e.g. H5N1). Sometimes, we rely heavily on the results of a single molecular test but we need to keep realizing that a single negative test is not always conclusive; it is thus advisable that alternative tests (second molecular test or better: alternative methods altogether such as virus isolation) should remain available as back up. Individual (reference) laboratories or consortia of laboratories should use alternative methods for pathogen detection, in order to track changes in pathogen quasispecies, and thus ensure that emerging variants are detected.

Ron A.M. Fouchier

Genetic diversity among various subtypes and genotypes of viruses and other pathogens is known to exist. The choice of diagnostic primer and probe sequences from highly conserved genomic regions favors the formation of specific amplification products; however genetic diversity presents a challenge in designing molecular assays for accurate diagnosis while also maintaining their sensitivity and specificity. Improved sharing of regional sequence data, the standardization of assay protocols including primers and probes and availability of commercial controls will extend our understanding of the impact of genetic variability with an increasing range of pathogens.

Lance Jennings

Careful design of conserved primers is required to detect as large a selection of closely related (highly similar) targets (quasispecies) as possible. The design of equally conserved probes within the same target region is sometimes impossible. In this case, sets of different probes, all for use with the same primer pair(s), enable detection of amplified targets, possibly in multiplexed assays. Guidelines for design and thermal characteristics of effective primers/probes would be helpful, since standard software packages do not address optimal design of primers for sets of aligned sequences.

Bernhard Kaltenboeck

Quasispecies complicate modern diagnostics. There is, for example, not yet one real-time PCR test on the market that is sensitive to all HIV clades. The situation is similar with HCV and many other pathogens. And the more we learn from genotyping species from new geographic areas the more variation we find. Of course, if using an inappropriate test the infectious agent can go undetected. The problem is particularly important in developing countries, where the abundant variants may not be known, and available tests may not be sensitive. What is very important to do, but unfortunately is hard to find financing for, is to genotype at least the most serious pathogens in afflicted developing countries so appropriate tests can be developed. It is also important that the sequence information is made available. For diagnostic tests the precision of quantification is not that important and one way to increase the sensitivity is by designing multiplex assays targeting more of the quasispecies.

Mikael Kubista

Routine diagnostics ignores this problem—until the first fatal case of a wrong diagnosis will be made public. I believe that "home-brew" assays make the situation safer; the worst case would be if all the world used only one kit—the fittest quasispecies would be selected and multiply very quickly. The future will show that some pathogens might escape from detection. From time to time we get reports about false results; one example—in the last ring trial for MRSA there was one MSSA isolate containing an error in a region of the mecA gene outsides the standard detection fragment—all results had been false positive.

Olfert Landt

Most studies of viral quasispecies focus on a handful of targets (e.g. HCV, HIV and poliovirus). The potential for quasispecies variation to interfere with assay design and performance has not been investigated for many other viruses. Hence those of us using PCR as an applied tool to investigate other microbial infections cannot predict what impact there may have been on historical prevalence values, nor on current values from studies that do not address the quasispecies problem. To date quasispecies studies have remained the domain of the basic virology research field but to investigate their scope for causing human illness, a better understanding of their role will need to filter down to the application-driven fields.

To my mind, many questions relating to the potential of quasispecies among other viruses need to be addressed:

- How many respiratory virus quasispecies co-exist in an infected host and does that vary with tissue? Is the number affected by the specific host's immune system and/or the speed of a developing infection?
- Do different patients infected with the same inoculum develop different quasispecies?
- How much variation differentiates one quasispecies from another? Do some genes exhibit more quasispecies variation than others?
- Do quasispecies render nucleotide sequencing an overly subjective method and how much weight can be attribute to sublineages and sub- sublineages etc, determined using sequencing?
- Is it correct to consider quasispecies variation to be more subtle than the variation we use to assign virus strains into lineages, sublineages, and as is becoming popular recently, sub- sublineages?
- Could the host's swarm of quasispecies include the innate potential among its variants to be better at, (i) causing disease *in situ*, (ii) transmitting, (iii) replicating in culture, (iv) causing different illness severity, (v) evading the immune system and, (v) even evading antiviral pressures?
- Is a swarm of quasispecies transmitted during infection, or only the predominating strain(s)?
- Are the predominant quasispecies in the host also present when a specimen is incubated during *in vitro* culture or does the "hierarchy" of quasispecies change to suit its environment?
- Have quasispecies confounded conventional and molecular assays to the extent that the prevalence of some viruses has been kept artificially low e.g. HCoV-229E, IFCV and HPIV-4?

Ian M. Mackay

Quasispecies have been investigated in particular in HCV and HIV research. Numerous publications deal with this subject, but still on a limited number of viral targets. This is currently still a research topic, of academic interest and worthwhile to pursue, but not of impact in day-to-day clinical diagnostics. The information as is, cannot be transferred to the individual patient in a timely manner.

H.G.M. Niesters

HIV, Hep C and Hep B infections. Recognize them as important clinically and a natural consequence of immune and drug pressure.

Michael D. Nissen

I would believe that this is primarily an issue where genotypic changes can result from antimicrobial resistance. One concern is the detection of low levels of organism with mutations may not be detected within a larger population of wild-type organisms. This is possibly only an issue when a consensus PCR is performed and sequencing or probe hybridization is used to characterize the amplification product. Low target levels of some quasispecies would remain undetected as amplification would bias towards higher copy number targets. This may be less of an issue if successive specimens are taken, especially in instances of treatment failure.

Astrid Petrich

Quasispecies are of emerging clinical relevance for HCV, HIV and potentially for any RNA virus with an error-prone polymerase. In HCV the quasispecies evolution is being studied as a predictor of antiviral response, but is not yet clearly of value in the clinical setting. We should be continuing with studies of quasispecies in treatable RNA virus infections (HCV, HIV, influenza), in well-described cohorts, and correlating such virological measures with clinical and treatment outcomes.

William Rawlinson and Gillian M. Scott

Even when bacteria first started to build up their own species millions of years ago during the harsh process of evolution, the genetic rearrangement by mutations (SNPs), insertion elements or hopping transposons was a permanent process leading to a number of slight genetically distinct variants within a given species. With respect to "global" assay specificity (encompassing all genetic variants of a given species present in clinical isolates) it is mandatory not only to screen type -trains of a given target organism but also as many clinical isolates of different geographic origin as possible. This kind of thorough evaluation should be requested for any diagnostic PCR assay. Testing a nice collection of static ATCC strains does not really address the problem of genetic variation or quasispecies.

Udo Reischl

Eigen had a precise mathematical notion of the quasispecies as the equilibrium frequencies of a pool of closely related variants that act together as an entity. Each entity cannot be characterized by a single molecular species so he used the term "quasispecies". Under this definition the individual HIV sequences within a single human are a quasi-species but the concept can be widened so that for example the *Staphylococcus aureus* clones associated with a human population can also be seen as a quasi-species. In the latter case the individuals do not merely differ in terms of nucleotide polymorphisms but also in presence/absence of hundreds of complete genes comprising the accessory genome. The quasi-species concept is very helpful in reminding us of what we are actually trying to achieve using molecular diagnostic methods. The aim might be to detect the presence of one or more classically defined species, a pathogenic subgroup or even a single clone. Once this has been decided and a suitable genetic target has been identified it is clearly essential to take account of quasi-species in the design of the particular assay. For example, by choosing essential genes or highly conserved sequence motifs. Where the level of genetic variation leads to a significant danger of false-negatives occurring in the diagnostic test applied in the relevant diagnostic setting, future molecular diagnostics should be able to build in parallel testing for multiple target sequences.

Nick Saunders

Entero- and rhinovirus genomes vary greatly and strains exist as swarms of microvariants or quasispecies. The variation is greatest in the VP1 protein coding gene and may reach 30% within a single enterovirus serotype. Even in the 5′ non-coding region, the most conserved region, commonly used for primer and probe locations in pan-enterovirus and pan-rhinovirus molecular detection, variation exists. In picornavirus diagnostics it is very important to take the variability of the genome into account when applying molecular methods of detection. Due to the quasispecies nature, one may miss several serotypes in detection, if primers or probes are designed only to match the consensus sequence. For instance, in a Finnish study several concurrent types of rhinoviruses were shown to circulate during the same epidemic season in a limited area. A molecular detection method should be prepared to identify all of them. It is also possible that new pathogenic and virulent variants evolve from the less common ones. Thus it would be important to recognize them. This of course created increased challenges for primer and probe design. Furthermore, when degenerate nucleotide positions are used in oligonucleotides, the quality of production and purification must be excellent. It is important to maintain the specificity of e.g. pan-enterovirus RT-PCR by revising primer sequences according to newly recognized serotypes and strains.

Carita Savolainen-Kopra

With respect to the field of newly detected and emerging respiratory viruses you have to optimize your detection methods, i.e. quasispecies frequently lead to the use of degenerate or, if possible, highly conserved, primers for PCR analysis/detection, as you otherwise will have too many false negative results.

Oliver Schildgen

Previous experimental results have conclusively demonstrated that many and varied genetic alterations may occur during RNA viral replication, due to the infidelity of genome replication and large population sizes. The variant genomes are selected by mechanisms that are still poorly understood, including environmental factors, to generate viral quasispecies. Variation, selection, and random sampling of genomes occur continuously and unavoidably during virus evolution.

The result is that numerous genetic variants of a viral species are circulating at any one time in a single population, and that these variants may genetically differ in different populations at any one geographic location.

The challenges for vaccine development and diagnostics are therefore significant. New generations of vaccines and diagnostic reagents (monoclonal antibodies, peptide antigens, oligonucleotides for PCR, etc.) may be adequate to prevent disease and detect some or even most of the circulating quasispecies of any given RNA pathogen, but the dynamics of quasispecies mandates careful consideration of those reagents to be incorporated into diagnostic assays.

In particular, the design of oligonucleotide primers and probes for real-time PCR needs to be carefully considered. The PCR efficiency is intimately dependent on the complementarity existing between primers and template, and sequence mismatches may result in diminished sensitivity, or at worse, false negative results. It is essential therefore that in the design of a real-time PCR assay, one has a detailed knowledge of the genetic composition of local quasispecies and that genetic variation of these viruses is reflected in the primer design. By definition, a reliance on sequence information available on public databases (Genbank, EMBL) which are often based on other geographical viral populations may lead to inadequate assay development. It is therefore of the utmost importance that a local genetic profile is determined by sequencing of an adequate number of local virus strains.

Also, once an assay has been developed and validated in the laboratory, and subsequently included in the routine diagnostic repertoire, it essential that further genetic monitoring of local virus strains is maintained to identify the emergence of dominant quasispecies that may not be detected by the assay developed. Of particular concern are the potential long-term effects of weak selective pressures that may initially go unnoticed. Variant viruses resulting from evolutionary pressure imposed by vaccines or drugs may insidiously and gradually replace previous dominant quasispecies.

Theo P. Sloots

All forms of PCR assays depend critically on a proper match between the templates and the primers. Some templates are more variable than others, and especially the genome of RNA viruses. However not all regions are equally variable, there are some hypervariable regions. In addition, quasispecies may cause heterogeneity in a fraction of the templates. Our approach for PCR design has been to develop consensus primers from sequence alignment of several genotypes, or even of different by closely related species. The identification of conserved regions helps to avoid hypervariable regions. Real-time PCR using an internal probe adds a layer of complexity, since heterogeneity and mismatches with the probe are also issues. PCR-based assays that can amplify more than one species require a step to distinguish them, which can be the use of several specific probes, but this, of course, requires careful validation. Quantitative real time PCR adds a further level of complexity: unless great care is taken with the design, internal probes may bind to different genotypes with different affinities, introducing biases in the quantification.

Raymond Tellier

Some quasispecies, such as those that confer antiviral resistance in influenza virus, have major clinical implications. These quasispecies may exist even in patients or poultry without prior exposure to antiviral agents. In general, if the primers are not specifically designed, the quasispecies will not be detected. Additional primers specific for the quasispecies have to be included in order to detect the quasispecies.

Patrick Woo

What do you perceive as the hurdles/obstacles to a microbiology laboratory wishing to introduce real-time PCR assays?

Up-front costs of equipment purchase are an issue. Most companies in the area provide a reasonable level of expert advice with respect to primer/probe design etc. However, for optimal performance there needs to be some scientific expertise in the laboratory to cover development and validation of new assays.

Chris Birch and Julian Druce

There should be few problems (other than cost) for a laboratory which has already been using conventional PCR and has staff who are experienced in the science behind the assays. However, I think caution should be exercised in introducing real-time PCR into laboratories with little molecular experience and no R&D backup. There needs to be a good understanding of the infrastructure required such as separate clean areas for setting up of reactions and practical experience of trouble-shooting PCR. In addition, there may be some trepidation felt by staff in implementation of the technology but it should be possible to overcome this with suitable training.

Patricia A. Cane

Training of staff, understanding the science, dealing with high C_T's, picking up contamination, choosing an extractor, keeping abreast of the best real-time kits, quality control of reagents, deciding on internal controls, finding decent run controls, choosing the best assay from the literature.

William Carman

In practice there are no real hurdles for implementing "standard" assays. Generating real-time PCR data will be no problem. However, experience is required for the proper laboratory organization and interpretation of results.

E.C.J. Claas

Labs have to choose between the variety of real-time PCR instrumentation, chemistry and kits that are available. Mistakenly buying inappropriate or soon to be superseded machines can be expensive. To start with, a diagnostic lab may be best advised to buy an established instrument for which there are kits for the virus or bacteria they're interested in, and for which a many others are available. Also, they should ensure that they can obtain standard reagents at known concentrations, for calibration and normalization of their assays. It is probably not advisable for a lab new to this technology and wanting to use it for diagnostic purposes to dive in and reinvent the wheel by buying an instrument, making their own assay and attempting to calibrate it with high titer cloned DNA. Better first to use what is already out there, and then later setup a development program. PCR carryover and contamination from cloned targets are as a big a problem as they ever were.

Jon Clewley

One obstacle is the need to have technicians with the adequate training, and experts to supervise their work. This is particularly necessary at the early stages of a project involving real-time PCR with new microbial agents. The people in charge should be knowledgeable in the design of the control amplifications to exclude artifacts in the interpretation of the runs with the real samples, and to ensure quantitative and reproducible results. In some countries, budget limitations may impede the purchasing of the required instrument and reagents.

Esteban Domingo and Cristina Escarmís

Cost. Space (multiple rooms required). Preventing sample contamination. Assay validation.

Ron A.M. Fouchier

In fact real-time PCR is easier to handle than conventional PCR, with a minimized risk of contamination. Availability of commercial test for an increasing number of pathogens facilitates the introduction of real-time PCR assays in labs without previous experience in molecular diagnostics.

Vincent Foulongne and Michel Segondy

We have had a problem introducing PCR assays—it is difficult to bill for assays that are not FDA approved and there may be some resistance to these.

Michael G. Ison

Microbiology laboratories wishing to introduce real-time PCR will need to address the same issues as those required for the introduction of any new test into routine diagnostic service and should follow recommended protocols. Adequate laboratory space for separated pre-analytical and analytical work areas, staff with some molecular expertise, accessing training, trouble-shooting and addressing quality assurance requirements, will all be major hurdles. For a laboratory moving from conventional/in-house PCR to real-time PCR, the hurdles are more likely to be related to work-flow management, the choice of instrumentation and whether to develop in-house methods or purchase commercial test kits. Laboratory management will require a business proposal including cost-effectiveness evaluations. The projected volume of test requests will determine whether it is cost effective to introduce commercial test kits which usually require the "batching" of tests, rather than develop in-house methods both for the diagnosis and quantitation of viruses and other pathogens. The validation of new tests whether in-house or commercial in comparison with existing technology is also a potential hurdle, especially for the smaller laboratory and access to R&D expertise should be considered essential.

Lance Jennings

Most users are not aware that real-time PCR is a quantitative method, as opposed to the strictly qualitative standard PCR method. As a quantitative method, real-time PCR requires much more stringent quality assurance than standard PCR.

Bernhard Kaltenboeck

Cost, lack of trained staff and ability to develop and validate "in-house" assays.

Michael D. Nissen

Laboratory space: Even though real-time PCR assays with their closed tube systems minimize the risk of amplicon contamination they do not eliminate it. Capillaries do break, we may want to open the tubes and run the products on an electrophoresis gel. With this in mind it is prudent to have at least two distinct areas for real-time PCR. One for reagent preparation and specimen extractions and the other for amplification and detection.

Resource allocation: To train and have dedicated staff that can become familiar and proficient in the assays. Even though molecular assays are becoming routine, they still are not as robust as many assays used in the laboratory. Senior staff need to be employed that are familiar with the design, troubleshooting and validation of molecular assays.
Platforms: Deciding on a platform that is suitable for as many assays as possible. Often laboratories have very limited budgets and they require a system that can be efficiently used. Issues to be considered are the number of samples that can be tested, reaction vessels, ability to support different chemistries, ease of use, ability to interface with the LIM, product support and of course, cost.

Another hurdle when introducing real-time PCR is trying to keep the molecular laboratory in close association with the traditional Microbiology laboratory. There is a push to centralize all molecular services. This may be entertained for other molecular assays, but molecular assays for a microbiology service need to be intimately linked with the microbiology service. Often tests are performed on cultures, which should not be transported around a Health Care Facility. Also many samples are tested by culture (bacterial, fungal, viral), serology and molecular techniques. These samples are best split in the one laboratory and distributed to the other sections in the Microbiology laboratory. There are distinct infection control measures that need to be taken when dealing with pathogenic organisms which are best handled and understood by microbiologists. The increased level of RNA testing for molecular infectious diseases precludes its integration within a general "genetic" molecular service.

Ian D. Kay and Silvano Palladino

The TATAA Biocenter supports laboratories in setting up real-time PCR diagnostics. The most common problem is contamination due to inadequate separation of pre-PCR, PCR and post-PCR working areas, and improper routines for sample handling, decontamination, sample disposal etc. Since it is very costly for laboratories to remove contamination we strongly advise laboratories to seek PCR consultancy when setting up real-time PCR.

Mikael Kubista

Reservations of the (older) personnel against using new technologies, the presence of existing instruments, necessary new investments and the requirement for additional rooms will all hinder introduction.

Olfert Landt

Much of the expertise gained from conventional PCR bodes well for a laboratory in the real-time PCR era. However a lot of the technical optimization of in-house assays has come from the observation of band intensity on gels and the change of mindset to crossing-points, efficiencies and fluorescent slopes can take some adaptation.

If a laboratory is new to all PCR then appropriate and ongoing training is paramount. While real-time PCR reduces many of the issues associated with conventional PCR (e.g. carry-over contamination) it may be wise to retain the same level of wariness of pre- and post-PCR areas. Interestingly, in training many people on PCR and real-time PCR, I have largely found that good microbiologists make good PCR operators—presumably due to the familiarity of the similar "aseptic technique" required.

John Mackay

Although PCR and even more, real-time PCR assays, are brilliant research and epidemiology tools, their unique sensitivity makes them a bit of a problem for diagnostic purposes. Technicians not specifically trained in such assays will not avoid embarrassment, either because of contamination or complete assay failure! Unusually strict measures against contamination should be taken and this may add to the routine problems of a microbiology laboratory. On the other hand, the clinical significance of a quantitative result is not clear in all cases, therefore moving into real-time PCR prematurely may reduce its significance. Cost is also an issue

Nikos Papadopoulos

Issues include:

- cost justification
- lack of cost recovery; molecular assays are often "add-on" assays and some health systems provide insufficient compensation or none at all
- obtaining and retaining suitably trained staff
- space issues
- instrumentation
- lack of commercial assays available (FDA approved and/or ASRs)
- being able to adequately validate assays (insufficient well-characterized specimens to work with or imperfect gold standards)
- difficulty demonstrating proficiency as no proficiency program is available for a many microbial targets.

Over time new regulations may be written and the experience of early users will be better distributed to inform others of their experiences and to provide expertise and/or possible solutions to these issues. Continue dialog, discussion and communication between laboratories.

Astrid Petrich

[WR] Expertise in molecular testing, laboratory physical set up, test cost per assay, ability to increase scale, and the need for regulation of in house assays.

[GS] Choice of technology appropriate to the laboratory and an ability to apply that technology to different microorganisms. We have found the Roche LightCycler® useful in the implementation of new tests for different pathogens, but within this platform the choice of detection systems needs to be looked at. We have found SYBR™ green detection to be less expensive, but on some specimen types (e.g. PMBC preparations) we have found non-specific DNA detection a problem affecting quantitation results. Use of hybridization probes for detection eliminates non-specific fluorescence but is more costly and there is some evidence of reduced sensitivity. Standardization and quality control of the assays is also necessary.

William Rawlinson and Gillian M. Scott

Among the main hurdles are the costs for the technical equipment, implementation of the molecular workflow, and the costs for establishing physically separate areas for sample processing and aliquotting, nucleic acid isolation (the so called "sample prep") and amplification. Since real-time PCR platforms have evolved as closed and easy-to-use systems, installing a separate amplicon detection area with negative air pressure seems to be no longer necessary.

Other obstacles can be the availability of personnel specifically trained in molecular diagnostics and a frequent absence of willingness by "traditional microbiologists" to approach their molecular colleagues for advice or help. Since the ego is hard to overcome, the latter point could be the highest hurdle.

Udo Reischl

Cost can be an issue both in terms of instrumentation and reagents however in most situations it should be possible to build a convincing business case. Acquiring the necessary laboratory expertise to implement the new tests may also be a hurdle but this is relatively minor compared with the most intractable problem. This is likely to lie in the need to find resources to do the work required to demonstrate the superiority of the new real-time assays.

Nick Saunders

In my opinion the major problem is of a financial nature. In general, conventional PCR techniques are cheaper since reagents and instruments are less expensive, and in most cases it is enough to know whether the pathogen is present or not. Nevertheless, the impact of and the need for quantification and also the speed of the techniques are of increasing importance.

Oliver Schildgen

In order of priority

1 Capital outlay. The purchasing of real-time systems is expensive, and equipment at best only has a practical "shelf-life" of 3 years. This limits the introduction of the technology to larger organizations (laboratories).
2 Wide scale acceptance is limited by the lack of availability of commercial real-time PCR diagnostic kits. These are generally available for the big volume tests (HCV, HBV gonorrhea) but not for the targets of "lesser" (commercial) importance. Also most commercial test kits are expensive with a single test costing between AUD50 to 200.
3 In-house assays are a common way of reducing costs. However, this requires staff with technical skill and awareness to develop and validate these assays and to support any technical difficulties experienced during their routine performance. Such staff are often difficult to recruit.
4 There must be a comprehensive quality dimension to the performance of diagnostic real-time PCR. This must include adequate and comprehensive validation of each assay using the appropriate specimen matrix, quality control (QC) measures to monitor the daily performance of the test, enrolment in an appropriate quality assurance program (QAP) for each assay to ensure continued performance, and the recording, monitoring and review of results to document and address any inadequacies in test performance long-term.

5 The availability of dedicated highly skilled technical staff to perform the routine diagnostic assays. Some staff are just not able to perform PCR at the required level. Therefore an educational/training program must be in place to raise proficiency standards, and from those the suitable candidates should be selected for the performance of the assays.
6 Physical environment. So often laboratories try to perform (real-time) PCR assays in facilities that are unsuitable for this technology. Even with real-tine PCR which is a closed system, some physical separation of extraction, loading and amplification/detection must be maintained together with unidirectional workflow.
7 A commitment from the laboratory's upper management to invest the appropriate resources for this technology, and a continued support for investment in capital equipment, appropriate laboratory facilities and staff training.

Theo P. Sloots

In our experience, familiarity with standard PCR and the underlying principles greatly facilitates the introduction of real time PCR. Another significant obstacle of real time PCR is the cost: capital cost of the instruments, but also cost of the commercial kits, much more costly than in-house assays.

Raymond Tellier

The main obstacles are (1) budget constraints, as the budgets of most clinical microbiology laboratories are very tight and both the machines and consumables for real-time PCR assays are very expensive; (2) lack of expertise, in microbiology laboratories other than those in referral centers, expertise in molecular techniques is still limited.

Patrick Woo

How will detection of multiple agents be best achieved? Is there a limit to the number of detections possible in a single tube, and what causes that limit?

There is definitely a limit to the number of reliable detections in a single tube. In a real-time situation this is likely due to the lack of suitable fluorogenic dyes currently available that can be reliably detected although other factors also contribute. Because real-time is essentially single-round, there is a likelihood that amplification kinetics will be compromised as more primers/probes are added.

Chris Birch and Julian Druce

Is there a limit to the number of detections possible in a single tube, and what is the cause of that limit? We for example detect 14 respiratory viruses in 5 tubes. I guess we can add one more to each, but it is getting close. The problem is going to be having fully automated systems which can choose primer/probe combinations in a random access format if system-specific samples are not put through in batches. I can see a number of integrated extraction/PCR machines, perhaps from different manufacturers, being required, one for each sample type, but hopefully not. Tm Biosciences can detect 15 viruses in one tube—not sure how they can do it, although not sure if data on multiple viruses in one sample are out there.

William Carman

With current technologies there is a possibility to multiplex many targets, but read-out requires labeling of these targets which is usually by PCR amplification. A question remains whether the intrinsic sensitivity is sufficient for detection of pathogens in microbiology without prior amplification.

E.C.J. Claas

Multiple detection of high copy number targets is perhaps not too much of a problem. The problems come when trying to detect low copy number targets, particularly when they come from viral genomes of different size and type (ssRNA, dsRNA, ssDNA, dsDNA etc.). This is one limitation. Another limitation comes when mixing lots of primers of different sequence together—it can be hard to balance between a highly sensitive, specific primer sequence, and that sequence being compatible with a standard reaction mixture and the presence of other interfering sequences. Another limitation is, of course, the number of dye chemistries that there are and which can be implemented on one instrument. It will be interesting to see how the Luminex® microsphere-based multiplexing system develops, as this has 100 colors.

Jon Clewley

The detection of multiple agents would be achieved with the use of different dyes, each displaying a distinguishable spectrum, and each bound to a probe for a specific agent. Different dyes with non-overlapping spectra should be developed. The limit of such an approach would largely depend on the number and type of agents to be detected and the specificity of each primer and probe for one of the agents present in the mixture. For a given mixture of agents, a preliminary screening of nucleotide sequence identities and similarities among the potential target sequences for primers and probes should provide valuable information before the initiation of preliminary trials with reconstructed mixtures of agents.

Esteban Domingo and Cristina Escarmís

I prefer individual detections in separate reactions, but have no problem with multiplex reactions. All multiplex assays I have seen are less sensitive than single reactions, and why would one want to give this up? Price could be an issue, but the difference in price is limited because reagents in multiple tubes only increase costs by a small amount in the western world.

Ron A.M. Fouchier

Detection of multiple agents can theoretically be achieved by combining multiple fluorophore-labeled oligoprobes and further melting curves analysis of the amplified products. There are published studies reporting more than five different targets detected in a multiplex assay. This seems to be actually the upper limit of the number of targets that can be discriminated in a single tube, because we are limited by: (i) the number and quality of the fluorochromes available (narrow spectra, quenchers that quench broad range wavelength...), (ii) the number and the quality of optical filters used, (iii) the use of a monochromatic exciting source and finally and, (iv) the system hardware.

Vincent Foulongne and Michel Segondy

Conventional multiplex PCR (using multiple primers directed at multiple templates within a single reaction) with end-point detection can discriminate 5 or more amplified sequences. The limit to the number of templates detected relates to the acceptable detection sensitivity of each template required for the diagnosis of a specific infectious disease. While for diagnosis there needs to be a balance between the clinical relevance of performing a specific test and the "screening" for multiple pathogens, the identification of multiple pathogens associated with an illness and a possible association with disease severity, will encourage the development of multiplex PCR. Limiting laboratory factors with in-house PCR tests relate to optimization, consistent test performance and the production of accurate results. Standardized protocols and commercial controls will help to improve this, however in our hands, multiplex PCR is less sensitive than conventional PCR with internal controlling. Multiple agent detection by real-time PCR currently has constraints related to the availability of fluorescent dye combinations, the capacity of the reaction and instrumentation.

Lance Jennings

Multiplex detection comes at the price of sensitivity, particularly in detection of the lowest-abundance target. It is therefore extremely important to verify sensitivity for each target at different concentrations of the other targets, in all permutations. In our experience, if quantification in real-time PCR is important, only duplexing will work. The higher the number of different targets, the higher the loss in sensitivity. Well designed duplex PCRs allow detectable and reasonably accurate amplification of single target copies. This is in our experience not possible for 3 or more targets. However, genetic typing of medium- to high-abundance targets works reasonably well in multiplexed PCRs.

Bernhard Kaltenboeck

Even though it would be economical and resource-efficient to detect multiple targets in the one reaction there is a realistic limit. This limit is determined by:

- Type of RT Machine—how many channels are available and at what wavelengths
- Sensitivity—multiplexing has the potential to decrease the sensitivity of detection of targets when compared to each reaction performed as a uniplex. This issue is exacerbated when more than one target may be amplified per reaction.
- Design of oligonucleotides—the more primers and probes that are combined in a reaction the harder it is to find sequences that do not have complementarity, hairpin loops and other conformational issues which may compromise the assay.

Ian D. Kay and Silvano Palladino

Multiplexing turned out to be more difficult than people expected. The largest number of parallel reactions in a single tube that have been separated are six. But the ability to separate colors is rarely the limiting factor in multiplexing. Rather, competition between assays is more serious problem. Techniques such as LATE PCR (J. Aquiles Sanchez *et al.* Linear-After-The-Exponential (LATE)–PCR: An advanced method of asymmetric PCR and its uses in quantitative real-time analysis. PNAS 101, 1933, 2004) may turn out to be valuable for multiplexing. Another possibility is to pre-amplify the test material in a limited number of cycles. This can be done for essentially any number of targets, and then assay each target separately.

Mikael Kubista

Multiple detection is most affected in cases where we have more than one amplification reaction in the same tube. Multiprimer PCR with only one target present is not a major problem. One limit in real-time PCR is the number of fluorophores which can be used; bead-based detection (flow cytometer) could allow detection of more products.

Olfert Landt

Often in multiplex PCR there is the focus on the number of channels an instrument possesses. Yet for most applications, multiplex PCR is the simultaneous detection of an internal control (perhaps better described as duplex PCR). With the possibility to resolve targets on the basis of T_M and melting curves, it is possible to detect a number of targets at the one wavelength. One author of this chapter has shown this for mycobacteria strains and perhaps this has been best demonstrated with the Septi*Fast* assay from Roche which distinguishes more than 20 of the most common bacteria involved in sepsis (plus internal control) using 4 wavelengths and multiple primer and probe sets.

Another approach to T_M differentiation of multiple targets has been the use of intercalating dyes (e.g. SYBR Green I) where primers have amplified a number of species of, e.g. Leptospira or Aspergillus; to determine an infection; before the species is resolved by melting curve analysis. However, the use of these dyes for diagnostic purposes is another issue!

A particular issue on the feasibility of multiplexing is whether multiple targets in the assay may be present together in the sample being tested (which may then be limited in sensitivity without exacting optimization) or whether one target of the multiplex will exist in any sample?

The optimization of in-house multiplex assays is another demand—the optimization required rises seemingly like PCR itself: exponentially for each target added into the reaction!

John Mackay

Microchips/arrays may hold the best hope of multiple agent detection without the pit falls of multiplexing.

Michael D Nissen

Multiplex assays frequently give good results and facilitate the diagnostic procedure, but this cannot always be generalized. I can't think of any unsurpassable theoretical reason why the number of parallel detections cannot increase considerably, although in practice reactions become more "unstable" with increasing number of primers.

Nikos Papadopoulos

There are two ways this can be achieved, multiplex amplification (i.e. use of multiple primer pairs and possibly multiple probes for detection) or the use of multiple assays each amplifying one or two targets. Ideally you perform one assay that will detect all desired targets. The technology to date is moving in this direction (TM BioScience, Genaco, etc.), however, the sensitivity does not yet appear to be equivalent to a uniplex assay. Perhaps there are applications where sensitivity can be sacrificed to detect multiple targets such as respiratory or gastrointestinal pathogens. The load of these organisms in samples may be sufficient that a slight loss in sensitivity may not be significant when testing clinical samples. However, in instances where the pathogen load is low, i.e. viral or tuberculous meningitis, a sacrifice in sensitivity may mean missing the target with significant clinical implications. Here uniplex assays may be more appropriate until technology demonstrates that the sensitivity of multiplexes can approach that of uniplexes.

Astrid Petrich

First we have to distinguish if there is only one possible pathogen present in the matrix or multiple. If we know there is only one dominant pathogen we can use as many primers as possible. There are examples in the literature of two dozen primers in food diagnostics. If there are more pathogens present in parallel, the limit in multiplex PCR is set by the number of dyes which can be differentially detected. Sixplex detection might be the maximum today.

Michael W. Pfaffl

This will change with time—in 5 years it is likely be via arrays (preferably on optical fibers as used in the Illumina® SNP detection system) which currently use multiplex amplification and multiplex probe techniques. The number is limited by primer stoichiometry—if this is overcome by using a fixed base for the primers or probes (such as bead linkages) then the number becomes limited only by physical constraints—effectively in the thousands of different targets.

William Rawlinson and Gillian M. Scott

One has to distinguish between simple and complex multiplex assay concepts. Simple multiplex assays are based on the amplification of suitable target regions by a single primer pair annealing to conserved regions but flanking the sequence regions of interest (e.g. 16S rDNA or other signature genes of bacterial or viral pathogens). The reaction mixture of such assays contains 2 or 3 primer oligonucleotides and a high number of probes which are chemically blocked against elongation. After an initial optimization of PCR buffer components and thermocycling conditions, big problems are usually not observed with this kind of assay concept. On the other hand, multiplex assays based on a complex mixture of specific oligonucleotides primers and probes in one reaction tube are very hard to establish and once established, usually not very robust. Minor variations in the input target DNA concentration can lead to misbalanced salt (Mg^{2+}) concentrations in the reaction mixture and non-specific elongation or mispriming of primer oligonucleotides can lead to incorrect polymerase actions. In general, a reduced analytical sensitivity is observed with most of these complex multiplex assay concepts.

Udo Reischl

Multiplex PCR comes with an inevitable cost to sensitivity. Whether the loss of sensitivity is significant or not depends on the design specification of the particular assay. If sensitivity is a crucial issue multiple separate PCRs are the best solution while if sufficient levels of sensitivity can be achieved multiplex PCR is obviously preferable. Apart from the sensitivity issue, the level of multiplexing possible in real-time PCR is really only determined by the instrument capabilities and the availability of compatible fluorescent labels that can be accurately resolved. The practical limit with existing technology is probably in the range of 4–6 labels.

Nick Saunders

The most sensitive technique is PCR for a single pathogen. The more pathogens, the less sensitive the technique is. The reason for this is simple competitive inhibition, as there can only be a limited number of polymerase molecules within a single reaction. Also primer binding may interfere if more than one pathogen and consequently more than two primers are in the reaction.

Oliver Schildgen

There is a clear limit for the number of detections caused by the number of optical channels in real-time PCR instruments and by the software to analyze the resulting data. Not a limit but a constraint is that several PCRs together in one single tube require design of primer and probes specific for the one aimed PCR product but without cross reaction with oligonucleotides from the other assays which becomes more and more effort with an increasing number of parallel assays.

In an additional dimension it is possible to get more information from a set of primers and probes by melting curve analysis. Thereby in a homogeneous assay the reaction is heated after PCR under constant fluorescence measurement resulting in distinct changes of the fluorescence signal indicating the melting of the probe off from PCR fragments. Sequence variations under the probe result in measurable differences of the fluorescence profile. In microbiology this information can be used to detect different strains or species with a single set of primers and probes by detecting characteristic sequence variations.

Rudolf Seibl

Both multiplex PCR and real-time PCR have limitations for the detection of multiple agents. Multiplex PCR is limited in the number of primers and probes that may be used in a single reaction, because interactions between these oligonucleotides may impair the efficiency of the PCR. In addition, in cases of co-infection the amplification of one target may compete with the second target thereby yielding a result only for the target in excess. These limitations also apply to real-time PCR assays, but these are further restricted by the number of fluorescent markers that may be detected during the reaction. In reality this is a function of the number of wavelengths any instrument can accommodate, and in current instrumentation is limited to 6.

An alternative approach would be the use of molecular beacons (or similar technology) with a range of fluorescent markers that may be distinguished and recorded in a fluorimeter capable of detecting a broad range of wavelengths.

Theo P. Sloots

Broad range primers can theoretically be used to detect a panel of "related" agents in one amplification reaction; however, when the test is designed to co-amplify several genomically variable pathogens, it is usually impractical to use broad range primers. Most of the time, the multiplex PCR becomes the only choice, and utilizes numerous primers within a single reaction tube in order to amplify nucleic acid fragments from different targets. But the conventional multiplex PCR has two major technical problems to be addressed: incompatible primer sets and high background amplification/detection. It is especially the case when real-time PCR formats are used since there are additional probes included in the PCR reactions. It is difficult to optimize the conventional multiplex PCR procedure because each amplification target has one pair of primers that dictates one optimal annealing condition. If a particular target has a PCR condition that conflicts with other targets, it will not amplify efficiently. Therefore, the range for "optimal" conditions, such as annealing temperature and salt concentration, is relatively narrow. On the other hand, the conventional multiplex PCR is limited by the number of primers, which have to be included in a single reaction, the primer-primer interference, and non-specific nucleic acid amplification. Different primers and probes can form primer–dimers and result in high non-specific background amplification. Modifications have been sought to improve the multiplex PCR procedure. One successful example is the Templex™ technology developed by Genaco Biomedical Products (Huntsville, AL, USA). The Templex assay uses nested primers to increase compatibility among loci. Four alternative products are generated to serve as templates; therefore, a common amplification condition for all the intended targets is more attainable. In addition, the Templex uses low concentrations of unlabeled, gene-specific primers to reduce background and to improve reproducibility. The main target amplification is completed subsequently by a high concentration of "SuperPrimers."

Yi-Wei Tang

One problem is that the template to be amplified should have similar GC content, and primers have similar T_M, otherwise it is difficult to find reaction conditions acceptable for all amplifications. It becomes much more difficult to design primers if one is to avoid interactions (primer–dimers, etc) between several pairs of primers. The possibility of multiple infections may lead to drop in sensitivity simply by competition of amplicon synthesis.

Raymond Tellier

What are the limitations of real-time PCR in relation to detection, sensitivity, usability etc?

Amount of input template is limited by small master mix volume. Number of suitable fluorogenic dyes limits the number of specific targets. Expertise and experience is needed to interpret some amplification plots.

Chris Birch and Julian Druce

Sensitivity tends to be as good as nested, although if there is a lot of secondary structure, a nested, or at least a two-step RT real-time assay, is done, e.g. enteroviruses. High C_T's are the bane of our existence—one way to address this may be to do a nested, with real-time being the second step. This tends to give much lower, clear-cut C_T's. One has to be careful not to over-interpret curves on screens without looking at raw data. Fluorescence change can be tricky—always look at the vertical axis. Some cross talk between probes can occur.

William Carman

Theoretically there are no limitations in detection or sensitivity. However, the major problem we are currently facing is the clinical relevance of the data that are generated. In for example, positive real-time PCR results with high Ct values of *Pneumocystis jiroveci* or Human rhinovirus, the clinical relevance can be considered questionable. However, it is very difficult to define a threshold of "clinically relevant" results.

E.C.J. Claas

I'm not sure that there are any real limitations other than those associated with any PCR. If you're going to setup a PCR for detection of a pathogen these days, you might as well do it by a real-time assay, unless you're doing it for a specialist research reason.

Jon Clewley

For the detection of RNA, current limits to obtain dose-dependent crossing points and a reliable quantification are in the range of 10^1 to 10^2 RNA molecules. It is conceivable that new enzymes (or proteins that could be added to the enzymes currently in use) may enhance the difference in crossing points resulting from different numbers of RNA molecules (a steeper slope in the standard curve is an advantage for relating crossing points with RNA dose)

Esteban Domingo and Cristina Escarmís

I see few limitations for routine diagnostics. It is fast and more sensitive than most other tests. Cost is still an issue. And it is possible we miss-diagnose infections because pathogen DNA/RNA may still be present while the real infection is already over. In addition, for some purposes it would be desirable to have the pathogen in hand; I am a bit afraid that most microbiology labs in the future will have available original specimens from patients and paper test results, but no reagents required for additional work that has proven critical in microbiology in the past (e.g. for pathogen characterization, fundamental research)

Ron A.M. Fouchier

When compared to conventional PCR, the sole limitation of real-time PCR remains the expense for platform acquisition.

Vincent Foulongne and Michel Segondy

Well designed real-time PCR will reliably detect single target copies and allow accurate quantification over at least 4 orders of magnitude.

Bernhard Kaltenboeck

Real-time PCR can detect a single molecule. The factors limiting sensitivity are therefore sample preparation and extraction rather than detection. Kits for sample prep and extraction from different suppliers are likely to have quite different yields and it is advisable to compare the yields of some different kits when setting up new diagnostic assays. For RNA samples reverse transcription may also be limiting. The reverse transcription yield can vary up to 200-fold depending on priming strategy and the reverse transcriptase used (A. Ståhlberg *et al.* Clin. Chem. 50, 509, 2004). A probe-based PCR does not detect primer–dimers and will report the presence of a single target molecule. However, the assay may not be quantitative if primer–dimers are formed, because the formation of primer–dimers will compete with the amplification of the target, which may affect the C_T values. The new dye BOXTO can be used as an internal control in probed based assays to detect formation of interfering primer–dimer products (K. Lind *et al.* Biotechniques 40, 315–18, 2006).

In diagnostics robustness is key factor. We have found that different assays can be inhibited to different degrees in the same sample (A. Ståhlberg *et al.* Clin. Chem. 49, 51–59, 2003). This differential inhibition can seriously complicate analysis. In many cases anomalous inhibition can be detected by analysis of the real-time PCR response curve (T. Bar *et al.* Nucl. Acids Res. 31, e105. 2003), and automatic solutions for quality assurance based on this approach are being developed by LabonNet (www.labonnet.com).

Currently most tests are based on a single marker. Sensitivity may be improved by tests for more markers. Such tests may also classify the diseases based on the expression patterns of the markers. MultiD Analyses (www.multid.se) has developed software called GenEx for such classifications.

Mikael Kubista

Typical limits are background of related organisms (if targeting conserved genes), absolute concentration and difficult matrices (TB in sputum, fungi in sepsis tests) and the RT reaction (in general). Optimized reactions reach a limit of 10 genome equivalents, which is enough for most pathogen detection assays.

Olfert Landt

Many of the limitations of real-time PCR are shared with all PCR methods—namely the template going into the reaction is only as good as the sample storage, extraction method and concentration of the target prior to the PCR. If RNA is being detected then there is the added complexity of the relatively low efficiency reverse transcription step that can introduce its own inhibitors into subsequent real-time PCR reactions. There can often be too much focus on the fluorescent probe(s) in the real-time reaction—perhaps forgetting that the sensitivity of any assay is still governed by appropriate primer design. While primer dimers and other non-specific products will not be detected using probe-based assays, they will adversely affect the sensitivity due to reagents being sequestered for these "invisible" amplicons. There is nothing currently better in terms of sensitivity for routine diagnostic use and usability often tends to be operator-specific.

John Mackay

Cost and currently limited to large labs.

Michael D. Nissen

A major limitation is that in many cases the clinical significance of the quantitative information has not been validated. This is also a major issue because it may differ between microbes, disease states, populations and types of samples. So usability is somehow limited by our restricted ability for interpretation. Practical problems, such as the fast pace of development and subsequent high cost can also be an issue; as in many other cases, platforms became really cost-effective and useful when they "mature" a little.

Nikos Papadopoulos

There are no "real" limitations, but there a few things to improve for the near future. New fluorescent dyes with high fluorescence readout may improve the next generation quantification. Better instruments with more sensitive fluorescence readers will also enhance the outcome.

qPCR is sometimes limited by inadequate sequence information from "unknown or new pathogens" for primer and probe design. This may impact not only on sensitivity but also on the ability to detect all potential targets.

Hardware (real-time cyclers and automation) and software at a reasonable price will also assist with bringing real-time PCR assays into all laboratories.

Michael W. Pfaffl

Still primarily a single target assay, though new fluorescent dyes and instruments that have multiple channels has alleviated this somewhat. Sometimes limited by not having adequate sequence information to choose primers and probes for the assays. This may impact not only on sensitivity but also on the ability to detect all potential targets. Automation at a reasonable price will also assist with bringing real-time PCR assays into all laboratories.

Astrid Petrich

[*WR*] Detection is predominantly probe and sensitivity limited, and can be overcome by appropriate design. [*GS*] Sensitivity is an issue, where we see specimens positive by qualitative PCR but below the detection limit for quantitative PCR. Again, our experience is with LightCycler technology, using plasmids containing the PCR product as standards to create a standard curve. However, the results obtained with plasmids do not necessarily reflect the added problems associated with amplification from clinical samples. This is presumably because the plasmids are more "pure" with less cellular and other interfering DNA, achieving real-time detection down to 100 copies per reaction. Therefore the problems with sensitivity are related to the specimen type used. See above re: problems with different detection systems (i.e. SYBR green vs. probes).

William Rawlinson and Gillian M. Scott

When identical PCR reaction volumes are applied, analytical sensitivity is almost is comparable between real-time PCR and traditional PCR assays. Since most of the real-time PCR devices are working with 10 or 20 µl reaction volumes, there are clear limitations with respect to sensitivity compared to the traditional 50 or 100 µl block cycler reactions. The usability mainly depends on the individual software.

Udo Reischl

Real-time PCR is often more sensitive and specific than conventional PCR (i.e. with a gel end-point) because of the use of probes that can detect the intended product in a mixture of non-specifically primed amplicons. This enhanced sensitivity depends on the efficiency of the probe system. Real-time PCR can be applied in any circumstances that are suitable for conventional PCR with the proviso that it may not be possible to design an efficient probe for use with all of the available detection chemistries. This particularly applies to methods based on hydrolysis (TaqMan®) probes in instances when the inter-primer sequence of the amplicon includes many sequence variations.

Nick Saunders

Until now I see only two limitations. First, as already mentioned, there is a financial limit. Second, the length of the amplified region is limited (around 150–300 base pairs for real time PCR vs. up to 10.000 or more base pairs for conventional PCR). Therefore, subsequent analysis methods like sequencing lead to rather limited additional information.

Oliver Schildgen

For nearly every problem a technical solution can be described using available instrumentation, chemistry and enzymology. The limitation is the development of robust standardized assays. The limiting factors in assay performance are often not related to real-time PCR. Instead they may relate to other aspects of the workflow for instance sample handling or nucleic acid preparation.

Rudolf Seibl

Real-time PCR has been widely used in the diagnostic molecular microbiology field, especially for microorganism quantitative detection; however, it is important to keep in mind that there is an intrinsic problem with sensitivity and specificity. The arithmetic, spontaneous increase in fluorescent background emissions interferes with the exponential, specific energy emissions during the simultaneous detection procedure, producing potentially false positive results. To overcome this non-specific issue, the system has to either increase the C_T cut-off value or decrease the cycle numbers during the amplification, which results in decreased sensitivity. Another limitation of real-time PCR is that it lacks an objective criterion to set up the cut-off C_T threshold. This could be a big headache when the real-time PCR assay is used for qualitative detection when the target microorganism load is low in the tested specimens.

Yi-Wei Tang

Do you believe that real-time PCR is inherently more sensitive than conventional PCR when using the same primer pair?

Yes, if it's a direct comparison between real-time with a fluorogenic probe and conventional single round PCR. But if the conventional PCR is nested then it may be more sensitive than real-time. There is also some evidence that the amplification of a small amplicon is favored in a real-time assay.

Chris Birch and Julian Druce

Yes.

William Carman

No, I do not believe this. Using probes will improve the sensitivity in comparison to read-out by gel electrophoresis, but probe hybridization in real-time PCR will not inherently be more sensitive than for example PCR-EIA. Improved sensitivity can be achieved by real-time PCR as there are better optimization features resulting in an increased efficiency and thus better sensitivity of real-time PCR. But an optimized assay will be as sensitive in conventional format as in real-time PCR.

E.C.J. Claas

I don't think it's inherently more sensitive; its greater sensitivity arises from the use of a smaller reaction volume and from the increased sensitivity of the detection system compared with ethidium bromide staining of an agarose gel.

Jon Clewley

In our experience, the sensitivity of the two methods is very similar. In both cases, the sensitivity is largely dependent on the enzymes (reverse transcriptase and DNA polymerase) used in the copying and amplification processes.

Esteban Domingo and Cristina Escarmís

No. When conventional PCR products are detected in combination with hybridization to specific probes, I think conventional PCR can be similar in sensitivity.

Ron A.M. Fouchier

In our experience the sensitivity of real-time PCR is equivalent to that of conventional PCR when using the same primers. An advantage of real-time PCR is the possibility of amplifying smaller DNA fragments by choosing alternative primers. This may allow a gain in sensitivity.

Vincent Foulongne and Michel Segondy

I'm not sure there is a significant increase in sensitivity but the turn-around time and data provided are greater.

Michael G. Ison

There is variability between pathogens. With some, where low copy numbers are present, conventional PCR, especially when nested, can be more sensitive than real-time PCR. This can relate to the volume of nucleic acid assayed, however, as with any new laboratory test, the in-house verification and validation of the test's performance in comparison with an existing test and with different clinical samples should be standard procedure.

Lance Jennings

No, it's equally sensitive given optimal design and execution of both methods.

Bernhard Kaltenboeck

There is no reason real-time PCR should be inherently more sensitive than conventional PCR. In fact, presence of dyes and probes complicates the reaction and may compromise sensitivity. But real-time PCR is easier and faster to optimize, because one has better control over the reaction. Therefore, real-time PCR assays usually turn out to be more sensitive than conventional PCR assays, particularly if they were developed as conventional assays and not tested in real-time PCR

Mikael Kubista

Believe? The amplification is made by the PCR-primers. Real-Time PCR might be able to show a curve where a gel picture might be unclear and foggy, but this is also at the detection limit of Real-Time PCR—and therefore the sensitivity is the same.

Olfert Landt

To address this we need to see data from experiments using the same primers, the same reaction conditions (reagents, volumes and cycling conditions) and the same template amounts. We also need to choose a detection method—agarose gel or PCR-ELISA for conventional PCR and non-specific or specific chemistries for real-time PCR.

If real-time PCR was shown to detect a higher dilution of template than single-round, conventional PCR under these sorts of conditions, then I'd be convinced. Unfortunately many simply say "real-time is more sensitive" without taking into account the inherent deficiencies of their comparison—if they even perform one. In my opinion (no data here either!) the amplification process is not likely to be different under conditions of direct comparison, but detection of fluorescence (be it from a fluorogenic probe or from SYBR) or color (from a PCR-ELISA) using an analytical instrument may be more sensitive than visualizing an ethidium bromide-stained gel. It would be nice if this could be addressed in the near future.

Ian M. Mackay

No. There is a limitation on fluorescent detection in that a certain amount of amplification must occur before the amplification curve emerges from the fluorescent "noise". Some comparisons using the same primer set have been done years apart and newer enzymes (e.g. hot-start) and buffer systems have seemingly advantaged real-time PCR. Also there is often the claim "but it has a probe in there" that provides the otherwise unfounded belief. Some assays may be more sensitive with real-time but some are just as likely more sensitive via agarose gel with SYBR Green (or even ethidium bromide) staining.

John Mackay

It's difficult to objectively and meaningfully compare when sensitivities in many assays are approaching single copy-levels. Furthermore, differences between sequences, platforms, detection methodologies etc are usually more relevant than the sensitivity of the assay per se.

Nikos Papadopoulos

From personal experience it appears so. It seems that the hybridization of an internal probe or probes may be more sensitively detected than a band by gel electrophoresis. We have seen this hold true for two different nested PCR assays. They were equivalent in sensitivity to a real-time PCR and more sensitive than a single round of amplification. This sensitivity is not always reflected in analytical sensitivity but appears to be a larger issue in clinical sensitivity.

Astrid Petrich

Both methods are more or less identical in all parameters, including sensitivity, variability and reproducibility, but in real-time PCR we have a fully quantitative answer!

Michael W. Pfaffl

No, not necessarily. It is dependent upon primer-template relationship, use of nesting for conventional PCR, and may be dependent upon secondary structure in complex templates.

William Rawlinson and Gillian M. Scott

Without any doubt, real-time PCR is NOT inherently more sensitive than conventional PCR. The physics behind SYBR Green and ethidium bromide are almost identical and the amplification rate is also not significantly higher in the real-time PCR reactions. It is all a matter of total reaction volume.

Udo Reischl

Real-time PCR is not inherently more sensitive than conventional PCR using the same primer pair. It depends entirely upon the sensitivity of the detection system.

Nick Saunders

No.

Oliver Schildgen

Due to the fact that real-time PCR results in Cps, target sequence numbers, etc which can be more easily and precisely used for statistical analysis, the statistically significant sensitivity with defined confidence levels is higher in comparison to a visual gel analysis of end-point PCR.

Rudolf Seibl

I do not think there is any experimental data to draw any firm conclusions on this. Clearly sensitivity may be related to a number of things.

1 The reaction parameters
2 Primer composition
3 Probe composition
4 The amplification system/method used
5 The detection method

If the same primer pair is used and reaction parameters have been optimized for both assays, then the detection system should determine the final sensitivity of the assay. Clearly a fluorescent signal has a lower limit of detection than visual gel-based systems. Similarly a colored end-product should have a lower limit of detection if a spectrophotometer is used, and should approach but not quite match the detection limits of fluorescence. However these issues may be confounded by the quality of the instrumentation used to perform the assay.

Theo P. Sloots

No. Some of the points have been addressed in my answers to the previous question. In contrast, conventional PCR is usually linked to an additional amplification product detection procedure such as the Southern blot, enzyme immunoassay, or direct sequencing. When additional signal amplification is incorporated into the detection procedure, such as an antigen-antibody chain, the sensitivity of the PCR assay can be further enhanced.

Yi-Wei Tang

I do not think that real-time PCR is inherently more sensitive. With current PCR enzymes and reagents, and a well designed pair of primers, detection of 1–10 molecules of template DNA can often be achieved with standard PCR. Real-time PCR with a probe, in my view, brings the advantage of template authentication by probing, and the possibility of quantification.

Raymond Tellier

Yes, in our experience, real-time PCR is more sensitive than conventional PCR when the same pair of primers is used.

Patrick Woo

How do microarrays and real-time PCR compare at the present time, and what is the future of arrays for routine versus research applications in microbiology?

Microarrays are a detection method at this time. Random generic amplification is needed up-front to make this technology more diagnostically relevant.

Chris Birch and Julian Druce

Still need to do PCR, so very time consuming and expensive. Needs a leap in technology and probably will not suit diagnostic labs in the medium term, but may be OK for expensive home-testing kits.

William Carman

Microarrays are very useful for high-throughput SNP analysis. For microbiology, its application may be found in screening isolates for markers (mutations) of antimicrobial or antiviral resistance. For pathogen detection in clinical specimens the sensitivity is still too low and real-time PCR is superior. In time, combining detection of single molecule interactions in an array-like environment may change the situation.

E.C.J. Claas

Arrays are still limited to a few laboratories. The practical problems of making them and using them have not yet all been solved. Perhaps the biggest issue with the use of arrays is the lack of a front-end amplification method that will amplify all sequences that are present in the sample. Then, of course, you have to have a suitable capture probe on the array—one that will hybridize to the "correct" amplified sequence. And you don't necessarily know what to expect, i.e. what is "correct", or how to interpret the result of any unanticipated signal from the array.

Jon Clewley

Microarrays are more suitable than real time PCR to detect multiple agents (or variants of one agent) in a given sample. However, microarray detection is a rather qualitative method, very difficult to adapt as a routine analytical procedure in a standard diagnostic laboratory. Improvements in instrumentation, bioinformatics and reproducibility of signals are needed for a routine implementation of microarray technology. Real time PCR is essentially a quantitative technique, reliable for quantification of a limited number of agents in a sample and generally more reproducible than microarrays at the present stage of development of the two techniques.

Esteban Domingo and Cristina Escarmís

I have not used microarrays myself for diagnostics. I THINK that real-time PCR is more sensitive at the moment, but I believe that microarrays based on gene detection or protein detection will dominate the microbiology lab in the future. They are more suitable for detecting multiple pathogens simultaneously (and at the same time other diseases). The detection technology is improving rapidly to ensure sufficient sensitivity.

Ron A.M. Fouchier

Real-time PCR is more suitable than microarrays for nucleic acid quantitation, whereas microarrays are more adapted for testing simultaneously a high diversity of microbial agents (e.g. respiratory pathogens) or for genotyping. The current cost of microarrays represents an obstacle for routine utilization.

Vincent Foulongne and Michel Segondy

PCR is superior and easier to do often than microarrays; you often answer different questions with these (PCR usually detects DNA/RNA while microarrays detect function or downstream proteins

Michael G. Ison

Microarray technologies currently require larger amounts of starting material than real-time PCR and have a limited dynamic range for quantitation. It is likely that in the future, arrays will be used for the high through-put screening of a large number of targets, with specific quantitation then being performed on the targets of interest.

Lance Jennings

Microarrays don't have the exponential amplification procedure so the signal is only linearly related to the target quantity. Therefore microarrays have low sensitivity and detect only the most abundant and easy targets. Unless exponential amplification can be combined with microarray detection, I do not foresee any great use of microarrays for sensitive detection with the use of PCR.

Another, frequently overseen problem in miniaturized arrays and "hot" microfluidics technology is that you simply cannot add a lot of sample volume to these devices. This severely limits the potential for low-abundance targets even if exponential signal amplification is included.

Bernhard Kaltenboeck

Today microarrays and real-time PCR are preferred for different applications. Microarrays are suited for genotyping, while real-time PCR is preferred for quantification of a single or a few selected markers. But the two techniques are merging. Companies like Biotrove (www.biotrove.com), Fluidigm® (www.fluidigm.com) and WaferGen (www.wafergen.com) are developing high throughput real-time PCR platforms for massive analysis of parallel reactions and these platforms will replace microarrays in many applications.

Mikael Kubista

Microarrays hold promise however it is still expensive for companies to manufacture chips for relatively small-volume users. Most routine laboratories do not have the facilities to manufacture chips so are reliant on another company to produce them. For many targets, e.g. antimicrobial resistance genes and RNA viruses, there is continual genetic drift. So this will require changing of the sequences on the array. This will need to be done by the commercial laboratory adding to cost and turn-around time. The clinical laboratories will have difficulties in adapting to the ever-changing microbial environment. Different arrays have different readers as some arrays are on glass slides, at the bottom of 1.5mL tubes etc. These readers can be expensive. Newer systems are now appearing on the market that hold the promise of addressing some of these concerns. For example the Nanogen™ system is a "generic" open system that provides the platform for array testing, and the user's diagnostic laboratory provides the specific primers and probes. These remove the necessity for a commercial lab to be commissioned to provide the specific oligonucleotides on the arrays. In validating arrays where there can be over a hundred individual probes, it is necessary to validate each of these probes with another method. This may be by PCR or by sequencing or a combination. This will require a significant amount of laboratory resources and may make it difficult for clinical laboratories to achieve. It is difficult to change the oligonucleotides on the chips if there is a change due to genetic drift. Also validating arrays means that each probe needs to have another assay performed to validate the array.

Ian D. Kay and Silvano Palladino

Arrays give limited or no quantification results. Arrays make use of multiprimer PCR or amplify conserved regions. By this arrays are superior for the identification of species, variants, types and resistances. The detection limit (for a single PCR) is identical. A typical scenario could be (1) array for identification and real-time PCR for verification and quantification or (2), real-time PCR followed by array-based typing.

Olfert Landt

A direct comparison would be unfair; combinations of the assays may occur, as arrays are "wide", while PCRs are "deep" and therefore may very well complement each other.

Nikos Papadopoulos

Microarrays are generally not used for routine clinical applications yet. I have not used them for microbial diagnostics myself. They have had a significant impact on research. They will likely end up playing a major role in pharmacological characterization of individuals and other genetic applications. They are less likely to play a role in detection of infectious agents, at least in the traditional sense of detection of genomic targets. This would require optimization of the multiplex assay for sensitivity. The future could possibly have multiplex assays followed by microarrays to characterize the amplification products; identification of the organism, antimicrobial resistance, toxin production, etc..

Astrid Petrich

Arrays will be too expensive for routine diagnostics, but for research the combined approach of screening with hybridization arrays and quantifying by qPCR is optimal. Hereby the advantages of both quantification systems can be added—the high throughput capacity of the microarray as well as sensitivity and specificity of the real-time RT-PCR.

Microarray technology is a powerful technique for analysis of the gene expression of thousands of genes (up to millions) in a very short time. Problems encountered include inconsistent fidelity, high variability, insufficient sensitivity for poorly expressed genes, discrepancy in fold-change calculations and lack of specificity for different isoforms or differentially expressed genes. The DNA microarray-based screening of tissue-specific gene expression and confirmation of putative candidate target genes by kinetic RT-PCR represents a powerful and optimal combination.

Michael W. Pfaffl

[*WR*] The arrays are currently expensive, the technology still requires considerable technical effort (tweaking), and interpretation is still questionable. [*GS*] Arrays definitely have great potential for routine screening for multiple pathogens, and multiple genotypes within a particular pathogen group e.g. detection of Hepatitis B virus antiviral resistant variants.

William Rawlinson and Gillian M. Scott

Microarrays as such are not competing with real-time PCR. Their dynamic range is very low and the composition of probe sequences is usually static. But they can be helpful for the pointed analysis of complex mixtures of amplicons generated in the course of a broad-range real-time PCR amplification. Since the multiplex real-time PCR is hampered by cross-hybridizations and cross-reactions within a complex mixture of hybridization probes, microarrays can be best used for subsequent product analysis and species identification.

Udo Reischl

Real-time PCR is currently more sensitive than microarray hybridization for measurement of either genomic nucleic acid or RNA transcript levels. This remains true even when PCR, or alternative amplification methods, are used to increase the levels of sequences prior to hybridization to the array. For quantitative applications there is good evidence that microarrays have a more restricted dynamic range than real-time PCR and that they underestimate differences between samples in the number of copies detected. Furthermore microarrays, microarray consumables and microarray equipment are more costly than equivalents for real-time PCR. Where microarrays score highly is in the ability to provide parallel data on many (i.e. from tens to hundreds of thousands) targets in one assay. Several routine applications in microbiology already make use of this ability. The genotyping of human papillomavirus is a good example.

Nick Saunders

We have no current experience with microarrays in our group. Nevertheless, I believe it will be a useful diagnostic tool in the future.

Oliver Schildgen

I think that the current applications of microarray technology are mainly as a detection system only. This extends into the routine detection of human and microbial (other than viral) genomic sequences, as long as these do not require pre-amplification by a targeted molecular method such as PCR. As such microarrays hold immense promise for the rapid and accurate detection of gene expression products and the expression of resistance markers.

However, during viral infection of the human host, the load of the organism present in clinical samples is generally very low and requires some form of amplification before detection. Amplification can also be affected by cell culture, but this is time-consuming and slow. If a molecular amplification system is used these are, by their present design, targeted to specific sections of the viral genome. To incorporate this into a multiarray system would require each element of the array to be coupled with a single PCR, and would not provide any major improvement over current technologies. Here real-time PCR has obvious advantages over microarrays.

However, the application of microarrays on a limited scale would be suitable for the detection of single nucleotide polymorphism (SNP) in viral genomes, and hence ensure the detection of viral quasispecies that may be circulating.

Theo P. Sloots

Microarrays will increasingly be important since they will allow for simultaneous detection of a large number of templates. There are some problems, chiefly the technical problems of the simultaneous, high yield amplification of several templates prior to probing with the arrays.

Raymond Tellier

At the moment, microarrays are even more expensive and expertise for microarrays even more limited than for real-time PCR. Moreover, even in good hands, the major hurdle for microarrays is nucleic acid enrichment in the clinical samples. In the long run, microarrays should be more powerful if the problem of nucleic acid enrichment can be overcome because the number of microbes that can be detected by multiplex PCR is incomparable to that for microarrays.

Patrick Woo

What is the life-span of real-time PCR and what are the "next big things" which will challenge the longevity of real-time PCR?

For at least the next 10 to 15 years unless microarrays or systems like Luminex are improved for microbial diagnostics.

M.G.H.M. Beld

What's limiting real-time PCR at the moment are the sheer number of pathogens that should be tested for in certain diagnostic situations. Improved microarray technology will definitely play a role in the future.

Chris Birch and Julian Druce

A major limitation of diagnostic PCR is that it relies on specific primers and probes so that one can only ever look for designated targets. Thus there is little scope for lateral thinking or microbe discovery. New approaches are being developed in response to the human genome project such as Solexa™'s technology which allows simultaneous analysis of hundreds of millions of individual molecules. Thus individual nucleic acid molecules can be analyzed and identified. Such technologies may one day find a place in the molecular diagnostic laboratory so one can ask "what bugs are in a sample?" rather than "is bug X or Y in the sample?."

Patricia A. Cane

The increasing sensitivity of microarrays or multitarget detection systems enabling detection of pathogens directly in clinical samples will challenge real-time PCR applications. If that is achieved, all microbiological targets may be challenged in a single experiment, including information on antimicrobial sensitivity.

E.C.J. Claas

Standard PCR has been around for over 20 years and shows no signs of going away. Although real-time PCR has been around for a shorter time, I expect it to be equally long lasting. Real-time sequencing would be a big thing.

Jon Clewley

We anticipate a long (but obviously unpredictable) life-span for real time PCR based on the same basic principles of the current instrumentation and chemistry. Improvements may come from the incorporation into the system of modified probes with higher specificity and of new combinations of compatible dyes. Also, improvements can come from new or modified enzymes that can increase the yield of the products of amplification. We feel at a loss trying to anticipate what the next, related "big things" might be.

Esteban Domingo and Cristina Escarmís

I think that for routine diagnostics, real-time PCR will soon be replaced with microarray technology. When the issues of sensitivity have been fully addressed (soon; few years), I believe that microarrays will FULLY replace routine diagnostics for simple pathogen detection, SNP detection, etc. On microarrays, it will be much easier to deal with genetic variability of pathogens, by simply including multiple targets in the array. I have also seen new diagnostic tests based on protein detection on chips that are almost as sensitive as

PCR without amplification of the target. This could reduce chances of sample contamination and issues like that.

Ron A.M. Fouchier

Real-time PCR is now widely spread in microbiology laboratories and there is still a large potential for further development of the fluorescent chemistries, real-time apparatus and software. Furthermore, in the future, new developments are awaited: combination of real time PCR and nucleic acid arrays technologies, solid phase or microfluidic amplification techniques and, perhaps, unit dose real-time PCR including in a single tube or on one array, different targets of interest, controls and calibrators.

Vincent Foulongne and Michel Segondy

Conventional PCR moved from the research laboratory into diagnostic laboratories during the late 1980s and early 1990s. Virology laboratories were a leading light in this respect. However, it has taken at least 10 years for real-time PCR to become accepted and its current use in diagnostic microbiology laboratories is largely focused on those applications associated with the larger acute admitting centers. Modern diagnostic services are often cost driven. With the increasing routine use of real-time technology, its cost-effectiveness for a wider range of tests will be recognized. However, the requirement for improved efficiencies, especially with high-volume tests, is likely to facilitate the introduction of automation of the pre-analytical, analytical and post-analytical steps, new instrumentation and with possible centralization. Staff safety and the need to reduce repetitive use injuries will become increasingly important as a driver towards automation. The move to the molecular quantitation of pathogens may also have an impact on the longevity of real-time PCR. Never-the-less, the current avian influenza outbreak has clearly identified that laboratory capacity for surveillance and diagnosis needs to be widely distributed globally, ensuring the usefulness of this technology for many years yet.

Lance Jennings

I foresee at least 2 decades of extensive use of real-time PCR until a better technology will replace it.

Bernhard Kaltenboeck

Real-time PCR is here to stay at least for some applications, such as quantitative analysis of a single target. For genotyping and assays for multiple targets other techniques will come. Particularly exciting are platforms based on oligomodified dyed particles, which makes it possible to separate the targets, as developed by Illumina (www.illumina.com). There is also very interesting development in mass spectrometry and chromatography that will allow for very sensitive detections.

Mikael Kubista

PCR is more than 20 years old, Real-Time PCR 10 years. NASBA and others could not compete and single molecule detection is (still) not working. We will live another 20 years with real-time PCR, with more automation over the next 10 years, and maybe a decrease after 2020 (if the oil is not used up before...). One should not forget that once installed, instruments usually cement a technology even if it is no longer efficient or up-to-date. I do not know what will be next (nobody knows—otherwise we would all invest in the right direction). Sequencing like that from 454 Life Sciences™ and the Luminex platform have some potentials, but usually I see ONE patient and ONE question, which cannot be solved efficiently with highly parallel methods. The next big things will be pandemic—AND—low-price parallel detection methods (One-Dollar-Array with 10–100 probes).

Olfert Landt

I believe the technology behind real-time PCR will remain largely the same but platforms will change to increase automation and provide the oft-requested 'sample-in, result-out" outcomes. It may be that this increased automation will allow increased parallel testing—even up to the thought of real-time PCR's inventor: a microarray of real-time PCR reactions!

John Mackay

This is very difficult to determine. Although one speculates about the developments of DNA chip technology, one also has to realize that real time technology is not yet fully implemented in diagnostics at all levels. Still, the costs are relatively high, making it almost impossible for introduction around the world. However, one should also remember that more classical technologies remain within a niche, and I don't have the impression that real-time PCR technology will be replaced by a different technology within the next 10 to 15 years. The manner in which we use it will change, more in an automated form (sample in, result out), but this also takes time. Easier said then done. However, with the likely discovery of new infectious agents and the emergence of pathogens, there will be enough work to be done. Remember that in the last 15 years, only the same viral targets have been introduced commercially, i.e. HBV, HCV and HIV-1.

The next big things include new technologies like DNA chips, genomics and proteomics. The possibility to perform multiplex, meaning really a large number of targets, is becoming closer to routine diagnostics. Also, the capability to perform these technologies easily at the bedside or emergency room will likely be introduced in the near future. However, it is also clear that there is a necessity to systematically investigate and search for still unknown targets. They must be around since we still have a gap between clinical disease and likely infectious agents.

H.G.M. Niesters

From a narrow point of view "real-time" may not last for long, but, as in the case of "conventional" PCR (or even more e.g. immunoassays), new varieties will probably continue to evolve for decades. The "next big thing" could be a combination of array and PCR technology.

Nikos Papadopoulos

Should a sensitive microarray system come along that can realistically compete with real-time PCR (multiplex included) for speed of turnaround, cost, and sensitivity, it is likely that this would replace real-time PCR, however, no technology is yet able to achieve these desired specifications. Many thought that PCR would dramatically change microbial diagnostics in a short period of time. This has not occurred due to a number of issues; inadequate template extraction, the need for isolates for further characterization (so amplification assays are "add-on" testing) and because its use has not improved clinical outcome etc. Some of these issues have been addressed with real-time PCR and new assays; however, it will still probably take a while for real-time PCR to be offered at all routine microbiology laboratories and even longer for any new technology to make inroads. It would therefore appear that real-time PCR may play a role for the next 20 years or so and perhaps longer for specific microbial targets that are not amenable to multiplexing.

Astrid Petrich

PCR is one of the basic methods in molecular biology and nucleic acid detection will therefore *"stay for ever"*. The new innovation of real-time PCR; combined with an RT step, has shown tremendous benefits for DNA and RNA diagnostics. With it's exponential amplification behavior it is hard to beat its sensitivity, specificity and therefore overall diagnostic value.

The next big things should include multiplex applications in quantitative real-time PCR to detect multiple strains or microbes and much better nucleic acid extractions for extreme natural (e.g. feces, urine, sputum, capsulated viruses and bacterial spores), or processed matrices (e.g. beverages and food).

Also, a bank of validated internal controls for spiking the sample and validated SOPs are needed e.g. CRM (certified reference material) from the EU.

Michael W. Pfaffl

Bead based microarrays (Genomics Illumina), other methods currently used for SNP genotyping only.

William Rawlinson and Gillian M. Scott

There is a permanent evolution of methods and technology in diagnostic and analytical PCR—mainly driven by competition between the diagnostic companies and their patent portfolio. Since no analytical process can be more rapid than "real-time", the next big things may happen "in-situ". Detection directly from the clinical or biological specimens will speed up the analytical process. Twenty years ago, Southern-blotting was up-to-date. Perhaps FISH will be out-competed by a more rapid, specific and sensitive technique one day.

Udo Reischl

This contributor is not far-sighted enough to see the demise of real-time PCR in all its current and future applications. However, it seems possible that for some applications it could be displaced by some types of nanoscale device. Possibly these will be based on the use of arrays (perhaps cantilever arrays). Alternatively, mass-spectrometry may eventually fulfill some of its potential and provide some real-time PCR-beating applications.

Nick Saunders

The next "big things" may be MassTag PCR and more sensitive protein methods (e.g. chips) provided it will be economic enough for routine diagnostics. Also "old-fashioned" techniques like cell culture isolation will be of importance, as demonstrated by the detection of SARS coronavirus, HMPV, and coronavirus NL63 in the last 5 years.

Oliver Schildgen

Real-time PCR is still at the early or middle of the technological S-curve. Increasing throughput, integrated workflows, the utilization of melting curve analysis, and evolutionary improvements in both enzymology and fluorescent labels will open up new applications and make existing applications faster, more convenient and cheaper.

Rudolf Seibl

What is the best fluorogenic dye and the best chemistry (incorporation of a fluorogenic dye into an oligoprobe or primer) and why?

FAM/fluorescein is the most common and may therefore be the best. A separately labeled probe improves specificity.

Chris Birch and Julian Druce

Most reproducible results are obtained with FAM dyes, but in multiplex PCR clearly more fluorophores are required. HEX has been found to be the dye giving most problems (in our lab) most likely because of its reduced stability. Initially our multiplex real-time PCR assays were designed using Molecular Beacons, as TaqMan probes used TAMRA as a quencher. With TAMRA being a fluorescent quencher, this reduced the multiplex possibilities in the filter measuring in the 600 nm emission range (e.g. for Texas Red probes). Now so called Dark quenchers are available, TaqMan probes are more economical than molecular beacons.

E.C.J. Claas

Since we currently work with a LightCycler platform, we prefer to use FRET chemistry when available. This presents the advantage—when compared to 5′ nuclease assays—of further specificity control and also genotyping.

Vincent Foulongne and Michel Segondy

The simplest chemistry is the best. In real-time PCR, quality requirements are much less stringent than in immunofluorescent microscopy. Cheap unpurified oligos 3′-labeled with dyes that are coupled right on the solid support to the 3′-nucleotide work perfectly well. We prefer FRET probes because essentially any sequence works as probe and no optimization is required. The 5′-labeled second FRET probe requires a second linkage step and is the most expensive oligo in the primer/probe set required for FRET PCR.

Bernhard Kaltenboeck

FRET hybridization probes allow for melting curve analysis. These provide a high degree of specificity which is a significant advantage for their use. However FRET hybridization probes require the design of two probes to span 40 to 50 bases. Finding a homogeneous region in some microbes is difficult especially for RNA viruses. TaqMan probes only require a stretch of 20 bases, so they are easier to design however there is no check of specificity by melting curve analysis.

Ian D. Kay and Silvano Palladino

Fluorescein or carboxy fluorescein (FAM) is presently the preferred dye. It has very high molar absorptive around 490, which is the favored excitation wavelength of essentially all PCR instruments, and it has high fluorescence quantum yield. However, fluorescein is pH sensitive and can assume several forms (R. Sjöback *et al.* Spetrochimica Acta Part A 51, L7-L21, 1995), which can give rise to some variation.

Mikael Kubista

Fluorescein is the less-expensive and most versatile dye with superior fluorescence properties. End-labeled probes are less expensive than (internal) labeled primers and there is a higher consumption of the primer part, making probes advantageous.

Olfert Landt

It depends on the application I think. If the aim is to amplify multiple related species and resolve the amplified targets then hybridization probes and melting curve analysis provide that without resorting to multiple assay sets. However if the detection of a single species in a mix of nucleic acid is required for a yes/no assay then hydrolysis probes perform very well. One concern is that quasi-species may be missed due to hydrolysis probes" increased intolerance of mismatches as compared with hybridization probes.

John Mackay

Not sure what the optimal dye system is; ideally you would like something that should provide optimal specificity and the opportunity to perform melting curve analysis. The choice of probe system may be limited by the DNA sequence within the amplification product. Where possible, we have chosen to use the FRET system, assuming there is an increase in specificity and possibly sensitivity (over SYBR Green), with the ability to perform melting curve analysis. I am unfamiliar with all the data surrounding differences in signal background levels and linearity over concentration ranges for the different fluorescent systems.

Astrid Petrich

[*GS*] We have only used SYBR green, but have problems with detection of non-specific DNA using this dye particularly with buffy coat specimens. We know this because we also run the real-time PCR products on an agarose gel and find multiple bands of differing size in the sample. Non-specific bands should be identified by the technology through analysis of melting temperatures, but these bands are often not detected by the LightCycler technology. SYBR green also permits the identification of primer–dimers and these are usually obvious from differing T_M.

William Rawlinson and Gillian M. Scott

I absolutely favor the LightCycler hybridization probe concept. It is most specific since the generation of a fluorescent signal is based on four independent hybridization processes

(2 primers and 2 labeled probes) and it allows for a post-PCR melting curve analysis. By melting curve analysis one can identify a number of mutations within the amplified region or distinguish between a number of variant target sequences without the need for any further manipulations or automated DNA sequencing.

But all of the currently available concepts and dyes have their specific pros and cons. No simple answer possible.

Udo Reischl

This is strongly dependent on the lab and the individual experiences, I do not believe that there is a gold standard.

Oliver Schildgen

There are many different strategies that can be used to link fluorescence changes of labeled oligonucleotides to their interaction with target amplicon in a homogeneous assay. A skilled practitioner is likely to be able to achieve his analytical goals using many of them. Nevertheless the simplicity of the synthesis and of the assay and the ease of assay development does vary considerably.

With respect to fluorescent labeling it is often the quality of oligo labeling and post-labeling purification that has a big impact on the performance of any given probe.

For gene expression analysis Hydrolysis Probes and HybProbe® Probes can be used equally well. For genotyping, HybProbe® Probes and SimpleProbe® Probes have the advantage of allowing a melting curve analysis disclosing additional information about the sequence under the probe. Thereby a single PCR in one tube provides a clear answer with an inherent control and the additional chance of identifying currently unknown or unexpected variations under the probe.

Rudolf Seibl

I would be partial to the use of fluorescent tagging of the probes rather than of the primer pair; probing provides an additional layer of specificity, and also if sequencing post PCR is required, you can use the PCR primers as sequencing primers; if they are not tagged the stock of primers can be used for both assays, which simplifies the laboratory logistics.

Raymond Tellier

Are molecular diagnostics cost-effective?

Certainly they are in a hospital environment, where the cost of the test usually represents a very small fraction of the overall cost of that patient's care. In other environments, thought needs to be given to whether the test is really necessary.

Chris Birch and Julian Druce

In comparison to conventional methods molecular tests are generally more expensive. However, cost-effectiveness will be reached by improved treatment and a reduced hospital stay.

E.C.J. Claas

In general, we believe they are.

Esteban Domingo and Cristina Escarmís

It depends on the assay—for CMV, for example—yes.

Michael G. Ison

Emphatically yes.

Bernhard Kaltenboeck

Yes.

Olfert Landt

I can recall three specific examples, and the children associated with them, whereby laboratory diagnoses that provided a result in a couple of hours prevented continuing illness, uncomfortable invasive testing and shotgun antimicrobial therapies. Let's not forget the human faces behind the specimens we receive—they usually don't get "costed" during effectiveness studies.

Ian M. Mackay

Why is this question raised for molecular diagnostics, and not for serology or virus culture? It has never been determined for every technology exactly what cost-effectiveness means. For sure, in screenings programs (like for HPV) this is a matter of intensive debate. Also for screening blood for the major blood-borne viruses; and this even depends on the continent where it is applied. Furthermore, costs for molecular diagnostics should be seen in relation to TOTAL costs. In the treatment of HIV-infected patients, cost of diagnostics is disappearing in relation to costs of medication, although people continue to discuss diagnostics, and never the overall costs. I believe that this is a different science, and once the costs are calculated (if possible for everything at this moment), the technology has renewed itself, and the process has to start over again. We should then discuss the costs of disease!

H.G.M Niesters

Which are the most difficult to produce and expensive chemistries?

Our impression is that scorpion probes are difficult. However, from the laboratory's perspective the degree of chemical difficulty is not an issue, since we're not producing the reagents. Obviously though this will have cost implications. Any of the probe-based assays are a bit more costly compared to conventional assays.

Chris Birch and Julian Druce

PNA is still difficult to make—and since Applied Biosystems overtook PerSeptive and terminated the licenses there is also no legal basis to continue with this. For DNA (LNA)-based probes the very long ones with more than two modifiers are the most difficult to make (Scorpions).

Olfert Landt

In general most difficult to produce equals most expensive since the cost of licenses appears to be fairly uniform. It is probably easiest to identify the least expensive chemistry first. The intercalators win easily. ResonSense is also relatively inexpensive as only one dye modification is needed. The most expensive probes are those that generate self-probing amplicons since they are larger molecules that require multiple (including internal) modifications, i.e. probes used in the Scorpions, Angler and IntraTaq chemistries. The use of LNA or PNA bases also ramps up the cost.

Nick Saunders

What is the next big chemistry?

Further developments in detection of Single Molecule Interactions.

E.C.J. Claas

At the moment it is LNA.

Olfert Landt

The "big chemistries" to date have been highly flexible and provided increased benefits in sensitivity or specificity—as opposed to just an alternative for consideration. The chemistries becoming increasingly interesting are Locked Nucleic Acids (LNA) which have been used in both hydrolysis and hybridization probe formats. This has shown greater sensitivity compared with standard DNA probes (hydrolysis) and increased melting resolution for variant analysis (hybridization probes).

New DNA-binding (intercalating) dyes are being used which offer increased resolution between multiple amplified targets and may provide target specificity benefits to allow a wider use of this arm of real-time chemistries

John Mackay

Improvements in dye chemistry and the performance of modern instruments mean that it is now possible to detect single base changes anywhere within a PCR product without the use of a specific probe. This opens up a range of applications such as isolate classification and the discovery of sequence variations *de novo* by performing a melting curve analysis after the amplification (Literature: Reed GH, Wittwer CT. Sensitivity and specificity of single-nucleotide polymorphism scanning by high-resolution melting analysis. Clin Chem. 2004 Oct; 50(10):1748–54. Epub 2004 Aug 12).

Rudolf Seibl

Are "intercalating" dye chemistries worth using in microbiology?

Yes, depending on how they are used. For example, they are useful for "generic" assays where the target is conserved. Also, where the target is less conserved and probe design is difficult there may be some applications. For example, we use SYBR chemistry to confirm our probe-based influenza H5 assay.

Chris Birch and Julian Druce

Confirmation of the PCR products by probes was already adding to the specificity of conventional PCR assays. For real-time PCR this is the same. Intercalating dyes, even though providing some confirmation by using melting curves, cannot provide the same specificity as probes.

E.C.J. Claas

Despite the opportunity of performing melting curves analysis, we would never recommend non-specific labeled dye chemistries for diagnosis in microbiology. These have to be reserved for set up or research applications.

Vincent Foulongne and Michel Segondy

The use of dyes such as SYBR green which intercalate with double-stranded DNA are the simplest and cheapest amplicon detection methods and in principle require the least specialist knowledge. As such they have a place, especially for laboratories moving from conventional to real-time PCR, as additional fluorescence-labeled oligonucleotides are not needed. However specific and non-specific PCR products are detected and the assays need careful optimization of PCR conditions and are best used in combination with melting-curve analysis. Awareness of the inherent aberrations in melting-curve analysis due to the occurrence of mutations is also needed.

Lance Jennings

We have long given up on dye-detection and only use FRET probe PCRs for exquisite specificity.

Bernhard Kaltenboeck

In Microbiology we are often looking for low copy number targets in a large concentration of genomic DNA. SYBR Green is a non-specific dye, which will bind to any double stranded DNA. This means in these samples there will be a high background which may swamp the target-specific signal. Some authors have suggested that these dyes be used as a quick screening tool, to be confirmed by another method. However this luxury is not usually available to diagnostic laboratories due to price and turnaround time.

Intercalating dyes may have a role when using pure bacterial colonies, however in our laboratory we mainly use probes

Ian D. Kay and Silvano Palladino

The probe cost is very small compared to the total cost of an assay, so there is no significant saving on using dyes in optimized assays. However, one can use dyes during the initial optimization of the assay, when testing different primers, and then change to probe in the optimized assay. As already mentioned, dyes can also be used for quality assessment of probe based assays (K. Lind *et al.* Biotechniques 40, 315–18, 2006).

Mikael Kubista

If it is diagnostics for patients we would prefer probes. If the test is for "flavivirus" and other non-conserved targets, intercalating dyes make sense.

Olfert Landt

SYBR can be used while setting up a new assay, but if one ultimately intends taking up an oligoprobe, conditions may have to change—so why not use that from the outset? I've seen some instances of primer–dimer with denaturation temperatures above those for larger, specific amplicons (presumably GC content-related, although not yet investigated)—so I remain unconvinced that melting analysis is the way to overcome the innate flaw of SYBR's non-specificity. Human gene studies seem to have found many uses for the non-specific chemistries—especially in SNP analysis, but this field rarely suffers from a paucity of reliable template—virology is a different beast. I'm not a fan.

Ian M. Mackay

I did not think so until I saw some work firsthand on detection and differentiation of Leptospira serovars using SYBR Green (Merien *et al.*, FEMS Microbiology Letters 249 (2005) 139–147). Using well-designed primers and careful evaluation, there was no non-specific amplification and the serovars could be differentiated by T_M. Melting temperatures offer an additional benefit over agarose gel detection (the common comparison) in that the T_M is largely dependent on GC content of the amplicon and not just length.

This use may be dependent on the likely complexity of the sample material being used and reinforces that assay validation is specific for the intended biological sample.

John Mackay

We have found them useful in the development of new real-time assays as a first examination of whether a chosen primer set gives a specific product and/or is prone to primer–dimer interactions. Other than this we have chosen to use other specific probe systems for detection as they give increased specificity and appear to give increased sensitivity.

Astrid Petrich

[GS] I don't know of any uses any more.

William Rawlinson and Gillian M. Scott

I think that "intercalating" dyes like ethidium bromide, SYBR Green or LCGreen® I are appropriate real-time PCR chemistries for the initial steps of assay development. Also, these dyes are useful for certain research applications, qualitative and quantitative real-time PCR or the newly developed approach of high-resolution melting analysis. Intercalating dyes are not the "golden choice" for the majority of diagnostic PCR assays.

Having a relatively high frequency of false-positive PCR results in mind when PCR product analysis was formerly accomplished by EtBr-stained agarose gel analysis, the solely size-dependent characterization of amplicons seems to be insufficiently specific for diagnostic purposes.

Although the real-time PCR products are not separated in agarose gels, the signal generated with intercalating dyes is dependent on the length of the amplicons (subsequent T_M-analysis or the respective T_M in the thermocycler profile where fluorescence was determined) but not on the detection of characteristic stretches of nucleotide sequences within the amplicons.

Due to the sometimes dramatic consequences on patient management or antibiotic treatment in the case of positive PCR results, all reasonable technological measures should be taken to avoid or at least to reduce the likelihood of false-positive results.

Recently the experts in Germany and Switzerland agreed to postulate a nucleotide sequence-specific detection of the corresponding amplicons when money is charged for a given PCR-based analysis in clinical diagnostics. Among the sequence-specific detection methods listed in the order of perfection (restriction enzyme digest, nested PCR, solid phase hybridization, real-time PCR with fluorescent probe(s), or automated DNA-sequencing), real-time PCR with hybridization probe-based amplicon detection is the favored approach with respect to specificity, turnaround time and usability.

Udo Reischl

The "intercalating" dye chemistries are certainly worth using as part of the assay development process when it is necessary to test the efficiency, specificity and sensitivity of a range of primers. This saves the expense of obtaining fluorescently labeled probes for each potential primer pair. In addition, the use of "intercalating" dyes allows the reaction kinetics of the elongation step and the amplicon T_M to be determined simply. Some of the existing "intercalating" dye chemistries include the use of specific probes (e.g. ResonSense) and are therefore as powerful as the methods that do not rely on an intercalating dye. However, if the question is taken to refer to the use of "intercalating" dye in diagnostic assays that rely on the generation of a signal due to the accumulation of dsDNA alone then the degree of specificity becomes a crucial consideration. In some cases the primer design combined with melting analysis may confer sufficient specificity on the assay. However, such situations are probably rare and consequently this contributor would not recommend that workers attempt to design or evaluate a real-time PCR intended for diagnostic microbiology that did not rely on a probe.

Nick Saunders

Yes, they are extremely useful, well known, and cheap.

Oliver Schildgen

Do target-specific chemistries exist that are better suited to microbial diagnostics and research than human studies?

We have found that molecular beacons appear to have optimal specificity and sensitivity compared to FRET probes, however, I am unfamiliar with a direct comparison with all of the chemistries to give a complete answer.

Astrid Petrich

[GS] We have found fluorogenic probes are more specific for target detection.

William Rawlinson and Gillian M. Scott

Due to the availability of well-tried yet robust real-time PCR platforms all of the fluorogenic chemistries may work well for applications in all areas of research and diagnostics irrespective of the target organism.

For the *de novo* establishment of an assay, however, it is wise to start with the "cheap" SYBR Green chemistry for primer specificity and sensitivity testing. Once a suitable primer pair and PCR conditions are identified, one should spend some efforts on the design of single TaqMan oligoprobes or pairs of LightCycler hybridization probes to enable a sequence-specific detection of the corresponding amplicons. It should be kept in mind that the most important aspects of analytical sensitivity and quantitative behavior of an assay are primarily linked to the basic amplification process—and the common or sometimes exotic fluorogenic labels used in the course of the detection process only "report" the qualitative or quantitative results to the user sitting in front of the computer screen. There are some myths about the pros and cons of different fluorogenic chemistries, but most of the storytellers are sitting at diagnostic companies (which are strictly dependent on their individual patent portfolio) and were paid to praise their own chemistries.

Anyhow, for some real-time PCR applications the pointed detection of nucleotide mutations within the amplicons may supersede laborious and costly DNA sequencing. Here the use of anchor and sensor LightCycler hybridization probe pairs covering the sequence regions of interest and the option for subsequent melting curve analysis is highly advisable.

Udo Reischl

In human studies most of the target sequences involved can be considered invariant with the obvious exception of any SNPs that may be the object of the assay. This means that primer and probe design for real-time PCR need not consider the occurrence of strain specific sequences. Since viral and bacterial genomes are generally far more variable than the human genome, probe and primer design is more complicated. Hydrolysis (TaqMan) probes work poorly when not perfectly matched to their target sequence and consequently may not be the best choice when the sequence may include SNPs. In contrast the probe chemistries that do not require hydrolysis, because signal is generated on hybridization, are far more robust to sequence variations.

Nick Saunders

What is the value of retrospective versus prospective microbial real-time PCR research-based studies?

One of the values would be to compare real-time PCR results with previous results as found by classical technologies.

M.G.H.M. Beld

Can be helpful in determining the epidemiology of a newly discovered pathogen if that's considered to be important. It may also shed light on clinical associations if the database is comprehensive. The retrospectivity should be sufficient to cover natural cyclic variation of these pathogens.

Chris Birch and Julian Druce

Retrospective studies allow selection of well-validated samples, but they may be negative or less positive after storage so need to be run through comparator test as well as standard test again. Prospective need larger numbers but fresh samples, freshly extracted are best.

William Carman

Analysis of a significant proportion of samples from patients with encephalitis or respiratory disease does not result in identification of a pathogen. It is highly likely that we still do not know about all pathogenic viruses and that new ones are emerging. For example, in the last 5 years human metapneumovirus, bocavirus, and several coronaviruses have been newly described and shown by retrospective molecular studies to have been with us for some time. In contrast, the SARS coronavirus was shown to be new to the human population. So, retrospective studies have an important role to play in some circumstances. The advantage of well planned prospective studies is that criteria can be defined for inclusion of patients and appropriate denominators included.

Patricia, A. Cane

Retrospective analysis is a valuable tool in validating real-time PCR assays and providing data on clinical relevance. In addition retrospective studies can provide markers for pathogenesis of an infection, which subsequently should be tested prospectively.

E.C.J. Claas

If you're doing a retrospective study you will presumably have a well characterized sample set with all the relevant clinical and epidemiological information. A problem with retrospective studies, though, is that the samples may not have been collected and stored in a way suitable for maintaining the integrity of the virus genome that you're interested in. Therefore, the copy number estimates you make based on real-time PCR may be underestimates. A prospective study may be expected to give more accurate results.

Jon Clewley

Most studies involving real time PCR must be considered retrospective in the sense that samples that are suspected of including one or several agents are tested with a quantitative-analytical aim. However, prospective studies (in the sense of anticipatory studies) are of potentially great value. For example, in the event of diagnosis of a pathogen in an infected host, it may be important to follow by real-time PCR the evolution of the level of the pathogen (or of variant forms of a pathogen) as a function of time. This does not seem to be applied at present on a routine basis, but it may be of value as a means to follow the evolution of disease or the response to treatment.

Esteban Domingo and Cristina Escarmís

As a "virus hunter" (Editors choice of terminology), I am a big fan of retrospective studies. It allows us to make rapid assessments of the impact of "new" pathogens, and compare it to others. There are obvious downsides, such as differences in the technology used to detect pathogens now and, say, 5 years ago. Retrospective sample collections may also have been stored imperfectly, and datasets are often incomplete, which can always affect results. As long as one realizes the downsides of retrospective studies, it is good to perform them to save time and money. However, I always greatly appreciate good prospective studies that are hypothesis-driven. I think we cannot do without both types of studies.

Ron A.M. Fouchier

Research projects which requires the collection of clinical data and human samples for microbiological analyses should be designed so that both prospective real-time PCR testing and retrospective testing as new assays are developed or become available, can be performed. Appropriate ethical approval for storage and future testing will be required. A specimen bank of appropriately stored, aliquoted and documented samples is an invaluable resource to a microbiology laboratory involved in research, assay development or evaluation. It allows the evaluation of the clinical sensitivity and specificity of new real-time tests for known pathogens, while at the same time permitting the application of new tests to establish the incidence and prevalence of novel pathogens. Quantitative real-time PCR can also be applied to such sample banks to establish the normal range of the viral load associated with an infection.

Lance Jennings

Important for understanding evolution and traveling/infection routes (see HIV-1, Spanish Flu or Black Death research). Both make sense. Retrospective studies give faster data.

Olfert Landt

We have to learn what the impact of these new technologies is on well-characterized patient material, mostly collected in a retrospective manner. If available, this will be very worthwhile to determine whether a newly detected virus, or low levels of viral nucleic acid, do have clinical impact. For sure, this is related to legal issues, but will provide the best opportunity for introducing advanced diagnostics for better health care.

H.G.M. Niesters

The issues of specimen and clinical data collection are best addressed by prospective studies. Valuable information in terms of incidence of the agent in question are still able to be determined by retrospective analysis.

Michael D. Nissen

Prospective studies are always necessary in order to avoid selection biases that always occur in retrospective ones. Relative stability of microorganisms may also affect retrospective approaches. On the other hand, retrospective studies are also needed for a variety of reasons, including speed, historical evaluations as well as cost.

Nikos Papadopoulos

Retrospective studies can allow you to look at new novel agents in older well-characterized populations, possibly different from those that would be obtained in a prospective study. If the samples are suitably anonymous there may not be the need to obtain ethics approval with retrospective samples.

Astrid Petrich

If all analytical steps between sampling and data interpretation were performed properly, and internal controls or valid standards were included, the diagnostic value would be maximal. If the scientist skips over any one step, the reduced efficiency, e.g. in extraction, in PCR efficiency, variability of the method, etc. renders the diagnostic value useless. Therefore validated internal controls for spiking the initial sample and validated SOPs are needed.

Michael W. Pfaffl

[*WR*] The availability of well-characterized cohorts, standard nature of specimen handling wrt freeze–thawing, extraction etc, same time of sample testing

[*GS*] We would assume that there is some loss in the DNA levels through freeze-thawing of samples (particularly repeated freeze-thawing), although it would be interesting to test whether this is the case.

William Rawlinson and Gillian M. Scott

Retrospective studies are well-suited for the development of a real-time PCR assay. But a thorough clinical evaluation needs to be based on prospective studies. Ideally in direct comparison with the current PCR assay or other analytical procedure which is to be substituted by the novel real-time PCR assay.

Udo Reischl

There are two obvious advantages of prospective studies: First, data-mining with respect to clinical information is in general better and more detailed. Second, a therapeutic intervention is possible in a (well-planned) prospective study, thus giving important information on a putative drug regimen for future episodes of the respective infection.

The major advantage of the retrospective studies is that the number of samples that can be investigated is normally higher and thus leads to statistically significant results, provided that the material is stored in an appropriate way (i.e. –70°C or liquid nitrogen) until the study is performed.

Oliver Schildgen

The best advantage of retrospective studies is that we can get an overall picture on the problem quickly, based on specimens collected over a period of time. For prospective studies, serial specimens and multiple specimens can be obtained from the same patient for monitoring of microbial load changes, the presence of the microbe in different specimens, etc.

Patrick Woo

Do molecular results impact on your treatment of patients? Would they have more of an impact if even faster result turn-around were possible?

Yes, mostly due to the time getting results in comparison to classical methods and due to the accuracy of real-time PCR.

M.G.H.M. Beld

Yes, definitely.

Their impact depends on how fast the existing turn-around-time is. We provide once-daily testing and are working on ways to make this twice daily for certain pathogens.

Chris Birch and Julian Druce

We cannot be sure. We know that clinicians phone us to get results so they must be interested. Some patients get taken off of antimicrobials if we have a PCR positive. Some patients get put on to antivirals of course. I doubt if they result in a shorter hospital stay. A test that took 2 hours would make a real difference as long as sensitivity and specificity were excellent. We tend to get most results out by the next day.

William Carman

Many molecular results do impact treatment of patients and obviously would even have more impact with faster turn-around times. However, an important issue to address at the moment, is proving the clinical value of molecular test results to some clinicians, especially in those cases where no or poor diagnostic possibilities were available.

E.C.J. Claas

There is talk of "bedside" PCR and PCR diagnostics in the doctor's surgery. These will impact on the treatment of patients, although there will always be a need for specialist labs providing confirmatory assays and resolving any problematic and ambiguous results. A fast turn-around is also useful when trying to locate the source of outbreaks (e.g. hospital infections) but this approach needs to incorporate some typing methodology to establish that the suspected source or index case and the patients have the same strain.

Jon Clewley

We do not deal with patients directly, but we collaborate with colleagues who do. The information we have gathered is that knowledge of molecular features of a pathogen (for example, specific nucleotide sequences and their proportion) is generally of value for the planning of treatment. Depending on the disease episode, its evolution, complications, etc. it may be crucial to obtain fast results with molecular information on the causative pathogen.

Esteban Domingo and Cristina Escarmís

We would stop such diagnostics if they would not. In principle, we only test for pathogens if treatment depends on the diagnosis. Sometimes pathogens can be tested for simply to exclude them; even that can be valuable. It is true that the faster the turn-around time, the better it is. For instance when it comes to outbreak prevention (e.g. nosocomial infections), and for treatment of certain pathogens (flu and neuraminidase inhibitors), time is very important.

Ron A.M. Fouchier

Sure. This is illustrated by the management of HIV-1 infected patients and also for HCV, HBV, CMV, EBV (lymphoma), or BKV (renal transplant recipients) infected patients. Furthermore, real-time PCR analyses that are rapid and sensitive with low risk of carry-over contamination are really compatible with emergency diagnostics. Those techniques are now well accepted as gold standards for HSV meningo-encephalitis diagnostics. In fact, faster result turn-around could be achieved by improving the critical limiting step for molecular diagnostics that is the sample preparation step through nucleic acid extraction.

Vincent Foulongne and Michel Segondy

There is a clear impact of molecular results in patient care (i.e. HSV encephalitis vs. meningitis; detection and management of CMV and adenovirus). The major limiting factor is turn-around time—fast is essential.

Michael G. Ison

It is well established that timely pathogen identification by conventional or molecular means can lead to fewer laboratory tests being ordered, timelier antiviral or antimicrobial therapy and a cost saving per patient hospitalized. One service in the hospital setting that does benefit from a fast turn-around time of conventional respiratory virus diagnosis, and would benefit from more timely molecular test results is infection control. During the influenza season the introduction of rapid influenza (point-of-care) testing for the Emergency Department, providing patient results within one hour, during the triaging period, can result in influenza positive patients either being discharged from hospital or being admitted to a single room. The application of real-time PCR with a similar turn-around-time and the ability to detect a range of respiratory viruses in addition to influenza would assist with the activation of infection control protocols in the hospital setting throughout the year, and lead to improved patient management and decreased nosocomial infection.

Lance Jennings

Both questions emphatically yes.

Bernhard Kaltenboeck

Since I am not a medical doctor...no. Elsewhere—I believe yes—but only in 2–5% of the cases (my estimation).

Olfert Landt

Some reports, as well as our friend "anecdotal evidence," suggest that despite relatively fast results (be they rapid antigen detection or PCR-based assays), we still don't provide results fast enough to prevent shotgun antibiotic therapies from being applied. Practically, such delays are more likely related to specimen processing and result reporting time than to actually performing a real-time PCR assay, nonetheless who could fault the clinician for wanting to cover all bases—especially when a patient's health is on the line? However that thinking will never challenge the paradigm. Surely there are some illnesses that are not so serious that they could wait and in so doing, prevent the unnecessary over-use of antibiotics in at least some instances. This would avoid side-effects experienced by the patients and importantly, reduce the production of drug-resistant mutants.

Ian M. Mackay

Without any doubt. The best examples are of course the treatment of patients infected with HIV-1 or 2, HBV and HCV. Determining viral load for use as a marker to create an optimal treatment regimen. But also the detection of EBV, CMV or adenovirus in transplant patients has reduced morbidity and mortality. The detection of viral targets in CSF, including HSV-1, enterovirus or parechovirus. Being able to detect an infectious agent will have impact on treatment options and choices to be made. Whether the virus is important or not, its presence always triggers options.

H.G.M. Niesters

Yes, Yes...good examples are meningococcal and herpes simplex PCRs.

Michael D. Nissen

Slowly but steadily the result turn-around and the validation of diagnostic methodologies are translated to clinically relevant information. Certainly, a clinician needs the response about the presence of a microbe in a patient as soon as possible and definitely before the patient is either cured or dead! On the other hand, we should not forget that even "classical" approaches in infectious disease diagnostics, i.e. antibody titers may need weeks to be completed.

Nikos Papadopoulos

Yes, in certain situations they do. Some examples are; *M. tuberculosis*-specific amplification assays performed directly on smear-positive respiratory samples, Herpes and Enterovirus PCR performed on CSF, Norovirus PCR vs. Electron microscopy, VRE, and MRSA. It is possible that even faster turnaround results would have an impact. However, other factors may minimize the affect of a rapid laboratory result such as delays in specimen acquisition and reporting of results, with physicians interpreting results and acting on them.

Astrid Petrich

Significant impact upon patient treatment, particularly in sorting out diagnosis. In the ideal world of rapid/near instant results, the diagnosis would be much easier as some common conditions could be ruled out immediately, and appropriate further diagnostic tests undertaken rapidly. For example in respiratory disease, immediate diagnosis of influenza has public health, isolation, therapeutic, and diagnostic benefits. Some of these are already realized with point-of-care tests, although these suffer considerably from poor performance characteristics.

Udo Reischl

Especially in the case of pre-treated patients or infections with slow-growing or unculturable bacterial pathogens, molecular results are very helpful. Like the eagerly awaited results of microscopy, an early and specific diagnosis employing automated sample prep combined with real-time PCR may also be very helpful in many other clinical situations. For example, we already practice a turn-around time of less than two hours for the direct detection of MRSA from nasal or wound swabs. This was well received by the clinicians since identifying whether a patient is positive or negative for a given pathogen within hours not only improves the options for specific treatment but also ensures that infection control measures are effective. In general, I think that we should aim at "same-day" PCR results. Reducing the total assay time from 60 minutes to 50 minutes is not so relevant in most applications.

William Rawlinson and Gillian M. Scott

Yes to both. In the case of our cohorts of children with respiratory infections superficial antibacterials can be avoided in many cases, if a respiratory virus was detected. Vice versa, therapy can be adapted to the pathogen if it is detected earlier. Furthermore, in cases of, e.g. HIV and HBV infection, molecular resistance testing influences the subsequent therapy.

Oliver Schildgen

In our institution we have established several assays that have had great impact on patient care and treatment. The main reasons have to do with greater sensitivity and faster turn-around-time (TAT). These assays have also greatly facilitated the diagnosis of agents difficult or currently impossible to culture. In Toronto, the SARS outbreak has also illustrated another point: because nucleic acid extraction quickly inactivates agents, it is possible to safely provide laboratory diagnosis of BSL-3 agents using BSL-2 installations and BSL-3 procedures, which would not be possible with culture-based methods. There are of course difficulties as well; when faced with new, emerging agents, primers of genome sequences will not be immediately available. TAT is a complicated question. The faster TAT of real time PCR, compared to a standard PCR assay followed by gel electrophoresis, is not necessarily the main determinant of TAT. In most laboratories the molecular assays will be done in batches, and then of course the main determinant of TAT will be how often you run batches during the week. There are new technologies (e.g. Cepheid) allowing for testing of samples "as they come" but currently these are expensive and difficult to scale up.

Raymond Tellier

Molecular results have a huge impact on treatment of our patients and they would have more impact if even faster result turn-around were possible. For example, in our hospital, tuberculosis was mainly diagnosed by AFB smear and culture in the past. However, AFB smear is very insensitive while culture takes 3–6 weeks, and clinicians are very reluctant to give empirical anti-TB treatment to patients, as TB drugs are associated with significant side effects, especially in older patients and patients with underlying diseases. After the introduction of TB-PCR, this has been one of the most valued molecular tests in the past few years. However, due to budget constraints, TB-PCR can only be performed once a week in our laboratory. If this can be performed 2–3 times a week, the impact would be markedly increased.

Patrick Woo

Do you think a wider range of microbial quantitative assays would help you diagnose and treat patients?

Yes, but there is the rule of diminishing returns!

Chris Birch and Julian Druce

Not apart from the ones we already do (CMV, HBV, HCV, HIV, EBV, BK, adeno). Time will tell if a lot of rhinovirus equates in a precise way with clinical outcome, for example.

William Carman

Not necessarily. For many pathogens their presence (i.e. qualitative diagnostic result) itself is sufficient for medical intervention and quantification is not necessary.

E.C.J. Claas

Yes, if there were a greater ease of access to them. Sometimes it's hard to follow up on hunches because an assay is only available in one or a few centers. And there is sometimes the problem of too little sample and too many tests.

Jon Clewley

According to the interaction we have had with colleagues who treat patients, a wider range of microbial quantitative assays would be very helpful for diagnosis and treatment planning.

Esteban Domingo and Cristina Escarmís

Yes. Even if determining quantity is not a requirement to assist in treatment directly, it provides information that can be of use. For instance, it could help to identify laboratory contamination (e.g. if all patients have high loads, a low load could be indicative of contamination). It is also well known that pathogen RNA/DNA can persist longer than the infectious pathogen itself, and also here a low load or declining load may be indicative. In the longer run, quantitative information on pathogens can be correlated with disease severity and final outcome, and studies on this issue should be welcomed.

Ron A.M. Fouchier

Real-time PCR already allows us to increase the number of pathogens of interest that are worth monitoring in some patients like transplant recipients.

Vincent Foulongne and Michel Segondy

Probably.

Michael G. Ison

The drivers for the provision of laboratory diagnostic services vary between health systems; however they are usually either a request from clinicians or recognition by the laboratory of an advance which would improve the diagnostic service. With new pathogens, this invariably means the introduction of an in-house qualitative molecular assay. Similar drivers for quantitative assays operate: as new treatments become available and our understanding of the pathogenesis of various pathogens and relevance of pathogen load with disease increases, quantitative information will be increasingly required.

Lance Jennings

Yes, by all means. Particularly if coupled with a 24 hr turnaround. If PCRs are designed to use the same thermal protocol and can run simultaneously then the rate-limiting step in this is nucleic acid extraction, limiting the number of real-time thermocyclers required.

Bernhard Kaltenboeck

Very often qualitative assays would be enough.

Olfert Landt

Yes....but only if basic work such as the significance of load can be linked to prognosis and or treatment.

Michael D. Nissen

At this moment in time, validation rather than quantity seems more important. Conceptually, quantitative molecular methods could become the gold standard in diagnostic microbiology. However, in several instances, quantitation is not completely necessary, while in others it may even become a problem when "levels" of presence cannot be interpreted. This will require some more work, before reaching the bedside.

Nikos Papadopoulos

I believe we are just at the starting point for determining the utility of quantitative microbial assays for monitoring the onset of disease in immune-compromised individuals (CMV, EBV, Adenovirus, BK), for monitoring therapeutic responses and antimicrobial resistance and for predicting therapeutic outcomes. This role of quantitative amplification assays will definitely grow in microbiological diagnostics.

Astrid Petrich

QPCR is very fast, much faster than classical microbiological approaches. If we can generate qPCR results in a short time, while the patient is waiting, an immediate and pathogen-related dose treatment is possible! The advantage for the patient is very clear!

Michael W. Pfaffl

Yes, it is apparent that quantitation has a significant impact on treatment and clinical prognostication in HIV, and increasingly in HCV. The availability for other diseases (CMV, respiratory diseases etc) is increasing as treatments for these become available. This notwithstanding, quantitation has enormous potential impact in understanding pathogenesis of viral, bacterial and fungal diseases.

William Rawlinson and Gillian M. Scott

Only in the field of diagnostic virology (see answer three questions below)

Udo Reischl

Yes. Yet, there is rather limited information on the correlation between viral load and clinical severity for a number of pathogens, e.g. respiratory viruses, but I'm sure that such correlations have to be taken into account in the future.

Oliver Schildgen

Quantitative viral load assays for chronic viral infections like HIV, HCV and HBV have proven to be very useful. I am not convinced however that a quantitative assay is always required for all agents tested. After all, for bacterial blood cultures we do not provide quantification; they would be meaningless most of the time given the changing titers and the fact that most bacteremias are transient. Also, a problem with quantitative assays often encountered in practice is that whereas considerable emphasis and effort is put on the quantification of the PCR, too often not enough care is taken to quantify accurately the sample (e.g. tissues, broncho-alveolar lavage) and without that measurement, the meaning of the numbers given by a quantitative PCR is unclear.

Raymond Tellier

This depends on the spectrum of antivirals available and data on correlation between microbial quantitative assays and treatment outcomes.

Patrick Woo

Should all C_T values, even high ones, be reported to the clinician?

No. The laboratory scientists need to know the limitations of the particular test and report the results accordingly.

Chris Birch and Julian Druce

If all controls are OK, a high C_T value reflects the presence of a pathogen. If that pathogen is not supposed to be there, these results should be reported but be accompanied by expert comments.

E.C.J. Claas

No, unless confirmed by gel electrophoresis.

Esteban Domingo and Cristina Escarmís

This is a very important point concerning the adverse effects of highly sensitive methods and the significance of low signals. This could be an illustration of the critical necessity of a dialogue between biologists and clinicians. Both sides need to advise each other of their own constraints. Picking up the phone is more constructive and efficient for all the actors, biologists, clinicians and most importantly, patients, than releasing raw data.

Vincent Foulongne and Michel Segondy

This depends on the pathogen. MRSA can be found everywhere (skin) and will be not significant. With HIV-1, everything is critical. Depending on the target... I would repeat the analysis before irritating the clinician.

Olfert Landt

In the "old" days, we reported all qualitative amplification results to the clinician and never thought about calling something weak positive or strong positive. Currently, with the advances of real-time PCR technology, we have the idea to be able to identify weak or strong positive signals, although we also realize that the clinical impact of these so-called high C_T values, are still unknown. However, a SARS coronavirus or H5N1 positive (with all correct controls) with a high C_T value still has impact! It is a matter of education from both sides. The laboratory side still has to determine what it really means, and whether we should maybe report a range of values. The clinicians also have to be educated to understand what low copy number values could mean. But the data are there, and should be interpreted together with other clinical data; never based on one high C_T value alone. We have to work together to learn to understand.

H.G.M. Niesters

How will progressive changes in regulatory issues impact upon use of real time PCR?

Regulation of real-time PCR in a diagnostic setting is important for maintaining quality of performance. A targeted quality assurance program would be a step forward.

Chris Birch and Julian Druce

Already severe impedances to use of in-house assays. Most labs may need to go with commercial assays with time.

William Carman

In time using "home brew" methods will become impossible due to strict regulations for the generation of *in vitro* diagnostic results.

E.C.J. Claas

There was recently an irresponsible piece of science journalism in a UK newspaper about the ease with which one could buy synthetic oligonucleotides. Around this the journalist concocted a fantasy about terrorists being able to make biological weapons (e.g. smallpox virus). This story lead to questions in parliament and a high level civil service committee deliberating the threat. Hopefully, it has been appreciated that the threat is minimal, but there remains the danger that ill-informed legislation might restrict the supply of molecular biological reagents and make it more difficult to make and use PCR and other assays.

Jon Clewley

Most regulatory issues refer to the implementation of some types of human interventions (use of early stage embryos, stem cells, etc.), or ethical issues regarding human or animal manipulations. If implemented rationally, these regulations should not affect at all (or only very indirectly) the applications of real-time PCR for diagnostic purposes. It is conceivable that sample preparation, with the aim of extracting nucleic acids of interest, may be limited when the biological material is subject to regulatory limitations.

Esteban Domingo and Cristina Escarmís

I am not a QA expert, but there is one regulatory issue that does affect the use of real-time PCR; biosafety levels for working with pathogens. At the present time, many microbiology laboratories do not have the facilities to work with all pathogens under the appropriate conditions (e.g. HPAI influenza, SARS-Coronavirus). The number of pathogens requiring high biosecurity laboratories will likely increase in the years to come (upgrading of old pathogens, discovery of new ones). Molecular detection is potentially the only way to perform diagnostics for those laboratories lacking high biosecurity facilities.

Ron A.M. Fouchier

Regulatory changes are required to ensure standardization of real-time PCR protocols and the use of positive, negative and internal controls in the routine diagnostic laboratory. The likely impact will be on the increased incentive for the smaller diagnostic laboratory to use commercial kits.

Lance Jennings

Mostly by requiring strict quality assurance, which is highly desirable. As we all know, there are huge disparities in the quality of PCR diagnostic (and research) labs.

Bernhard Kaltenboeck

Cost and delays in time to clinical testing release

Michael D. Nissen

The new IVD regulations for class 1 to 3 formalize what most good laboratories should already be doing. That is, validation of the assay, determination of sensitivity, specificity (both analytical and clinical) and reliability. However the regulations for Class 4 such as HCV and HIV ensure that any assay whether commercial or home-brew must meet strict criteria. This will make it difficult for many home-brew users to provide sufficient data and validation to complete the dossier. Commercial companies will probably be dominant in this area as they have the resources to complete the studies required to meet the new regulations.

Ian D. Kay and Silvano Palladino

PCR will get more relevance.

Olfert Landt

The cost differential between commercial and in-house assays will decrease as the cost of validating in-house assays increases. In addition, separate regulations often make it difficult for laboratories to generate in-house controls (e.g. permits for plasmid cloning) that mean it can be difficult to even get the controls to evaluate basic assay performance in terms of sensitivity and reproducibility.

John Mackay

The significant workload for initial validation and then for ongoing QC of the molecular assays and instrumentation will have a large impact on laboratories especially in situations where molecular assays are add-on tests and not replacements to conventional testing.

Astrid Petrich

[*WR*] Significant, as in-house assays are the only ones available for many of these. It is unlikely commercial assays will become available because of cost considerations in development of new assays.

[*GS*] I believe increased regulation will not decrease the use of real-time PCR, and will promote the use of and belief in this technology as a useful and accurate diagnostic tool.

William Rawlinson and Gillian M. Scott

Any increase in the level of standardization will lead to an overall increase of reliability of the results. Since one of the main arguments against the widespread reimbursement of real-time PCR testing in infectious diseases is the lack of reliability among the various in-house protocols, progressive changes in regulatory issues and national or international quality control initiatives may invalidate this argument.

Udo Reischl

The gradual tightening of the regulatory framework for laboratories involved in diagnostic assays is probably one of the drivers behind movement away from traditional PCR toward real-time PCR. This is because real-time PCR can be highly specific and it produces objective end-points that can be interpreted automatically. Over time it is inevitable that laboratories will move toward the use of commercially prepared real-time PCR kits that have the best evidence base. This will be driven by the increasing regulatory risk of relying on in-house assays. However, in-house methods will continue to play an essential role in epidemiological studies, as confirmatory tests, in quality assessment and for research.

Nick Saunders

I don't know; this is a rather political question.

Oliver Schildgen

I think that regulatory controls proposed by the regulators will primarily have an impact on the development and performance of "in-house" assays. It is proposed that strict validation criteria be met before an assay may be used for diagnostics. This should lead to the development of high quality assays and improved quality of service. Laboratories that are not able to meet these criteria will be forced to use commercial assays if these are available, which may drive up the cost of medical testing.

Theo P. Sloots

The elimination of post-amplification processing combined with rapid thermocycling makes the real-time PCR procedure a desirable point-of-care (POC) test that can be implemented in a doctor's office, providing real-time service. Cepheid (Sunnyvale, CA, USA) ingrates a real-time PCR protocol with specimen processing and manufactures GeneXpert®, the first simple, integrated real-time PCR analyzer. However, regulation agencies, such as the Food and Drug Administration (FDA) in the United States, have not approved any PCR tests for a POC setting and Cepheid has just received FDA clearance for its group B streptococcus assay at a moderate-complexity designation. It will take time for a change of concept in regulations.

Yi-Wei Tang

One concern that I have with regulatory issues is enforcement of commercial assays approved for clinical diagnostic use. The problem here is that they are often considerably more expensive than in-house assays. In principle a well designed in-house assay can be as accurate. I would like to see a regulatory emphasis on validation and proficiency monitoring rather than enforcing only a few "approved" commercial tests.

Raymond Tellier

I don't think there will be much impact because the use of real-time PCR is not regulated in most circumstances, except in situations like HIV and HBV.

Patrick Woo

Do non-specialist real-time PCR labs produce quality results that will be reliable in times of pandemic disease?

They should, although specimen numbers often determine skill levels. Non-specialist laboratories may not necessarily be in the best position to decide what is the best commercially available test.

Chris Birch and Julian Druce

Only if they are given the appropriate primers, probes, standard operating procedure (SOP) and have been trained. So, that almost makes them specialist!!

William Carman

This should not be recommended because of an increased risk for false negative results. This may be circumvented by very clear protocols with sufficient controls, provided that the lab has ample experience in real-time PCR in their field. Nevertheless, in times of true pandemic disease as for example influenza, diagnostic results will not be required as clinical symptoms will induce a rapid treatment protocol.

E.C.J. Claas

I hope so.

Jon Clewley

It depends on how we define a "non-specialist" real-time PCR lab. It is inappropriate to give a diagnostics lab the responsibility of carrying out real-time PCR analyses just because of the need to respond to a disease emergence. Any lab involved in real-time PCR (as in any other diagnostic procedure) must have shown a minimal reliability in performance. Reliability should be tested by carrying out "blind" assays (with samples of a composition unknown to the practitioner) involving candidate laboratories during times prior to the onset of an emergency. Some central laboratory should be responsible for the preparation of a set of samples with known amounts of pathogens (or mixtures of pathogens) of interest. The different laboratories should be provided with protocols and reagents to test their capability of producing correct quantitative diagnosis.

Esteban Domingo and Cristina Escarmís

I think not. It takes some time to get the technology "in the fingers," and the logistics for rapid testing sorted out. The large numbers of samples that can be expected in times of pandemic would make it difficult for the non-specialist labs to produce the desired quality results at once. Once testing is ongoing, problems may arise with the tests (especially for genetically variable pathogens), and it would be difficult for non-expert laboratories to recognize such phenomena. It will be tough for expert labs to rely on and verify the results of non-expert labs, and guide the non-experts when such problems arise. An additional problem for the high volume testing is that many laboratories (in particular non-experts) may not be fully prepared for this sort of testing in the first place; availability of personnel to provide continuous service (24/7?), maintenance of sufficient stocks of reagents and equipment when there is (global?) shortage are difficult enough to deal with for expert labs, let alone non-experts.

Ron A.M. Fouchier

Yes. We think that is one of the advantages of real-time PCR that is easy to set up in any laboratory. We can report the really successful set-up of home-brew real-time PCR assays for both HIV-1 and HSV-2 quantification in an African laboratory with low resources and little experience in molecular techniques.

Vincent Foulongne and Michel Segondy

Molecular diagnostics will play a pivotal role in surveillance to assist the public health response during the early phases of the next pandemic. Following a pandemic alert, "border management" and "cluster management" pandemic phases, will require the identification of individual cases or clusters of infection with a novel influenza virus. The most sensitive and specific tests with a fast turn-around-time will need to be used and the real-time PCR assay is currently the most appropriate method for the rapid detection and sub-typing of a novel influenza viruses. This assay will need to be widely available and is likely to be performed by laboratories with varying expertise. However; confirmatory testing by a second real-time PCR laboratory should be required in all pandemic influenza laboratory protocols, to ensure a high level of reliability of the results. Adherence to these protocols will be essential as a single identification is likely to lead to substantial media attention and hysteria. Once a pandemic virus becomes established, laboratory surge capacity is likely to be rapidly reached and laboratory testing will be focused on supporting the management of hospitalized cases. Now, during the inter-pandemic period is when every effort possible should be made by local and regional laboratory networks to address quality issues between laboratories through the introduction of quality assurance program and pandemic exercises.

Lance Jennings

See answer to question above. My major concern is false negative results; less false positive results if probe technologies are used.

Bernhard Kaltenboeck

The European EQUAL study compared real-time PCR results obtained from a large number of different laboratories. When provided cDNA the results where in good agreement, while when provided cells the variation was not acceptable (EQUAL-quant: An International External Quality Assessment Scheme for Real-Time PCR. Clin. Chem., 2005). This supports the notion that variability is primarily due to sample extraction and reverse transcription (A. Ståhlberg *et al.* Clin. Chem. 50, 509, 2004).

Mikael Kubista

We have seen very bad "specialist's" results in ring trials... and yes, non-specialists can also achieve reliable results.

Olfert Landt

In posing this question I had in my mind the definition of a non-specialist lab as one with little or no experience in designing, optimizing, evaluating and validating in-house real-time PCR assays. Such a lab is probably unable to take on board someone else's in-house assay for a pandemic strain of virus because of a lack of hands-on experience with such systems. These labs would usually employ kit-based real-time PCR diagnostics, and/or receive few specimens because of their distance from large population centers.

Should we try to include smaller labs in pandemic testing at some level? Surely they will play a role in specimen collection and transport of these potentially lethal pathogens anyway. Are dedicated laboratories realistically prepared for the surge in specimens resulting from a pandemic and wouldn't we be better served by planning ahead of time to "annex" other laboratories wherever possible? Perhaps I should have rephrased my question to ask whether reference laboratories and advisory panels are considering training small pandemic-ready teams of staff, located at smaller laboratories, in all aspects of real-time PCR to ensure preliminary screening results are both useful and conducted at more sites.

Ian M. Mackay

Yes as long as suitable validation procedures and controls are maintained.

Michael D. Nissen

Probably not. Setting up the method and particular protocols requires time, effort and regular quality control measures. Times of pressure, such as in the case of a pandemic, are rather ideal for messing up without being able to afford the luxury of continuously controlling results. High throughput, specialist labs should do the job.

Nikos Papadopoulos

I'm not sure. If it was a laboratory that routinely performed PCR for other purposes (i.e. a genetics laboratory) there might be sufficient expertise to perform the assay (i.e. setting up the reaction mixture and programming the thermal cycler), however, the ability of the laboratory to interpret the results correctly and to recognize the limitations of the assay may be imperfect. There may also be some differences in extraction methods and safe treatment of samples, such as biohazard level 2 vs. biohazard level 1.

Astrid Petrich

Difficult to say—QA programs such as the RCPA and Virology Division SEALS (and others) are performing for HPAI will inform these questions

William Rawlinson and Gillian M. Scott

The true quality of PCR results is not only influenced by the available technology and the proper use of prefabricated diagnostic kits but also by a number of pre- and post analytical steps. The sample workup in the individual lab as well as the critical and qualified interpretation of PCR results in the context of the clinical picture may have an equally important impact on overall diagnostic performance. Such aspects can be hardly assured in a non-specialist real-time PCR lab.

Udo Reischl

This depends strongly on the lab and on the (commercial) assay they use, but why not?

Oliver Schildgen

The broad answer is who knows? Apart from the National association of testing authorities in Australia (NATA) there is no other accreditation mechanism for these laboratories, and NATA certainly does not have the expertise to make a realistic and in-depth assessment of molecular diagnostic methods. Certainly the current evidence is that there are many (specialist) laboratories that are performing real-time PCR diagnostics that either have no idea of the fundamentals, or choose to ignore them. These regularly pass NATA inspection. My personal experience regarding the threat of influenza H5N1 to the Queensland population is that most laboratories (private and public) as well as the public health laboratories are ill-prepared and do not know the performance characteristics of their real-time assays that may be performed in these situations.

Theo P. Sloots

Laboratories with little experience in molecular techniques can quickly run into problems. The consequence of false negatives or false positives are magnified when dealing with new emerging pathogens that may have a pandemic potential.

Raymond Tellier

I don't think so.

Patrick Woo

Is it possible to do truly quantitative PCR in microbiology?

Yes, depending on the availability of agreed copy number standards. Quantitation (as opposed to detection) at low copy number is inherently less reliable. A problematic issue is that the nucleic acid extracted is fractured, so there may not be a truly 1 to 1 recovery.

Chris Birch and Julian Druce

I think so. At least, from a clinical perspective. Are you asking if the numbers are right—probably not, but who cares? The results appear to be consistent and useful and help to lead to better outcomes.

William Carman

Yes, if proper standards are available.

E.C.J. Claas

This is hard to do. You have to ask what it is that you're measuring. There will be normal fluctuations in the amount of the RNA or DNA molecule that you're interested in. Then you have to try to extract that without losing any through degradation or sticking to surfaces etc. And you should (although people rarely do) normalize it against some other housekeeping molecule. Then there's the question of which quantitative PCR assay to use, and whether different assays or techniques are directly comparable. So all in all you end up with a ball park figure that you need to interpret with caution.

Jon Clewley

It is possible provided the know-how, adequate instruments, and reliable reagents are available to trained experts and technical staff.

Esteban Domingo and Cristina Escarmís

As good as any other test, maybe better. I don't see why other tests would be more quantitative than PCR. Most steps in the procedures can be spiked to validate the quantitative PCR. If this spiking does not solve the problems in the quantitation, the problems cannot generally be solved in other microbiological tests. The only thing one cannot quantify easily by PCR is the amount of pathogen that is still infectious, whereas one can if one uses isolation techniques. As I indicated above, I think this remains an important issue.

Ron A.M. Fouchier

Yes but this is mainly true for the microbial agents showing high genetic stability. The exactness of the quantitative result may be impaired by genetic variability where HIV-1 is an ideal example.

Vincent Foulongne and Michel Segondy

This will become very important as many pathogens may be present in low numbers when colonizing but in larger numbers when true infection is underway (i.e. Aspergillus)

Michael G. Ison

Emphatically yes.

Bernhard Kaltenboeck

I believe it is. In digital real-time PCR the samples are diluted into aliquots that contain at most 0.1 molecules in average. Most positive samples had one target molecule initially. Some had two and few had three. The distribution is given by the Poisson equation. From the fraction of positive samples the absolute number of the initial target molecules can be determined.

Mikael Kubista

No—all we can do is pseudo-quantitative tests. But this is more than enough for the practice

Olfert Landt

Not in the sense that we think of quantification now. Unfortunately our current concepts are often reinforced by real-time PCR companies whose biggest customers work in human gene quantification, where specimen types and collection methods are much less complicated. We need to redefine what "quantitative" real-time PCR really means to us today and we need to respect that it will be easier to perform for agents associated with certain infectious diseases (e.g. due to blood-borne infections) than for others (e.g. due to respiratory infections). For respiratory infections, the production of viral load data that can be compared between laboratories and studies is not currently possible unless we relax our expectations.

There is likely to be no clinically significant difference between 10^3 genome equivalents (ge) per ml of a particular respiratory virus in a swab from a patient in Brisbane, and 10^5 ge of that virus in a nasal wash from a patient in Berlin. Probably our best chance of linking virus concentration with clinical implications is to define ranges of virus concentration, e.g. less than 10^3 ge/mL, 10^3 to 10^6 ge/mL and more than 10^6 ge/mL. These ranges can then be married to clinical data and associated with clinical outcome. The challenge is in defining the ranges. But we like challenges, we're working in infectious disease.

Ian M. Mackay

In order to do what with the result? It is relatively easy to get a copy number from a machine but if this number should be compared between laboratories to give a prognosis then there must be a level of standardization that seems near-impossible with in-house assays (e.g. sample collection, nucleic acid extraction and consistent standards).

John Mackay

In the cases where microbes are "evenly" distributed within a sample and the sample easy to handle, then a major step towards true quantitation can be achieved. Other details, such as intracellular or extracellular presence, different developmental stages, viability etc. may complicate quantitation. Nevertheless, a major problem occurs when obtaining or handling of the clinical sample is not straightforwardly quantitative, as in the case of nasal secretions or stool.

Nikos Papadopoulos

Possibly not, but it does provide relative quantitation. If the quantity obtained for a particular PCR using specific standards can be correlated to clinical condition and over time for a patient, this quantitation would probably be sufficient. Truly quantitative PCR from multiple laboratories requires that they use the same amplification standards (as a minimum, this may also include similar extraction and amplification conditions.

Astrid Petrich

We have to distinguish between dead and alive cells, not to overestimate the viable pathogens present in the matrix. Therefore a live marker gene or better a live marker mRNA is much of value.

Michael W. Pfaffl

[*WR*] Yes, with appropriate controls and appropriate validation. [GS] Yes. The use of QPCR over time is very helpful for the monitoring of patients for increasing load, rather than determination of load at single time points, as these do not necessarily mean anything without comparison unless they are very high (10^6–10^8). Here I am referring to viral loads, and my particular experience is with CMV, so I'm not sure how this might relate to bacteria and other pathogens. Increasing CMV loads can be an indication of development of antiviral resistant CMV strains in transplant recipients, for example. Some patients will have a constant low-mid load of $\sim 10^3$ with no evidence of clinical disease or other complications, whereas some patients with this load may have disease, hence the need for detection of INCREASING loads. Therefore, while I believe the viral load quantitation is "true", it is these mid-range levels that may not be indicative of disease.

William Rawlinson and Gillian M. Scott

Without any doubt, the current real-time methodology allows for a truly quantitative PCR. The highly sophisticated and permanently improved software of the current real-time PCR systems and platforms in combination with internal or external quantification standards represents a robust basis for reliable quantification of target molecules in the PCR reaction mixture. It should be considered, however, that quantification of target molecules or target organisms only makes sense when the investigated sample is obtained in a quantitative manner. In diagnostic virology you often receive quantitatively obtained specimens like blood, serum, or CSF which can be reliably investigated for the corresponding pathogens by quantitative real-time PCR. The majority of clinical specimens in diagnostic bacteriology are biopsies, swabs, catheter tips, stool, BAL, sputum, or other biological material which cannot be obtained or processed in a standardized way—here a quantitative analysis does not make sense at all.

Udo Reischl

There are many potential obstacles to true quantification by real-time PCR in microbiology. The first and often insurmountable problem is that of obtaining a representative sample. This may be relatively simple (e.g. in blood borne infection) but can be difficult or impossible in many other situations (e.g. respiratory infections). If a suitable sample is available the next difficulty lies in quantitative extraction of the intact nucleic acid. In some cases true quantification of viable target organisms may require that they be separated from target nucleic acid derived from non-viable cells. This may be impossible within the framework of a diagnostic assay. A further complication is that many tissue samples contain nucleases which must be inhibited once the nucleic acid has been released from the bacterial cell or viral particle. Fortunately, conditions that inhibit nucleases can usually be found. A particular problem for some bacteria (e.g. some gram-positive cells) is that efficient cell lysis may require the use of lytic enzymes (e.g. *Staphylococcus aureus*) or strongly denaturing conditions. These need to be applied rigorously since different strains or even cells from within different clinical specimens may vary in the degree of resistance to lysis. Finally, many samples contain materials that inhibit thermostable polymerases (e.g. heme) and if they cannot be completely removed from the extracted nucleic acid the efficiency of amplification will be reduced affecting the accuracy of quantification if samples are not equally contaminated. Provided all of these problems can be addressed truly quantitative real-time PCR is possible even in clinical microbiology.

Nick Saunders

Yes, it is. The problem is the interpretation of the data. Although the number of nucleic acid copies may be truly quantitative, it is no parameter for the number of infectious particles.

The detection of a nucleic acid in a clinical sample does not mean that it is infectious. This fact has to be taken into account in the interpretation, as in general the number of infectious particles is the quantitative parameter measured.

Oliver Schildgen

It is not possible to perform ABSOLUTE quantitation in microbiology, because we are dealing with biological systems which are complex and highly diverse. However, it is possible to approach a measure of quantitation based on recognizing all the variables that may affect the results of PCR. At best then the final results can be expressed as a quantitative estimate together with a measure of the degree of confidence for the result. That is, to determine the confidence limit. This is currently very topical in microbial serology, where the measurement of uncertainty is proposed, and I think that could be applied to quantitative PCR.

Theo P. Sloots

Would you trust in-house/home-brew real-time PCR assays as much, more, or less than commercial real-time PCR assays?

Both assay types have to be validated comparatively. There are definite examples of commercial assays being inferior to in-house. But to some extent there's less that can go wrong with a commercial assay.

Chris Birch and Julian Druce

CE marking in Europe and its equivalent elsewhere may provide some reassurance of the quality of a commercial assay. One practical disadvantage of commercial assays is that they can be so expensive that fewer controls are included by the user and repeats of assays are less readily undertaken than with an in-house assay. A good optimized in-house assay with all reagents subject to QC and appropriate controls included in every run can provide a very good or better alternative to commercial assays. However, it would be wise to compare assay performance between laboratories both for sensitivity and specificity. For example, the Clinical Virology Network in the UK has carried out a performance assessment of in-house assays for various targets using panels of samples with the aim of determining both whether there is variation between the assays and to enable recommendations to be made to laboratories seeking to initiate such testing.

Patricia A. Cane

I would trust them less, unless they came from a lab I trusted and I had tried them myself. When commercial assays work well, they work well, but they go wrong more often than they ought to

William Carman

Reliability of a test is not dependent on whether it is home-brew or commercial, but on how well the test is validated.

E.C.J. Claas

Commercial assays should be better validated and quality controlled than in-house assays: that is what you are paying for. But there's no reason why a lab shouldn't be able to setup, monitor and run accurate in-house assays, there's just more investment required.

Jon Clewley

We would trust in-house/home-brew assays provided the laboratory carrying out the experiments proves its competence in a series of controlled, blind tests of the type suggested in our answer regarding quality results by a non-specialist laboratory. In this case, the blind samples should be supplied to both, laboratories using in-house reagents and those using standard, commercial reagents. Results should be evaluated by an independent panel of experts. These procedures based on contrasted, blind tests could be applied also to assays with new instruments or any other novelty to be introduced in the entire cycle of procedures involved in a real-time PCR assay.

Esteban Domingo and Cristina Escarmís

In our own lab I trust the in-house assays better. They all have been evaluated carefully, and we have full control over all variables. It is a mistake to think that the commercial tests cannot fail, or loose reactivity. As much quality control as a commercial supplier may do (and some do less than others), there is never a guarantee that it will keep working. One advantage if one has multiple home-brew tests is that if one of the components of the test goes bad, one knows sooner because it will result in bad test results with other tests as well. If one has 5 different commercial tests from different companies, this would be much harder. In fact, even if they were from the same manufacturer, the use of different lot numbers could result in test A running fine, and test B failing without seeing it. A second advantage of home-brew tests is that they have usually been fully optimized in the context of a particular lab and particular machines and procedures. A third advantage of home-brew tests is that they can easily be updated. For instance, some of our H5N1 flu tests have been updated 3 times in the last 10 years, because the virus changed. I think we do this faster than commercial companies can do it. It is always a good idea to compare home-brew tests with commercial tests. If a commercial test is better than your own; don't be stubborn. We use some commercial tests because they simply were found to work better than our own.

A problem with home-brew tests is if they are designed and optimized by non-experts. So perhaps a beginner-level lab should start with commercial tests or team up with expert labs.

Ron A.M. Fouchier

In theory, commercial tests are more reliable than in-house tests because industrial controls on the reagents are more stringent. However we have participated in an external European control for quantitative EBV PCR by submitting results obtained using both commercial and in-house assays. Results were more satisfactory with the in-house test. The in-house tests might probably be as reliable as the commercial ones in labs with extensive experience in molecular techniques.

Vincent Foulongne and Michel Segondy

If appropriately validated, yes. It would be difficult to compare from center to center.

Michael, G. Ison

The issue is not so much with trust but with the level of the laboratories understanding of test validation, standardization and the use of controls. Commercial real-time assays can be used inappropriately. Participation in approved proficiency program and the inclusion of requirements for accreditation and certification will lead to improved standardization within the diagnostic laboratory.

Lance Jennings

We have given up use of commercial brews because our home-brew consistently outperforms any of them.

Bernhard Kaltenboeck

TATAA Biocenter has developed many home-brew assays for hospitals and we have also developed assays that were commercialized by companies. Much more time and resources are devoted to the commercial assays. Our experience is that the commercial assays are more reliable than the home-brews. In particular the home-brews are much more sensitive to changes in protocols and other conditions, including sample matrix, which may not always be under the control of the laboratory.

Mikael Kubista

Well-controlled in-house assays have (with few exceptions) the same or better quality compared to commercial assays. The latter might be better controlled (panels, references) but once they are established it is harder to change them in sequence and composition.

Olfert Landt

Most (but not all) commercial assays offer validated systems and are therefore trustworthy. Validation and quality control varies much more among in-house assays. So in respect to an in-house method, trust will vary in parallel with the focus, experience and expertise of a laboratory.

Nikos Papadopoulos

I would trust them as much, provided that the proper validation, training and QC were performed on either type of assay by the user laboratory. Some commercial assays are ASRs with no more that primers and probes being provided. This is not very different to an in-house assay where the primers and probes have been QCed but the rest of the assay needs to be put together.

Astrid Petrich

Home-brew is fine, but for diagnostics we need highly standardized assays performed with standard operating protocols (SOPs). Both can only come from ISO certified diagnostic companies.

Michael W. Pfaffl

[*WR*] This depends upon the source, and the agent being tested. No for HIV, HCV, CMV, but yes for less common, more variable RNA viruses.

[*GS*] I believe they are equivalent as long as standard controls are applied. In-house assays allow for more flexibility in assay parameters and interpretation which is better for research purposes, but may be a problem in the diagnostic setting unless strict regulation is ensured. I could understand if outsiders would be less trusting of in-house assays.

William Rawlinson and Gillian M. Scott

I know a number of home-brew real-time PCR assays which clearly outperform any of the commercial kits—but I have also experienced a number of well-published in-house assays with very poor performance in routine practice. After a thorough evaluation with respect to robustness and analytical specificity and sensitivity using local microbial strains and clinical specimens, I would trust any in-house assay. A clear advantage of commercial kits is the use of quality controlled components. On the other hand, FDA approval or a CE certificate only indicates that the assay has passed the formal and judicial hurdles—but does not guarantee a robust and idiot-proof performance in the local laboratory. Once the manufacturer has obtained the FDA approval after months or years of evaluation studies, the willingness to change any component or procedure is nearly zero. The concept of home-brew assays is open for necessary changes and, for example, can be immediately adopted to cover some previously unknown clinical variants of the target organisms or for using improved real-time PCR reagents.

Udo Reischl

Realistically, in-house assays are always likely to be less reliable than commercial real-time tests. This applies even though an in-house test may have performance characteristics that are clearly superior. The problem lies in production and validation of the reagents required. Most non-commercial laboratories simply do not have the ethos or inclination required to produce and validate the same quality in multiple batches over time. Even commercial laboratories can sometimes slip-up.

Nick Saunders

Depending on the experience in the individual lab that performed the PCR, I would trust the commercial assays less than the homebrew.

Oliver Schildgen

This really depends on the laboratory that has developed, validated and continues to support the assay. Many laboratories should stick to commercial assays only because they lack the technical awareness to run real-time assays within a quality framework. Personally I think that in-house assays are better if they are developed/validated by a skilled lab, and provided that these assays are run in an environment of continuous quality improvement. Sadly not many diagnostic laboratories have those abilities. When a commercial assay is purchased you rely on the fact that the manufacturer has targeted the best gene sequences, has optimized the reaction parameters, and provided appropriate quality control reagents. Experience tells us that often this is not the case.

Theo P. Sloots

Well designed in-house assays that are well validated should be as reliable, with one important exception: quantitative assays. Reliable and consistent quantitative assays require extensive quality control, not only of the reagents and PCR primers, and probes but also of the several quantitative controls required, and ensuring that they all work together from batch to batch. I believe that this is a level of quality control that may be too taxing for most clinical laboratories and may be best left to the industry.

Raymond Tellier

In all, there is no difference in quality between home-brew and commercial real-time assays. If a home-brew assay was well developed and fully validated, I would trust it more than other commercial products even if the latter received FDA or similar regulation agents" clearance.

Yi-Wei Tang

In good hands, in-house assays are as good as commercial real-time PCR assays.

Patrick Woo

Are we doing enough to implement the appropriate controls for quantification in microbiology? What should we use as gold/reference standards?

In some areas (e.g. blood borne viruses such as HIV) it's well controlled through a QA program.

Chris Birch and Julian Druce

Lots of initiatives going on. I suspect cries for "clinical standards" will go nowhere, as one patient on immunosuppressives or with a wonky immune system will have a different clinically relevant level to a young healthy child

William Carman

WHO standards as available for Parvovirus B19, hepatitis B and hepatitis C viruses would be the standard of choice to calibrate quantitative assays. Although this results in IU/ml which is not equivalent to copies, it is the best way to standardize quantitative real-time PCR.

E.C.J. Claas

It is useful to have at least three different types of reagent. First, absolute standards that define e.g. an International Unit. In the UK, the National Institute for Biological Standards and Controls (NIBSC) make some of these standards. These are used to calibrate other controls. Second, run controls, which are positives and negatives that can be used to ensure that the assay has actually worked on a day-to-day basis. These are sometimes supplied as part of a kit, but if in-house methods are being used they must be prepared separately. Third, quality control panels, which should be used blind. Labs doing diagnostic PCR should participate in QC schemes whenever they are available. More could probably be done to facilitate the production of all these types of reagents. The importance of them should be drawn to the attention of funding agencies whenever possible.

Jon Clewley

As stated in previous answers, reliable quantification in microbiology requires an adequate planning with sets of standard reagents and detailed protocols prepared by a reference laboratory. Any interested or candidate laboratory should participate in comparative tests to document reliable performance in the required real-time PCR procedure. This is particularly important if the aim is to gather a number of different laboratories that could contribute analyses of one or multiple pathogens in the event of the emergence of microbial disease, or to monitor environmental (or other) samples suspected of microbial contamination.

Esteban Domingo and Cristina Escarmís

No. Each step of the procedure should be monitored, and this is often not done. So spiking the original material with something of similar genomic make-up is crucial (if you test for an RNA virus; spike with an RNA virus). It is not hard to quantify, one just needs to do some work to find the appropriate controls and set up tests for it. With respect to gold standards, this is a tough call. Classical microbiology is often used as the gold standard because classical microbiologists still run the show, and because we simply often have nothing better. I think that for many pathogens (certainly viruses), classical tests fail to serve as the gold standard. In my opinion, the molecular diagnostic labs need to do more to generate reference standards, perhaps through some companies/institutes that can help on a global scale. At the moment there are many national and international schemes for standardization of tests, but it usually involves selected labs and selected pathogens. Laboratories should be able to contact one or a few companies/institutes and get all the reagents needed for standardization and quantitation.

Ron A.M. Fouchier

This is a crucial point. Programs for quality control in molecular diagnostic are now available in Europe (QCMD; www.qcmd.org) and we recommend registration for laboratories involved in the field. Furthermore, use of internationally standardized controls would be desirable.

Vincent Foulongne and Michel Segondy

No—there are few reference standards. Additionally, there are few groups that certify the quantitative standards and local performance.

Michael G. Ison

Controls for quantitation are in general non-standardized, time consuming to produce in the laboratory, and should be more widely available from commercial manufacturers. Leadership needs to be shown in this area

Lance Jennings

No, we don't do nearly enough. The real gold standard is the purified genomic DNA/RNA of the target organism.

Bernhard Kaltenboeck

Well, it depends on how sensitive and accurate the test should be, how much it may cost and how complicated it can be. These factors vary from test to test and also on whether the test should be used in research or in routine. Essentially, 3–5 steps in the process of QPCR testing contribute to variation and should be controlled: sampling, extraction, storage, reverse transcription, and real-time PCR. Since real-time PCR determines the number of molecular copies, while we usually are interested in a concentration measure, we should normalize to the amount of relevant sample used. Since many samples are heterogeneous, identifying a good normalizer for sample amount can be challenging. Possible normalizers are sample volume, amount of tissue, cell count, genomic DNA (which should be equivalent to cell count), total RNA or expression of reference genes. Next is the extraction step. Extraction efficiency is expected to depend on target size, target conformation, target structure, target compartmentation, and target sequence. It will also depend on other factors present in the cell that bind to the target. The extraction control should be as similar to the target as possible. A DNA plasmid, for example, is not likely to extract with the same efficiency as an RNA virus. Reverse transcription contributes to the variation of RNA targets. It can be controlled using an RNA spike. Storage has not been addressed very much in the literature, but is expected to cause substantial variation of RNA targets. RNA degrades rapidly in a sequence dependent way by enzymatic processes and also by pH dependent self-degradation. Best precaution is to minimize degradation by rapid cooling of the sample or placing it in a medium that inhibits RNases and then, as soon as possible, reverse transcribe the sample. Reverse transcription is robust reaction (A. Ståhlberg *et al.* Clin. Chem. 50, 509, 2004), but the yields vary with priming strategy, the reverse transcriptase, and target sequence (A. Ståhlberg *et al.* Clin. Chem. 50, 1679, 2004). As long as the same protocol is used the data are nicely comparable, but it is not possible to compare, for example, results from two labs that use different protocols even if they use the same normalizer. The real-time PCR reaction is highly reproducible, but is prone to inhibition. Presence of inhibitors can vary substantially among the same type of samples and can affect different assays differently. One way to assess sample specific inhibition is by *in situ* calibration or standard additions (A. Ståhlberg *et al.* Clin. Chem. 49, 51–59, 2003).

Mikael Kubista

There are many different ways to realize positive controls and internal standards. Place for a whole chapter.

Positive controls—positive (native samples) and/or standardized plasmids. Plus blind (positive) samples included from time to time. *Internal controls*—One can amplify anything which is present in the sample, like a human DNA target which is co-extracted (disadvantage—various amounts of target could inhibit the analytical PCR; advantage—control of extraction included) or one can add a standard DNA and separate primers and probes (our preference) or one can construct a reference, using the same primers but different probes (Stöcher/Berg method;—the disadvantage is that you have design one control per

PCR, you have to control one more component and there is a competition between the primers).

Olfert Landt

I'm not even convinced that we're doing enough to define what we *should* use and disseminate those definitions let alone actually implement a process.

Ian M. Mackay

Plasmids or electron microscopy-enumerated microorganisms can be used, depending upon the microorganism tested.

Nikos Papadopoulos

We currently use cloned controls to generate standards for quantitation and use spectrophotometry to determine the concentration performing calculations to determine the copy number of the target nucleic acid. For quantitative RNA real-time assays we generate transcripts from cloned target and measure the OD of the RNA to determine copy number. Ideally there would be universal standards that all laboratories could use to generate their standard curve, thus allowing a laboratory to compare its numbers directly to those obtained from another laboratory. This would allow for universal clinical correlation and interpretation. Unfortunately these are not routinely available and in some cases where they are available they are quite expensive triggering laboratories to generate their own standards and their own interpretation of assay results.

Astrid Petrich

For the really important pathogens, certified reference materials (CRM) should be produced by the EU and should be the basis for a reliable and identical quantification EU-wide.

Michael, W. Pfaffl

[*WR*] No, there should be a microbiology community wide attempt to share reagent controls.

[*GS*] I agree. With the increasing use of this technology in the diagnostic setting, there needs to be greater assurance that QA across laboratories is maintained.

William Rawlinson and Gillian M. Scott

Actual well-characterized clinical specimens are the gold standard but good quantitative controls of this type are very difficult to manufacture since they rely on the availability of clinical material that can be stabilized and used to produce multiple batches over a period of time. Such material is usually scarce. Simulated specimens offer a route to avoid this problem but they usually have the obvious limitation that they are not equivalent to true specimens. Despite the problems, many laboratories and working groups worldwide are involved in trying to provide materials to satisfy the need for appropriate controls. More resources should probably be channeled into this effort.

Nick Saunders

In general, yes. Nevertheless, whenever available, the most sensitive technique is a good cell culture. Only by the combination of classical and modern method we can reach a "gold standard".

Oliver Schildgen

In addition to the above, careful independent validation by academic institutions is required, and especially with an eye to testing several different isolates/genotypes.

Raymond Tellier

What is required for better standardization of qPCR?

Why standardize everything in the world? To do so is annoying and also a risk. This does not mean we should leave out the validation of (new) assays.

Olfert Landt

As mentioned in other answers, it is important to distribute protocols, reagents and samples prepared in some reference laboratory, to carry out blind assays to ensure that different laboratories have the adequate instrumentation and know-how to perform reproducible qPCR assays.

Esteban Domingo and Cristina Escarmís

Full descriptions of protocols used provided with the subsequent performance data so that laboratories can measure the performance of an assay. Absolute detection limits as opposed to clinical sensitivities and specificities would assist here.

John Mackay

Standardization issues are often focusing on targets that have commercial value, like HIV-1, HBV and HCV. Since we as a scientific and diagnostics community have and are developing assays ("in-house" as the process is called) with a greater speed then standards can be introduced by official organizations, I feel that our scientific organizations should support the initiatives of developing reference preparations more strongly, and take the lead in this. Some initiatives are present, but as a group together, we could do more to develop these reference preparations and make them available. Furthermore, organizations dealing with proficiency testing (NEQUAS, INSTAND, QCMD) already have a lot of data to support these initiatives.

H.G.M. Niesters

If you use synthetic plasmid controls, do you use them circularized or linearized and why?

True pathogens should be preferred over plasmids. IF plasmids are used, circular or linear are not important as it is a surrogate standard anyway.

E.C.J Claas

Plasmids must be used linearized. A circularized plasmid may not amplify or may amplify in an artefactual way. Bacteria containing recombinant plasmid DNA and the plasmid DNA itself should be handled with care. Both usually exist at very high titers and can easily cause PCR contamination.

Jon Clewley

We don't. You can use plasmid DNA to test if a PCR works, but it is not the best control for most diagnostic purposes.

Ron A.M. Fouchier

We commonly use plasmids for controls and calibrators. We use circularized form since enzymatic digestion adds one more critical step with no enhancement in performance of our assays. We however recommend the use of a plasmid containing a large fragment of DNA that overlaps the PCR target for at least 200 bp on each side.

Vincent Foulongne and Michel Segondy

TATAA Biocenter has tested both and we have also developed extraction controls for industry. A circular molecule is supercoiled and interacts quit differently with proteins and other molecules in the cell than a linear molecular. A circular plasmid is therefore the appropriate control for a circular target, while a linearized plasmid is the preferred control for linear targets.

Mikael Kubista

Not important to linearize them.

Olfert Landt

No evidence to suggest linearization is necessary—except anecdotal of course. A common reason given is that linearization will remove secondary structure that exists solely in the circularized plasmid. I would like to see solid data before I believe that. An oligonucleotide can have significant predicted secondary structure and it's only 25 nucleotides long (and very linearized!).

Ian M. Mackay

I have used them circularized for real-time PCR. The only time I needed to linearize a plasmid was for a relatively long PCR (conventional PCR and a near 3kb product). With no evidence to back it up, I would suggest that circularized plasmids might also be more stable and at less risk of any exonucleases in a tube. Comparisons reported on the web describe no difference between the two but perhaps it may be sequence specific.

John Mackay

We do use them and we use them circularized. It has not proven to be an issue in our hands. Our targets are generally quite short (< 200 bp) and perhaps this is why it does not appear to impact on the amplification.

Astrid Petrich

Linearized DNA material is stable for years we have used it for over 10 years in an absolute quantification assay.

Michael W. Pfaffl

[GS] We use them circularized, only because we are not aware of any benefit in linearization.

William Rawlinson and Gillian M. Scott

I prefer to use them linearized, because a long time ago I had the experience that linearized DNA is behaving more accurate during denaturation and annealing. It seems that formation of disturbing secondary structures within the plasmid template is reduced in the linearized form and it behaves more similar to a huge chromosomal DNA. But this is only an hypothesis.

Udo Reischl

I use them circularized, as they are more stable. I do not believe that it really makes a difference. In my opinion it is most important to use uniform controls.

Oliver Schildgen

I am not aware of studies showing that this is a systematic problem. Conceptually, for a PCR the circular molecule is "locally linear" so unless the targeted amplicons is a large fraction of the whole plasmid I do not see a problem. There may be instances however where the primers may be used by the reaction in the other direction and synthesize the near complete plasmid instead, if PCR conditions are such that synthesis of large (> 3 kb) amplicons is possible.

Raymond Tellier

What is the biggest change in technology you have seen in microbiological research and diagnostics?

Real-time PCR and the use of UNG have changed a lot in terms of accuracy of results and in avoiding cross-contamination.

M.G.H.M. Beld

The introduction of PCR, surely?

William Carman

Definitely NAT testing. In the area of virology it has made diagnostics relevant to treatment for the first time.

Chris Birch and Julian Druce

Implementation of the polymerase chain reaction.

E.C.J. Claas

PCR.

Jon Clewley

PCR has had an immense impact on research in virology although transfer of its benefits to routine diagnostics has been painfully slow. A major spin-off from PCR, combined with the development of automated nucleotide sequencing, has been the enabling of detailed molecular epidemiology studies of microbes. For example, it is now possible to completely sequence a new isolate of influenza virus within days. The use of nucleotide sequencing to guide patient treatment, for example to undertake genotypic drug resistance testing of HIV, is an example of transfer of high-tech to the clinical laboratory which would have been inconceivable 20 years ago.

Patricia, A. Cane

A very important development in microbial research has been the advent of the whole-genome, shotgun sequencing procedure based in the sequential, enzymatic release of pyrophosphate, in what is now termed pyrosequencing technology. This new procedure is the basis of the Genome Sequencer 20 System, commercialized by Roche Applied Science, and used already in several recent research studies. It can increase the speed of nucleotide sequencing by 100 times, as compared with current procedures. It requires advanced expertise in bioinformatics and very specialized personnel. At this stage it is not realistic to consider this development for a diagnostic laboratory, but one has the impression that this new technology will first revolutionize microbial (and general cellular) genomics, and will eventually impact diagnosis, due to the possibility of simultaneous analysis of genomes contained in a biological specimen.

Esteban Domingo and Cristina Escarmís

You could have just asked how old I am. I am old enough to say that for research, the availability of restriction enzymes and sequencing technology has been crucial. Without these, there would not be much molecular biology. For diagnostics, I would certainly say PCR (but of course this would not be available/possible without sequence info).

Ron A.M. Fouchier

In the area of microbiological diagnostics, automation in molecular approaches has allowed high-throughput and short turnaround times.

Vincent Foulongne and Michel Segondy

The introduction of molecular techniques, specifically PCR in the late 1980s has impacted on the microbiology laboratory in a number of ways. It has been the one technology which has been able to improve both the sensitivity and specificity of diagnostic assays especially in the area of virology and at the same time has facilitated the detection of important pathogens which could not have been, or would have been difficult to, detect by conventional microbiological methods. The precision required for infectious disease diagnostics and the frequent need for careful result interpretation has increased the awareness of the importance of controls during the pre-analytical and analytical stages. While the concern over standardization has driven improved quality assurance programs for the diagnostic laboratory. It has also been the single most important technology for driving microbiology research, by extending our knowledge of the pathogenesis of existing infectious diseases and of an ever-increasing range of new infectious diseases, and facilitating a resurgence of previously dormant areas of applied microbiology research and the evolution of new research disciplines such as molecular epidemiology.

Lance Jennings

Well, the introduction of real-time PCR has certainly been a major break-through in routine diagnostics. Supported by UNESCO, and hopefully more welfare organizations in the future, TATAA Biocenter trains people from developing countries in real-time PCR and assists them in setting up the technique for routine work. The advantage of real-time PCR is that students with a reasonable amount of training can set up their own home-brew assays, which saves them a lot of money.

Mikael Kubista

The change to (Real-time) PCR (complementing microscopy and culture) is still not established everywhere. Downscaling volumes does not make sense because this would lower the detection limits—so we all are part of slow change.

Olfert Landt

Apart from coffee pods and easy to clean tiny espresso machines? I'm a fan of easy-to-order fluorogenic oligoprobes. These permitted us to move away from PCR-ELISA methods. We employed our in-house version, the enzyme-linked, amplicon hybridization assay (ELAHA) to increase the specificity of our PCRs, permit objective multiplex detection of five or more amplified microbial targets and for competitive quantification. Let me emphasize that the quantitative potential of real-time PCR was greatly appreciated because of the likelihood that I would never have to perform conventional qcPCR again!

Ian M. Mackay

It is for sure that the development of amplification technologies, and in particular PCR, has increased the possibilities to detect targets at a low level, although it is not always clear what these low levels mean for patient care. However, in a large number of viral targets, this is obvious, like the detection of HSV or enteroviruses in CSF. But there are more examples.

With the development of real-time PCR technologies, one is able to quantify viral load more clearly, and to determine the relation between viral load and clinical outcome, or significance. Since the technology has now advanced in such a way that results can be obtained in a timely manner, this has indeed impacted on day-to-day patient management. The ability to determine mutants or variants of viruses very quickly has also improved patient management and care. Sequencing, for instance, is becoming very easy.

Furthermore, the advancements to detect unknown viruses, have been enormous. Like the detection of HCV, but also hMPV, SARS coronavirus, NL63 coronavirus or the recently discovered bocavirus. And there will be more to come. But also the understanding that a pandemic of H5N1 could be handled easier without growing the virus, but by being able to quickly detect and determine the nature of the virus in a diagnostic manner.

H.G.M. Niesters

Cost and occasionally no valid references to assess clinical implications of testing

Michael D. Nissen

Nucleic acid molecular diagnostics are revolutionizing microbiology research and diagnosis, as their high sensitivity allows detection of very low numbers of microorganisms, making it necessary to understand the role and/or natural history of some of them in cases where they were not found before.

Nikos Papadopoulos

The biggest change is the movement of molecular methods from research to diagnostics and the increased stringency and standardization involved in using these assays for clinical purposes.

Astrid Petrich

NA amplification techniques available widely. Availability of routine sequencing of isolates for confirmation.

William Rawlinson and Gillian M. Scott

After so many years in the PCR business I have seen two really big changes: The first change was the introduction of thermostable enzymes and automated thermocyclers instead of using three water bath and adding a drop of Klenow enzyme after each denaturation step. The second change was the introduction of the LightCycler real-time PCR device (the improved one from Boehringer Mannheim Inc.).

Udo Reischl

This contributor has been around for a long time and so the biggest change has been the introduction of standard PCR. PCR is now a cornerstone technology involved in many laboratory activities. For research the most important impact has been on the ease with which it is now possible to perform sequencing leading to the determination of many complete genome sequences. In microbial diagnostics PCR provides the key that unlocks the power of nucleic acid hybridization as a tool for sensitive and specific detection of microorganisms (e.g. by real-time PCR or on microarrays).

Nick Saunders

I speak as a molecular epidemiologist working in a research laboratory. I mainly analyze samples retrospectively, but my working experience deals also with some patient diagnostics as a reference laboratory molecular virologist. I work with human enteroviruses, including polioviruses, and human rhinoviruses. The common feature of these viruses is their vast genetic variation, hampering development of uniform methods for molecular diagnosis and genetic characterization. The biggest change I have been witnessing in microbiological research and diagnostics has been the introduction of automatic sequencing to everyday laboratory life. In the enterovirus field the biggest change from my perspective has been the possibility for genetic typing of all strains of human enteroviruses using a single primer pair in RT-PCR, introduced by Oberste et al., in 1999. This method is based on facts that in picornaviruses the VP1 protein gene is the most variable part of the genome, it contains several antigenic epitopes and the sequence is thus strongly correlated to serotype of a given strain. Together with the easiness of automatic sequencing this approach has given a name to countless strains of enteroviruses previously classified as untypeable when isolation in cell culture and conventional typing with neutralization using hyperimmune antisera did not provide an unequivocal serotype definition. The criteria for sequence homology were defined by Oberste *et al.*: "VP1 nucleotide sequence identity of more than 75% to a certain reference strain indicates that the serotype of an unknown sample is of homologous serotype, provided that the second-highest identity is less than 70%." Along with names for untypeable strains, use of this method has also revealed strains not fitting any pre-existing serotype. This in turn has revived interest in taxonomical issues and as a consequence the number of official separate types of human enteroviruses is now exceeding 100. New enteroviruses have been given a type number based on the VP1 sequence. Of course, one may ask what is the purpose of knowing the name of a virus and how does it help a physician or a distinct patient? Classification of viruses is still considered important and surely in time specific features can be connected to species and/or serotypes. The increase of sequence data has also enabled molecular epidemiological studies of several emerging enteroviruses, e.g. enterovirus 71 and echovirus 30. There have also turned up obstacles. One of the goals of molecular epidemiological studies and phylogenetic analyses has been to identify genetic determinants of certain differences in clinical disease. Unfortunately, distinct genetic lineages of enterovirus serotypes have not been connected to disease entities. Though single amino acid changes may have been seen and held responsible for certain manifestation of illness, the same change may have been absent in the next epidemic with similar clinical picture. This is not a surprise as it is likely that primary structure does not explain all properties of viruses, but the characteristic features are generated with conformations of secondary structure and even beyond that. However, this gives new challenges to sequence analysis.

Carita Savolainen-Kopra

In 2001 Saborio *et al.* established a method for cyclic amplification of prions. This was really impressive and may be a future technology. (Saborio GP *et al.* Nature. 2001;411(6839):810–3)

Oliver Schildgen

Without doubt, PCR. Initially for research, and as instrumentation developed and technical awareness increased the technology was applied to diagnostics in those areas where the technical knowledge was available. However it is real-time PCR that has made the technology more freely available as integrated instrumentation. Further advancements in this area are expected with full integration of extraction, amplification and detection into one instrument.

Theo P. Sloots

We now have access to several assays for a large number of pathogens present at high titers (mostly viruses). There are still considerable difficulties for molecular assays targeting agents present at very low titer, as is the case in most instances of bacteremia or fungemia with a titer of 1–10 CFU/ml. The difficulty is then the ability to extract DNA from a large volume of sample containing a very low amount of the targeted template and which may contain a large amount of human DNA.

Raymond Tellier

I think the most important technological advance in diagnostic microbiology in past decades is PCR, whereas the most important technological advances in research microbiology in the past decades are PCR, DNA sequencing, bioinformatics tools and technologies in large-scale biology.

Patrick Woo

What applications of real-time PCR do you consider to be lacking in the microbiological literature?

The ability to multiplex (including an internal control) while retaining sensitivity remains an issue.

Chris Birch and Julian Druce

There are not many, but we need good comparison of available assays with recognized panels of samples. We could also do with a qPCR for adeno that is validated on all types and a simple papillomavirus typing test. A way to detect HIV minor populations would be good.

William Carman

It would be important to improve efforts to analyze environmental samples in a quantitative way to define fluctuations in the frequency of different microbial forms.

Esteban Domingo and Cristina Escarmís

None??

Ron A.M. Fouchier

Large volume real-time PCR and other approaches to improve sensitivity by analyzing larger sample volumes, pre-amplification strategies for multiplexing, and real-time immuno-PCR for protein detection (K. Lind and M. Kubista. J. Immunol. Methods 304, 107–116, 2005; Barletta *et al.* 2004. Am J. Clin. Pathol. 122: 20–27).

Mikael Kubista

Multiple channels on real-time instruments have existed for years—and despite most people claiming they need all these channels for multiplexing; there has been very little in this area. Perhaps the demand for the overriding sensitivity precludes this?

It would be interesting to see more comparisons on extraction methods since a higher quality extraction is typically required for fluorescent applications in order to provide data unaffected by amplification inhibitors or fluorescent quenchers.

John Mackay

Single cell quantitation—more literature needed.

William Rawlinson and Gillian M. Scott

I think the field of possible real-time PCR applications is almost completely covered. You never know what innovations are in the pipeline at the present moment—but after more than 10 years of intensive research and development everyone in natural sciences is already aware of the considerable advantages of real-time PCR. I would appreciate to see any critical reflections of published real-time PCR protocols to prevent friendly colleagues from making similar frustrating experiences after investing a lot of money in poorly evaluated but quickly published real-time PCR hybridization probes.

Udo Reischl

In research tasks real-time PCR may be used as a fast and time saving tool to quantify virus titers. This may complement or at least replace plaque titrations as it saves time and more often is easier to perform.

Oliver Schildgen

Quantitative PCR is still not widely accepted or applied, probably because of a lack of understanding of the underlying technical issues. Also the detection of SNP or point mutations in drug resistance typing in both bacteriology and virology is not widely addressed. However, the greatest lack is in the application of quality systems. Appropriate and comprehensive management of quality systems is poorly addressed in the literature.

Theo P. Sloots

Is there a need for unifying our existing abbreviations for quantitative PCR, quantitative real-time PCR, real-time PCR and reverse-transcription PCR?

I didn't think so until I read this question!

Chris Birch and Julian Druce

Why don't you suggest something?? [Happy to—Ian M.]

William Carman

Yes, currently the use of RT-PCR for both real-time and reverse transcriptase is very confusing. In addition, qPCR which refers to quantitative PCR, is also used for qualitative applications. RNA-PCR, RT-PCR and qRT-PCR could be suggested.

E.C.J. Claas

No.

Jon Clewley

In our experience this is not essential. If deemed important, a panel of recognized experts should circulate proposals to many laboratories involved in real-time PCR assays, to obtain feed-back from the experts. If a list of abbreviations and definitions of terms is to be approved and used by the scientific community, the accepted proposal should be amply publicized through a consensus article in a widely read, specialized journal (Nucleic Acids Res. or similar).

Esteban Domingo and Cristina Escarmís

Yes.

Ron A.M. Fouchier

That is not really a priority; we are supposed to be aware of what we are talking about.

Vincent Foulongne and Michel Segondy

The current terminology can be confusing to those outside the field. Because RT-PCR is well established in the literature, changing the term "real-time" might then lead to more logical abbreviations and consensus.

Lance Jennings

Definitely! Presently it's very confusing for the non-expert.

Mikael Kubista

QPCR is a stupid abbreviation—full-length names are preferred.

Olfert Landt

I admit to an obsession with acronyms. Despite that, it seems unfair to me that multiplex-, quantitative-, reverse transcriptase-, nested-, semi-nested- and other methods of PCR have their own modifier, and yet real-time PCR does not. Why not use "rtPCR"? Even without this term in the popular literature, I have had students present draft theses to me with "RT-PCR" referring to real-time PCR (but never twice). It will happen despite our best efforts so why not try and standardize the lingo? My choices would be:

Quantitative PCR	qPCR
Multiplex PCR	mPCR
Nested PCR	nPCR
Semi-nested PCR	snPCR
Asymmetric PCR	aPCR
Competitive quantitative PCR	qcPCR
Conventional PCR	cPCR
Reverse transcription PCR	RT-PCR
Real-time PCR	rtPCR

When combining the modifier with real-time PCR, then I'd propose that the variant remains shortened but real-time is spelt out:

Real-time quantitative PCR	Real-time qPCR
Real-time RT-PCR	As written
Real-time nested PCR	Real-time nPCR
Real-time quantitative competitive PCR	Real-time qcPCR

Ian M. Mackay

I dislike real-time PCR acronyms—one of the main reasons being that there are so many of them published that you only need remember for the length of the paper before forgetting it—safe in the knowledge you are unlikely to see it again.

I prefer "real-time PCR"—to avoid confusion with the "other" RT-PCR (reverse transcription PCR) and because the other common acronym (qPCR for quantitative PCR) does not always apply to diagnostics when a yes/no result is required.

The original term from the inventor was kinetic PCR (kPCR)—perhaps a return to this is timely?

John Mackay

There is some confusion but provided that abbreviations are clearly indicated following the first use of the complete wording it does not appear to be a significant issue. I am not against a standardized set of acronyms. Yes its confusing and trying to explain the differences to trainees is beyond the written literature and how terms are being used currently. Please let me know if we standardize! I don't want to miss the new terminology due to busyness.

Astrid Petrich

Yes of course there should be a unique nomenclature:

Reverse-transcription PCR = RT-PCR
Quantitative PCR = qPCR
Quantitative real-time PCR = real-time qPCR
Quantitative real-time reverse-transcription PCR = real-time qRT-PCR

Michael W. Pfaffl

Yes

William Rawlinson and Gillian M. Scott

It is O.K.—and who should (or who can) force the colleagues to change their habits and terminology?

Udo Reischl

Yes, absolutely.

Oliver Schildgen

RT-PCR = reverse-transcription PCR
Q-PCR = quantitative PCR
No abbreviation for real-time PCR

Theo P. Sloots

Yes, of course, clarity is important. Too often I see junior colleagues confusing reverse transcriptase PCR and real time PCR.

Raymond Tellier

What are the issues with qPCR assays described in publications?

Many of them are not well-selected (poor sensitivity, false detections...), some are incorrectly published (reviewers seem not to BLAST/check them)—by the way—I do not like the term q-PCR.

Olfert Landt

Perhaps this is best shown with the number of errata published in journals to correct primer and probe sequences. BLAST searches should be performed on all sequences and even an amplicon sketch drawn up to be sure that the two primers are pointing 5′ to 3′ and the position of the probe(s) noted. Some assays have relatively poor primers (but high yielding fluorescent probes) from previous studies which may limit their sensitivity and cycling conditions should be eyed somewhat critically on the basis of the sequences being used (especially annealing temperatures).

John Mackay

What advice can you offer to someone wishing to publish their study, in order to pre-empt reviewer issues?

The validation should be as rigorous as possible. If there is insufficient clinical material available this should be mentioned and discussed. Suitable internal controls (controlling extraction, reverse transcription and amplification) should be included. Because the amplicon size for real-time PCR is small, it might be useful to confirm at least a sub-set of results with a PCR directed at an alternative sequence.

Chris Birch and Julian Druce

Keep it detailed, make it clear what the analytical sensitivity is, what range of samples from which geographical areas have been tested and keep the clinical sample numbers high, best done on prospective samples. Use appropriate comparison tests. Compare at least 2 extractors. Obvious stuff, really.

William Carman

The most important thing is to use any real-time PCR assay in the context of an actual clinical or epidemiological study, rather than just describing putting together an assay and then using it on a few samples plucked from the freezer. Then, when using the assay for a proper study, it is important to have appropriate controls. This might include measuring the "background" levels of the virus in the normal, healthy population, and so defining some minimum level of viremia necessary for a pathogenic effect. Sequencing some or all of the positives may also be important to show their relatedness, and to confirm that they're not all the same, which would be a sign of PCR contamination.

Jon Clewley

If this question refers to aiming at getting a favorable review upon submission of a manuscript, the advice is to emphasize the novelty of the real-time PCR used, and to provide in the manuscript all pertinent experimental details so that experts can easily reproduce the procedure described in the manuscript.

Esteban Domingo and Cristina Escarmís

Describe how well the primers were designed, optimize and quantify the test, and spike. Tests will be as good (or bad) as your primers. If this starts off wrong, it will never be a good test. Then, you try to get the primers to work as well as they can under your test conditions, and you state what the end result is. Finally, you make sure you control your test subsequently as well as you can. The issues that I have raised most as a reviewer are (1) your primers are not optimal to detect a pathogen, (2) what is the sensitivity of the assay, (3) why did you not control for sample processing and reactions?

Ron A.M. Fouchier

Make a nice story, include some technical advances.

Olfert Landt

Provide experimental details clearly and in sufficient detail.

Nikos Papadopoulos

Ensure that you describe what validation has occurred for the assay used in the study and that it was sufficient and indicate why the assay was chosen in the first place. Indicate clearly what controls were used and how they were generated/characterized prior to use. Also indicate the initial treatment of the specimen including extraction and whether an internal control or other method was used to assess the presence of inhibitors. It is also important to specify what the gold or reference standard employed was and justify why it was chosen and deemed adequate. If there are any discordant samples indicate what methodology was chosen to resolve the discordant results and why this was deemed suitable. There are differences between clinical and analytical sensitivity and if their values are addressed, this can prove valuable for the evaluation of a new assay in a manuscript.

Astrid Petrich

Consider the work as a critical reviewer—get your colleagues to read particularly with respect to the need for controls

William Rawlinson and Gillian M. Scott

A good question—please let me know about the "golden answer" (☺)

Udo Reischl

All studies should be well planned and thoroughly thought, so that the person performing the experiments and interpreting results knows what he/she is doing and why. Then it is also easier to foresee critical points and decide what to do if everything does not go as planned. To be able to publish the study and the results, one must know what has been done. This underlines the importance of notes and reports of experiments. As quality systems for laboratory practice call for, what is not documented has not been done. The most important advice I can think in writing is to be straight and honest. If all data is out in the open and discussed already by authors, several questions can be avoided as they have already been answered. Of course, especially as a young scientist it is impossible to predict all issues. Therefore the same open line should be continued when points raised by reviewers" are addressed. Discussion and critical judgment of one's own methods should be clear, but one should also defend chosen lines.

Carita Savolainen-Kopra

This is a good question. From my personal experience I would advise using a uniform methodology and look for enough parameters. Finally, if your manuscript is not accepted, send it to another Journal, that also helps sometimes.

Oliver Schildgen

In general, writing an attractive scientific article and getting it published is not easy, and I hope the following "tips" can help you when you are preparing your manuscript. Firstly, it is never too early to think about a journal to submit to. You should submit your manuscript to a journal that you presumably read regularly or at least follow closely. When you are preparing a manuscript, you should keep in mind Aristotle's golden rules: Every piece of writing should have a beginning, a middle, and an ending. You should have a clear idea of what you are doing, in terms of bringing your field of learning "forward", and never send a manuscript that is too long. Finally, it is a good idea that you get a colleague, friend, spouse or professor to read your semi-final draft. My next "two-cents" advice is specific for studies related to the development and validation of real-time PCR-based diagnostic procedures. (1) While the real-time PCR techniques themselves are no longer novel, it is important now to focus your study on clinical applications rather than the development and optimization of the technique. (2) A panel of well-characterized clinical specimens needs to be included for procedure validation. It is important to set up unbiased criteria ahead of time for including and excluding the validation specimens, and a constant standard to determine the status of the included specimens. (3) In addition to analytical sensitivity and specificity of the technique itself, it is critical to have clinical sensitivity and specificity information as the main results of the study.

Yi-Wei Tang

Index

E

F

G

H

Hydrolysis probes *See* fluorogenic chemistries

I

J

L

M

N

T

U

V

W

Y

Z